Handbook of Biochemistry

Volume I

Handbook of Biochemistry
Volume I

Edited by **Bernard Wilde**

R CALLISTO REFERENCE

New York

Published by Callisto Reference,
106 Park Avenue, Suite 200,
New York, NY 10016, USA
www.callistoreference.com

Handbook of Biochemistry: Volume I
Edited by Bernard Wilde

International Standard Book Number: 978-1-63239-371-5 (Hardback)

Printed in the United States of America.

Contents

Preface

Among all the inventions and discoveries, it was obvious that man would soon get curious about himself and life around him and what separated them from the non-living world. These questions led to discovering chemical processes in life and its materials which in turn paved way for a new line of study called biochemistry. Biochemistry is broadly defined as the branch of science concerned with the chemical and physico-chemical processes and substances which occur within living organisms.

The scope and pervasiveness of the subject acquired immense importance over a period of time and now from botany to medicine, biochemistry research is applicable. Biochemists, as the practitioners of this field are known as; are committed to exploring the possibilities of applying biochemistry to answer further questions related to complexity of living organisms.

The advantages of biochemistry are for all of us to see. Since the discovery of the first enzyme to understanding complex biochemical processes inside and outside cell to identifying gene's role in transferring information in the cell, the subject has grown by leaps and bounds. While such pioneering research has been honored by top science awards like the Nobel Science prize, it has brought in a new era in other fields like criminology and agriculture.

However in an economy-centric world, the practical applications and employment opportunities of a research subject are equally important. Biochemistry has opened up new avenues in the fields of pharmaceuticals, genetics, education, cosmetics industries, etc., hence making it a viable career option for future students.

This book is an honest attempt to focus on the various facets of biotechnology and highlight its importance and contribution to the world at large.

Editor

Proteomics Shows New Faces for the Old Penicillin Producer *Penicillium chrysogenum*

Carlos Barreiro, Juan F. Martín, and Carlos García-Estrada

Proteomics Service of INBIOTEC, Instituto de Biotecnología de León (INBIOTEC), Parque Científico de León, Avenida. Real, no. 1, 24006 León, Spain

Correspondence should be addressed to Carlos Barreiro, c.barreiro@unileon.es

Academic Editor: Tanya Parish

Fungi comprise a vast group of microorganisms including the Ascomycota (majority of all described fungi), the Basidiomycota (mushrooms or higher fungi), and the Zygomycota and Chytridiomycota (basal or lower fungi) that produce industrially interesting secondary metabolites, such as β-lactam antibiotics. These compounds are one of the most commonly prescribed drugs world-wide. Since Fleming's initial discovery of *Penicillium notatum* 80 years ago, the role of *Penicillium* as an antimicrobial source became patent. After the isolation of *Penicillium chrysogenum* NRRL 1951 six decades ago, classical mutagenesis and screening programs led to the development of industrial strains with increased productivity (at least three orders of magnitude). The new "omics" era has provided the key to understand the underlying mechanisms of the industrial strain improvement process. The review of different proteomics methods applied to *P. chrysogenum* has revealed that industrial modification of this microorganism was a consequence of a careful rebalancing of several metabolic pathways. In addition, the secretome analysis of *P. chrysogenum* has opened the door to new industrial applications for this versatile filamentous fungus.

1. Introduction: Evolution of Fungi in Penicillin Production

Life comprises three domains: the bacteria, the archaea, and the eukaryota. Within the last one, the fungi kingdom forms a monophyletic group of the eukaryotic crown group, which collects the largest group of organisms [1]. Fungi consist of a heterogeneous group including yeasts, moulds, and mushrooms characterized by their lack of photosynthetic pigment and their chitinous cell wall [2]. Hawksworth and Rossman [3, 4] described that the number of fungal species ranged between 72,000 and 120,000, which supposes less than 10% of the theoretically estimated 1.5 million existing fungal species. In contrast, when environmental samples based on metagenomics data are evaluated, the species number increases to as high as 3.5 million [5]. The discovery of the new fungal species described along the last years was evaluated by Hibbett and coworkers [6], which averages 223 species per year, mostly Ascomycota. These data present a huge number of unrecognized and unidentified fungal species (more than 90%), which could be discovered in associations with plants, insects, and animals, as lichen-forming fungi or in undisturbed areas [4].

Traditionally, the four main fungal phyla were Ascomycota (majority of all described fungi) and Basidiomycota, which embody the mushrooms or higher fungi, and the Zygomycota and Chytridiomycota, which represent the basal or lower fungi. The first three phyla theoretically diverged from the last one (Chytridiomycota) approximately 550 million years ago [1, 7, 8]. This was an evolutionary step previous to the land invasion of the plants. The fungal taxonomy is in continuous movement; thus, Zygomycota has been replaced by several subphyla [9], and the arbuscular mycorrhizal fungi Glomeromycota has been included as a new phylum [10]. This evolution of the fungi and fungal-like microorganisms can be observed in the complex phylogenetic tree based on a Bayesian inference analysis reviewed by Voigt and Kirk [8].

The fungal environmental impact is remarkable due to the central role played in the organic matter decomposition process because of their ability to degrade recalcitrant compounds such as lignin; thereby the organic material utilization is enhanced by the microbial community. In addition, some species are involved in disease interactions with humans, plants, or animals, either by means of their direct action as disease agents or through the production of secondary metabolites (e.g., host-specific toxins). Related to this, filamentous fungi produce a diverse array of secondary metabolites and enzymes [11], which have a tremendous impact on society because they are exploited for their antibiotic (penicillins, cephalosporins, etc.) or pharmaceutical (cyclosporin and other immunosuppressants) activities and their industrial applications in white biotechnology (beverage industries). Therefore, fungi are the second most important group (after Actinobacteria) of secondary metabolites producers with industrial application [12].

Among secondary metabolites, antibiotics (and more precisely β-lactam antibiotics) are especially relevant. The discovery of β-lactam antibiotics is one of the most significant milestones of the human history and entailed a revolution in modern chemotheraphy. Members of this family of antibiotics are commonly prescribed due to their high activity and low toxicity and have helped medicine to reduce dramatically the mortality rate. The history of these compounds started up in 1928 after Sir Alexander Fleming's accidental discovery of the antimicrobial activity generated by a fungus culture contaminating a Petri dish cultured with *Staphylococcus* sp. Fleming, who worked at the St. Mary's Hospital in London, initially identified the mould responsible for the antibacterial effect as *Penicillium rubrum* [13]. However, it was not until 1932 when the Fleming's isolate was correctly identified as *Penicillium notatum* and the active compound inhibiting the bacterial growth was dubbed penicillin [14]. Fleming did not extend his work to clinical study due to the low amounts and instability of the penicillin purified from culture broths. It was in 1940 when a group of workers from the Sir William Dunn School of Pathology at Oxford University (H. W. Florey, E. B. Chain, N. Heatley, C. M. Fletcher, A. D. Gardner, M. A. Jennings, J. Orr-Ewing, A. G. Sanders, and E. Abraham) were able to undertake detailed studies on penicillin [15]. However, massive production could not be accomplished in England due to World War II. Therefore, Florey and Heatley went to the United States in June 1941 and convinced the U.S. Department of Agriculture (USDA) and several pharmaceutical firms (Charles Pfizer and Co., E. R. Squibb and Sons, and Merck and Co., among others) to produce penicillin [16]. Heatley remained for a period at the Northern Regional Research Laboratory of the USDA, Peoria, Illinois, USA. The isolation of new strains of *Penicillium chrysogenum*, industrial improvement programs (see below), and medium modifications favored the development of penicillin high-producing strains. The tremendous cooperative effort among universities and industrial laboratories in England and the United States during the war led to multiple large-scale clinical trials to treat those wounded in battle in England and in the United States during 1942 and 1943 [16]. In 1945, Ernst B. Chain, Howard W. Florey, and Sir Alexander Fleming were awarded with the Nobel Prize for Physiology and Medicine. A year later, penicillin was finally available in the open market.

Since the early days of penicillin development, it was rather clear that several different penicillins were being produced, depending on the composition of the medium, the fungal strain, and the fermentation conditions. Penicillins contain a bicyclic "penam" nucleus formed by fused β-lactam and sulphur-containing thiazolidine rings and an acyl side chain bound to the amino group at C-6. The side chain depends on the precursors present in the culture medium. Side chain attachment is relatively nonspecific, and therefore, natural penicillins, such as penicillin F (3-hexenoic as side chain), and K (octanoic acid as side chain), are synthesized under natural conditions. However, feeding the cultivation media with phenylacetic or phenoxyacetic acids directs the biosynthesis mainly towards benzylpenicillin (penicillin G) or phenoxymethylpenicillin (penicillin V), respectively. Biosynthesis, regulation, and evolution of the penicillin pathway have been reviewed extensively [17–19].

Another important milestone that opened a new era of chemotherapy was the detection and isolation of 6-aminopenicillanic acid (6-APA) in fermented broths in the decade of 1950s [20–23]. This penicillin precursor is the basis of semisynthetic penicillins, which are achieved through the addition of different side chains to 6-APA by a chemical process [24].

2. Classical Strain Improvement and Molecular Genomic-Transcriptomic Basis of the Increased Penicillin Productivity

One of the main problems that researchers initially had to face was the low penicillin titers (about 2 i.u. (international units)/mL or 1.2 μg/mL) produced by the original Fleming's *P. notatum* strain (NRRL 1249B21). This, together with the increasing demand of antibiotics derived from World War II and subsequent conflicts during the 20th century, led scientists to isolate new strains with more productivity (Figure 1). Thus, the strain improvement began in 1943 after the isolation of *P. chrysogenum* NRRL 1951 from a mouldy cantaloupe in a local market in Peoria, IL, USA [25, 26]. This strain was more suitable than *P. notatum* for penicillin production in submerged cultures (60 μg/mL). Selection of single spores allowed the isolation of the higher-producing lineage NRRL 1951-B25, which produced up to 150 μg/mL (250 i.u./mL). This strain was X-ray treated by Demerec at the Carnegie Institution (Cold Spring Harbor, NY, USA) giving rise to the X-1612 mutant, which yielded 300 μg/mL of penicillin. This strain was mutated by 275 nm ultraviolet irradiation of spores at the University of Wisconsin, and the Q-176 mutant was obtained. It was tested in 300-liter fermentors, yielding over 550 μg/mL (900 i.u./mL) [27]. This strain represents the origin of the Wisconsin family of superior strains, some producing over 1800 μg/mL (3000 i.u./mL) [28]. After ultraviolet mutagenesis of strain Q-176, the pigmentless strain BL3-D10 was obtained. This was desirable due to commercial purposes, since extraction of the yellow pigment produced naturally by *P. chrysogenum* gave rise to antibiotic losses during the purification process. Therefore, this strain was selected to produce directly the white commercial product although it involved a 25% reduction in penicillin yield [28]. The BL3-D10 strain was subjected to successive colony selection

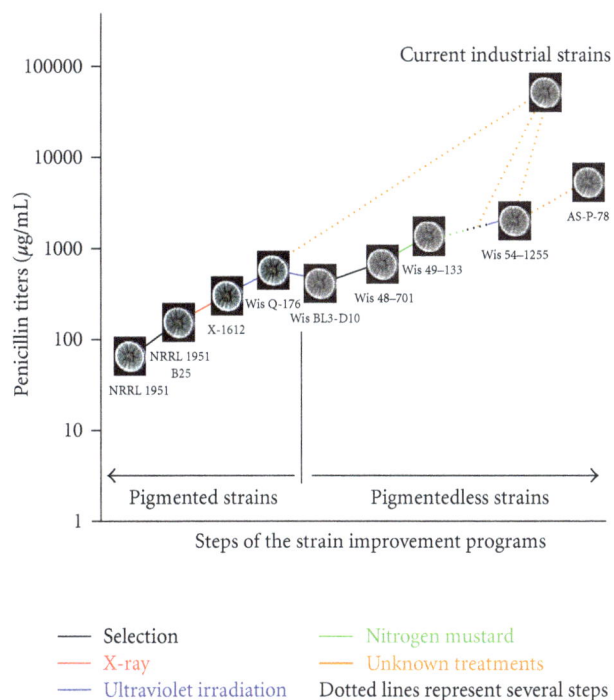

FIGURE 1: Scheme showing the industrial strain improvement program of *P. chrysogenum*, Barreiro et al., 2011.

rounds without mutagenesis to stabilize the mutant strains. In this way, 47–638, 47–1564, and 48–701 were obtained. The 48–701 strain produced around 650 μg/mL of penicillin and was mutated with nitrogen mustard, thus giving rise to the Wisconsin 49–133 strain. This strain represented the greatest single improvement among the Wisconsin pigmentless series, with penicillin yields 100% above that of the 48–701 parent, only 50% as much mycelium, and an "exceptionally efficient utilization of phenylacetic acid" [29]. After several steps of nitrogen mustard mutagenesis followed by several steps of selection and one step of ultraviolet irradiation, the improved-producer Wisconsin 54–1255 was obtained and became the laboratory reference strain [23, 30, 31]. Besides the University of Wisconsin, some companies developed their own industrial strain improvement programs in order to obtain penicillin high-producer strains, but few data are available about these programs. For example, Panlabs Inc. (Taiwan) started the program with strains P1 and P2, which are derived from the Q-176 strain [32], DSM (The Netherlands) developed the DS04825 strain from the Wisconsin 54–1255 strain [33], Wyeth Lab (West Chester, PA, USA) developed the M strains (obtained from the Wisconsin 51–20 strain, one of the ancestors of Wisconsin 54–1255), and Antibioticos S.A. (Spain) produced the AS-P-78, AS-P-99, and E1 strains from the Wisconsin family [34] (Figure 1). Overproducer mutants currently used for the industrial production of penicillin can reach titers of more than 50 mg/mL (83300 i.u./mL) in fed-batch cultures [32, 35].

As a consequence of the strain improvement programs, numerous modifications leading to impressive penicillin titers were introduced in *P. chrysogenum*. Some were

characterized in detail, such as the phenomenon of amplification that occurs in the genomic region containing the three penicillin biosynthetic genes, *pcbAB, pcbC,* and *penDE* (Figure 2(a)). These genes are arranged in a single cluster located on a DNA region present as a single copy in the genome of the wildtype NRRL 1951 and Wisconsin 54–1255 strains. This region undergoes tandem amplification in penicillin overproducing strains, giving rise to several copies of the biosynthetic gene cluster in those strains [34, 36]. Hence, the AS-P-78 strain was reported to contain 5 to 6 copies, the E1 strain 12 to 14 copies [34], and the DS0485 strain 5 to 6 copies [37]. The mechanism of gene amplification is intriguing; it has been suggested that the conserved hexanucleotides located at the borders of the amplified region may be hot spots for site-specific recombination after mutagenesis [34]. It is noteworthy that the amplified region contains other ORFs in addition to the three biosynthetic genes [37, 38]. Since these ORFs are amplified together with the penicillin biosynthetic cluster, a role in penicillin biosynthesis, regulation or secretion might be expected for those ORFs. However, it was demonstrated that the role of those ORFs was not essential for penicillin biosynthesis (although they may contribute to high penicillin production), since the presence of the three penicillin biosynthetic enzymes alone was sufficient to restore β-lactam synthesis in a mutant lacking the complete region [37, 39]. No penicillin pathway-specific regulators have been found in the genome region that contains the penicillin gene cluster, and the biosynthesis of this antibiotic is controlled by global regulators, such as LaeA [40], the velvet complex [41], or StuA [42]. Other genes encoding enzymes important for the biosynthesis of penicillin are also absent from the amplified region. This is the case of the *ppt* gene encoding the phosphopantetheinyl transferase that posttranslationally activates the δ-(L-α-aminoadipyl)-L-cysteinyl-D-valine synthetase (the first enzyme of the penicillin biosynthetic pathway) [43], or the *phl* gene encoding the phenylacetyl-CoA ligase, which activates the aromatic side chain of benzylpenicillin [44]. Interestingly, the amplification of either the *laeA, ppt,* or *phl* genes gave rise to increased penicillin titers [40, 43, 44].

Although high-producing strains contain several copies of the penicillin cluster, it has been reported that penicillin production does not follow a linear correlation with the biosynthetic gene copy number, the transcript, or enzyme levels [36, 45–47], indicating that other modifications are also playing an important role. One of those modifications has also been characterized in detail and is related to the catabolism of phenylacetic acid (the benzylpenicillin side chain precursor). Phenylacetic acid is a weak acid that is toxic to cells depending on the concentration and the culture pH. This compound can be metabolized in *P. chrysogenum* through at least two routes; incorporation to the benzylpenicillin molecule or catabolism via the homogentisate pathway [48–51] (Figure 2(b)). The first step of the homogentisate pathway consists of 2-hydroxylation of phenylacetic acid ring yielding 2-hydroxyphenylacetate. This reaction is catalyzed by a microsomal cytochrome P450 monooxygenase (phenylacetic acid 2-hydroxylase; phenylacetate, NAD(P): oxygen oxidoreductase (2-hydroxylating); EC number 1.14.13),

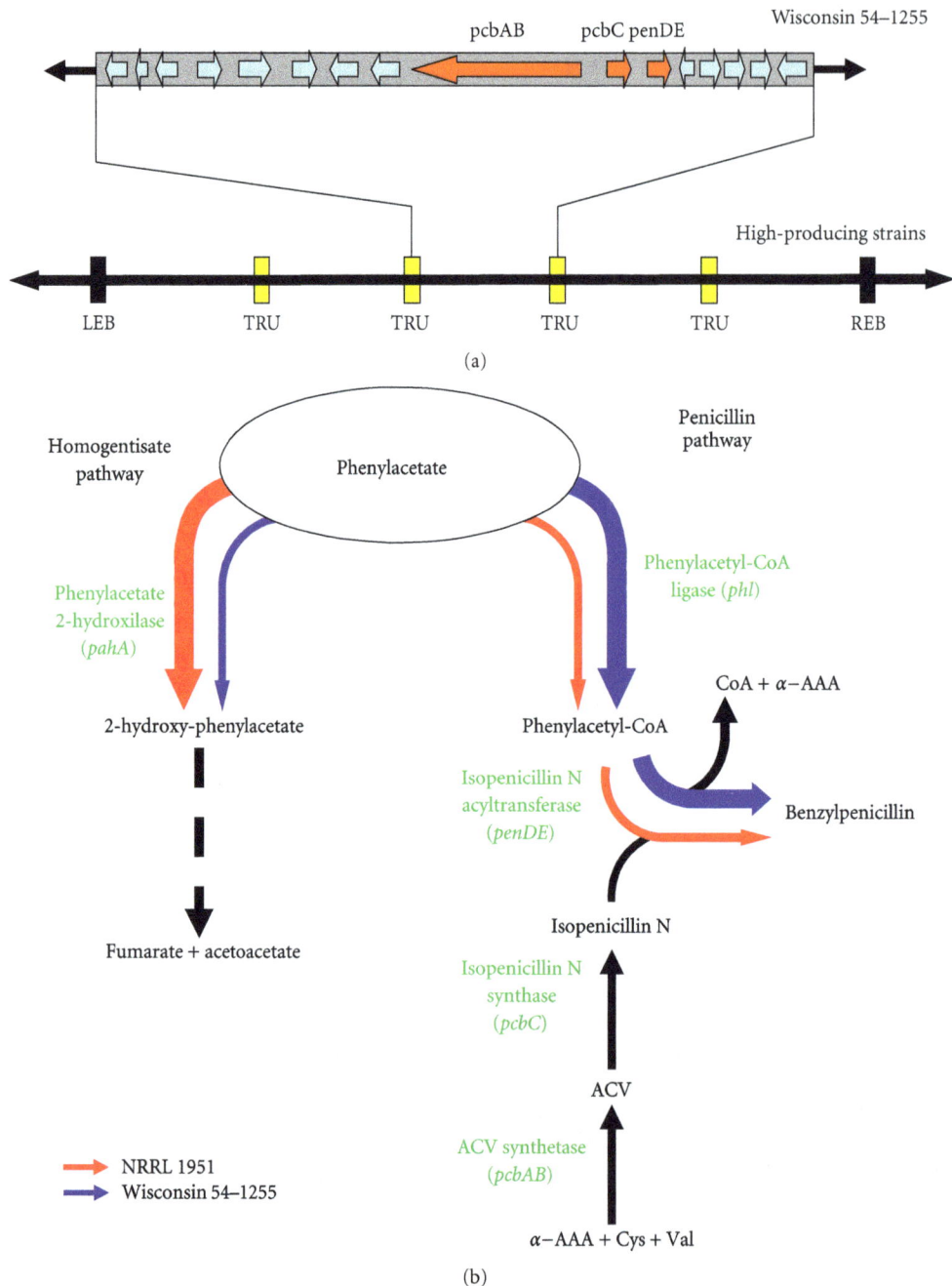

FIGURE 2: Examples of two well-known modifications that occurred during the strain improvement program. (a) Amplification of the DNA region containing the penicillin biosynthetic gene cluster in high-producing strains. LEB: left end border; REB: right end border; TRU: tandem repetition union. (b) Modification in the metabolic fluxes through the homogentisate pathway for phenylacetic acid catabolism and the penicillin biosynthetic pathway. Thickness of arrows indicates the flux rate through a specific enzyme. α-AAA: α-aminoadipic acid; ACV: δ-(L-α-aminoadipyl)-L-cysteinyl-D-valine; CoA: coenzyme A; Cys: cysteine; Val: Valine, Barreiro et al., 2011.

which is encoded by the *pahA* gene. Sequencing of the *pahA* gene from several strains of *P. chrysogenum* revealed that whereas the wildtype strain (NRRL 1951) contains a C at position 598 of the gene, the 49–133 strain (and likewise its descendant Wisconsin 54–1255) replaces this nucleotide with a T [52]. This modification in the gene sequence originates an L181F mutation in the protein, which is responsible for the reduced function of the enzyme. Therefore, the catabo-

lism of phenylacetic acid through the homogentisate pathway is diminished in Wisconsin 54–1255 and, presumably, in derived strains as well, leading to a reduced degradation of phenylacetic acid and to penicillin overproduction [52] (Figure 2(b)). The importance of this enzyme in penicillin production was also highlighted after the comparative analysis of the *pahA* gene of *P. notatum* (the Fleming's isolate) and *P. chrysogenum* NRRL 1951 (the wildtype strain isolated in Peoria). The latter strain shows a C1357T (A394V) mutation in this gene that is conserved in Wisconsin 54–1255 (and presumably, in derived strains as well). This mutation is responsible for a loss of function in the phenylacetic acid 2-hydroxylase, which is directly related to penicillin overproduction and supports the historic choice of *P. chrysogenum* as the industrial producer of penicillin [53]. Therefore, the consecutive accumulation of point mutations in the phenylacetic acid 2-hydroxylase is another reason that made *P. chrysogenum* a good penicillin producer.

A direct relation between microbody (peroxisome) abundance and increased penicillin titers was also observed [54]. Microbodies are organelles involved in the final steps of the penicillin pathway, which are catalyzed by the acyl-coenzyme A: isopenicillin N acyltransferase (IAT) and phenyl-acetyl-CoA ligase [18, 55, 56]. In fact, it has been reported that microbody volume fractions are enhanced in high-producing strains [57]. Compartmentalization of the penicillin pathway between cytosol and microbodies has been crucial for productivity. *P. chrysogenum* seems to have lost its capability to synthesize penicillins in the cytosol [58], and peroxisomes are required for efficient penicillin biosynthesis [59].

In addition to these observations, some modifications remained hidden until the arrival of the "omics" era. The publication of the genome sequence of *P. chrysogenum* Wisconsin 54–1255 [57] represented the ultimate springboard for several studies aimed to decipher the genetic secrets of the industrial production of penicillin. Some modifications that have been reported are the introduction of a 14 bp repeat in a cephalosporin esterase homolog between Q-176 and Wisconsin 54–1255 (disturbing the ORF), which suggests a reduction in unwanted β-lactam degradation, or a mutation in an ABC transporter homolog between Q-176 and Wisconsin 54–1255, suggesting a change in transport capabilities [60]. Transcriptomics data revealed that expression of genes involved in biosynthesis of the amino acid precursors for penicillin biosynthesis (cysteine, valine, and α-aminoadipic acid) was higher in the high-producing strains. The same trend was followed by several genes encoding microbody proteins [57]. Interestingly, the transcription of the penicillin biosynthesis genes *pcbAB*, *pcbC*, and *penDE* was only two-to fourfold higher in the high-producing strain [57] suggesting that changes in other metabolic pathways related to penicillin biosynthesis have contributed as much to penicillin biosynthesis as the amplification of the three direct biosynthetic genes. Another interesting study dealing with transcriptomics was carried out by Harris and coworkers [61]. These authors analysed the expression responses of *P. chrysogenum* to the addition of the side-chain precursor phenylacetic acid and found that the homogentisate pathway genes, as well as those genes involved in nitrogen and sulfur metabolism, were upregulated strongly in those cultures supplemented with the side chain precursor.

Although the above-mentioned reports on genomics and transcriptomics provided a global overview of some mechanisms contributing to increased penicillin production, full exploitation of *P. chrysogenum* still required the integration of other "omics," such as proteomics. In fact, as we will show in next sections, the study of the proteome was crucial

to decipher the molecular mechanisms responsible for the improved productivity of *P. chrysogenum*.

3. Methodology of the Intracellular Protein Extraction in *P. Chrysogenum*

As previously described, the vast knowledge accumulated along decades about the *P. chrysogenum* metabolism, and penicillin production contrasts with the scare information generated about methodology related with the "omics" approach, particularly with proteomics. Thus, the most basic procedures such as good-quality protein extraction for bidimensional analysis should be updated. A bibliographic search shows different procedures for protein extraction, precipitation, or protein solubilization that should be adapted for *P. chrysogenum*. These extraction procedures, in the case of free-living microorganisms, have some common steps, such as the mechanical breaking, the use of protease inhibitors, reducing agents, and TCA precipitation for intracellular proteomes that help in the protocol update.

The process to obtain cytoplasmic proteins should include an initial step to discard the media, unbroken mycelia, cell wall, and membrane contaminants in order to present the intracellular proteome as clean as possible. Thus, when cultures reach the desired conditions, the most common mycelia-collection process is filtration, since centrifugation of mycellial microorganisms does not generate compact pellets that make the subsequent washing steps difficult. This procedure can be done by filtering through a Whatman 3 MM paper (Whatman, Maidstone, England) or nylon filters (Nytal Maissa, Barcelona, Spain) [62–65]. Hence, mycelia is collected and washed, and the media can be stored for further analysis of secreted proteins (Figure 3). Washing steps, which allow the media elimination that can interfere in the protein purification, are usually done at 4°C so as to diminish the protein lysis by intracellular proteases. The most commonly employed washing solutions are (i) water [64, 66, 67]; (ii) phosphate buffered saline (PBS) [68], or (iii) combination of 0.9% sodium chloride and water [69]. In order to prevent protein degradation, samples collected at different time points or conditions are washed, paper-dried, and immediately stored at −20°C to −80°C for several months.

Regarding the intracellular protein extraction, keeping the temperature as low as possible and the addition of protease inhibitors are two common steps included in all the protocols. Thus, the use of liquid nitrogen to decrease warming of the disruption systems is a widely used method. The main breaking system is the traditional prechilled mortar grinder due to its efficiency against the fungal cell wall [68, 70, 71] although waring blender machines [62], or glass bead-beating systems, either combined with a 10 mM Tris-HCl buffer [72] or with a phenol buffer [64, 70], have been successfully applied. After the breaking step, the protein solubilization buffer always includes a protease inhibitor (e.g., protease inhibition cocktail for fungi and yeast (Sigma), COMPLETE (Roche)) [63, 69], in addition to a reducing agent (e.g., 2-mercaptoethanol (BME), Dithiothreitol (DTT)) that reduces the disulfide linkages between two

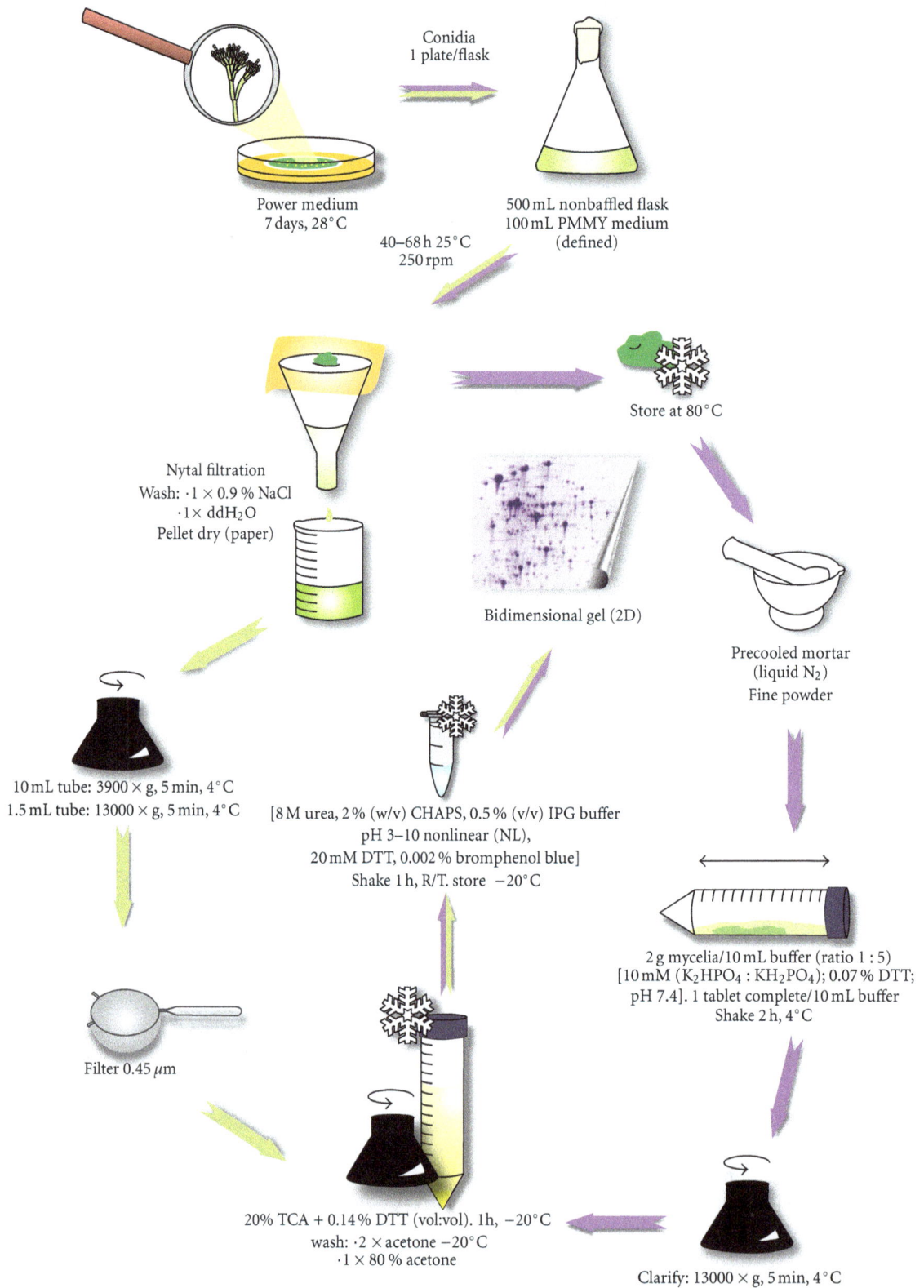

FIGURE 3: Schematic representation of the optimized method for mycelia and secreted proteins collection of *P. chrysogenum* in order to analyze the intracellular and extracellular proteomes. Green/purple-coloured arrows show the common steps for both methods. Green arrows represent the specific method for secreted proteins isolation. Purple arrows show those steps specific for intracellular proteome isolation, Barreiro et al., 2011.

cysteines [63, 66]. Frequently, the reducing agent is added to the precipitation solution (0.093% BME, 0.14% DTT). This precipitation step (previous to resuspension in sample solution) is employed to selectively purify proteins from contaminants such as salts, nucleic acids, detergents, or lipids, which interfere the final bidimensional analysis. Although the combination of trichloroacetic acid (TCA) and acetone is usually more effective than either TCA or acetone alone [73], other mixtures, such as methanol/ammonium acetate, have been described for fungi [70]. Optionally, a nuclease mix (0.5 mg mL^{-1} DNase, 0.25 mg mL^{-1} RNase, and 50 mmol L^{-1} MgCl$_2$; or commercially available, for example, Benzonase (Merck)) can be used as additional treatment before sample precipitation [68, 74]. When the precipitation step is omitted, direct solubilization in homogenization buffer is done, which includes fungal DNase/RNase as described by Oh and coworkers [72].

Proteins may be difficult to resolubilize and may not resolubilize completely after TCA precipitation ("2-D Electrophoresis. Principles and Methods"; GE Healthcare). Thus, residual TCA must be removed intensively by cold acetone washing steps, for example, (i) 2x acetone plus 0.07% DTT and 1x 80% acetone [69]; (ii) 2x acetone containing 1% BME [66, 67]; (iii) 3x acetone/0.3% DTT (wt/vol) [70]. A helpful tip after TCA precipitation and washing is to avoid long drying periods of time in order to improve protein resuspension (1-2 minutes is enough). The main components of the final sample buffer or homogenization buffer described in fungal bibliography have urea (7–9 M), thiourea (0–2 M), CHAPS (1–4% w/v), and ampholyte (0,5–2%) combined as follows:

(i) 8 M urea, 2 M thiourea, 1% (w/v) CHAPS, 20 mm DTT, and 0,5% (v/v) ampholyte 3–10 [65];

(ii) 8 M urea, 2% CHAPS, 20 mM DTT, and 0.5% (v/v) biolytes [63];

(iii) 7 M urea, 2 M thiourea, 4% CHAPS (w/v), 20 mM DTT, and 1.0% IPG buffer (v/v) [67];

(iv) 7 M urea, 2 M thiourea, 4% (w/v) CHAPS, 0.8% (v/v) ampholytes, 20 mM Tris, and 20 mM DTT [66];

(v) 7 M urea, 2 M thiourea, 2% (w/v) CHAPS, 1% (w/v) Zwittergent 3–10, and 20 mM Tris [70];

(vi) 7 M urea, 2 M thiourea, 4% (w/v) CHAPS, 1% (w/v) DTT, 20 mM Tris, and 1% (v/v) Pharmalyte pH 3–10 [68];

(vii) 8 M urea, 4% CHAPS, 40 mM Tris-pH 7.4, 100 mM DTT, and 0.2% ampholyte [72].

A protocol described by Kniemeyer and coworkers [66] for protein extraction of Aspergillus fumigatus summarized part of the steps described above, since after the mycelia homogenization in a pre-cooled (in liquid nitrogen) mortar they did the extraction, cleaning, and precipitation steps together, using the Clean-Up kit (GE Healthcare). This procedure produced quite good results when it was applied to P. chrysogenum (Figure 4(a)) showing a good protein yield. However, some problems, such as the poor representation of large proteins (>75 kDa), stripping in the basic region, and the improper cleaning of the middle-sized protein region (50–100 kDa), were observed. Notwithstanding the fact that some problems are described, this procedure can be applied for fungal protein extraction.

Based on the methods described by Fernández-Acero and coworkers [63] for Botrytis cynera, the mortar grinding approach combined with a phosphate buffer extraction was used for P. chrysogenum. This method improved the final bidimensional gel results (Figure 4(b)), replacing the lacks observed in the direct Clean-Up kit method. In addition, this procedure yielded large protein amounts, which allowed blue silver Coomassie colloidal staining [76], thus simplifying the 2D gels handling.

As a conclusion of the protocols and the experimental work found in literature, the best protein extraction method for P. chrysogenum (Figure 3) was the one described in our laboratory by Jami and coworkers [69], which consisted of the following.

(i) Filtrate through nylon filters (Nytal membrane), 1x 0.9% NaCl washing, 2x ddH2O washing, paper dried and stored at −80°C.

(ii) Grinding of two grams of mycelia to a fine powder in a pre-cooled mortar using liquid nitrogen.

(iii) Stirring the homogenized powder at 4°C for 2 h in 10 mL (ratio: 1 g/5 mL) of phosphate buffer (10 mM potassium phosphate buffer (K$_2$HPO$_4$:KH$_2$PO$_4$) (pH 7.4) containing 0.07% (w/v) DTT and supplemented with tablets of protease inhibitor mixture complete (one tablet/10 mL of buffer)). Clarify by centrifugation at 13000 × g for 5 min twice.

(iv) Protein precipitation for 1 h at −20°C after addition of 1 volume of 20% TCA in acetone containing 0.14% (w/v) DTT. Harvest the protein pellet at 4,000 × g, 5 minutes at 4°C. Washing steps: 2x acetone and 1x 80% acetone (salt elimination). Solubilization of the final pellet was done by shaking 1 h at room temperature in 500 μL of sample buffer (8 M urea; 2% (w/v) CHAPS; 0.5% (v/v); IPG buffer pH 3–10 nonlinear (NL) (GE Healthcare); 20 mM DTT; 0.002% bromophenol blue).

(v) Centrifugation at 13,200 rpm for 1 min to discard the insoluble fraction. The supernatant was collected, and protein concentration was determined according to the Bradford method [77].

As a result of the protein extraction method optimization, an interactive reference map suitable for protein location of P. chrysogenum Wisconsin 54–1255 (reference strain) was uploaded at the website: http://isa.uniovi.es/P_chrysogenum_proteome/.

4. Industrial Strain Evolution of P. chrysogenum from the Proteomics Point of View

The evolution of the penicillin producer strains, as previously described (Section 2), has been a complex process of random

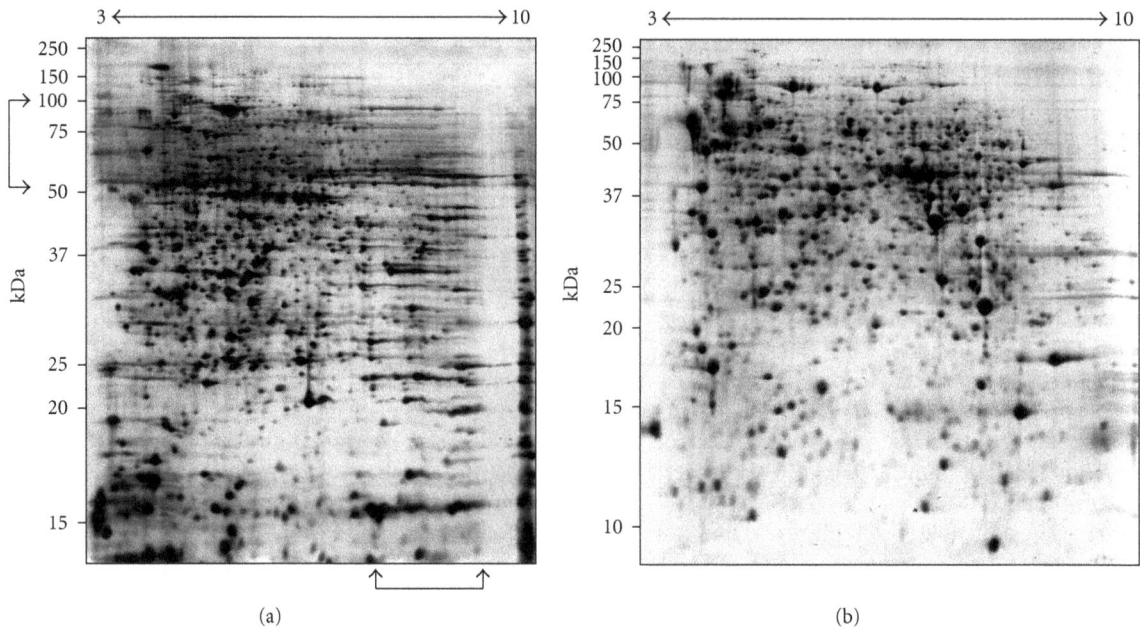

(a)   (b)

FIGURE 4: Optimization of the protein extraction protocol from mycelia of *P. chrysogenum*. (a) Extraction and purification of protein samples with "Clean-Up Kit" (GE Healthcare), based on Kniemeyer and coworkers [66] and silver stained. As sample buffer was used, 7 M urea, 2 M thiourea, 4% (w/v) CHAPS, 0.8% (v/v) IPG buffer pH 3–10 nonlinear (NL) (GE Healthcare), 40 mM Tris, 1 mM EDTA, and 20 mM DTT [75]. (b) The extraction is based on the method described by Fernández-Acero and coworkers [63] by using mortar gridding and phosphate buffer, plus blue silver Coomassie colloidal staining [76]. As sample buffer was used, 8 M urea, 2% (w/v) CHAPS, 0.5% (v/v) IPG buffer pH 3–10 NL (GE Healthcare), 20 mM DTT, and 0.002% bromophenol blue [69]. Precision plus protein standards (Bio-Rad) were used as markers. The molecular mass is indicated in kilo-Daltons (kDa). Note the problematic regions observed with the extraction method A, which are highlighted by arrowed square brackets (left and bottom), Barreiro et al., 2011.

mutagenesis and further screening. This system accumulates beneficial unknown mutations, which increased the final titers of penicillin, but does not provide clues about the modified genes or proteins. The recent application of proteomic techniques has allowed the identification of several of these modifications, which opens the option for the genome breeding approach to fungi following the steps developed in other microorganisms [78]. Along the strain improvement process, several metabolic pathways have been redirected to obtain (i) more precursors, (ii) energy-supplying molecules for the penicillin synthesis in parallel to (iii) decrease in virulence or pigment production (Figure 5). Comparison among three different strains of *P. chrysogenum*: (i) NRRL 1951 (wildtype strain), (ii) Wisconsin 54–1255 (model strain; moderate production), and (iii) AS-P-78 (high-producing strain) showed a clear picture of the industrial improvements [69] (Figure 1). Thus, the three precursor amino acids (L-α-aminoadipic acid, L-cysteine and L-valine) involved in the nonribosomal condensation, which is the initial step of penicillin synthesis, presented a decrease of their catabolic pathways: (i) cysteine catabolism (cystathionine β-synthase); (ii) valine catabolism (methylmalonate-semialdehyde dehydrogenase); (iii) fatty acid degradation (phytanoyl-CoA dioxygenase family protein) and an increase of their biosynthetic pathways involved in the penicillin synthesis: (i) cysteine synthesis (cysteine synthase), (ii) valine synthesis (branched-chain amino acid aminotransferase), and (iii) branched amino-acid synthesis (dihydroxy-acid

dehydratase). Surprisingly, the high-producer strain (AS-P-78) does not present a significant overrepresentation of the proteins directly involved in the penicillin production, despite the fact that AS-P-78 contains five or six copies of the amplified region that includes the penicillin biosynthetic genes [34, 38] (see Section 2). Similar results were obtained with transcriptomic analysis using other high-producing strains [57]. In fact, the lack of a linear relationship between the number of copies of the penicillin gene cluster and the penicillin titers was previously observed [36]. In summary, it was concluded that penicillin gene cluster amplification is not the main mechanism that transformed the wildtype strain (NRRL 1951) into a penicillin overproducer (AS-P-78), even when it had a clear impact on production.

The penicillin titers are also increased in high-producing strains by the downregulation of penicillin modification or degradation-related proteins, such as an esterase, which contains a β-lactamase domain possibly involved in penicillin degradation. Besides, there are proteins involved in other secondary metabolite production pathways underrepresented in the high-producing strain, which decreased their expression throughout industrial strain evolution. A 3-hydroxy-3-methylglutaryl-CoA synthase and a coproporphyrinogen oxidase III involved in terpenoid synthesis and porphyrin metabolism, respectively, are clear examples of this redirection of metabolism [69].

The wildtype strain (NRRL 1951) presents a variety of proteins involved in pigment formation, in agreement with

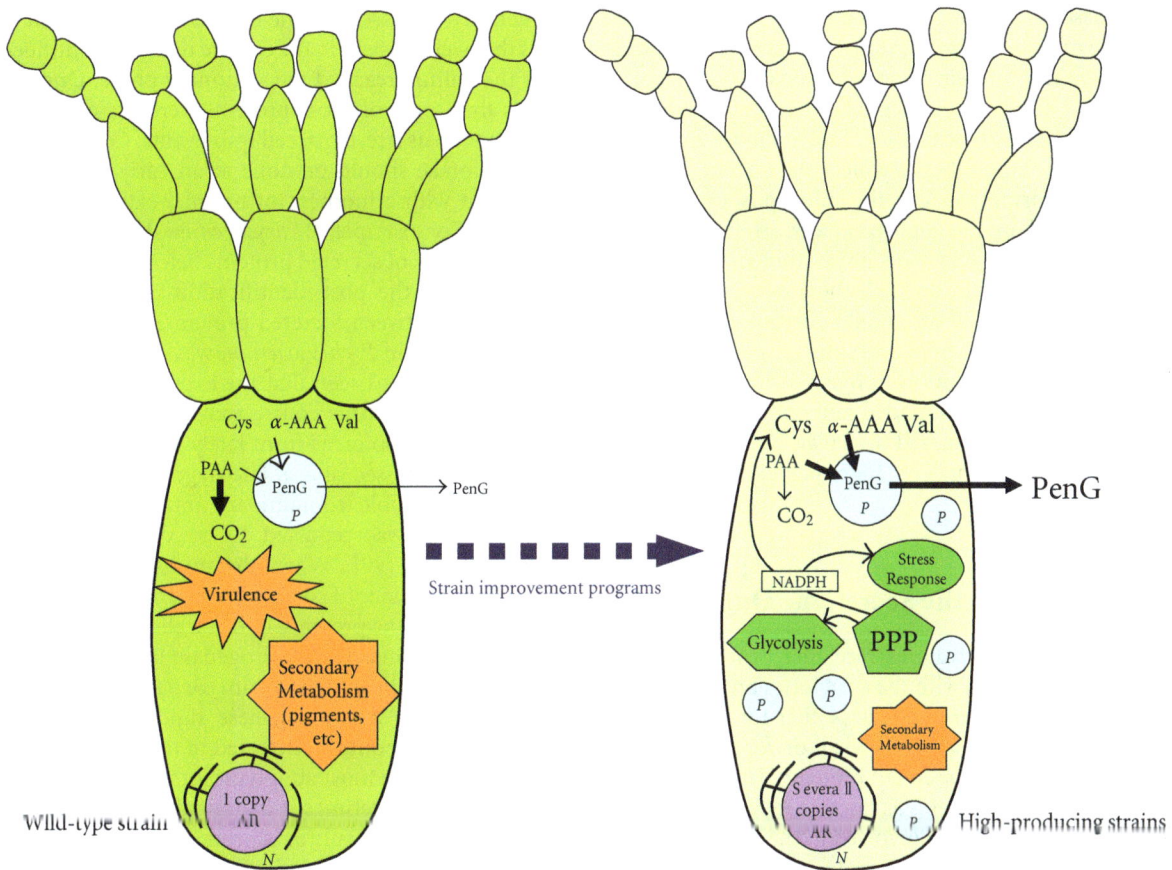

FIGURE 5: Pathways and networks modified during the strain improvement program. Font size is in concordance with concentration and differences; the thickness of arrows indicates the flux through a specific pathway. α-AAA: α-aminoadipic acid; Cys: cysteine; N: nucleus; NADPH: reduced form of nicotinamide adenine dinucleotide phosphate; P: peroxisome; PAA: phenylacetic acid; PenG: benzylpenicillin; PPP: pentose phosphate pathway; Val: valine, Barreiro et al., 2011.

the well-known ability of this strain to synthesize large amounts of pigment, which decreased or disappeared in the industrial strains. Thus, a zeaxanthin epoxidase involved in natural xanthophyll pigment production is 11-fold downrepresented in the AS-P-78 strain, and a scytalone dehydratase (involved in the biosynthesis of melanin protecting hyphae and spores from UV light, free radicals, and desiccation in addition to serving as a virulence factor [79]) is only present in the wildtype strain. As it was mentioned above, during the mutagenic treatments (from Q-176 to BL3-D10 strains), pigmentless strains were selected (Figure 1).

The analysis of the cytoplasmatic proteome has also shown a rearrangement of the internal energetic fluxes through the glycolysis and pentose phosphate pathways to the penicillin biosynthetic pathway, which constitutes a major burden on the supply of NADPH. In the one hand, the AS-P-78 (high production) strain overproduces enzymes of the pentose phosphate pathway like ribose-5-phosphate isomerase or transketolase [69]. It is likely that a high concentration of ribulose 5-phosphate, together with reducing power in the form of NADPH, is formed in the high-producer strain. Thus, this sugar phosphate excess is converted by the ribose-5-phosphate isomerase into ribulose 5-phosphate (precursor of nucleic acids and nucleotides) and glycolysis

precursors by the transketolase, which connects both metabolic pathways (pentose phosphate and glycolysis) [69]. On the other hand, the NADPH level is strongly correlated to β-lactam production (8–10 mol of NADPH are required for the biosynthesis of 1 mol of penicillin) [80]. In addition, the ATP level and cysteine concentration have been reported to have a great influence over the penicillin production [81]. Because NADPH is required for cysteine biosynthesis, a positive relationship among cysteine, NADPH levels, and penicillin production can be established [81]. In addition, the increase in cysteine biosynthesis could be also related to an increased demand for glutathione biosynthesis to cope with higher levels of the oxidative stress related to high aeration rates in the high penicillin producer strain. Besides, the oxidized glutathione during oxidative stress requires NADPH to reduce it. Therefore, the connection between NADPH, oxidative stress, and pentose phosphate pathway has been well established by means of the proteomics approach in *P. chrysogenum* [69] (Figure 5).

In concordance with the underrepresentation of the proteins involved in pigment formation, as well as other secondary metabolites in the high-producer strains (less energy consumption), the proteins involved in virulence are also underrepresented. This situation has been observed with a glucose oxidase, which has been described as involved in the pathogenicity of *Penicillium expansum* in apples [82] and with a host infection protein (Cap20), which plays a significant role in the fungal virulence on avocado and tomato fruits [83]. Besides, UDP-galactopyranose mutase involved in galactofuranose biosynthesis, which plays a crucial role in cell envelope formation and the infectious cycle [84], is downrepresented in the high-producer strain AS-P-78. Therefore, the decrease of proteins involved in virulence and plant cell wall degradation through the strain evolution is substantiated by the specialization on industrial raw material consumption by the mutagenized strains.

In summary, the proteomics approach through the intracellular proteome analysis has added more relevant information about the events that led to create "domesticated" strains, which are the current high penicillin producer strains.

5. How to Guarantee the Secretome Quality?

Even when the analysis of the extracellular proteome does not suggest a clear relation with the antibiotics production in *P. chrysogenum*, it can offer new clues either for the general understanding of the fungal metabolism or for possible industrial applications of the secreted proteins. This is the case of other secreted fungal proteins as those of *Aspergillus oryzae* or *Aspergillus niger*, which are involved in the production of fermented foods and beverages or used for the production of organic acids, food ingredients, and industrial enzymes, respectively [85, 86]. In addition, white rot fungi (the degradation of hemicelluloses and lignin by basidiomycetes and ascomycetes) and brown rot fungi (the attack of cellulose and hemicelluloses, exclusively, accomplished by basidiomycetous) are involved in the degradation of lignocelluloses, which represent the world's largest renewable carbon source [87]. This huge biomass coming from plant primary metabolism, which is about 140 billion tons annually [88], is proposed as a possible substitute of the limited availability of fossil fuels and can be utilized easily and bioconverted with the help of fungal white biotechnology.

The most crucial step for the secreted protein analysis is to distinguish the proteins really secreted from those contaminant proteins coming from cell lysis events. This capital process can be tackled sequentially from three different points of view: (i) sample collection; (ii) post-identification protein analysis; (iii) "moonlighting proteins" and unconventional secreted protein analysis.

Some tips in the collection of secreted protein are shown in Figure 3, such as the intensive mycelia elimination process (the use of nylon filters, rigorous centrifugation, and filtration through 0.45 μm filters) or the use of low temperatures to avoid mycelia degradation. Sample collection is also crucial to discard those non-natural secreted proteins. It is described in the scientific literature how the presence of extracellular proteins in the culture medium is correlated directly to the growth phase [89]. Jami and coworkers [90] showed the linear correlation existing between the presence of proteins in the culture medium and the culture growth (biomass

formation). Interestingly, the increase in the amount of proteins that were present in the culture medium continued even after the culture reached the stationary phase. One explanation is that some of those proteins were present in the culture medium as a result of cell lysis events. Consequently, the sample uptake should be done at an early time point, but coincident with a moment of optimal extracellular protein secretion. For example, *P. chrysogenum* does not show significant amounts of secreted proteins before 24 hours [90].

Related to the post-identification protein analysis, the comparison between secreted proteins at 68 h versus those secreted at 40 h in *P. chrysogenum* was done as an additional control [90]. It was seen that the expression differences observed during the time course were mainly due to the presence at 40 h of isoforms from lately expressed proteins and to proteins expressed only at late stages. As a conclusion, the amount of possible contaminant intracellular proteins in the secretome was reported to be very low (6.09%) in *P. chrysogenum* [90]. When the intracellular proteome is known, the presence of contaminant proteins due to cell lysis can be tested by the presence of intracellular high-abundance proteins. Thus, malate dehydrogenase and glyceraldehydes-3-phosphate dehydrogenase in *Botrytis cinerea* or the flavohemoglobin and manganese superoxide dismutase in *P. chrysogenum* should be detected in the case of lysis events, since these are the most abundant intracellular proteins [69, 91]. However, sometimes it is difficult to ascertain whether a protein is secreted truly or is present in the culture medium as a consequence of cell lysis. Recently, new bioinformatics tools have been developed to characterize the secreted proteins either in a classical or non-classical way by means of prediction of secretion signal motifs. SignalP (for classical secretion signal motifs) and SecretomeP (for nonclassical signal motifs) are useful prediction softwares [92, 93] that can be used for such purpose. The availability of several fungal genomes in databases together with the use of those prediction programs for secretory proteins has allowed the recent development of platforms for the annotation of fungal secretomes, such as the Fungal Secretome Database [94] or the Fungal Secretome KnowledgeBase (FunSecKB [95]), which allow the proper identification of secreted proteins.

Even when the above-mentioned information is taken into account and you have the certainty of working with secreted proteins, some unconventional proteins, dubbed "moonlighting proteins" [96], can be detected. These proteins present unconventional protein secretion systems [97–99], and then, it is difficult to define their authenticity as secreted proteins. One meaning of moonlighting, according to The Compact Oxford English Dictionary [100], is "to do paid work, usually at night, in addition to one's regular employment"; therefore, "moonlighting proteins" is a quite appropriate designation for the collective of multifunctional proteins, which are widespread among organisms ranging from bacteria to mammals [101]. In spite of this ubiquity, more attention has been traditionally given to the "moonlighting proteins" found in higher eukaryotes, but recently the yeasts have caught the attention for lower eukaryotes [101, 102]. Thus, in many cases, the fusion of two genes that initially encoded proteins with single functions has been the origin of that duality. On the other hand, a significant number of proteins can perform dissimilar functions [101, 102].

Another phenomenon that has been observed in the *P. chrysogenum* secretome is the fact that some extracellular proteins identified in the reference map were also identified previously by Kiel and coworkers [103] in the microbody matrix of this fungus. Therefore, an alternative explanation given to the presence of intracellular proteins in the culture broth was the selective autophagic degradation of peroxisomes (pexophagy), which can give a new sense to the strange "moonlighting proteins" as we explained in *P. chrysogenum* [90]. It is well known that peroxisome abundance can be decreased rapidly through autophagic pathways, which selectively degrade peroxisomes by fusion to lysosomes or vacuoles [104]. Integration of peroxisomes into vacuoles may lead to secretion of the proteins located in the peroxisomal matrix by exocytosis, a mechanism that has been discussed as an alternative route for the release of penicillin from peroxisomes to the culture medium [18].

Taking into account all these factors, the analysis of the extracellular proteome of *P. chrysogenum* has been completed fully [90]. It can be consulted in an interactive way by means of the reference map present at the website: http://isa.uniovi.es/P_chrysogenum_secretome/.

6. *Penicillium chrysogenum*: A Versatile Cell Factory for Biotechnology

Genome sequence of *P. chrysogenum* revealed the presence of several nonribosomal peptide synthetases (NRPSs), polyketide synthases (PKSs), and hybrid NRPS-PKS enzymes [57], highlighting the potential of this microorganism to produce other secondary metabolites different from penicillin. Even more interesting than the production of homologous secondary metabolites is the potential use of *P. chrysogenum* as a host for the biosynthesis of heterologous secondary metabolites. In fact, due to the good metabolic fluxes of the penicillin biosynthetic pathway, *P. chrysogenum* was engineered to produce cephalosporin intermediates [105–107]. The available information about fermentation conditions and optimization of *P. chrysogenum* to industrial conditions has promoted the development of host strains that lack the penicillin gene cluster [61, 108], which could be optimal for the production of heterologous metabolites.

In addition to secondary metabolites, secreted enzymes also reflect the versatility of *P. chrysogenum*. The analysis of the secretome of this fungus [90] has also suggested the potential interest of nonpenicillin producer strains for other biotechnological uses, such as food and beverage industries. Some proteins identified in the culture medium of *P. chrysogenum* are of interest for this biotechnological area. This is the case of a probable isoamyl alcohol oxidase, an extracellular enzyme in *A. oryzae* catalyzing the formation of isovaleraldehyde, which is the main component of mureka that gives sake an off-flavor [109]. Another example is sulphydryl oxidase, which may be used for treatment of bakery products or for removal of off-flavour from UHT-treated

milk or beer [110]. A relevant enzyme for its application in food industry is dihydroxy-acid dehydratase, which removes vicinal diketones. These compounds are responsible for the off-flavor of diacetyl (butter flavor or sweaty flavor in beer, "tsuwari-ka," or nauseating flavor in sake) in brewed alcoholic beverages. Overproduction of dihydroxy-acid dehydratase in *S. cerevisiae* has been patented to produce alcoholic beverages with superior flavor [111]. It is also important the secretion of a prolyl aminopeptidase during the ripening of the Camembert cheese since it decreases the amount of bitter tasting peptides [112]. Other proteins, such as pectinolytic enzymes, which play an important role in beverage and juice industries, were also found in the secretome of *P. chrysogenum*. Additional relevant secreted enzymes are ferulic acid esterases, which release ferulic acid (a naturally occurring hydroxycinnamic acid derivative with beneficial medical effects [113]) from plant cell walls. It has been proposed that ferulic acid esterases have the ability not only to degrade plant biomasses, but also to synthesize novel bioactive components [114].

Although there are other systems for the production of useful proteins, such as bacteria, insects, mammalian cell culture, transgenic animal, plant cell cultures, and transgenic plants, the versatility of *P. chrysogenum*, its plasticity, its ability to grow on different media and conditions, and the availability of several mutants and tools for its genetic manipulation may confer some advantages on *P. chrysogenum* over other systems to produce and secrete high amounts of homologous and heterologous proteins.

7. Concluding Remarks

The new "omics" era has allowed scientists, 80 years after the Fleming's fortuitous discovery of the penicillin-producing fungus *Penicillium notatum*, to start to comprehend the molecular basis of the effect of six decades of classical strain improvement. Besides, the advances in microbiology, biochemistry, genetics, and molecular biology have contributed to the knowledge of structural, ancillary, and regulatory biosynthetic genes and enzymes, together with the characterization of the subcellular compartments involved in the biosynthesis of β-lactam antibiotics. Therefore, genomics, transcriptomics, proteomics, metabolomics, and fluxomics studies have revealed that major modifications in primary and secondary metabolism (phenylacetic acid catabolism, peroxisome biogenesis, pentose phosphate pathway, redox metabolism, biosynthesis of precursor amino acids, penicillin biosynthetic enzymes, etc.) occurred during the improvement programs, rebalancing several cellular processes that led to the impressive penicillin production reached by the current industrial strains. As conclusion, these findings will help the scientific community from industry and academia not only to continue the exploitation of the successful synergy existing between *P. chrysogenum* and β-lactam antibiotic production, but also to explore the production of other compounds using this versatile fungus as a cell factory.

Acknowledgments

This work was supported partially by Grants from the European Union (Eurofung Grant QLRT-1999-00729 and Eurofungbase). C. Barreiro was supported by the European Union program ERA-IB (BioProChemBB project (EIB.08.008)). The authors acknowledge the technical assistance of the INBIOTEC staff: B. Martín, J. Merino, A. Casenave, and A. Mulero.

References

[1] J. Guarro, J. Gené, and A. M. Stchigel, "Developments in fungal taxonomy," *Clinical Microbiology Reviews*, vol. 12, no. 3, pp. 454–500, 1999.

[2] D. L. Hawksworth, "The fungal dimension of biodiversity: magnitude, significance, and conservation," *Mycological Research*, vol. 95, pp. 641–655, 1991.

[3] D. L. Hawksworth and A. Y. Rossman, "Where are all the undescribed fungi?" *Phytopathology*, vol. 87, no. 9, pp. 888–891, 1997.

[4] D. L. Hawksworth, "The magnitude of fungal diversity: the 1.5 million species estimate revisited," *Mycological Research*, vol. 105, no. 12, pp. 1422–1432, 2001.

[5] H. E. O'Brien, J. L. Parrent, J. A. Jackson, J. M. Moncalvo, and R. Vilgalys, "Fungal community analysis by large-scale sequencing of environmental samples," *Applied and Environmental Microbiology*, vol. 71, no. 9, pp. 5544–5550, 2005.

[6] D. S. Hibbett, A. Ohman, and P. M. Kirk, "Fungal ecology catches fire," *New Phytologist*, vol. 184, no. 2, pp. 279–282, 2009.

[7] M. L. Berbee and J. W. Taylor, "Ascomycete relationships: dating the origin of asexual lineages with 18S ribosomal RNA gene sequence data," in *The Fungal Holomorph: Mitotic, Meiotic and Pleomorphic Speciation in Fungal Systematics*, D. R. Reynolds and J. W. Taylor, Eds., pp. 67–78, CAB International, Wallingford, UK, 1993.

[8] K. Voigt and P. M. Kirk, "Recent developments in the taxonomic affiliation and phylogenetic positioning of fungi: impact in applied microbiology and environmental biotechnology," *Applied Microbiology and Biotechnology*, vol. 90, no. 1, pp. 41–57, 2011.

[9] D. S. Hibbett, M. Binder, J. F. Bischoff et al., "A higher-level phylogenetic classification of the Fungi," *Mycological Research*, vol. 111, no. 5, pp. 509–547, 2007.

[10] A. Schüßler, D. Schwarzott, and C. Walker, "A new fungal phylum, the Glomeromycota: phylogeny and evolution," *Mycological Research*, vol. 105, no. 12, pp. 1413–1421, 2001.

[11] J. M. Palmer and N. P. Keller, "Secondary metabolism in fungi: does chromosomal location matter?" *Current Opinion in Microbiology*, vol. 13, no. 4, pp. 431–436, 2010.

[12] C. Olano, F. Lombo, C. Mendez, and J. A. Salas, "Improving production of bioactive secondary metabolites in actinomycetes by metabolic engineering," *Metabolic Engineering*, vol. 10, no. 5, pp. 281–292, 2008.

[13] A. Fleming, "On the antibacterial action of cultures of a penicillium, with special reference to their use in the isolation of B. influenzae," *British Journal of Experimental Pathology*, vol. 10, pp. 226–236, 1929.

[14] P. W. Clutterbuck, R. Lovell, and H. Raistrick, "Studies in the biochemistry of micro-organisms. The formation from glucose by members of the *Penicillium chrysogenum* series of a pigment, an alkali-soluble protein and penicillin—the antibacterial substance of Fleming," *Biochemical Journal*, vol. 26, no. 6, pp. 1907–1918, 1932.

[15] H. W. Florey, E. B. Chain, N. G. Heatley et al., *Antibiotics*, vol. 2, Oxford University Press, London, UK, 1949.

[16] K. F. Kong, L. Schneper, and K. Mathee, "Beta-lactam antibiotics: from antibiosis to resistance and bacteriology," *Acta Pathologica, Microbiologica et Immunologica Scandinavica*, vol. 118, no. 1, pp. 1–36, 2010.

[17] A. A. Brakhage, M. Thön, P. Spröte et al., "Aspects on evolution of fungal β-lactam biosynthesis gene clusters and recruitment of trans-acting factors," *Phytochemistry*, vol. 70, no. 15-16, pp. 1801–1811, 2009.

[18] J. F. Martín, R. V. Ullán, and C. García-Estrada, "Regulation and compartmentalization of beta-lactam biosynthesis," *Microbial Biotechnology*, vol. 3, pp. 285–299, 2010.

[19] C. García-Estrada, F. Fierro, and J. F. Martín, "Evolution of fungal β-lactam biosynthesis gene clusters," in *Current Research, Technology and Education Topics in Applied Microbiology and Microbial Biotechnology*, A. Mendez-Vilas, Ed., vol. 1, pp. 577–588, Formatex Research Center, Badajoz, Spain, 2010.

[20] K. Kato, "Occurrence of penicillin-nucleus in culture broths," *The Journal of Antibiotics*, vol. 6, no. 3, pp. 130–136, 1953.

[21] A. L. Demain, "The mechanism of penicillin biosynthesis," *Advances in Applied Microbiology*, vol. 1, no. C, pp. 23–47, 1959.

[22] F. R. Batchelor, F. P. Doyle, J. H. C. Nayler, and G. N. Rolinson, "Synthesis of penicillin: 6-aminopenicillanic acid in penicillin fermentations," *Nature*, vol. 183, no. 4656, pp. 257–258, 1959.

[23] A. L. Demain and R. P. Blander, "The β-lactam antibiotics: past, present, and future," *Antonie van Leeuwenhoek*, vol. 75, no. 1-2, pp. 5–19, 1999.

[24] C. García-Estrada and J. F. Martín, "Penicillins and cephalosporins," in *Comprehensive Biotechnology*, M. Moo-Young, M. Butler, C. Webb et al., Eds., Elsevier, Amsterdam, The Netherlands, 2nd edition, 2011.

[25] K. B. Raper, D. F. Alexander, and R. D. Coghill, "Penicillin: II. Natural variation and penicillin production in *Penicillium notatum* and allied species," *The Journal of Bacteriology*, vol. 48, no. 6, pp. 639–659, 1944.

[26] K. B. Raper, "The development of improved penicillinproducing molds," *Annals of the New York Academy of Sciences*, vol. 48, pp. 41–56, 1946.

[27] M. P. Backus, J. F. Stauffer, and M. J. Johnson, "Penicillin yields from new mold strains," *Journal of the American Chemical Society*, vol. 68, no. 1, pp. 152–153, 1946.

[28] M. P. Backus and J. F. Stauffer, "The production and selection of a family of strains in *Penicillium chrysogenum*," *Mycologia*, vol. 47, pp. 429–463, 1955.

[29] R. F. Anderson, L. M. Whitmore, W. E. Brown et al., "Production of penicillin by some pigmentless mutants of the mold, *Penicillium chrysogenum* Q176," *Industrial & Engineering Chemistry*, vol. 45, pp. 768–773, 1953.

[30] R. P. Elander, "Strain improvement and preservation of beta-lactam producing microorganisms," in *Antibiotics Containing the Beta-Lactam Structure I*, A. L. Demain and N. Solomon, Eds., pp. 97–146, Springer, New York, NY, USA, 1983.

[31] R. P. Elander, "University of Wisconsin contributions to the early development of penicillin and cephalosporin antibiotics," *SIM News*, vol. 52, pp. 270–278, 2002.

[32] J. Lein, "The Panlabs Penicillium strain improvement program," in *Overproduction of Microbial Metabolites*, Z. Vanek

and Z. Hostalek, Eds., pp. 105–140, Butterworths, Stoneham, Mass, USA, 1986.

[33] R. J. Gouka, W. van Hartingsveldt, R. A. L. Bovenberg, C. A. M. J. J. van den Hondel, and R. F. M. van Gorcom, "Cloning of the nitrate-nitrite reductase gene cluster of Penicillium chrysogenum and use of the niaD gene as a homologous selection marker," Journal of Biotechnology, vol. 20, no. 2, pp. 189–199, 1991.

[34] F. Fierro, J. L. Barredo, B. Díez, S. Gutierrez, F. J. Fernández, and J. F. Martín, "The penicillin gene cluster is amplified in tandem repeats linked by conserved hexanucleotide sequences," Proceedings of the National Academy of Sciences of the United States of America, vol. 92, no. 13, pp. 6200–6204, 1995.

[35] M. A. Peñalva, R. T. Rowlands, and G. Turner, "The optimization of penicillin biosynthesis in fungi," Trends in Biotechnology, vol. 16, no. 11, pp. 483–489, 1998.

[36] R. W. Newbert, B. Barton, P. Greaves, J. Harper, and G. Turner, "Analysis of a commercially improved Penicillium chrysogenum strain series: involvement of recombinogenic regions in amplification and deletion of the penicillin biosynthesis gene cluster," Journal of Industrial Microbiology and Biotechnology, vol. 19, no. 1, pp. 18–27, 1997.

[37] M. A. van den Berg, I. Westerlaken, C. Leeflang, R. Kerkman, and R. A. L. Bovenberg, "Functional characterization of the penicillin biosynthetic gene cluster of Penicillium chrysogenum Wisconsin54-1255," Fungal Genetics and Biology, vol. 44, no. 9, pp. 830–844, 2007.

[38] F. Fierro, C. García-Estrada, N. I. Castillo, R. Rodriguez, T. Velasco-Conde, and J. F. Martín, "Transcriptional and bioinformatic analysis of the 56.8 kb DNA region amplified in tandem repeats containing the penicillin gene cluster in Penicillium chrysogenum," Fungal Genetics and Biology, vol. 43, pp. 618–629, 2006.

[39] C. García-Estrada, I. Vaca, M. Lamas-Maceiras, and J. F. Martín, "In vivo transport of the intermediates of the penicillin biosynthetic pathway in tailored strains of Penicillium chrysogenum," Applied Microbiology and Biotechnology, vol. 76, pp. 169–182, 2007.

[40] K. Kosalková, C. García-Estrada, and R. V. Ullán, "The global regulator LaeA controls penicillin biosynthesis, pigmentation and sporulation, but not roquefortine C synthesis in Penicillium chrysogenum," Biochimie, vol. 91, pp. 214–225, 2009.

[41] B. Hoff, J. Kamerewerd, C. Sigl et al., "Two components of a velvet-like complex control hyphal morphogenesis, conidiophore development, and penicillin biosynthesis in Penicillium chrysogenum," Eukaryotic Cell, vol. 9, no. 8, pp. 1236–1250, 2010.

[42] C. Sigl, H. Haas, T. Specht, K. Pfaller, H. Kurnsteiner, and I. Zadra, "Among developmental regulators, StuA but not BrlA is essential for penicillin V production in Penicillium chrysogenum," Applied and Environmental Microbiology, vol. 77, pp. 972–982, 2011.

[43] C. García-Estrada, R. V. Ullán, T. Velasco-Conde et al., "Posttranslational enzyme modification by the phosphopantetheinyl transferase is required for lysine and penicillin biosynthesis but not for roquefortine or fatty acid formation in Penicillium chrysogenum," Biochemical Journal, vol. 415, pp. 317–324, 2008.

[44] M. Lamas-Maceiras, I. Vaca, E. Rodríguez, J. Casqueiro, and J. F. Martín, "Amplification and disruption of the phenylacetyl-CoA ligase gene of Penicillium chrysogenum encoding an aryl-capping enzyme that supplies phenylacetic acid to the isopenicillin N-acyltransferase," Biochemical Journal, vol. 395, no. 1, pp. 147–155, 2006.

[45] D. J. Smith, J. H. Bull, J. Edwards, and G. Turner, "Amplification of the isopenicillin N synthetase gene in a strain of Penicillium chrysogenum producing high levels of penicillin," Molecular and General Genetics, vol. 216, no. 2-3, pp. 492–497, 1989.

[46] J. G. Nijland, B. Ebbendorf, M. Woszczynska, R. Boer, R. A. L. Bovenberg, and A. J. M. Driessen, "Nonlinear biosynthetic gene cluster dose effect on penicillin production by Penicillium chrysogenum," Applied and Environmental Microbiology, vol. 76, no. 21, pp. 7109–7115, 2010.

[47] R. Smidák, M. Jopcík, M. Kralovicová et al., "Core promoters of the penicillin biosynthesis genes and quantitative RT-PCR analysis of these genes in high and low production strain of Penicillium chrysogenum," Folia Microbiologica, vol. 55, pp. 126–132, 2010.

[48] J. M. Fernández-Cañón and M. A. Peñalva, "A fungal model for inborn errors in human phenylalanine metabolism," Proceedings of the National Academy of Sciences of the United States of America, vol. 92, pp. 9132–9136, 1995.

[49] J. M. Mingot, M. A. Peñalva, and J. M. Fernández-Cañón, "Disruption of phacA, an Aspergillus nidulans gene encoding a novel cytochrome P450 monooxygenase catalyzing phenylacetate 2-hydroxylation, results in penicillin overproduction," Journal of Biological Chemistry, vol. 274, no. 21, pp. 14545–14550, 1999.

[50] E. Arias-Barrau, E. R. Olivera, J. M. Luengo et al., "The homogentisate pathway: a central catabolic pathway involved in the degradation of L-phenylalanine, L-tyrosine, and 3-hydroxyphenylacetate in Pseudomonas putida," The Journal of Bacteriology, vol. 186, pp. 5062–5077, 2004.

[51] F. Ferrer-Sevillano and J. M. Fernández-Cañón, "Novel phacB-encoded cytochrome P450 monooxygenase from Aspergillus nidulans with 3-hydroxyphenylacetate 6-hydroxylase and 3,4-dihydroxyphenylacetate 6-hydroxylase activities," Eukaryotic Cell, vol. 6, no. 3, pp. 514–520, 2007.

[52] M. Rodríguez-Sáiz, J. L. Barredo, M. A. Moreno, J. M. Fernández-Cañón, M. A. Peñalva, and B. Díez, "Reduced function of a phenylacetate-oxidizing cytochrome p450 caused strong genetic improvement in early phylogeny of penicillin-producing strains," The Journal of Bacteriology, vol. 183, pp. 5465–5471, 2001.

[53] M. Rodríguez-Sáiz, B. Díez, and J. L. Barredo, "Why did the Fleming strain fail in penicillin industry?" Fungal Genetics and Biology, vol. 42, no. 5, pp. 464–470, 2005.

[54] J. A. K. W. Kiel, I. J. van der Klei, M. A. van den Berg, R. A. L. Bovenberg, and M. Veenhuis, "Overproduction of a single protein, Pc-Pex11p, results in 2-fold enhanced penicillin production by Penicillium chrysogenum," Fungal Genetics and Biology, vol. 42, no. 2, pp. 154–164, 2005.

[55] W. H. Müller, T. P. van der Krift, A. J. J. Krouwer et al., "Localization of the pathway of the penicillin biosynthesis in Penicillium chrysogenum," The EMBO Journal, vol. 10, no. 2, pp. 489–495, 1991.

[56] L. Gidijala, I. J. van der Klei, M. Veenhuis, and J. A. K. W. Kiel, "Reprogramming Hansenula polymorpha for penicillin production: expression of the Penicillium chrysogenum pcl gene," FEMS Yeast Research, vol. 7, no. 7, pp. 1160–1167, 2007.

[57] M. A. van den Berg, R. Albang, K. Albermann et al., "Genome sequencing and analysis of the filamentous fungus Penicillium chrysogenum," Nature Biotechnology, vol. 26, pp. 1161–1168, 2008.

[58] W. H. Müller, R. A. L. Bovenberg, M. H. Groothuis et al., "Involvement of microbodies in penicillin biosynthesis," Biochimica et Biophysica Acta, vol. 1116, no. 2, pp. 210–213, 1992.

[59] W. H. Meijer, L. Gidijala, S. Fekken et al., "Peroxisomes are required for efficient penicillin biosynthesis in *Penicillium chrysogenum*," *Applied and Environmental Microbiology*, vol. 76, no. 17, pp. 5702–5709, 2010.

[60] M. A. van den Berg, "Functional characterisation of penicillin production strains," *Fungal Biology Reviews*, vol. 24, no. 1-2, pp. 73–78, 2010.

[61] D. M. Harris, Z. A. van der Krogt, P. Klaassen et al., "Exploring and dissecting genome-wide gene expression responses of *Penicillium chrysogenum* to phenylacetic acid consumption and penicillinG production," *BMC Genomics*, vol. 10, article 75, 2009.

[62] D. Lim, P. Hains, B. Walsh, P. Bergquist, and H. Nevalainen, "Proteins associated with the cell envelope of *Trichoderma reesei*: a proteomic approach," *Proteomics*, vol. 1, no. 7, pp. 899–910, 2001.

[63] F. J. Fernández-Acero, I. Jorge, E. Calvo et al., "Two-dimensional electrophoresis protein profile of the phytopathogenic fungus *Botrytis cinerea*," *Proteomics*, vol. 6, supplement 1, pp. S88–S96, 2006.

[64] J. V. F. Coumans, P. D. J. Moens, A. Poljak, S. Al-Jaaidi, L. Pereg, and M. J. Raftery, "Plant-extract-induced changes in the proteome of the soil-borne pathogenic fungus *Thielaviopsis basicola*," *Proteomics*, vol. 10, no. 8, pp. 1573–1591, 2010.

[65] V. Yıldırım, S. Ozcan, D. Becher, K. Buttner, M. Hecker, and G. Ozcengiz, "Characterization of proteome alterations in Phanerochaete chrysosporium in response to lead exposure," *Proteome Science*, vol. 9, pp. 1–12, 2011.

[66] O. Kniemeyer, F. Lessing, O. Scheibner, C. Hertweck, and A. A. Brakhage, "Optimisation of a 2-D gel electrophoresis protocol for the human-pathogenic fungus *Aspergillus fumigatus*," *Current Genetics*, vol. 49, no. 3, pp. 178–189, 2006.

[67] M. Shimizu, T. Fujii, S. Masuo, K. Fujita, and N. Takaya, "Proteomic analysis of *Aspergillus nidulans* cultured under hypoxic conditions," *Proteomics*, vol. 9, no. 1, pp. 7–19, 2009.

[68] X. Lu, J. Sun, M. Nimtz, J. Wissing, A. P. Zeng, and U. Rinas, "The intra- and extracellular proteome of *Aspergillus niger* growing on defined medium with xylose or maltose as carbon substrate," *Microbial Cell Factories*, vol. 9, article 23, 2010.

[69] M. S. Jami, C. Barreiro, C. García-Estrada, and J. F. Martín, "Proteome analysis of the penicillin producer *Penicillium chrysogenum*: characterization of protein changes during the industrial strain improvement," *Molecular and Cellular Proteomics*, vol. 9, no. 6, pp. 1182–1198, 2010.

[70] M. Vödisch, K. Scherlach, R. Winkler et al., "Analysis of the *Aspergillus fumigatus* proteome reveals metabolic changes and the activation of the pseurotin a biosynthesis gene cluster in response to hypoxia," *Journal of Proteome Research*, vol. 10, no. 5, pp. 2508–2524, 2011.

[71] R. Cobos, C. Barreiro, R. M. Mateos, and J. R. Coque, "Cytoplasmic- and extracellular-proteome analysis of *Diplodia seriata*: a phytopathogenic fungus involved in grapevine decline," *Proteome Science*, vol. 8, article 46, 2010.

[72] Y. T. Oh, C. S. Ahn, J. G. Kim, H. S. Ro, C. W. Lee, and J. W. Kim, "Proteomic analysis of early phase of conidia germination in *Aspergillus nidulans*," *Fungal Genetics and Biology*, vol. 47, no. 3, pp. 246–253, 2010.

[73] V. Bhadauria, W. S. Zhao, L. X. Wang et al., "Advances in fungal proteomics," *Microbiological Research*, vol. 162, no. 3, pp. 193–200, 2007.

[74] C. Barreiro, E. Gonzalez-Lavado, S. Brand, A. Tauch, and J. F. Martín, "Heat shock proteome analysis of wild-type *Corynebacterium glutamicum* ATCC 13032 and a spontaneous mutant lacking GroEL1, a dispensable chaperone," *The Journal of Bacteriology*, vol. 187, no. 3, pp. 884–889, 2005.

[75] D. W. Kim, K. Chater, K. J. Lee, and A. Hesketh, "Changes in the extracellular proteome caused by the absence of the *bldA* gene product, a developmentally significant tRNA, reveal a new target for the pleiotropic regulator AdpA in *Streptomyces coelicolor*," *The Journal of Bacteriology*, vol. 187, no. 9, pp. 2957–2966, 2005.

[76] G. Candiano, M. Bruschi, L. Musante et al., "Blue silver: a very sensitive colloidal Coomassie G-250 staining for proteome analysis," *Electrophoresis*, vol. 25, no. 9, pp. 1327–1333, 2004.

[77] M. M. Bradford, "A rapid and sensitive method for the quantitation of microgram quantities of protein utilizing the principle of protein dye binding," *Analytical Biochemistry*, vol. 72, no. 1-2, pp. 248–254, 1976.

[78] J. Ohnishi, M. Hayashi, S. Mitsuhashi, and M. Ikeda, "Efficient 40 degrees C fermentation of L-lysine by a new *Corynebacterium glutamicum* mutant developed by genome breeding," *Applied Microbiology and Biotechnology*, vol. 62, no. 1, pp. 69–75, 2003.

[79] B. L. Gómez and J. D. Nosanchuk, "Melanin and fungi," *Current Opinion in Infectious Diseases*, vol. 16, no. 2, pp. 91–96, 2003.

[80] R. J. Kleijn, F. Liu, W. A. van Winden, W. M. van Gulik, C. Ras, and J. J. Heijnen, "Cytosolic NADPH metabolism in penicillin-G producing and non-producing chemostat cultures of *Penicillium chrysogenum*," *Metabolic Engineering*, vol. 9, no. 1, pp. 112–123, 2007.

[81] U. Nasution, W. M. van Gulik, C. Ras, A. Proell, and J. J. Heijnen, "A metabolome study of the steady-state relation between central metabolism, amino acid biosynthesis and penicillin production in *Penicillium chrysogenum*," *Metabolic Engineering*, vol. 10, no. 1, pp. 10–23, 2008.

[82] Y. Hadas, I. Goldberg, O. Pines, and D. Prusky, "Involvement of gluconic acid and glucose oxidase in the pathogenicity of *Penicillium expansum* in apples," *Phytopathology*, vol. 97, no. 3, pp. 384–390, 2007.

[83] C.-S. Hwang, M. A. Flaishman, and P. E. Kolattukudy, "Cloning of a gene expressed during appressorium formation by *Colletotrichum gloeosporioides* and a marked decrease in virulence by disruption of this gene," *Plant Cell*, vol. 7, no. 2, pp. 183–193, 1995.

[84] S. M. Beverley, K. L. Owens, M. Showalter et al., "Eukaryotic UDP-galactopyranose mutase (GLF Gene) in microbial and metazoal pathogens," *Eukaryotic Cell*, vol. 4, no. 6, pp. 1147–1154, 2005.

[85] M. Machida, K. Asai, M. Sano et al., "Genome sequencing and analysis of *Aspergillus oryzae*," *Nature*, vol. 438, no. 7071, pp. 1157–1161, 2005.

[86] H. J. Pel, J. H. de Winde, D. B. Archer et al., "Genome sequencing and analysis of the versatile cell factory *Aspergillus niger* CBS 513.88," *Nature Biotechnology*, vol. 25, no. 2, pp. 221–231, 2007.

[87] H. Bouws, A. Wattenberg, and H. Zorn, "Fungal secretomes—nature's toolbox for white biotechnology," *Applied Microbiology and Biotechnology*, vol. 80, no. 3, pp. 381–388, 2008.

[88] A. Steinbüchel, "Nachwachsende rohstoffe für die weiße biotechnologie," in *Weiße Biotechnologie-Industrie im Aufbruch*, S. Heiden and H. Zink, Eds., pp. 76–91, Biocom AG, Berlin, Germany, 2006.

[89] J. F. Peberdy, "Protein secretion in filamentous fungi—trying to understand a highly productive black box," *Trends in Biotechnology*, vol. 12, no. 2, pp. 50–57, 1994.

[90] M. S. Jami, C. García-Estrada, C. Barreiro, A. A. Cuadrado, Z. Salehi-Najafabadi, and J. F. Martín, "The *Penicillium*

chrysogenum extracellular proteome. Conversion from a food-rotting strain to a versatile cell factory for white biotechnology," *Molecular and Cellular Proteomics*, vol. 9, no. 12, pp. 2729–2744, 2010.

[91] P. Shah, J. A. Atwood, R. Orlando, H. E. Mubarek, G. K. Podila, and M. R. Davis, "Comparative proteomic analysis of botrytis cinerea secretome," *Journal of Proteome Research*, vol. 8, no. 3, pp. 1123–1130, 2009.

[92] J. D. Bendtsen, H. Nielsen, G. von Heijne, and S. Brunak, "Improved prediction of signal peptides: signalP 3.0," *Journal of Molecular Biology*, vol. 340, no. 4, pp. 783–795, 2004.

[93] J. D. Bendtsen, L. J. Jensen, N. Blom, G. von Heijne, and S. Brunak, "Feature-based prediction of non-classical and leaderless protein secretion," *Protein Engineering, Design and Selection*, vol. 17, no. 4, pp. 349–356, 2004.

[94] J. Choi, J. Park, D. Kim, K. Jung, S. Kang, and Y. H. Lee, "Fungal secretome database: integrated platform for annotation of fungal secretomes," *BMC Genomics*, p. 105, 2010.

[95] G. Lum and X. J. Min, "FunSecKB: the fungal secretome knowledgeBase," *Database*, vol. 2011, pp. 1–10, 2011.

[96] C. J. Jeffery, "Moonlighting proteins," *Trends in Biochemical Sciences*, vol. 24, no. 1, pp. 8–11, 1999.

[97] W. Nickel, "The mystery of nonclassical protein secretion: a current view on cargo proteins and potential export routes," *European Journal of Biochemistry*, vol. 270, no. 10, pp. 2109–2119, 2003.

[98] W. Nickel, "Unconventional secretory routes: direct protein export across the plasma membrane of mammalian cells," *Traffic*, vol. 6, no. 8, pp. 607–614, 2005.

[99] W. Nickel and C. Rabouille, "Mechanisms of regulated unconventional protein secretion," *Nature Reviews Molecular Cell Biology*, vol. 10, no. 2, pp. 148–155, 2009.

[100] J. A. Simpson and E. S. C. Weiner, *The Compact Oxford English Dictionary*, Oxford University Press, New York, NY, USA, 2nd edition, 1992.

[101] C. Gancedo and C. L. Flores, "Moonlighting proteins in yeasts," *Microbiology and Molecular Biology Reviews*, vol. 72, no. 1, pp. 197–210, 2008.

[102] C. L. Flores and C. Gancedo, "Unraveling moonlighting functions with yeasts," *IUBMB Life*, vol. 63, no. 7, pp. 457–462, 2011.

[103] J. A. K. W. Kiel, M. A. van den Berg, F. Fusetti et al., "Matching the proteome to the genome: the microbody of penicillin-producing *Penicillium chrysogenum* cells," *Functional and Integrative Genomics*, vol. 9, no. 2, pp. 167–184, 2009.

[104] M. Oku and Y. Sakai, "Peroxisomes as dynamic organelles: autophagic degradation," *The FEBS Journal*, vol. 277, no. 16, pp. 3289–3294, 2010.

[105] C. A. Cantwell, R. J. Beckmann, J. E. Dotzlaf et al., "Cloning and expression of a hybrid *Streptomyces clavuligerus* cefE gene in *Penicillium chrysogenum*," *Current Genetics*, vol. 17, no. 3, pp. 213–221, 1990.

[106] R. V. Ullán, S. Campoy, J. Casqueiro, F. J. Fernandez, and J. F. Martín, "Deacetylcephalosporin C production in *Penicillium chrysogenum* by expression of the isopenicillin N epimerization, ring expansion, and acetylation genes," *Chemistry & Biology*, vol. 14, pp. 329–339, 2007.

[107] D. M. Harris, I. Westerlaken, D. Schipper et al., "Engineering of *Penicillium chrysogenum* for fermentative production of a novel carbamoylated cephem antibiotic precursor," *Metabolic Engineering*, vol. 11, pp. 125–137, 2009.

[108] J. M. Cantoral, S. Gutiérrez, F. Fierro, S. Gil-Espinosa, H. van Liempt, and J. F. Martín, "Biochemical characterization and molecular genetics of nine mutants of *Penicillium chrysogenum* impaired in penicillin biosynthesis," *Journal of Biological Chemistry*, vol. 268, no. 1, pp. 737–744, 1993.

[109] N. Yamashita, T. Motoyoshi, and A. Nishimura, "Purification and characterization of isoamyl alcohol oxidase ("Mureka"-forming enzyme)," *Bioscience, Biotechnology, and Biochemistry*, vol. 63, pp. 1216–1222, 1999.

[110] H. E. Swaisgood, V. G. Janolino, and P. J. Skudder, "Continuous treatment of ultrahigh-temperature sterilized milk using immobilized sulfhydryl oxidase," *Methods in Enzymology*, vol. 136, no. C, pp. 423–431, 1987.

[111] Y. Kodama, Y. Nakao, and T. Shimonaga, "Dihydroxy-acid dehydratase gene and use thereof," USPTO Patent Application 20090148555, 2009.

[112] Y. Fuke, S. Kaminogawa, H. Matsuoka, and K. Yamauchi, "Purification and properties of aminopeptidase I from *Penicillium caseicolum*," *Journal of Dairy Science*, vol. 71, pp. 1423–1431, 1988.

[113] M. Srinivasan, A. R. Sudheer, and V. P. Menon, "Ferulic acid: therapeutic potential through its antioxidant property," *Journal of Clinical Biochemistry and Nutrition*, vol. 40, no. 2, pp. 92–100, 2007.

[114] T. Koseki, S. Fushinobu, Ardiansyah, H. Shirakawa, and M. Komai, "Occurrence, properties, and applications of feruloyl esterases," *Applied Microbiology and Biotechnology*, vol. 84, pp. 803–810, 2009.

Mouse Ficolin B Has an Ability to Form Complexes with Mannose-Binding Lectin-Associated Serine Proteases and Activate Complement through the Lectin Pathway

Yuichi Endo,[1] Daisuke Iwaki,[1] Yumi Ishida,[1] Minoru Takahashi,[1] Misao Matsushita,[2] and Teizo Fujita[1]

[1] *Department of Immunology, Fukushima Medical University School of Medicine, 1-Hikarigaoka, Fukushima 960-1295, Japan*
[2] *Department of Applied Biochemistry, Tokai University, Hiratsuka, Kanagawa 259-1292, Japan*

Correspondence should be addressed to Yuichi Endo, yendo@fmu.ac.jp

Academic Editor: Nobutaka Wakamiya

Ficolins are thought to be pathogen-associated-molecular-pattern-(PAMP-) recognition molecules that function to support innate immunity. Like mannose-binding lectins (MBLs), most mammalian ficolins form complexes with MBL-associated serine proteases (MASPs), leading to complement activation *via* the lectin pathway. However, the ability of murine ficolin B, a homologue of human M-ficolin, to perform this function is still controversial. The results of the present study show that ficolin B in mouse bone marrow is an oligomeric protein. Ficolin B, pulled down using GlcNAc-agarose, contained very low, but detectable, amounts of MASP-2 and small MBL-associated protein (sMAP) and showed detectable C4-deposition activity on immobilized *N*-acetylglucosamine. These biochemical features of ficolin B were confirmed using recombinant mouse ficolin B produced in CHO cells. Taken together, these results suggest that like other mammalian homologues, murine ficolin B has an ability to exert its function *via* the lectin pathway.

1. Introduction

Ficolins (FCN/Fcn) are a family of proteins comprising a collagen-like and a fibrinogen-like domain [1] the latter binding specifically to *N*-acetyl compounds such as *N*-acetylglucosamine (GlcNAc) [2–4]. Three types of ficolin have been identified in humans: L-FCN [1, 5], M-FCN [6, 7], and H-FCN [8]. Transcripts for L-FCN and H-FCN are mainly produced in the liver, and the proteins circulate as serum ficolins, whereas the mRNA for M-FCN is expressed mainly in peripheral monocytes and the protein is present in the serum at low concentrations [9]. Two ficolins have been identified in mice: ficolin A (FcnA) and ficolin B (FcnB) [10, 11]. FcnA mRNA is mainly expressed in Kupffer cells in the liver [12], and the protein is present in serum. FcnB mRNA is mainly expressed in cells of myeloid cell lineage within the bone marrow [12]. The location of FcnB protein is still unclear, although it is reported to localize within the lysosomes of activated macrophages [13].

Our phylogenetic analyses show that FcnB is the murine orthologue of human M-FCN, that FcnA and L-FCN were independently diverged in the murine and primate lineages, respectively, from the ancestral FcnB/M-FCN, and that the *H-FCN* gene is a pseudogene in the murine lineage [14]. In addition, our previous ontogenetic study showed that the spatial-temporal expression pattern was different for FcnB and FcnA, suggesting that each of the ficolins might have a specific role in the prenatal and postnatal stages [12].

Thus, ficolins are roughly classified into two groups: a serum type (plasma type), which includes L-FCN, H-FCN and FcnA produced mainly in the liver and present in the circulation as serum lectins [15] and a nonserum type (nonplasma type), which includes M-FCN and FcnB hardly detectable in the serum. The latter group, particularly murine FcnB, has not been studied in detail at the protein level, because of difficulties in identifying and isolating a sufficient amount of the protein. There are no reports regarding the biochemical features of native FcnB.

Mouse Ficolin B Has an Ability to Form Complexes with Mannose-Binding Lectin-Associated Serine Proteases and
Activate Complement through the Lectin Pathway

17

To date, we have shown that mammalian ficolins, including three human ficolins and mouse FcnA, associate with mannose-binding lectin (MBL)-associated serine proteases (MASPs) and activate the lectin pathway [5, 16–18]. We also reported that recombinant mouse FcnB produced in *Drosophila* S2 cells does not associate with MASP-2 and small MBL-associated protein (sMAP) [18]. Recently, however, it was reported that rat recombinant FcnB associates with MASPs and activates the lectin pathway by binding to PAMPs [19]. In the present study, we carefully examined the biochemical properties of FcnB using both native FcnB isolated from mouse bone marrow fluid, and recombinant mouse FcnB produced in CHO cells. The results show that like rat FcnB, mouse FcnB has the ability to form complexes with MASPs and sMAP.

2. Materials and Methods

2.1. Preparation of FcnB from Mouse Bone Marrow.
To avoid complications resulting from the co-presence of FcnA and FcnB, the bone marrow tissue used as the source of FcnB was collected from FcnA-deficient mice generated on a C57BL6 background by gene targeting (manuscript in preparation). The bone marrow fluid and cells were collected as supernatant and precipitate, respectively, by centrifugation of the pooled tissue at 10,000 ×g for 10 min. The supernatant was then subjected to affinity chromatography on a GlcNAc-agarose column. The bound fraction was eluted sequentially with 0.3 M mannose and then with 0.3 M GlcNAc. The recovered eluate was dialyzed against 50 mM Tris-HCl, at pH 7.5, containing 0.15 M NaCl and 2.5 mM CaCl$_2$ (TBS-Ca), concentrated in a centrifugal filter (Amicon Ultra-4, Millipore, Billerica, MA, USA) and stored at −80°C until required. The GlcNAc-eluate was used as a source of native FcnB in further study.

2.2. Preparation of Recombinant FcnB (rFcnB).
Full-length mouse FcnB cDNA was constructed in a pIRCMV vector and cotransfected with a pFerH vector encoding a transposase into CHO cells as previously described [20]. The rFcnB-producing CHO cells were screened by culturing in the presence of 0.5 mg/mL neomycin G-418. After several passages in DMEM medium containing 10% FCS, the neomycin-resistant CHO cells were cultured in a serum-free medium (CHO-S-SFM, GIBCO, Grand Island, NY, USA). The rFcnB secreted into the culture medium was purified on a GlcNAc-agarose column. Briefly, after washing the column with TBS containing 0.05% Tween-20, 0.1 M mannose, 0.1 M galactose, and 0.1 M glucose, the bound fraction was eluted with 0.3 M GlcNAc. The eluate was dialyzed against TBS-Ca, concentrated, and stored at −80°C until use. The N-terminal amino acid sequence of rFcnB was determined using a Procise cLC Protein Sequencing System (Applied Biosystems, Poster City, CA, USA). The rFcnB produced by CHO cells was simply termed rFcnB in this study, while rFcnB produced previously in *Drosophila* S2 cells was termed rFcnBs2-1 [18]. The recombinant protein concentration was

determined using a BCA protein assay kit (Pierce, Rockford, IL, USA) with BSA as a standard protein.

2.3. Preparation of Recombinant Mouse MASP-1, MASP-2, MASP-3, and sMAP.
Two recombinant forms of mature mouse MASP-2 were produced in *Drosophila* S2 cells with a histidine (His)-tag as previously described [21]: one comprised the normal sequence with protease activity (rMASP-2a) and the other a mutated sequence (Ser632Ala) with no activity (rMASP-2i). Recombinant mouse MASP-1 (rMASP-1i) was prepared in a His-tagged form using a *Baculovirus* expression system (Invitrogen, Carlsbad, CA, USA), and is an inactive form harboring a mutated catalytic site (Ser646Ala) [22]. Recombinant mouse MASP-3 (rMASP-3) [23] and sMAP (rsMAP) [21] were prepared as His-tagged forms in *Drosophila* S2 cells. All recombinants (rMASP-1i, -2i, -2a and -3, and rsMAP) were purified by affinity chromatography on Ni-NTA agarose columns (Qiagen Inc., Valencia, CA, USA) followed by elution with imidazole. The recombinant proteins were dialyzed against TBS, concentrated, and stored at −80°C until use. The protein concentration was determined as described above.

2.4. Western Blotting for FcnB, MBLs, MASPs, and sMAP.
SDS-PAGE was performed on 10% polyacrylamide gels under reducing conditions according to the method of Laemmli. After electrophoresis, the proteins were transferred to a polyvinylidene difluoride membrane filter (Millipore, Billerica, MA, USA). The membrane filter was blocked with Blocking One reagent (Nacalai Tesque Inc., Kyoto, Japan) and probed with 500~2000-fold-diluted polyclonal antibodies (Abs) against mouse FcnB and MASP-2/sMAP [18, 21] and 500-fold-diluted monoclonal Abs against MBL-A and MBL-C (clones 8G6 and 16A8, resp., Hycult Biotechnology, Uden, The Netherlands) in 10 mM phosphate buffer, at pH 7.4, containing 137 mM NaCl and 2.7 mM KCl (PBS) containing 0.1% Tween-20 (PBS-T). For the detection of MASP-1 and -3, a monoclonal anti-penta-His-tag Ab (Qiagen) was used as the primary Ab. After washing, the filters were further incubated with either HRP-conjugated secondary Abs or biotinylated secondary Abs (Dako Cytomation, Glostrup, Denmark) followed by an avidin-biotinylated HRP complex (Vector Lab., Burlingame, CA, USA). Finally, the membranes were developed using a chemiluminescent substrate (ECL, Amersham Biosciences, Buckinghamshire, UK). The chemiluminescent image was analyzed using an LAS-3000 (Fuji film, Tokyo, Japan).

2.5. Treatment of FcnB with Endoglycosidases.
The N-linked carbohydrates expressed on FcnB were removed by treatment with endoglycosidase F (EMD Biosciences Inc., La Jolla, CA, USA) as previously described [24]. Selective removal of O-linked glycans was achieved by treatment with 0.1 U neuraminidase (Wako Pure Chemicals, Osaka, Japan) and 20 mU endo-α-N-acetylgalactosaminidase (Seikagaku Co., Tokyo, Japan) at 37°C for 16 h.

2.6. Gel Filtration Chromatography of FcnB. To estimate the size distribution of oligomeric FcnB, the FcnB preparations were subjected to gel filtration chromatography using a Superose 6 10/300GL column equilibrated with PBS and connected to an ÄKTA purifier system (Amersham Biosciences, Uppsala, Sweden). An aliquot of each recovered fraction (0.5 mL/fraction) was assessed for FcnB by western blotting under reducing conditions.

2.7. Complex Formation of rFcnB with rMASPs and rsMAP. rFcnB was incubated with rMASPs and rsMAP at a molar ratio of 3 : 1 : 8 (rFcnB : rMASPs : rsMAP) overnight at 4°C in TBS containing 2.5 mM $CaCl_2$, 3% BSA, and 0.05% Tween-20 as previously described [18]. The above molar ratio was chosen by reference to the concentrations of FcnA, MASP-2, and sMAP in the mouse serum [18]. The mixture was further incubated with a GlcNAc-agarose slurry (50%, 40 μL) at 4°C for 3 hr to pull down rFcnB, and the bound fraction was eluted with 0.3 M GlcNAc. The eluate was dialyzed against TBS-Ca and the final sample subjected to western blotting and a C4-deposition assay. For western blotting, rMASP-2i was used as a source of MASP-2 to ensure clear results, since it is known that rMASP-2a is converted, in part, into its active form, comprising the heavy and light chains connected *via* a disulfide bond, during purification. For the C4-deposition assay, MASP-2a was used as the source of MASP-2 instead of rMASP-2i. Similar autoactivation is also seen with rMASP-1; therefore, rMASP-1i was used as the source of MASP-1 for western blotting to detect complex formation with rFcnB.

2.8. C4-Deposition Assay. C4-deposition activity was determined by ELISA as previously described [18]. Briefly, the GlcNAc eluates prepared from bone marrow fluid or the rFcnB/rMASP-2a/sMAP complexes, were incubated in 100 μL of TBS-Ca at 37°C for 10 min in a GlcNAc-BSA-coated microtiter plate. The plate was then incubated with human C4 on ice for 30 min, followed by washing with PBS-T. The C4b generated on the plate was detected with an HRP-conjugated sheep anti-human C4b Ab (Biogenesis, Poole, UK) and color developed using TMB (KPL Co., Gaithersburg, MD, USA) and H_2O_2 as substrates. After termination of the reaction with 0.5 M H_3PO_4, the plates were read at 450 nm in a Multimode detector DTX880 (Beckman Coulter Inc., Brea, CA, USA).

3. Results

To detect the FcnB protein in the bone marrow, the tissue supernatants and precipitates were subjected to western blotting. As shown in Figure 1(a), a 38 kDa band was observed in the supernatant under reducing conditions, suggesting that FcnB is secreted into the mouse bone marrow fluid as a soluble protein. FcnB was also detected as a 37 kDa band at high levels in the precipitate. This suggests that FcnB in bone marrow cells is slightly small due to incomplete processing prior to secretion. When FcnB in the supernatant was treated with endoglycosidase F, its molecular weight

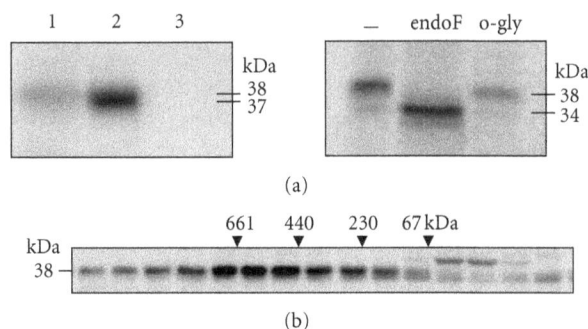

FIGURE 1: Characterization of FcnB in the mouse bone marrow. (a) Western blotting of FcnB in the supernatant (lane 1) and precipitate (lane 2) of bone marrow tissue, and plasma (lane 3) (left panel). Two μL of the supernatant/precipitate equivalent to the original tissue and 2 μL of plasma from $FcnA^{-/-}$ mice were subjected to western blotting under reducing conditions. FcnB in the supernatant (4 μL equivalent to the original tissue) was treated with endoglycosidase F (endoF) or neuraminidase plus endo-α-N-acetylgalactosaminidase (o-gly) and subjected to western blotting (right panel).—, not treated. (b) Superose 6 gel chromatography of FcnB in the bone marrow supernatant from $FcnA^{-/-}$ mice. An aliquot of each fraction was subjected to western blotting under reducing conditions. Arrowheads depict the eluted positions of the molecular weight markers (661 kDa, thyroglobulin; 440 kDa, ferritin; 230 kDa, catalase; 67 kDa, BSA).

reduced from 38 to 34 kDa, whereas treatment with endo-α-N-acetylgalactosaminidase resulted in either no or a smaller reduction in molecular weight (Figure 1(a)). To determine the size distribution of oligomeric FcnB, the supernatant was subjected to gel chromatography. FcnB was recovered from fractions corresponding to the elution positions of marker proteins ranging from 100 to >1000 kDa with a peak around 600 kDa (Figure 1(b)), indicating a heterogeneous structure composed mainly of 12–18-mers.

Next, the bone marrow supernatant was subjected to GlcNAc-agarose affinity chromatography to purify FcnB. As shown in Figure 2(a), FcnB was recovered in the GlcNAc eluate, whereas MBL-A and MBL-C (MBLs) were recovered in the mannose eluate. The mannose eluate included significant amounts of the MASP-2 pro-enzyme, MASP-2 heavy chain, and sMAP. This suggests that MBLs are present in the bone marrow fluid as complexes with MASP-2 and sMAP. Interestingly, trace amounts of MASP-2 and sMAP were also detected in the GlcNAc eluate, suggesting that at least a part of FcnB also exists in complex with MASP-2 and sMAP. This GlcNAc eluate showed C4-deposition on GlcNAc-coated microplates (Figure 2(b)), although the level was very low compared with that of the mannose eluate. This activity was significantly decreased by passage of the eluate through anti-FcnB Ab-coupled Sepharose 4B. These results suggest that the FcnB/MASPs complexes can activate complement component C4 through the lectin pathway.

To confirm the above results, rFcnB was produced in CHO cells and purified by GlcNAc-agarose chromatography. Western blotting showed that rFcnB consisted of a monomer with a molecular weight of 37 kDa; slightly smaller

Mouse Ficolin B Has an Ability to Form Complexes with Mannose-Binding Lectin-Associated Serine Proteases and
Activate Complement through the Lectin Pathway

19

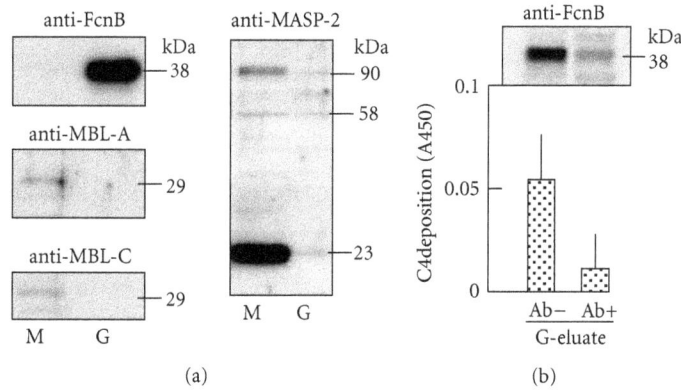

FIGURE 2: : GlcNAc-agarose chromatography of bone marrow fluids from FcnA$^{-/-}$ mice. (a) Western blot of FcnB, MBL-A, MBL-C, and MASP-2/sMAP in the mannose (M) and GlcNAc (G) eluates from GlcNAc-agarose chromatography under reducing conditions. Right panel: 90 kDa, 58 kDa, and 23 kDa bands represent the proenzyme form of MASP-2, the heavy chain of MASP-2, and sMAP, respectively. For each sample, 60 μL equivalent to the original volume of bone marrow tissue was loaded per lane. (b) C4-deposition activity of the GlcNAc eluate on GlcNAc-BSA-coated microplates. Before assessment, the GlcNAc eluate was passed through anti-FcnB Ab-coupled Sepharose 4B (Ab+) or not (Ab−). The activity of a 30 μL sample equivalent to the original bone marrow tissue was determined in quadruplicate (mean ± SD). Inset: western blot of FcnB in the eluates used for C4 deposition.

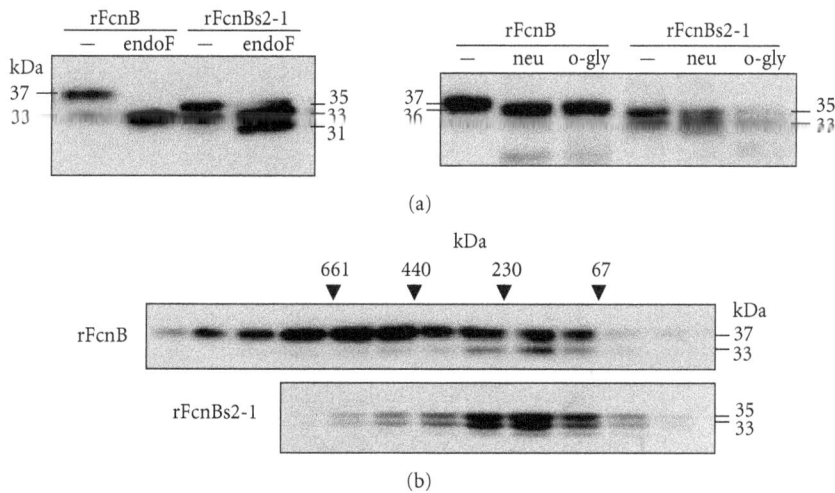

FIGURE 3: Structural characterization of rFcnB. (a) Left panel: Western blot of rFcnB and rFcnBs2-1 treated with endoglycosidase F (endoF). Right panel: western blot of rFcnB and rFcnBs2-1 treated with neuraminidase (neu) or neuraminidase plus endo-α-N-acetylgalactosaminidase (o-gly). Western blotting was performed under reducing conditions. —, not treated. (b) Gel chromatography of rFcnB (upper panel) and rFcnBs2-1 (lower panel). rFcnB or rFcnBs2-1 (400 μL; 5–10 μg) was applied to a Superose 6 column (1 cm φ× 30 cm) and fractionated into 0.5 mL/fractions. An aliquot of each fraction was subjected to western blotting for FcnB under reducing conditions.

than native FcnB, but larger than rFcnBs2-1 (33–35 kDa) (Figure 3(a)). Upon treatment with endoglycosidase F, the molecular weights of rFcnB and rFcnBs2-1 were reduced to 33 kDa and 31–33 kDa, respectively. Treatment of rFcnB with endo-α-N-acetylgalactosaminidase resulted in a slight reduction in the molecular weight to 36 kDa, while treatment of rFcnBs2-1 had no effect. The N-terminal amino acid sequence of rFcnB was T^{20}CPELKV, indicating that the preceding 19 amino acids were removed as a signal peptide by the host CHO cells. The N-terminal sequences of the 35 kDa

and 33 kDa bands of rFcnBs2-1 were RSPWPGVFV^{15}HAAG and A^{18}GTCPEL, respectively, indicating that the 35 kDa band corresponded to our designed rFcnBs2-1 product containing the eight plasmid-derived amino acids (underlined) at the N-terminal [18], and the 33 kDa band was another rFcnBs2-1 product with a different N-terminal, which was processed by *Drosophila* S2 cells.

Oligomeric rFcnB was subjected to gel chromatography to determine its size distribution. It was found that, like native FcnB, the main rFcnB species was eluted in the

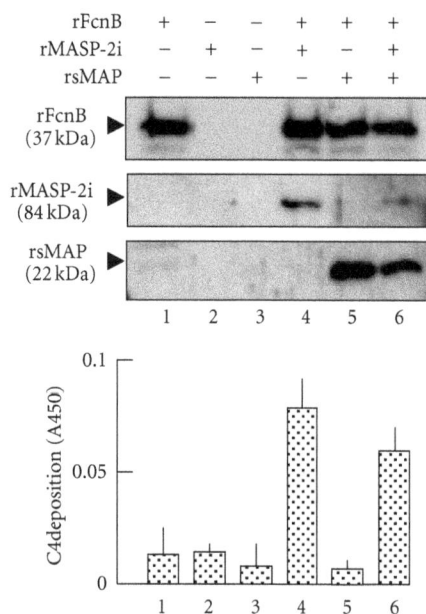

FIGURE 4: Complement activation by rFcnB/rMASP-2/rsMAP complexes. Upper panel: western blot of FcnB and MASP-2/sMAP. After incubation of the recombinant proteins as shown in the table, the generated complexes were pulled down by GlcNAc-agarose and subjected to western blotting. Lower panel: C4-deposition activity of the similar pull-down samples on GlcNAc-BSA-coated microplates. The samples used for C4 deposition were prepared the same as those for western blotting except for employment of rMASP-2a instead of rMASP-2i. The activity was determined in quadruplicate (mean ± SD).

range corresponding to 100 to >1000 kDa with a peak around 600 kDa (Figure 3(b)). A minor band was observed at 33 kDa in the rFcnB preparation eluted between 100 and 200 kDa. These results indicate that the rFcnB preparation contains a major 12–18-mer made up of 37 kDa monomers, and a minor 3–6-mer made up of 33 kDa monomers. Gel chromatography of rFcnBs2-1 showed that this protein ranged from 100 to 300 kDa, suggesting that it is a 3–9-mer composed of 33–35 kDa monomers.

To confirm the interaction between rFcnB and MASPs and sMAP, rFcnB was incubated with rMASP-2i and rsMAP and then subjected to FcnB pull down using GlcNAc-agarose. As shown in Figure 4, rMASP-2i and rsMAP were coprecipitated only in the presence of rFcnB. Coincubation of rMASP-2i and rsMAP resulted in reduced binding to rFcnB compared with incubation with each alone, suggesting their competitive bindings to rFcnB. The rFcnB/rMASP-2a complex activated C4 on GlcNAc-coated microplates. This C4-deposition activity was inhibited in part by rsMAP. In addition, rFcnB bound to rMASP-1i and rMASP-3, and this binding was partially inhibited by coincubation with rsMAP (Figures 5(a) and 5(b)), suggesting that FcnB associates with all types of MASP and sMAP in a similar manner. No activation of rMASP-3 was observed under these experimental conditions, even when it was complexed with rFcnB on GlcNAc.

4. Discussion

The present study clearly indicates that, like rat recombinant FcnB, both native and recombinant forms of mouse FcnB associate with MASPs and sMAP. It also demonstrates that FcnB/MASPs/sMAP complexes activate C4 on immobilized GlcNAc. Taken together with the results of Girija et al. [19], these results suggest that at least a part of murine FcnB essentially executes its function through the lectin pathway.

In the present study, it was observed that the monomer size of FcnBs2-1 was smaller than that of the native FcnB and rFcnB proteins, largely due to the N-linked carbohydrate content. It was also found that FcnBs2-1 formed smaller oligomers (3–9-mers), in contrast to the highly oligomeric forms of the native FcnB and rFcnB (12–18-mers). These results simply suggest that the processing of proteins in insect cells is different from that in mammalian cells. To confirm this in the present study, we prepared a third form of recombinant mouse FcnB, termed rFcnBs2-2, in *Drosophila* S2 cells. The N-terminal residue of rFcnBs2-2 was adjusted to Thr[20] as same as that in rFcnB, which was performed by ligation of a FcnB cDNA containing an extra six bases into a pMT/Bip/V5-His A vector. The generated rFcnBs2-2 showed a molecular weight of 31 kDa under reducing conditions, and treatment with endoglycosidase F reduced this to 30 kDa (data not shown). It was also found that rFcnBs2-2 was less oligomeric, existing mainly as 3–6-mers, and that it failed to associate with rMASP-2i or rsMAP (data not shown). These results clearly indicate that the processing of recombinant mouse FcnB in *Drosophila* S2 cells is different from that in mammalian cells.

An interesting result was observed in gel chromatography of the rFcnB preparation, which contained a major and highly oligomeric species comprising 37 kDa monomers and a minor and poorly oligomeric species comprising 33 kDa monomers (Figure 3(b)). In a preliminary study, we observed that culture of CHO cells in the presence of tunicamycin, an inhibitor of N-linked glycosylation, resulted in the preferential production of rFcnB comprising 33 kDa-monomers, which was also less oligomeric (3–6-mers) (data not shown). These results suggest that full N-linked glycosylation of FcnB is essential for high level of oligomerization.

We recently observed that *Drosophila* S2 cells produced trimers of human H-FCN, while CHO cells produced highly oligomeric H-FCN (~18-mers), which were structurally similar to native H-FCN in human serum (data not shown). However, *Drosophila* S2 cells do not always produce less oligomeric ficolin. For example, this cell line successfully produced highly oligomeric forms of mouse FcnA and human M-FCN (~600 kDa) [17, 18]. At present, the reason why ficolin molecules are processed differently in *Drosophila* S2 cells it is not known. One possibility might be a small difference among these ficolins in the amino acid sequence that forms the N-linked glycosylation site, for example, the surrounding sequence around Asn-X-Ser/Thr.

Our studies, including the present study, demonstrate that, regardless of the type of host cell used, recombinant ficolins expressed as highly oligomeric forms associate with

Mouse Ficolin B Has an Ability to Form Complexes with Mannose-Binding Lectin-Associated Serine Proteases and Activate Complement through the Lectin Pathway

21

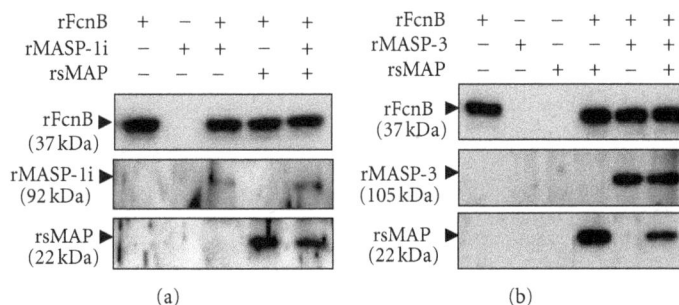

rFcnB	+	−	+	+	+
rMASP-1i	−	+	+	−	+
rsMAP	−	−	−	+	+

rFcnB	+	−	−	+	+	+
rMASP-3	−	+	−	−	+	+
rsMAP	−	−	+	+	−	+

(a) (b)

FIGURE 5: Complex formation between rFcnB and rMASP-1i (a) and rMASP-3 (b). After incubation of the recombinant proteins as outlined in the table, samples pulled down with GlcNAc agarose were subjected to western blotting. Western blotting for rMASP-1i, rMASP-3, and rsMAP was performed using an anti-penta-His tag Ab.

MASPs/sMAP, while less oligomeric forms do not. Thus, it is suggested that full processing including N-linked glycosylation is essential for the formation of highly oligomeric ficolin and that high oligomerization is in turn essential for association with MASPs and sMAP. At least, it can be concluded from our studies that CHO cells are available to produce highly oligomeric recombinant for all types of ficolin.

The molecular mechanism underlying the activation of the lectin pathway by mouse FcnB appears to be similar to that of other mammalian ficolins, at least in terms of complex formation with MASPs, sMAP, and probably Map44 (MAP-1) [25] and in activating C4, possibly C3 and C2 on targets. However, it is unclear as to how strongly FcnB exerts its activity through the lectin pathway *in vivo* and as to how FcnB shares its activity with FcnA and MBLs. It is noteworthy that, in the present study, rMASP-3 was not activated when complexed with rFcnB on GlcNAc. The rFcnB/rMASP-3 complex may be useful for identifying native target molecule(s) recognized by FcnB.

5. Conclusions

FcnB was identified in mouse bone marrow fluid as a highly oligomeric protein. GlcNAc-agarose chromatography of bone marrow fluid showed that FcnB was present in a complex with MASP-2 and sMAP. This complex also exhibited C4-deposition activity on immobilized GlcNAc. These results were confirmed using rFcnB produced in CHO cells. Comparison of the biochemical features of the three types of the recombinant ficolin, rFcnB, rFcnBs2-1, and rFcnBs2-2, suggests that full processing, including N-linked glycosylation, is essential for oligomerization of FcnB and its association with MASPs/sMAP. Taken together, we conclude that mouse FcnB acts as a recognition molecule working through the lectin pathway. To further understand the role of murine FcnB, its location site and real target(s) and the precise stoichiometry of the FcnB complex need to be clarified.

Acknowledgments

The authors thank Ms N. Nakazawa and K. Kanno for technical assistance. This work was supported in part by grants from the Ministry of Education, Culture, Sports, Science and Technology of Japan.

References

[1] M. Matsushita, Y. Endo, S. Taira et al., "A novel human serum lectin with collagen- and fibrinogen-like domains that functions as an opsonin," *Journal of Biological Chemistry*, vol. 271, no. 5, pp. 2448–2454, 1996.

[2] V. Garlatti, N. Belloy, L. Martin et al., "Structural insights into the innate immune recognition specificities of L- and H-ficolins," *EMBO Journal*, vol. 26, no. 2, pp. 623–633, 2007.

[3] V. Garlatti, L. Martin, E. Gout et al., "Structural basis for innate immune sensing by M-ficolin and its control by a pH-dependent conformational switch," *Journal of Biological Chemistry*, vol. 282, no. 49, pp. 35814–35820, 2007.

[4] M. Tanio, S. Kondo, S. Sugio, and T. Kohno, "Trivalent recognition unit of innate immunity system: crystal structure of trimeric human M-ficolin fibrinogen-like domain," *Journal of Biological Chemistry*, vol. 282, no. 6, pp. 3889–3895, 2007.

[5] M. Matsushita, Y. Endo, and T. Fujita, "Complement-activating complex of ficolin and mannose-binding lectin-associated serine protease," *Journal of Immunology*, vol. 164, no. 5, pp. 2281–2284, 2000.

[6] Y. Endo, Y. Sato, M. Matsushita, and T. Fujita, "Cloning and characterization of the human lectin P35 gene and its related gene," *Genomics*, vol. 36, no. 3, pp. 515–521, 1996.

[7] J. Lu, P. N. Tay, O. L. Kon, and K. B. M. Reid, "Human ficolin: cDNA cloning, demonstration of peripheral blood leucocytes as the major site of synthesis and assignment of the gene to chromosome 9," *Biochemical Journal*, vol. 313, no. 2, pp. 473–478, 1996.

[8] R. Sugimoto, Y. Yae, M. Akaiwa et al., "Cloning and characterization of the Hakata antigen, a member of the ficolin/opsonin p35 lectin family," *Journal of Biological Chemistry*, vol. 273, no. 33, pp. 20721–20727, 1998.

[9] C. Honoré, S. Rørvig, L. Munthe-Fog et al., "The innate pattern recognition molecule Ficolin-1 is secreted by monocytes/macrophages and is circulating in human plasma," *Molecular Immunology*, vol. 45, no. 10, pp. 2782–2789, 2008.

[10] Y. Fujimori, S. Harumiya, Y. Fukumoto et al., "Molecular cloning and characterization of mouse ficolin-A," *Biochemical and Biophysical Research Communications*, vol. 244, no. 3, pp. 796–800, 1998.

[11] T. Ohashi and H. P. Erickson, "Oligomeric structure and tissue distribution of ficolins from mouse, pig and human," *Archives*

of Biochemistry and Biophysics, vol. 360, no. 2, pp. 223–232, 1998.

[12] Y. Liu, Y. Endo, S. Homma, K. Kanno, H. Yaginuma, and T. Fujita, "Ficolin A and ficolin B are expressed in distinct ontogenic patterns and cell types in the mouse," *Molecular Immunology*, vol. 42, no. 11, pp. 1265–1273, 2005.

[13] V. L. Runza, T. Hehlgans, B. Echtenacher, U. Zähringer, W. J. Schwaeble, and D. N. Männel, "Localization of the mouse defense lectin ficolin B in lysosomes of activated macrophages," *Journal of Endotoxin Research*, vol. 12, no. 2, pp. 120–126, 2006.

[14] Y. Endo, Y. Liu, K. Kanno, M. Takahashi, M. Matsushita, and T. Fujita, "Identification of the mouse H-ficolin gene as a pseudogene and orthology between mouse ficolins A/B and human L-/M-ficolins," *Genomics*, vol. 84, no. 4, pp. 737–744, 2004.

[15] Y. Endo, M. Matsushita, and T. Fujita, "Role of ficolin in innate immunity and its molecular basis," *Immunobiology*, vol. 212, no. 4-5, pp. 371–379, 2007.

[16] M. Matsushita, M. Kuraya, N. Hamasaki, M. Tsujimura, H. Shiraki, and T. Fujita, "Activation of the lectin complement pathway by H-ficolin (Hakata antigen)," *Journal of Immunology*, vol. 168, no. 7, pp. 3502–3506, 2002.

[17] Y. Liu, Y. Endo, D. Iwaki et al., "Human M-ficolin is a secretory protein that activates the lectin complement pathway," *Journal of Immunology*, vol. 175, no. 5, pp. 3150–3156, 2005.

[18] Y. Endo, N. Nakazawa, Y. Liu et al., "Carbohydrate-binding specificities of mouse ficolin A, a splicing variant of ficolin A and ficolin B and their complex formation with MASP-2 and sMAP," *Immunogenetics*, vol. 57, no. 11, pp. 837–844, 2005.

[19] U. V. Girija, D. A. Mitchell, S. Roscher, and R. Wallis, "Carbohydrate recognition and complement activation by rat ficolin-B," *European Journal of Immunology*, vol. 41, no. 1, pp. 214–223, 2011.

[20] H. Nakanishi, Y. Higuchi, S. Kawakami, F. Yamashita, and M. Hashida, "PiggyBac transposon-mediated long-term gene expression in mice," *Molecular Therapy*, vol. 18, no. 4, pp. 707–714, 2010.

[21] D. Iwaki, K. Kanno, M. Takahashi et al., "Small mannose-binding lectin-associated protein plays a regulatory role in the lectin complement pathway," *Journal of Immunology*, vol. 177, no. 12, pp. 8626–8632, 2006.

[22] M. Takahashi, Y. Ishida, D. Iwaki et al., "Essential role of mannose-binding lectin-associated serine protease-1 in activation of the complement factor D," *Journal of Experimental Medicine*, vol. 207, no. 1, pp. 29–37, 2010.

[23] D. Iwaki, K. Kanno, M. Takahashi, Y. Endo, M. Matsushita, and T. Fujita, "Mannose-binding lectin-associated serine protein 3 (MASP-3) induces activation of the alternative complement pathway," *The Journal of Immunology*, vol. 187, pp. 3751–3758, 2011.

[24] N. Okada, R. Harada, T. Fujita, and H. Okada, "A novel membrane glycoprotein capable of inhibiting membrane attack by homologous complement," *International Immunology*, vol. 1, no. 2, pp. 205–208, 1989.

[25] S. E. Degn, A. G. Hansen, R. Steffensen, C. Jacobsen, J. C. Jensenius, and S. Thiel, "MAp44, a human protein associated with pattern recognition molecules of the complement system and regulating the lectin pathway of complement activation," *Journal of Immunology*, vol. 183, no. 11, pp. 7371–7378, 2009.

3

The Glycosylation of AGP and Its Associations with the Binding to Methadone

Jennifer L. Behan,[1] Yvonne E. Cruickshank,[1] Gerri Matthews-Smith,[2] Malcolm Bruce,[2] and Kevin D. Smith[1,3]

[1] School of Life, Sport and Social Sciences, Edinburgh Napier University, Sighthill Campus, Edinburgh EH11 4BN, UK
[2] School of Nursing, Midwifery and Social Care, Edinburgh Napier University, Edinburgh EH11 4BN, UK
[3] Community Drug Problem Service, Spittal Street Centre, Edinburgh EH3 9DU, UK

Correspondence should be addressed to Kevin D. Smith; k.smith@napier.ac.uk

Academic Editor: Viness Pillay

Methadone remains the most common form of pharmacological therapy for opioid dependence; however, there is a lack of explanation for the reports of its relatively low success rate in achieving complete abstinence. One hypothesis is that *in vivo* binding of methadone to the plasma glycoprotein alpha-1-acid glycoprotein (AGP), to a degree dependent on the molecular structure, may render the drug inactive. This study sought to determine whether alterations present in the glycosylation pattern of AGP in patients undergoing various stages of methadone therapy (titration < two weeks, harm reduction < one year, long-term > one and a half years) could affect the affinity of the glycoprotein to bind methadone. The composition of AGP glycosylation was determined using high pH anion exchange chromatography (HPAEC) and intrinsic fluorescence analysed to determine the extent of binding to methadone. The monosaccharides galactose and N-acetyl-glucosamine were elevated in all methadone treatment groups indicating alterations in AGP glycosylation. AGP from all patients receiving methadone therapy exhibited a greater degree of binding than the normal population. This suggests that analysing the glycosylation of AGP in patients receiving methadone may aid in determining whether the therapy is likely to be effective.

1. Introduction

It has been reported that the number of people who could be considered as being involved in opioid abuse, commonly coadministered with the benzodiazepines (BDZs), is between 12.8 and 21.8 million individuals [1]. Opioid dependence represents a universal problem to society and methadone remains as the most common form of pharmacological therapy. Similar to other drugs, the desired action of methadone requires that it must attain a certain *in vivo* concentration (the minimally effective concentration, MEC) at its site of action in order to provide a therapeutic effect. Unfortunately the success rate of methadone in achieving complete abstinence is poor and reportedly as low as 7% [2].

According to the free drug hypothesis, only a drug which is in a free unbound state *in vivo* is active. As most drugs access their target site of action through the bloodstream, they come into contact with numerous plasma proteins including albumin and α-1-acid glycoprotein (AGP), which are capable of binding the drug and rendering it inactive. After administration, methadone is strongly associated with AGP [3], therefore, alterations in the level and structure of the protein could greatly affect the efficacy of the drug. When doses of methadone are below the MEC, withdrawal symptoms may arise, including nausea, anxiety, tremors and dissipation of the euphoric "high" [4, 5]. Conversely, if concentrations exceed the threshold too rapidly or by too much, the effects of an overdose are displayed.

During the acute phase response to infection and disease, levels of AGP are known to increase two- to fivefold [6]. The increased production of AGP and any other drug binding plasma proteins causes an immediate shift in the binding equilibrium between drug and proteins upon the administration of the drug [7]. Consequently, the plasma level of bioactive (unbound) drug available to the target site of action or receptor is reduced, alongside its efficacy [8]. Rostami-Hodjegan and colleagues [9] noted that in heroin-dependent individuals with signs of opiate withdrawal, levels of AGP were elevated and were thought to reduce the level of active unbound methadone, promoting withdrawal symptoms.

Like many hepatically synthesised proteins, AGP is also modified through the addition of oligosaccharide or glycan chains (glycosylation). The process of glycosylation is an ordered and functionally significant process which results in a high degree of structural heterogeneity. The aforementioned variability is essentially a result of different cell types possessing a diverse array of enzymes thus catalysing a range of reactions to produce distinct glycans [10, 11]. AGP is extensively glycosylated (45%) with five oligosaccharide chains and exists *in vivo* as a heterogeneous population of variants (glycoforms) owing to differing occupancy of the five glycosylation sites by different oligosaccharide structures. Heterogeneity arises through subtle structural differences in monosaccharide sequence and linkages, degree of branching [bi-(2), tri-(3), tetra-(4) antennary], and the extent and number of charged sialic acid groups (sialylation). During several physiological and pathological conditions, the oligosaccharide "fingerprint" of AGP is altered [12, 13].

It is now widely accepted that AGP is an important drug-binding protein in the serum as it binds both endogenous and exogenous ligands. The drug-binding capabilities of AGP and the binding site of human AGP have been well characterised [14]. Muller [15] reported that AGP glycans are not critical in the interaction with drugs at the binding site, and instead it is solely dependent on the formation of the correct tertiary structure, describing the folding of the protein into its three-dimensional structure which is essentially determined by the sequence of amino acids (primary structure). However, although the drug binding site of AGP is peptide in nature, the hydrodynamic mass and surface location of the oligosaccharide chains are likely to affect the conformation, and correct folding, of both the protein in general and the binding site in particular. Therefore an alteration in the glycosylation of AGP is likely to affect the extent of binding to drugs.

This study sought to investigate whether differences in the relative level and glycosylation of AGP existed between that isolated from opioid-dependent individuals and a "normal" healthy population, which could correlate with variations in the binding to methadone. The structure of glycans expressed by AGP was analysed using high pH anion exchange chromatography (HPAEC). Subsequently, fluorescence quenching data was utilised as an indicator of binding to methadone.

2. Materials and Methods

2.1. Materials. Cibacron Blue 3GA, Dulbecco's phosphate-buffered saline, dimethyl sulphoxide Glacial acetic acid, (±)-methadone, polyethylene glycol (PEG) 3350, potassium chloride, potassium thiocyanate, Red Sepharose CL-6B, sodium acetate, sodium azide, sodium chloride, theophylline, and Trizma base were supplied by Sigma-Aldrich (Poole, UK). Ethanol was purchased from Bamford Laboratories Ltd.,

(Norden Rochdale, UK). HPLC-grade water was obtained from Rathburn Chemicals Ltd. (Walkerburn, UK). The poly prep disposable 10 mL columns were purchased from Bio-Rad, (Hemel Hempstead, UK). Amicon Ultra-4 centrifugal filter devices with a MW cutoff of 10,000 from Millipore (UK) Ltd. sodium hydroxide (50% w/v) were supplied by VWR International Ltd. (Lutterworth, UK). New England Biolabs Inc., (Hertfordshire, UK) provided the peptide-N-Glycosidase F (PNGase F) purified from *Flavobacterium meningosepticum*, 10% NP-40, and NE Buffer G7. HPAEC-PAD equipment, including the DX500 system and Carbopac PA-100 column, was purchased from Dionex, Camberley, UK. AGP N-linked glycan library was supplied by Prozyme (Europa Bioproducts Ltd., Cambridgeshire, UK). Nunc (Germany) supplied the 96-well microtitre plates, and A BMG Labtech Optima fluorimeter (Germany) was used to analyse the intrinsic fluorescence.

2.2. Methods. Ethical approval was granted by Lothian NHS and Edinburgh Napier University Ethics committees which allowed 2–5 mL blood samples to be obtained from consenting adults over 18 years of age ($n = 20$) undergoing various stages of treatment for opioid addictions at the Community Drug Problem Service (CDPS) in Edinburgh. Patient demographics are summarised in Table 1. Individuals were undergoing several phases including initial two-week titration (T), harm reduction (HR), or long-term methadone treatment (LT) and had no concomitant conditions present. Heparinised blood samples from "normal" healthy individuals were pooled by the blood transfusion service (BTS).

2.2.1. Isolation of AGP from Plasma. AGP was isolated from all blood samples using protocols similar to those reported by Elliott et al. [16]. SDS-PAGE indicated that the two-column isolation technique generated relatively pure AGP. Salt introduced during the isolation phase was removed using centrifugal filter devices and HPLC-grade water. The levels of isolated AGP were determined spectrophotometrically.

2.2.2. Monosaccharide Analysis. Preparation of the glycans was similar to methods reported by Fan et al. [17]. AGP (50 μg) was hydrolysed with 4 M HCl (six hours) and 2 M TFA (four hours) separately; we also found that the use of HCl rendered the neutral monosaccharides unstable over long periods of time. Also, the TFA was thought to be too weak to cleave N-acetylglucosamine (GlcNAc) residues from the core—as demonstrated by a higher level of GlcNAc in the presence of HCl compared to TFA. A Dowex cation-exchange resin was used to purify the neutral monosaccharides. HPAEC was used to separate and quantify the monosaccharide components.

2.2.3. Oligosaccharide Analysis. Structural analysis was also performed using HPAEC. Typically 100 μg of AGP was denatured at 100°C for three hours and the dried sample then treated with a reaction mixture of HPLC-grade water (79% v/v), NE Buffer G7 (10% v/v), NP-40 (10% v/v), and PNGaseF (5 U). The mixtures were incubated at 37°C for 24 hours and

TABLE 1: Patient demographics.

Stage of therapy	Identity	Age	Male/female	Methadone dose (mg)
Titration	T1	32	F	95
	T2	23	F	65
	T3	37	F	55
	T4	33	M	40
	T5	20	M	60
Harm reduction	HR1	37	F	90
	HR2	28	M	85
	HR3	23	F	150
	HR4	23	F	150
Long term	LT1	35	M	95
	LT2	33	M	150
	LT3	36	F	40
	LT4	44	F	90
	LT5	30	M	105
	LT6	51	M	85
	LT7	34	M	150
	LT8	42	M	50
	LT9	34	M	150
	LT10	31	M	200
	LT11	30	M	90

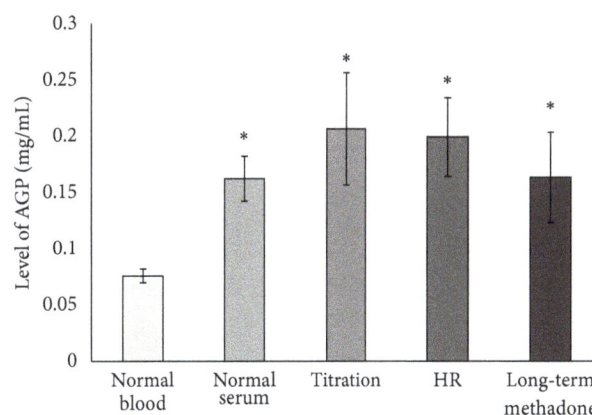

FIGURE 1: Relative level of AGP isolated from treatment groups. Summary of the mean level of AGP isolated from individuals at different stages of methadone therapy. *Statistically different to "normal" blood ($P < 0.05$).

then a further 5 U of PNGase F was added. After a final 24-hour incubation at 37°C, ethanol precipitation was performed at 3 : 1 ratio with the reaction mixture. The samples were then stored at −20°C overnight and analysed by HPAEC.

2.2.4. Drug Binding Analysis.
The binding of methadone to the isolated AGP preparations was investigated using intrinsic fluorescence studies at the microtitre level. Commercial AGP (5 mg/mL) was dissolved in d-PBS and plated in $10\,\mu L$ aliquots ($n = 3$). A range of theophylline (positive control) and methadone concentrations ($0\,\mu M$–$2500\,\mu M$) were added in $10\,\mu L$ volumes—having been dissolved in DMSO. A final reaction volume of $100\,\mu L$ was generated through the addition of D-PBS. Plates were excited at 280 nm and fluorescence emitted at 340 nm (AGP maximum emission) recorded. The experiment was repeated with AGP concentrations of 0.5 m/mL to allow for the low levels of AGP isolated from patient groups and ensures that the technique was adaptable to this level. Commercial and isolated AGP was plated (0.5 mg/mL) and methadone standards added to include values representing the therapeutic level of methadone (~ 4000 ng/mL) and others out-with this; all were diluted tenfold upon plating.

2.2.5. Statistical Analysis.
Two sample t-tests and one-way ANOVAs (with Tukey's post hoc) were performed using Minitab, version 15, to determine the statistical significance of all quantitative data generated during the various analyses.

3. Results and Discussion

3.1. AGP Levels.
Using a set of standards, an AGP calibration curve was generated and used to determine the relative level of AGP isolated from the individual samples (Figure 1). Statistical analysis was performed with two sample t-tests to compare the relative difference between each patient group and heparinised blood from a "normal" healthy population. Significantly lower levels of AGP were isolated from the "normal" blood population ($P < 0.05$) compared to those from all other treatment groups. The levels of AGP were found to be at least twofold higher which is comparable with the increase in amounts reported after the acute phase response [6].

3.2. Monosaccharide Analysis.
The levels of monosaccharides in the commercially purchased AGP were found (using one-way ANOVA) to be greater ($P < 0.05$) than those in AGP isolated from "normal" heparinised blood. However, direct comparisons were deemed invalid because the methods implemented in the isolation of the glycoprotein differed. It was considered more relevant to use the "normal" blood values as comparisons, having been isolated by the same technique.

There was a significant change in the glycosylation of AGP in individuals receiving methadone as an opioid replacement therapy compared to that from a "normal" healthy population. The levels of galactose (Gal) and galactosamine (GlcN; GlcNAc in the *in vivo* structure) were significantly greater in patients undergoing methadone therapy when compared to the "normal" group (refer to Figure 2). Relative consistency was displayed in the level of mannose (Man) between all samples which is consistent with its presence in the complex N-linked oligosaccharide chains of AGP (three mannose residues per chain). The most significant differences in monosaccharide composition occurred with respect to the levels of galactose (Gal) and glucosamine (GlcN) which suggests that there is an increase in the number of branches on the oligosaccharide chains present. AGP has five N-linked chains (so called because they are attached to the side chain of

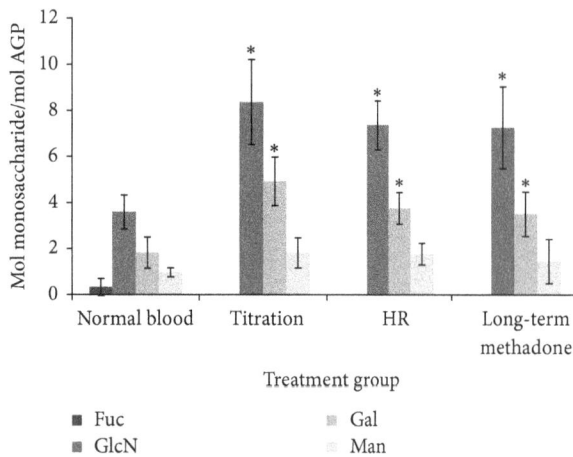

FIGURE 2: Mean level of monosaccharides in AGP glycans isolated from treatment groups. "Normal blood" represents heparinised blood from the Blood Transfusion Service. *Statistically significant difference compared to normal heparinised blood sample ($P < 0.05$) HR: harm reduction; LT: long-term treatment. Standard deviation calculated from triplicate analysis of a single sample while others were calculated from data from more than one patient ($n = 4/+$).

an asparagine amino acid) which can have a variable number of branches attached to a core sequence. Since galactose and the majority of glucosamine (as its acetylated form, N-acetylglucosamine) are only present in the outer branches of N-linked chains, an increase would tentatively suggest that the degree of branching of a chain has also increased. Our data suggests the occurrence of increased branching on the chains of the AGP from the methadone replacement patients. However, in the number of patients analysed, these changes did not appear to be dependent upon stage or types of opioid replacement therapy. The monosaccharide fucose (Fuc), whose presence on oligosaccharide chains is generally correlated with pathophysiological conditions such as inflammation and cancer [18, 19] was not detected in any patient sample.

3.3. Oligosaccharide Analysis.

Oligosaccharide analysis using HPAEC generated chromatograms which allowed qualitative analysis. The technique initially separates enzymatically cleaved glycans based on the degree of sialylation which provided the main negative charge. Glycans were separated into charge bands, those expressing a single SA residue eluted first (10–20 minutes), followed by bisialylated (20–30 minutes), trisialylated (30–40 minutes), and finally tetrasialylated, eluting between 40 and 50 minutes. An AGP N-linked glycan library was used to allow the corresponding charge groups to be identified (Figure 3).

The glycosylation of AGP isolated from "normal" heparinised blood samples was found to express numerous structures of different sialylation. Preference in this case appeared to be for bisialylated forms. The traces generated for the oligosaccharides of titration patients indicated that bi- and trisialylated structures were most common although not the only type present. This was similar to the "normal"

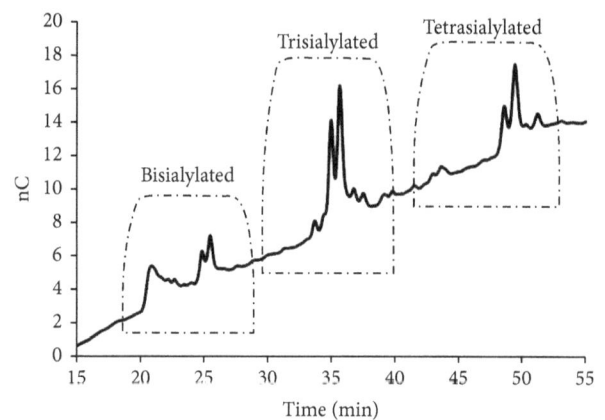

FIGURE 3: Oligosaccharide library trace generated during separation by HPAEC. Separation based on the charge of oligosaccharide chains as measured using nano-Coulomb (nC). Degree of sialylation highlighted in a dashed line.

and commercial samples, implying low levels of branching. Long-term patient AGP was also found to show a greater proportion of bi- and trisialylated structures, however, unlike the titration phase individuals, peaks in the tetrasialylated region were more defined (e.g., LT4, LT9, and LT10). In many patients, the number of peaks was fairly low but they were of a large size suggesting that there were many of the same glycan expressed.

Overall, it was shown that similarities existed between the treatment and nontreatment groups. Titration and "normal" appeared to have a high proportion of trisialylated structures. Variation existed, particularly within the LT group; some AGP displaying more highly branched structures (e.g., LT4 and LT10) while others having a greater proportion of less-sialylated structures (e.g., LT3).

3.4. Drug Binding.

The degree of fluorescence quenching increased as the concentration of methadone added was increased, reaching a plateau at concentrations above $\sim 100~\mu$M in both patient AGP and that isolated from "normal" heparinised samples. Values in the absence and presence of $1.07~\mu$M and $250~\mu$M of methadone are recorded in Table 2 to highlight the similarity of participant AGP with that from "normal" hepatrinised blood in the absence of drug. The rate of reduction in fluorescence was most pronounced at low concentrations indicating that binding became saturated relatively quickly. However, the drug did not completely quench the intrinsic fluorescence; some binding sites remained available. In general, the overall reduction in fluorescence was greater for all patient AGP samples analysed when compared to "normal" AGP, however, there were no clear differences between patients. Further studies are required to allow quantification and statistical analysis of binding, and this study was to allow preliminary comparison.

4. Conclusion

The success of drug therapies relies on the presence of specific concentrations of its bioactive form (MEC) at the

TABLE 2: Intrinsic fluorescence of AGP in absence and presence of 1.07 μM and 250 μM methadone.

Sample	Fluorescence in the absence of methadone (RFU)	Fluorescence in the presence of 1.07 μM methadone (RFU)	Fluorescence in the presence of 250 μM methadone (RFU)
Commercial (sigma)	41636	38236	30620
"Normal" 1	39943.5	39021	35476
"Normal" 2	37784.5	37032	36138
T1	39859	34006	30712
T2	35992	33371	30840
T3	36715	33020	30284
T4	37571	31089	29182
T5	38882	33959	29412
HR1	39107	33280	29289
HR2	38438	33215	29025
HR3	33531	30023	27072
LT1	31770	30775	27196
LT2	33324	30252	28346
LT4	38688	37189	31008
LT5	39937	36241	31893
LT6	39691	31681	27137
LT7	38484	31538	28658

corresponding site of action. Numerous factors determine whether this can be achieved; however, often overlooked is the binding to plasma proteins. Of primary interest to this study was whether the level and glycoform expression of AGP isolated from patients undergoing various stages and types of substitute therapy for opioid dependence differed to a "normal" healthy population. It was supposed that alterations may correlate to variations in the binding of the glycoprotein to methadone. It was hypothesised that AGP isolated from patients would exhibit higher levels and structural changes causing increased binding to methadone when compared to a "normal" AGP sample. Greater binding would at least partially explain why high doses of the drug are required—accounting for that which is bound and therefore inactivated.

Similar to research undertaken by Rostami-Hodjegan et al. [9], the current study detected increased levels of AGP expressed in individuals who displayed signs of withdrawal from heroin (represented by the titration group). The levels isolated from the patients in the remaining treatment groups—excluding the individual receiving heroin—were also greater than that isolated from the "normal" population. Our study seeks to extend this research by examining the effect of the glycosylation pattern of AGP, and any changes and correlating to the extent of binding to methadone.

Altered glycosylation of AGP has been widely studied in a number of pathophysiological conditions (see reviews in [6, 14]) and compared with the pattern present in healthy subjects which tends towards chains with either two (bi-) or three (tri-) branches. A decrease in the aforementioned number of branches has been demonstrated in acute inflammatory conditions while increased branching is observed to

be associated with chronic inflammations, liver diseases, and poor prognosis in cancer after surgery. The data from the current study suggests that AGP isolated from all phases of the methadone treatment programme is associated with a shift towards increased branching. The mixture of isolated glycan chains is separated, by HPAEC, according to the number of terminal negatively charged sialic acids possessed which can normally be directly correlated with the size of the chains. In other words a bi-sialylated structure contains two sialic acid residues which equates to one terminating each of two branches; a tri-sialylated structure has three sialic acids on typically three branches, and a tetra-sialylated chain has four branches each terminated with a sialic acid. Our results indicate that AGP isolated from subjects in the titration and long-term phases both had increases in the tri-sialylated/three branch population, and the latter also contained significant quantities of tetra-sialylated chains with four branches. The significance of these alterations with respect to the drug binding is likely to be related to changes in the conformation of the drug binding site of AGP resulting in either decreased or increased binding affinity for methadone. An increased binding affinity for methadone would result in a larger percentage of the free drug being bound therefore reducing or completely abolishing the desired therapeutic response.

Although it is well understood that structural changes to AGP glycosylation can significantly affect the functions of the biomolecule, there is currently a paucity of research data pertaining to this effect in patients undergoing opioid-replacement therapy. It is well understood that AGP generally exhibits a degree of selectivity for the ligands to which it binds, and, in terms of drugs, they are generally of the neutral or basic variety [8, 14]. If binding to methadone is altered by structural modifications to the glycan chains present on AGP, the free active concentrations and subsequent pharmacological effect could be changed. A reduction in efficacy would become apparent when the affinity increases or vice versa if the affinity was reduced.

Currently treatment is based on a titration phase where individuals are given low doses of drug to allow for its slow accumulation. Finally, a maintenance dose is determined. However, with investigations like these it may be possible to determine the relative level of glycoforms expressed by an individual's AGP to identify those where it is less likely to be effective or where higher doses will be required. Also, another consequence of i.v heroin use is collapsed veins, complicating venipuncture. However, this preliminary study has shown that potential correlation exists between the structure of AGP glycans and its ability to bind the basic drug methadone although a larger patient population is required to address other influential factors and determine whether the interaction could significantly affect the efficacy of methadone in the treatment of opioid dependencies.

The variability in the binding of methadone to AGP may partly explain why some individuals require more drug than others to produce the effect. That is not to forget the roles of metabolism and elimination which display interindividual variability due to differences in gene expression. Further work is required to quantify the degree of methadone binding to AGP with specific glycosylation patterns. Fortunately, the ability to quantify the interaction between a plasma protein such as AGP and a drug has recently improved through the use of surface plasma resonance (SPR) [20] which overcomes the limitations of previous methods such as equilibrium dialysis, ultrafiltration, and ultracentrifugation. The latter methods are associated with problems of cost, throughput, and quality of information obtained, as well as being equilibrium based and, therefore, providing no kinetic information. The use of SPR not only overcomes these limitations but also allows the more accurate statistical analysis of the correlation between the degree of binding and drug efficacy.

Conflict of Interests

There is no conflict of interests with any of the authors (nor did any of them gain financially).

Acknowledgment

The authors acknowledge the support of the Carnegie Trust for the Universities of Scotland in the award of a Caledonian research scholarship to Jennifer L. Behan.

References

[1] UNODC, *World Drug Report 2010*, Sales No.E.10.XI.13, United Nations Publication, 2010.

[2] N. McKeganey, M. Bloor, M. Robertson, J. Neale, and J. MacDougall, "Abstinence and drug abuse treatment: results from the drug outcome research in Scotland Study," *Drugs*, vol. 13, no. 6, pp. 537–550, 2006.

[3] M. K. Romach, K. M. Piafsky, and J. G. Abel, "Methadone binding to orosomucoid (α1-acid glycoprotein): determinant of free fraction in plasma," *Clinical Pharmacology and Therapeutics*, vol. 29, no. 2, pp. 211–217, 1981.

[4] O. Drummer and M. Odell, *The Forensic Pharmacology Drugs of Abuse*, Edited by O. H. Drummer, Hodder Arnold, 2001.

[5] M. M. Scimeca, S. R. Savage, R. Portenoy, and J. Lowinson, "Treatment of pain in methadone-maintained patients," *Mount Sinai Journal of Medicine*, vol. 67, no. 5-6, pp. 412–422, 2000.

[6] F. Ceciliani and V. Pocacqua, "The acute phase protein α1-acid glycoprotein: a model for altered glycosylation during diseases," *Current Protein and Peptide Science*, vol. 8, no. 1, pp. 91–108, 2007.

[7] Y. Kuroda, S. Matsumoto, A. Shibukawa, and T. Nakagawa, "Capillary electrophoretic study on pH dependence of enantioselective disopyramide binding to genetic variants of human α1-acid glycoprotein," *Analyst*, vol. 128, no. 8, pp. 1023–1027, 2003.

[8] J. M. H. Kremer, J. Wilting, and L. H. M. Janssen, "Drug binding to human alpha-1-acid glycoprotein in health and disease," *Pharmacological Reviews*, vol. 40, no. 1, pp. 1–47, 1988.

[9] A. Rostami-Hodjegan, K. Wolff, A. W. M. Hay, D. Raistrick, R. Calvert, and G. T. Tucker, "Population pharmacokinetics of methadone in opiate users: characterization of time-dependent changes," *British Journal of Clinical Pharmacology*, vol. 48, no. 1, pp. 43–52, 1999.

[10] A. Varki, R. D. Cummings, J. D. Esko et al., *Essentials of Glycobiology*, Cold Spring Harbour Laboratory Press, 2nd edition, 2009.

[11] M. E. Taylor and K. Drickamer, *Introduction to Glycobiology*, Oxford University Press Limited, 2003.

[12] W. van Dijk, G. A. Turner, and A. Mackiewicz, "Changes in glycosylation of acute-phase proteins in health and disease: occurrence, regulation and function," *Glycosylation and Disease*, vol. 1, no. 1, pp. 5–14, 1994.

[13] K. Higai, Y. Aoki, Y. Azuma, and K. Matsumoto, "Glycosylation of site-specific glycans of α1-acid glycoprotein and alterations in acute and chronic inflammation," *Biochimica et Biophysica Acta*, vol. 1725, no. 1, pp. 128–135, 2005.

[14] Z. H. Israili and P. G. Dayton, "Human alpha-1-glycoprotein and its interactions with drugs," *Drug Metabolism Reviews*, vol. 33, no. 2, pp. 161–235, 2001.

[15] W. E. Muller, "Stereoselective plasma protein binding of drugs," in *Drug Stereochemistry: Analytical Methods and Pharmacology*, I. W. Wainer and D. E. Drayer, Eds., p. 277, Marcel Dekker, New York, NY, USA, 1988.

[16] M. A. Elliott, H. G. Elliott, K. Gallagher, J. McGuire, M. Field, and K. D. Smith, "Investigation into the concanavalin A reactivity, fucosylation and oligosaccharide microheterogeneity of α1-acid glycoprotein expressed in the sera of patients with rheumatoid arthritis," *Journal of Chromatography B*, vol. 688, no. 2, pp. 229–237, 1997.

[17] J.-Q. Fan, Y. Namiki, K. Matsuoka, and Y. C. Lee, "Comparison of acid hydrolytic conditions for Asn-linked oligosaccharides," *Analytical Biochemistry*, vol. 219, no. 2, pp. 375–378, 1994.

[18] G. A. Turner, A. W. Skillen, and P. Buamah, "Relation between raised concentrations of fucose, sialic acid, and acute phase proteins in serum from patients with cancer: choosing suitable serum glycoprotein markers," *Journal of Clinical Pathology*, vol. 38, no. 5, pp. 588–592, 1985.

[19] J. J. Listinsky, G. P. Siegal, and C. M. Listinsky, "Alpha-L-fucose: a potentially critical molecule in pathologic processes including neoplasia," *American Journal of Clinical Pathology*, vol. 110, no. 4, pp. 425–440, 1998.

[20] S. S. Gustafsson, L. Vrang, Y. Terelius, and U. H. Danielson, "Quantification of interactions between drug leads and serum proteins by use of 'binding efficiency'," *Analytical Biochemistry*, vol. 409, no. 2, pp. 163–175, 2011.

4

Factors Affecting Poly(3-hydroxybutyrate) Production from Oil Palm Frond Juice by *Cupriavidus necator* (CCUG52238[T])

Mior Ahmad Khushairi Mohd Zahari,[1,2] Hidayah Ariffin,[3] Mohd Noriznan Mokhtar,[1] Jailani Salihon,[2] Yoshihito Shirai,[4] and Mohd Ali Hassan[1,3]

[1] *Department of Process and Food Engineering, Faculty of Engineering, Universiti Putra Malaysia, Serdang, 43400 Selangor, Malaysia*

[2] *Faculty of Chemical and Natural Resources Engineering, Universiti Malaysia Pahang, Lebuhraya Tun Razak, Kuantan, 26300 Pahang, Malaysia*

[3] *Department of Bioprocess Technology, Faculty of Biotechnology and Biomolecular Sciences, Universiti Putra Malaysia, Serdang, 43400 Selangor, Malaysia*

[4] *Department of Biological Functions and Engineering, Graduate School of Life Science and Systems Engineering, Kyushu Institute of Technology, 2-4 Hibikino, Wakamatsu-ku, Kitakyushu, Fukuoka 808-0196, Japan*

Correspondence should be addressed to Hidayah Ariffin, hidayah_a@biotech.upm.edu.my

Academic Editor: Anuj K. Chandel

Factors in uencing poly(3-hydroxybutyrate) P(3HB) production by *Cupriavidus necator* CCUG52238[T] utilizing oil palm frond (OPF) juice were clari ed in this study. Effects of initial medium pH, agitation speed, and ammonium sulfate $(NH_4)_2SO_4$ concentration on the production of P(3HB) were investigated in shake asks experiments using OPF juice as the sole carbon source. The highest P(3HB) content was recorded at pH 7.0, agitation speed of 220 rpm, and $(NH_4)_2SO_4$ concentration at 0.5 g/L. By culturing the wild-type strain of *C. necator* under the aforementioned conditions, the cell dry weight (CDW) and P(3HB) content obtained were 9.31 ± 0.13 g/L and 45 ± 1.5 wt.%, respectively. This accounted for 40% increment of P(3HB) content compared to the nonoptimized condition. In the meanwhile, the effect of dissolved oxygen tension (DOT) on P(3HB) production was investigated in a 2-L bioreactor. Highest CDW (11.37 g/L) and P(3HB) content (44 wt.%) were achieved when DOT level was set at 30%. P(3HB) produced from OPF juice had a tensile strength of 40 MPa and elongation at break of 8% demonstrated that P(3HB) produced from renewable and cheap carbon source is comparable to those produced from commercial substrate.

1. Introduction

Poly(3-hydroxybutyrate), P(3HB) is a biodegradable thermoplastic polyester accumulated intracellularly by many microorganisms under unfavorable growth conditions [1]. The high production cost of P(3HB) can be decreased by strain development, improving fermentation and separation processes [2–4], and/or using a cheap carbon source [5]. In P(3HB) production, about 40% of the total production cost is contributed by the raw material, whereby the cost of carbon feedstock alone accounts for 70 to 80% of the total raw material cost [6, 7]. Therefore, the utilization of renewable and sustainable substrates for the production of P(3HB) has

become an important objective for the commercialization of bioplastics. A lot of research have been carried out to discuss and propose the utilization of renewable biomass to replace commercial sugars as carbon source in order to reduce the production cost of P(3HB) [8–12].

Recently, we reported on the use of oil palm frond (OPF) juice as the novel and renewable feedstock for the production of P(3HB) [13]. We demonstrated that OPF juice is a good substrate for the production of P(3HB) from wild-type *Cupriavidus necator* (CCUG52238[T]), with better yield of product formation in comparison to technical grade sugars. This can be explained by the presence of minerals and nutrients in the OPF juice which are essential for bacterial growth during fermentation. Apart from contributing to higher product formation and microbial growth, the use of OPF juice is advantageous compared to the other lignocellulose-based sugars due to the ease in its processing wherein no harsh pretreatment steps and enzymatic treatment will be needed in order to obtain the sugars.

In our report, 32 wt.% of P(3HB) accumulation was successfully obtained under nonoptimized fermentation condition [13]. C. necator is well known as polyhydroxyalka-noate (PHA) producer and its ability to accumulate PHA more than 50 wt.% has been previously reported [8, 14]. In general, P(3HB) accumulation is favored by an excess carbon source and inadequate supply of macrocomponents such as nitrogen, phosphate, and dissolved oxygen or micro-components such as magnesium, sulphate, iron, potassium, manganese, copper, sodium, cobalt, tin, and calcium [7, 15]. Moreover, it was also reported that the accumulation of P(3HB) in microorganisms were influenced by several physical parameters including pH and agitation speed [16–19].

In order to make the production of P(3HB) feasible for industrial application, it is crucial to have high P(3HB) production yield. In this study, we investigated the effect of initial medium pH, agitation speed, and ammonium sulfate $(NH_4)_2SO_4$ concentration on P(3HB) production from C. necator (CCUG52238^T) utilizing OPF juice in shake asks fermentation with the aim to clarify the effect of each fermentation parameter on the microbial growth and P(3HB) formation. The effect of dissolved oxygen tension (DOT) level on cell growth and P(3HB) production was investigated by conducting batch fermentation in 2-L-bioreactor. P(3HB) produced from this study was then characterized for its thermal and mechanical properties.

2. Materials and Methods

2.1. Bacterial Strain.
In this study, C. necator (CCUG52238^T) was obtained from the Culture Collection, University of Goteborg, Sweden and used for the production of P(3HB). The culture was maintained on slants of nutrient agar at 4°C. The inoculum preparation, media, and cultivation conditions for C. necator (CCUG52238^T) are similar to those reported by Zahari et al. [13], unless otherwise stated.

2.2. Biosynthesis of P(3HB) in Shake Flask.
P(3HB) biosynthesis was carried out through one-stage cultivation fermentation in shake asks. OPF juice in this study was obtained by pressing fresh OPF following the method described earlier [13]. OPF juice which comprises fructose, glucose and sucrose was diluted from stock (55 g/L) to 16-17 g/L of total initial sugars and used as carbon sources throughout the study period. In order to study the effect of culture medium initial pH on biosynthesis of P(3HB), the initial pH value of each MSM and OPF juice was adjusted to pH 6.0–8.0 using 2 M NaOH prior to autoclaving. Another set of experiment was conducted to study the effect of agitation on P(3HB) production by testing several agitation speed at 180, 200, 220, 240, and 260 rpm. For the effect of ammonium sulfate

concentration, various concentrations of $(NH_4)_2SO_4$ in the range of 0–2.0 g/L were tested. The cultures were incubated at 30°C under aerobic condition, and all experiments were conducted in duplicates.

2.3. Biosynthesis of P(3HB) in 2-L-Bioreactor.
In order to study the effect of dissolved oxygen tension (DOT) on cell growth and P(3HB) production pro le under the optimized condition obtained from the shake ask study, batch experiment was conducted in 2-L-bioreactor (1 L working volume) at different DOT levels of 20, 30, 40, and 50%. 100 mL of pregrown cells from growing stage were transferred into 900 mL MSM in 2L bioreactors (Sartorius, Germany) supplemented with OPF juice at 30% (v/v) dilution. The stock of OPF juice with 55 g/L of initial total sugars concentration was autoclaved separately prior to addition with the MSM medium. The MSM compositions were prepared as previously reported by Zahari et al. [13], except that 0.5 g/L of $(NH_4)_2SO_4$ was used in this study. The temperature inside the bioreactor was set at 30°C, while DOT level was set at various concentrations of saturation throughout the fermentation using cascade mode and supplied with air at 1.0 vvm. The pH value during fermentation was controlled at pH 7.0 ± 0.05 by 2 M NaOH/H_2SO_4. Samples were withdrawn every 5 h for the period of 50 h for the determination of CDW, P(3HB) concentration, residual sugars, and ammoniacal nitrogen $(NH_3\text{-}N)$ content.

2.4. Analytical Procedures

2.4.1. Biomass and Culture Medium Separation.
Residual sugars concentration, cell dry weight measurement, and P(3HB) analysis were done as previously described by Zahari et al. [13]. The samples from the bacterial fermentations were taken at the end of the cultivation period to measure the total dry weight and P(3HB) content. Each sample was centrifuged at $11,000 \times g$ for 5 min at 4°C (Thermo Fisher Scienti c, NC, USA) and the solids were washed with distilled water and centrifuged for two consecutive times.

2.4.2. Determination of Cell Dry Weight and P(3HB) Content.
Dry weight measurements were carried out by drying the solids at 50°C and cooling in a desiccator to constant weight. The P(3HB) content and composition in the lyophilized cell were determined using the gas chromatography (Shimadzu GC-2014). Approximately, 20 mg of lyophilized cells were subjected to methanolysis in the presence of methanol and sulfuric acid [85% : 15% (v/v)]. The organic layer containing the reaction products was separated, dried over Na_2SO_4, and analyzed by GC according to the standard method [20] using an ID-BP1 capillary column, 30 m × 0.25 mm × 0.25 μm lm thickness (SGE).

2.4.3. Determination of Residual Ammoniacal Nitrogen $(NH_3\text{-}N)$ Content.
The supernatant was then analyzed for residual sugars and ammoniacal nitrogen content. Residual ammoniacal nitrogen $(NH_3\text{-}N)$ content analysis was done using

TABLE 1: Effect of initial pH value on the biosynthesis of P(3HB)[a].

Initial pH	CDW (g/L)	Total P(3HB) (g/L)	P(3HB) content (wt.%)[b]
6.0	6.42	1.28	20
6.5	7.12	1.99	28
7.0	8.57	2.91	34
7.5	6.89	1.72	25
8.0	4.02	0.40	10

[a]MSM containing 16 g/L of total sugars in OPF juice and supplied with 1.0 g/L of $(NH_4)_2SO_4$, incubated at 30°C for 48 h with agitation at 200 rpm.
[b]Determination by GC from freeze dried samples.
*Values obtained herewith are means of two independent experiments.

Nessler method according to standard procedures (HACH, USA) which was previously described by Zakaria [21]. Samples with appropriate dilution factor were lled to 25 mL in the sampling bottles. Three drops of mineral stabilizer were added into solution and the bottle was inverted for several times. Three drops of polyvinyl alcohol also were added into the solution and mixed well with inversion several times. Lastly, 1.0 mL of Nessler reagent was added to the mixtures and mix thoroughly by inversion. The standard solution was prepared by replacing the samples with deionised water as blank sample. The sample solution was determined at the wavelength (λ) 425 nm using DR/4000 spectrophotometer by following the manufacturer, instructions (HACH, USA).

2.4.4. Determination of Residual Sugars. Residual sugars were determined by a high performance liquid chromatography (HPLC) (Agilent Series 1200, USA) using the Supelcosil LC-NH2 column (Sigma Aldrich) (25 cm × 4.6 mm ID, 5 μm particles) with a RI detector operated at 30°C. The mobile phase was acetonitrile : water (75% : 25%) at a o w rate of 1.0 mL/min. The components were identi ed by comparing their retention times with those of authentic standards under analytical conditions and quanti ed by external standard method [22].

2.5. Extraction of P(3HB). Solvent extraction method as described by Zakaria et al. [23] was carried out in order to extract the P(3HB) produced from fermentation. P(3HB) lm was then prepared by solvent casting using chloroform.

2.6. Characterization of P(3HB). Thermal properties of the polymer were determined by differential scanning colorimetry (DSC) (TA Instruments). For DSC analysis, 5–7 mg of homopolymer samples were weighed and heated from 20 to 200°C at heating rates 10°C/min and held for 1 min. The rst scan was conducted to eliminate the polymer history. The samples were then fast cooled from 200°C to −30°C. The second scan was used in reheating the samples at the same heating rates and was used in evaluating the thermal properties of the biopolymer. The tensile strength, Young's modulus and elongation to break were determined by using Instron Universal Testing Machine (Model 4301) at 5 mm/min of crosshead speed [21]. Mechanical tensile data were calculated from the stress-strain curves on average of v e specimens.

3. Results and Discussion

3.1. Biosynthesis of P(3HB) in Shake Flask Experiment

3.1.1. Effect of Initial Medium pH. The effect of initial medium pH on biosynthesis of P(3HB) from OPF juice was studied by varying the pH between pH 6.0 and 8.0 due to the fact that *C. necator* can tolerate and produce PHA at the aforementioned pH range [24]. Suitable initial medium pH is crucial for the cell growth and P(3HB) accumulation by *C. necator* (CCUG52238[T]). As shown in Table 1, increasing the initial medium pH value at intervals of 0.5 units affected both the cell growth and P(3HB) production. Both the cell growth and P(3HB) content were increased when the initial medium pH was increased from pH 6.0 to pH 7.0, that is, from 6.42 g/L to 8.57 g/L for CDW and 20 wt.% to 34 wt.% for P(3HB) content, respectively. However, further increase of initial medium pH above pH 7.0 decreased both the CDW and P(3HB) content. From the results, it can be concluded that pH 7.0 was the optimum initial medium pH for the growth and biosynthesis of P(3HB) by *C. necator* (CCUG52238[T]) in which, 8.57 g/L of CDW and 34 wt.% of P(3HB) accumulation was recorded. The optimal pH for the cell growth and P(3HB) accumulation in this study was similar to those reported in the literature. It was reported that the optimum pH for growth and P(3HB) production by *A. eutrophus* was pH 6.9 and that a pH of 5.4 inhibited its growth [16].

On the other hand, lowest CDW and P(3HB) content, 4.02 g/L and 10 wt.%, respectively, were obtained at pH 8.0. Lowest cell growth and P(3HB) accumulation at this initial pH value were obtained might due to alkaline condition which could affect the P(3HB) production. These results corroborate with other be previous ndings. For instance, N. J. Palleroni and A. V Palleroni. [25], recommended a pH range of between 6.0 to 7.5 for microbial growth and P(3HB) production. Although P(3HB) production can be controlled by precisely manipulating the medium pH, it has been reported that pH values other than 7.0 affected P(3HB) production [26]. These results suggested that P(3HB) production is sensitive to the pH of cultivation.

3.1.2. Effect of Agitation Speed. Table 2 displays the effect of agitation speed on biosynthesis of P(3HB) using OPF juice as substrate in shake asks experiment. It is interesting to note that both cell growth and P(3HB) production

TABLE 2: Effect of agitation speed on the biosynthesis of P(3HB)[a].

Agitation speed (rpm)	CDW (g/L)	Total P(HB) (g/L)	P(3HB) content (wt.%)[b]
180	7.37	1.62	22
200	8.30	2.66	32
220	9.42	3.77	40
240	6.37	1.72	27
260	5.19	1.09	21

[a]MSM containing 16 g/L of total sugars in OPF juice and supplied with 1.0 g/L of $(NH_4)_2SO_4$, incubated at 30°C for 48 h (initial pH medium adjusted at 7.0 ± 0.1).
[b]Determination by GC from freeze dried samples.
*Values obtained herewith are means of two independent experiments.

TABLE 3: Effect of $(NH_4)_2SO_4$ concentration on biosynthesis of P(3HB)[a]

$(NH_4)_2SO_4$ concentration (g/L)	CDW (g/L)	Total P(3HB) (g/L)	P(3HB) content (wt.%)[b]
0.0	5.25	2.31	44
0.5	8.31	3.49	42
1.0	8.65	2.94	34
1.5	9.05	2.62	29
2.0	10.15	2.33	23

[a]MSM containing 16 g/L of total sugars in OPF juice, incubated at 30°C for 48 h with agitation at 200 rpm (initial pH medium adjusted at 7.0 ± 0.1).
[b]Determination by GC from freeze dried samples.
*Values obtained herewith are means of two independent experiments.

showed an increasing trend with the agitation speed up to 220 rpm. For the agitation speed of more than 220 rpm, the cell biomass and P(3HB) content was decreased. This result suggests that agitation speed plays an important role in the fermentation process. Agitation not only provides mixing and homogeneous cell and heat dispersion in the fermentation broth, but also better aeration for the cells by increasing the oxygen transfer rate throughout the fermentation medium. Generally, slower agitation speed may cause the possibilities of cells aggregation, making the culture medium more heterogeneous. This may cause the cell growth to be decreased and thus affecting the production of P(3HB). On the other hand, increasing agitation speed higher than its optimal level may reduce the P(3HB) formation, and hence, the CDW. This is due to the fact that PHA is only produced and stored as granules in the cell cytoplasm by microorganisms when they are under stress conditions, for example when there is limitation of nutrient or electron acceptor such as oxygen [27].

In our study, the best condition for the biosynthesis of P(3HB) is at moderate agitation speed which is at 220 rpm with the highest CDW and P(3HB) content reaching up to 9.42 g/L and 40 wt.%, respectively.

3.1.3. Effect of $(NH_4)_2SO_4$ Concentration. Nitrogen is an essential element for cell growth and P(3HB) accumulation. $(NH_4)_2SO_4$ has been widely used as the inorganic nitrogen source for the biosynthesis of P(3HB) by *C. necator*. It is important to optimize nitrogen content in fermentation medium as P(3HB) accumulation in the microorganisms can be triggered when one of the nutrients (N, P, Mg, and O_2) in the mineral salt is limited in the presence of excess carbon source [14, 15].

The effect of different $(NH_4)_2SO_4$ concentrations on biosynthesis of P(3HB) by *C. necator* (CCUG52338^T) from OPF juice is summarized in Table 3. In overall, it was observed that CDW was increased when $(NH_4)_2SO_4$ concentration increased from 0 to 2.0 g/L. On the other hand, P(3HB) accumulation decreased with the increase of $(NH_4)_2SO_4$ concentration. Highest P(3HB) accumulation at 44 wt.% was achieved when there was no addition of $(NH_4)_2SO_4$, in the culture medium. However, unsatisfactory cell growth that is, 5.25 g/L of CDW was observed in the experiment.

Based on the results, it was also found that $(NH_4)_2SO_4$ concentration at 0.5 g/L was the optimal concentration for P(3HB) accumulation and CDW formation, giving 42 wt.% and 8.31 g/L, respectively. Further increasing nitrogen concentration slightly improved the cells growth; however the accumulation of P(3HB) was restricted. This may be due to excess nitrogen concentration that limited the P(3HB) accumulation. These results corroborate to the literature, which reported that P(3HB) formation predominantly occurs under-nitrogen and oxygen-limited conditions [14, 15, 28, 29]. It was discussed that excess nitrogen source may restrict acetyl-CoA from entering P(3HB) production pathways and otherwise channelling into TCA cycle for biomass production [15, 21, 28].

3.1.4. Biosynthesis of P(3HB) under Optimized Condition in Shake Flask. Biosynthesis of P(3HB) was then carried out in shake ask under the optimized conditions: initial pH medium, 7.0; agitation speed, 220 rpm and $(NH_4)_2SO_4$ concentration, 0.5 g/L. Under these conditions, the maximum cell dry weight obtained was 9.31±0.13 g/L with 45±1.5 wt.% of P(3HB) accumulation. The P(3HB) produced from this

(a)

(b)

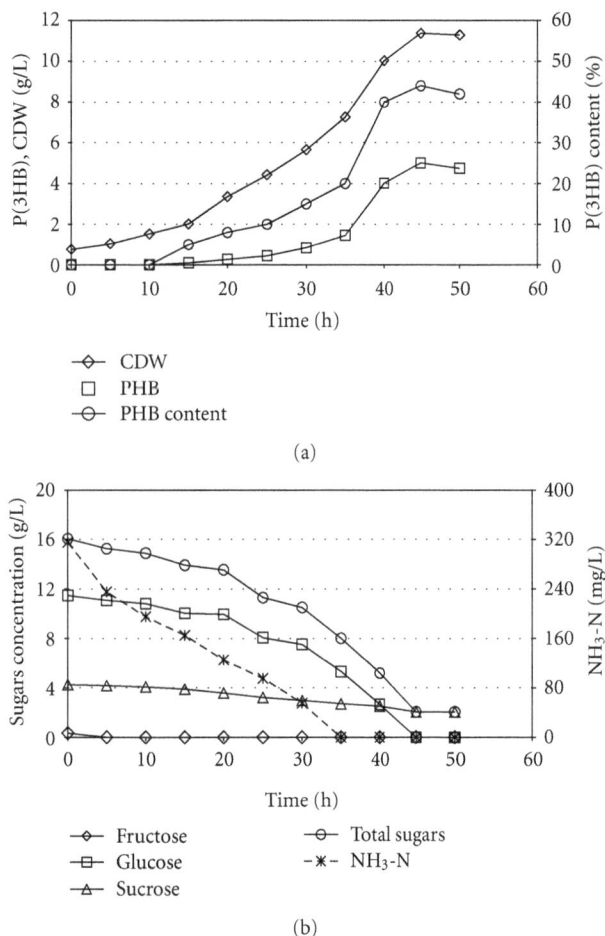

FIGURE 1: (a) Time pro le of cell growth and P(3HB) production by *C. necator* (CCUG52238[T]) using OPF juice in 2-L bioreactor at 30% DOT level. (b)Time pro le of sugars and ammonical nitrogen utilization by *C. necator* (CCUG52238[T]) using OPF juice in 2-L bioreactor at 30% DOT level.

study is 40% higher compared to the P(3HB) produced under nonoptimized condition as shown in our previous study [13]. The results presented herewith demonstrated that pH, agitation speed, and nitrogen concentration indeed plays an important role in the P(3HB) production by *C. necator* (CCUG52338[T]) utilizing OPF juice.

3.2. Biosynthesis of P(3HB) in 2-L Bioreactor

3.2.1. Effect of Dissolved Oxygen Tension (DOT). Biosynthesis of P(3HB) from OPF juice by *C. necator* (CCUG52338[T]) was carried out through batch cultivation process in 2-L bioreactor. The effect of DOT level in the bioreactor was studied for DO concentrations of 20 to 50% and the results are shown in Table 4. It was observed that CDW was increased when DOT level increased from 20 to 50%. On the other hand, P(3HB) accumulation was decreased with the increase in DOT level. Highest CDW (12.81 g/L) and P(3HB) content (46 wt.%) were achieved at 50 and 20% DOT level, respectively. Based on the

result, it was found that dissolved oxygen concentration in the fermentation medium improved the cell growth; however, P(3HB) accumulation was found to be increased towards oxygen limitation. This result suggested that appropriate level of oxygen is needed for cell development, and oxygen depletion was favorable for P(3HB) accumulation.

As shown in Table 4, P(3HB) accumulation was tripled at lower dissolved oxygen concentration (20%) compared to the higher ones (50%). This might be due to the fact that insufficient supply of oxygen to the bacteria may decrease oxidation of NADH and lead to P(3HB) biosynthesis [15, 21, 28]. A similar observation was obtained in our previous study on the effect of different $(NH_4)_2SO_4$ concentration on P(3HB) production using OPF juice in shake ask experiment. These results indicate that both nitrogen and oxygen limitation do not improve cell biomass development, but markedly improve the P(3HB) accumulation. Therefore, it can be suggested that besides nitrogen depletion, oxygen limitation is also important in getting the optimal level of P(3HB) accumulation.

3.2.2. Cell Biomass and P(3HB) Production Profile. In order to study cell biomass and P(3HB) production pro le by *C. necator* (CCUG52238[T]), batch cultivation process was carried out using OPF juice in 2-L bioreactor with aeration supplied at 30% DOT level and the results were depicted in Figures 1(a) and 1(b). It was observed that the culture entered the exponential phase after a lag of 15 h, and nitrogen was completely consumed within 35 h. Highest CDW (11.37 g/L) and P(3HB) content (44 wt.%) were achieved at 45 hr cultivation period. The biomass yield ($Y_{x/s}$) and P(3HB) yield ($Y_{p/s}$) were 0.81 g biomass/g sugars consumed and 0.36 g P(3HB)/g sugars consumed, respectively. The maximum P(3HB) productivity was 0.11 g/L/h.

Almost similar P(3HB) content with some improvement in cell growth was obtained in this study compared to the shake ask experiment under optimal condition. Higher CDW (11.37 g/L) and biomass yield (0.81 g biomass/g sugars consumed) obtained in fermentor compared to shake ask were due to different conditions which prevail in the shake asks and fermentor; some of these conditions include aeration, agitation, and temperature. In fermentor, aeration was supplied via air sparging, and agitation is provided by an impeller or by the motion imparted to the broth (liquid phase) by rising gas bubbles [30]. Temperature is maintained at a constant and uniform value by circulation of cooling water through coils in the vessel or in a jacket surrounding the vessel [31]. Compared to our previous studies in shake asks using technical grade sugars [13], batch studies in 2-L bioreactor using renewable sugars from OPF juice showed superior results in *C. necator* CCUG52238[T] probably due to the additional components in the OPF juice that improve the fermentation performance. An almost similar observation was reported by Koutinas et al. [12] when WH and FE were used as renewable feedstock for P(3HB) production. It was reported that the consumption of various carbon sources (carbohydrates,

TABLE 4: Effect of DOT (%) level on biosynthesis of P(3HB)[a].

DOT (%)	Maximum CDW (g/L)	Maximum total P(3HB) (g/L)	Maximum P(3HB) content (wt.%)[b]
20	9.55	3.93	46
30	11.37	4.78	44
40	12.52	3.38	25
50	12.81	2.37	15

[a]Experiments were conducted in 2-L bioreactor (1 L working volume) by batch mode using OPF juice with initial total sugars of 16 g/L as substrates.
[b]Determination by GC from freeze dried samples.
*Values obtained herewith are means of duplicate sample.

TABLE 5: Comparison of thermal and mechanical properties of P(3HB) obtained in this study with literature.

Microorganisms	Carbon sources	T_m	Tensile strength (MPa)	Elongation to break (%)	References
C. necator CCUG52238^T	OPF juice	162.2	40	8	This study
R. eutropha	Fructose	177	43	5	Doi, 1990 [28]
A. latus	Maple sap	177	—	—	Yezza et al., 2007 [10]

amino acids, peptides) presented in the feedstock resulted in high growth yields (up to 1.07 g cells/g glucose) as related to glucose.

As shown in Figures 1(a) and 1(b), the microbial growth is mainly associated with ammoniacal nitrogen consumption. For the rst 35 h, lower sugars consumption by C. necator was observed. The sugars consumption within the time range was only 8.03 g/L which is half of the total sugars in the culture broth. On the other hand, the NH$_3$-N was found to be decreased drastically from initial and completely exhausted after 35 h of cultivation period. This result indicates that at initial, the microbial growth was mainly attributed by the consumption of nitrogen sources from (NH$_4$)$_2$SO$_4$ supplied earlier as one of the medium composition in bioreactor. In addition to that, other organic compounds such as amino acids, carbohydrates, and other minerals which were previously characterized in the OPF juice could be used as supplementary growth substrates by the bacterium [13]. After that, the cell growth was mainly contributed by the cell expansion due to P(3HB) accumulation inside the cells. It can be seen that the P(3HB) accumulation was doubled that is, 20 wt.% to 40 wt.%, from 35 h to 40 h of cultivation period. From sugars consumption and P(3HB) pro les, it can be observed that the detectable depletion of sugars in the medium from 35 h onwards can be associated with P(3HB) accumulation. These results are in agreement with the nding s of other researchers that reported P(3HB) accumulation is favored by an excess of carbon source and limited supply of macrocomponents such as nitrogen and dissolved oxygen [7, 16, 29]. It is interesting to note that C. necator CCUG52238^T completely utilized the glucose in the OPF juice. Regardless of fructose (due to too low concentration of fructose in the medium), it seems like the bacterium preferred to consume glucose compared to sucrose. Glucose consumption rate by C. necator CCUG52238^T was much higher at 0.33 g/L/h, compared to that of sucrose that is, 0.049 g/L/h. This shows that C. necator CCUG52238^T prefers monosaccharide than disaccharide as its carbon source.

3.2.3. Characterization of Homopolymer P(3HB). The mechanical and thermal properties of the homopolymer produced in 2-L bioreactor are shown in Table 5. The mechanical and thermal properties of P(3HB) obtained in this study showed an almost similar properties to those reported in the literature. For instance, the tensile strength and elongation to break for P(3HB) produced in this study were 40 MPa and 8%, respectively, and it was comparable to the P(3HB) produced from pure fructose [28]. The melting temperature, T_m of P(3HB) obtained from OPF juice (T_m = 162.2°C), was slightly lower compared to the melting point 177°C reported for P(3HB) produced from pure fructose [28] and other renewable sugars such as maple sap [10]. This could be in uenc ed by other properties of the P(3HB) such as molecular weight. It has been reported that the molecular weight of P(3HB) produced is mainly in uenc ed by the type of bacterial strain, substrate, growth rate, and production temperature [15, 29, 32].

4. Conclusions

This study demonstrated that higher cell growth and P(3HB) accumulation can be obtained by culturing Cupriavidus necator strain CCUG52238^T at optimized condition using OPF juice as the sole renewable carbon source. Under the optimal conditions, the highest cell weight was 9.31 ± 0.13 g/L with 45 ± 1.5 wt.% of P(3HB) contained in the cells, accounts of 40% increment for P(3HB) content compared to the nonoptimized condition. Cultivation in a 2-L bioreactor with 30% DOT yielded CDW of 11.37 g/L and P(3HB) content of 44 wt.%. In the meanwhile, thermal and mechanical characterization of the P(3HB) obtained from OPF juice showed almost similar properties to those reported in the literature. It is worth to mention that this study may contribute to the process development for P(3HB) production from renewable OPF juice in pilot and industrial scale. Furthermore, since OPF is an abundant solid waste at oil palm plantation and is currently underutilized, it has a great potential to be used as sustainable, renewable, and cheap fermentation feedstock for the production of P(3HB).

Acknowledgments

The authors would like to acknowledge the Federal Land Development Authority, (FELDA) Malaysia, the Ministry of Science Technology and Innovation (MOSTI), Malaysia, and the Japan Society for the Promotion of Science (JSPS) for funding this research and giving the technical support during the study period. Their heartiest gratitude also goes to Universiti Malaysia Pahang (UMP) for providing the study leave. M. A. K. M. Zahari is a recipient of an academic training scholarship from the Ministry of Higher Education, Malaysia.

References

[1] S. Y. Lee, "Review bacterial polyhydroxyalkanoates," *Biotechnology and Bioengineering*, vol. 49, pp. 1–14, 1996.

[2] Beom Soo Kim, Seung Chul Lee, Sang Yup Lee, Ho Nam Chang, Yong Keun Chang, and Seong Ihl Woo, "Production of poly(3-hydroxybutyric acid) by fed-batch culture of *Alcaligenes eutrophus* with glucose concentration control," *Biotechnology and Bioengineering*, vol. 43, no. 9, pp. 892–898, 1994.

[3] B. S. Kim and H. N. Chang, "Production of poly(3-hydroxybutyrate) from starch by *Azotobacter chroococcum*," *Biotechnology Letters*, vol. 20, no. 2, pp. 109–112, 1998.

[4] Sei Kwang Hahn, Yong Keun Chang, Beom Soo Kim, and Ho Nam Chang, "Communication to the editor optimization of microbial poly(3- hydroxybutyrate) recovery using dispersions of sodium hypochlorite solution and chloroform," *Biotechnology and Bioengineering*, vol. 44, no. 2, pp. 256–261, 1994.

[5] W. J. Page, "Production of poly-β-hydroxybutyrate by Azotobacter vinelandii strain UWD during growth on molasses and other complex carbon sources," *Applied Microbiology and Biotechnology*, vol. 31, no. 4, pp. 329–333, 1989.

[6] J. M. B. T. Cavalheiro, M. C. M. D. de Almeida, C. Grand ls, and M. M. R. da Fonseca, "Poly(3-hydroxybutyrate) production by *Cupriavidus necator* using waste glycerol," *Process Biochemistry*, vol. 44, no. 5, pp. 509–515, 2009.

[7] M. Koller, A. Atlić, M. Dias, A. Reiterer, and G. Braunegg, "Microbial production from waste raw materials," *Microbiology Monograph*, vol. 14, pp. 85–119, 2010.

[8] D. Rusendi and J. D. Sheppard, "Hydrolysis of potato processing waste for the production of poly-β-hydroxybutyrate," *Bioresource Technology*, vol. 54, no. 2, pp. 191–196, 1995.

[9] B. S. Kim, "Production of poly(3-hydroxybutyrate) from inexpensive substrates," *Enzyme and Microbial Technology*, vol. 27, no. 10, pp. 774–777, 2000.

[10] A. Yezza, A. Halasz, W. Levadoux, and J. Hawari, "Production of poly-β-hydroxybutyrate (PHB) by Alcaligenes latus from maple sap," *Applied Microbiology and Biotechnology*, vol. 77, no. 2, pp. 269–274, 2007.

[11] R. Haas, B. Jin, and F. T. Zepf, "Production of poly(3-hydroxybutyrate) from waste potato starch," *Bioscience, Biotechnology and Biochemistry*, vol. 72, no. 1, pp. 253–256, 2008.

[12] A. A. Koutinas, Y. Xu, R. Wang, and C. Webb, "Polyhydroxybutyrate production from a novel feedstock derived from a wheat-based biore ner y," *Enzyme and Microbial Technology*, vol. 40, no. 5, pp. 1035–1044, 2007.

[13] M. A. K. M. Zahari, M. R. Zakaria, H. Ariffin et al., "Renewable sugars from oil palm frond juice as an alternative novel fermentation feedstock for value-added products," *Bioresource Technology*, vol. 110, pp. 566–571, 2012.

[14] S. Khanna and A. K. Srivastava, "Statistical media optimization studies for growth and PHB production by *Ralstonia eutropha*," *Process Biochemistry*, vol. 40, no. 6, pp. 2173–2182, 2005.

[15] A. J. Anderson and E. A. Dawes, "Occurrence, metabolism, metabolic role, and industrial uses of bacterial polyhydroxyalkanoates," *Microbiological Reviews*, vol. 54, no. 4, pp. 450–472, 1990.

[16] M. Beaulieu, Y. Beaulieu, J. Melinard, S. Pandian, and J. Goulet, "In uenc e of ammonium salts and cane molasses on growth of *Alcaligenes eutrophus* and production of polyhydroxybutyrate," *Applied and Environmental Microbiology*, vol. 61, no. 1, pp. 165–169, 1995.

[17] F. Tabandeh and E. Vasheghani-Farahani, "Biosynthesis of poly-β-hydroxybutyrate as a biodegradable polymer," *Iranian Polymer Journal*, vol. 12, no. 1, pp. 37–42, 2003.

[18] M. S. Baei, G. D. Najafpour, H. Younesi, F. Tabandeh, and H. Eisazadeh, "Poly(3-hydroxybutyrate) synthesis by *Cupriavidus necator* DSMZ 545 utilizing various carbon sources," *World Applied Science Journal*, vol. 7, no. 2, pp. 157–161, 2009.

[19] S. Philip, S. Sengupta, T. Keshavarz, and I. Roy, "Effect of impeller speed and pH on the production of poly(3- hydroxybutyrate) using *Bacillus cereus* SPV," *Biomacromolecules*, vol. 10, no. 4, pp. 691–699, 2009.

[20] G. Braunegg, B. Sonnleitner, and R. M. Lafferty, "A rapid gas chromatographic method for the determination of poly β hydroxybutyric acid in microbial biomass," *European Journal of Applied Microbiology and Biotechnology*, vol. 6, no. 1, pp. 29–37, 1978.

[21] M. R. Zakaria, *Biosynthesis of poly(3-hydroxybutyrate-co-hydroxyvalerate) copolymer from organic acids using Comamonas sp. EB172 [Ph.D. thesis]*, Faculty of Biotechnology & Biomolecular Sciences, Universiti Putra Malaysia, 2011.

[22] E. Kafkas, M. Koşar, N. Türemiş, and K. H. C. Başer, "Analysis of sugars, organic acids and vitamin C contents of blackberry genotypes from Turkey," *Food Chemistry*, vol. 97, no. 4, pp. 732–736, 2006.

[23] M. R. Zakaria, H. Ariffin, N. A. Mohd Johar et al., "Biosynthesis and characterization of poly(3-hydroxybutyrate-co-3-hydroxyvalerate) copolymer from wild-type *Comamonas* sp. EB172," *Polymer Degradation and Stability*, vol. 95, no. 8, pp. 1382–1386, 2010.

[24] L. P. A. Paladino, *Screening, optimization and extraction of polyhydroxyalkanoates and peptidoglycan from Bacillus megaterium [M.S. dissertation]*, Michigan Technological University, 2009.

[25] N. J. Palleroni and A. V. Palleroni, "*Alcaligenes latus*, a new species of hydrogen-utilizing bacteria," *International Journal of Systematic Bacteriology*, vol. 28, no. 3, pp. 416–424, 1978.

[26] Y. H. Wei, W. C. Chen, C. K. Huang et al., "Screening and evaluation of polyhydroxybutyrate-producing strains from indigenous isolate *Cupriavidus taiwanensis* strains," *International Journal of Molecular Sciences*, vol. 12, no. 1, pp. 252–265, 2011.

[27] L. S. Sera m, P. C. Lemos, M. G. E. Albuquerque, and M. A. M. Reis, "Strategies for PHA production by mixed cultures and renewable waste materials," *Applied Microbiology and Biotechnology*, vol. 81, no. 4, pp. 615–628, 2008.

[28] Y. Doi, "Structure and properties of poly(3-hydroxybutyrate)," in *Microbial Polyesters*, VCH, New York, NY, USA, 1990.

[29] L. L. Madison and G. W. Huisman, "Metabolic engineering of poly(3-hydroxyalkanoates): from DNA to plastic,"

Microbiology and Molecular Biology Reviews, vol. 63, no. 1, pp. 21–53, 1999.

[30] H. S. Fogler, *Elements of Chemical Reaction Engineering*, Prentice-Hall, Upper Saddle River, NJ, USA, 2nd edition, 1992.

[31] H. W. Blanch and D. S. Clark, *Biochemical Engineering*, Marcel Dekker, New York, NY, USA, 1997.

[32] M. A. Hassan, Y. Shirai, H. Umeki et al., "Acetic acid separation from anaerobically treated palm oil mill effluent by ion exchange resins for the production of polyhydroxyalkanoate by *Alcaligenes eutrophus*," *Bioscience, Biotechnology and Biochemistry*, vol. 61, no. 9, pp. 1465–1468, 1997.

5

Potential Role of Kringle-Integrin Interaction in Plasmin and uPA Actions (A Hypothesis)

Yoshikazu Takada

Department of Dermatology, and Biochemistry and Molecular Medicine, University of California Davis School of Medicine, Research III Suite 3300, 4645 Second Avenue, Sacramento, CA 95817, USA

Correspondence should be addressed to Yoshikazu Takada, ytakada@ucdavis.edu

Academic Editor: Edward F. Plow

We previously showed that the kringle domains of plasmin and angiostatin, the N-terminal four kringles (K1–4) of plasminogen, directly bind to integrins. Angiostatin blocks tumor-mediated angiogenesis and has great therapeutic potential. Angiostatin binding to integrins may be related to the antiinflammatory action of angiostatin. We reported that plasmin induces signals through protease-activated receptor (PAR-1), and plasmin-integrin interaction may be required for enhancing plasmin concentration on the cell surface, and enhances its signaling function. Angiostatin binding to integrin does not seem to induce proliferative signals. One possible mechanism of angiostatin's inhibitory action is that angiostatin suppresses plasmin-induced PAR-1 activation by competing with plasmin for binding to integrins. Interestingly, plasminogen did not interact with $\alpha v\beta 3$, suggesting that the $\alpha v\beta 3$-binding sites in the kringle domains of plasminogen are cryptic. The kringle domain of urokinase-type plasminogen activator (uPA) also binds to integrins. The uPA-integrin interaction enhances uPA concentrations on the cell surface and enhances plasminogen activation on the cell surface. It is likely that integrins bind to the kringle domain, and uPAR binds to the growth factor-like domain (GFD) of uPA simultaneously, making the uPAR-uPA-integrin ternary complex. We present a docking model of the ternary complex.

1. The Kringle Domains of Plasmin Interact with Integrins

The integrins are a superfamily of cell adhesion receptors that bind to extracellular matrix ligands, cell-surface ligands, and soluble ligands. They are transmembrane $\alpha\beta$ heterodimers and at least 18 α and eight β subunits are known in humans, generating 24 heterodimers [1]. The α and β subunits have distinct domain structures, with extracellular domains from each subunit contributing to the ligand-binding site of the heterodimer. The sequence arginine-glycine-aspartic acid (RGD) was identified as a general integrin-binding motif, but individual integrins are also specific for particular protein ligands. Immunologically important integrin ligands are the intercellular adhesion molecules (ICAMs), immunoglobulin superfamily members present on inflamed endothelium and antigen-presenting cells. On ligand binding, integrins transduce signals into the cell interior; they can also receive intracellular signals that regulate their ligand-binding affinity.

Angiostatin, a proteolytic fragment of plasminogen, contains either the first three or four kringle domains of plasminogen and is a potent inhibitor of tumor-induced angiogenesis in animal models [2, 3]. Angiostatin has promising therapeutic potential and is now in clinical trials. Plasminogen is first converted to the two-chain serine protease plasmin by cleavage of a single Arg561-Val562 peptide bond by urokinase-type plasminogen activator (uPA), and plasmin serves as both the substrate and enzyme for the generation of angiostatin [4]. Several other mechanisms have been proposed for the generation of angiostatin from the plasminogen molecule [5]. The antiangiogenic functions of plasminogen kringles have been extensively studied using recombinant plasminogen kringles and kringle fragments produced by elastolytic processing of native plasminogen. Smaller fragments of angiostatin display differential effects on the suppression of endothelial cell growth [6].

We found that bovine arterial endothelial (BAE) cells adhere to angiostatin in an integrin-dependent manner and that integrins $\alpha v\beta 3$, $\alpha 9\beta 1$, and to a lesser extent $\alpha 4\beta 1$, specifically bind to angiostatin. $\alpha v\beta 3$ is a predominant receptor for angiostatin on BAE cells, since a function-blocking antibody to $\alpha v\beta 3$ effectively blocks adhesion of BAE cells to angiostatin, but an antibody to $\alpha 9\beta 1$ does not. ε-Aminocaproic acid, a Lys analogue, effectively blocks angiostatin binding to BAE cells, indicating that an unoccupied Lys-binding site of the kringles may be required for integrin binding. It is known that other plasminogen fragments containing three or five kringles (K1–3 or K1–5) have an antiangiogenic effect, but plasminogen itself does not. We found that K1–3 and K1–5 bind to $\alpha v\beta 3$, but plasminogen does not. These results suggest that the anti-angiogenic action of angiostatin may be mediated via interaction with $\alpha v\beta 3$. Angiostatin binding to $\alpha v\beta 3$ does not strongly induce stress-fiber formation, suggesting that angiostatin may prevent angiogenesis by perturbing the $\alpha v\beta 3$-mediated signal transduction that may be necessary for angiogenesis [7].

Plasmin, the parent molecule of angiostatin and a major extracellular protease, induces platelet aggregation, migration of peripheral blood monocytes, and release of arachidonate and leukotriene from several cell types [8]. We found that plasmin specifically binds to $\alpha v\beta 3$ through the kringle domains and induces migration of endothelial cells. In contrast, angiostatin does not induce cell migration. Notably, angiostatin, anti-$\alpha v\beta 3$ antibodies, RGD-peptide, and a serine protease inhibitor effectively block plasmin-induced cell migration. These results suggest that plasmin-induced migration of endothelial cells requires $\alpha v\beta 3$ and the catalytic activity of plasmin and that this process is a potential target for the inhibitory activity of angiostatin [9].

We found that plasmin specifically interacts with integrin ($\alpha 9\beta 1$) and that plasmin induces migration of cells expressing recombinant $\alpha 9\beta 1$ ($\alpha 9$-Chinese hamster ovary (CHO) cells). Migration was dependent on an interaction of the kringle domains of plasmin with $\alpha 9\beta 1$ as well as the catalytic activity of plasmin. Angiostatin, representing the kringle domains of plasmin, alone did not induce the migration of $\alpha 9$-CHO cells, but simultaneous activation of the G protein-coupled protease-activated receptor (PAR)-1 with an agonist peptide induced the migration on angiostatin, whereas PAR-2 or PAR-4 agonist peptides were without effect. Furthermore, a small chemical inhibitor of PAR-1 (RWJ 58259) and a palmitoylated PAR-1-blocking peptide inhibited plasmin-induced migration of $\alpha 9$-CHO cells. These results suggest that plasmin induces migration by kringle-mediated binding to $\alpha 9\beta 1$ and simultaneous proteolytic activation of PAR-1 [10]. It is likely that other integrins that bind to plasmin may exert similar effects on plasmin signaling.

We propose a model (Figure 1) in which (1) upon plasminogen activation, integrin-binding site in plasmin is exposed. Note that plasminogen does not bind to integrins $\alpha v\beta 3$ or $\alpha 9\beta 1$. (2) Once activated, plasmin is able to bind to integrins on the cell surface through the kringle domains (since integrin-binding sites are exposed) and proteolytically activates PAR-1, which induces intracellular signaling. Plasmin is concentrated to the cell surface through integrin binding, and this process is probably critical since plasmin has much lower affinity to PAR-1 than thrombin. Angiostatin,

in contrast, binds to integrins, but does not activate PAR-1. Angiostatin is expected to suppress plasmin action by competing with plasmin for binding to integrins.

It has been reported that integrins $\alpha M\beta 2$ [11], $\alpha D\beta 2$ [12], and $\alpha 5\beta 1$ [13] bind to plasminogen, while we did not detect binding of $\alpha v\beta 3$ or $\alpha 9\beta 1$ to plasminogen. One possibility is that integrins $\alpha M\beta 2$, $\alpha D\beta 2$, and $\alpha 5\beta 1$ recognize plasminogen in the ways different from those of $\alpha v\beta 3$ or $\alpha 9\beta 1$. Another possibility is that integrin-binding sites in plasminogen (perhaps kringle domains) are exposed in partially denatured plasminogen. Supporting the second possibility we observed that freshly prepared plasminogen did not significantly bind to $\alpha v\beta 3$, but plasminogen binding to $\alpha v\beta 3$ appeared to increase as plasminogen preparations aged (data not shown). This issue should be clarified in future studies.

In conclusion, the kringle domains in plasmin are involved in direct integrin binding, in addition to binding to the C-terminal Lysine residues of many proteins, and playing a role in inducing intracellular signals through proteolytic activation of PAR-1. The kringle-integrin interaction may enhance the cell surface concentration of plasmin, or directly induce intracellular signals through outside-in integrin signaling. Interestingly, plasminogen does not interact with integrins $\alpha v\beta 3$ or $\alpha 9\beta 1$ (possibly the integrin-binding sites are cryptic in plasminogen) (Figure 1) Based on our results on the plasmin kringle-integrin interaction, we hypothesized that the kringle domains of other serine proteases may interact with integrins and the interaction may play a role in their functions. Consistent with this idea, kringle domains from other proteins such as tissue-type plasminogen activator (tPA) [14] and apolipoprotein [15] have been reported to interact with integrins. This suggests that kringle-integrin interaction is a common mechanism in kringle-containing proteins.

2. uPA Kringle-Integrin Interaction

uPA is a highly restricted serine protease that converts the zymogen plasminogen to active plasmin. uPA binds with high affinity to a cell-surface uPA receptor (uPAR) that has been identified in many cell types. uPAR is a glycosylphosphatidylinositol- (GPI-) anchored 35–55 kDa glycoprotein. This system mediates pericellular proteolysis of extracellular matrix proteins including fibrin degradation (fibrinolysis) and plays an important role in cancer, inflammation, and immune responses [16–19]. The single chain form of uPA has three independently folded domains: the growth factor-like domain (GFD) (residue 1–46), kringle (residue 47–135) domain, and serine protease domain (residue 159–411). Enzymatic digestion of single chain-uPA yields an amino terminal fragment (ATF), which consists of the GFD and kringle domains, and the low molecular weight fragment (LMW-uPA), which consists of the serine protease domain. The uPAR-binding site of uPA is located in the GFD domain [20]; this binding is stabilized by the kringle [21]. It has generally been accepted that uPA signaling involves its binding to uPAR through its GFD [22].

uPA binding to uPAR on the cell surface facilitates activation of plasminogen to plasmin in vitro by increasing the rate

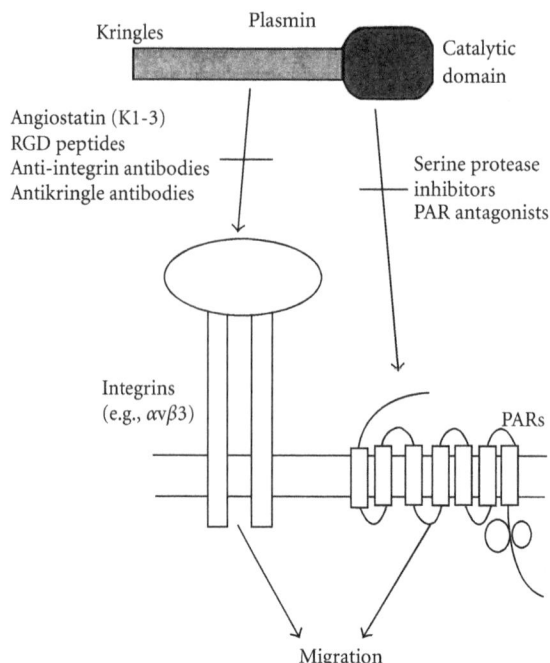

FIGURE 1: A model of plasmin-induced cell migration and the potential mechanism of angiostatin action. uPA activates plasminogen to plasmin pericellularly. Plasmin is accumulated on the cell surface by binding to integrins and stabilized. Free plasmin would be rapidly inactivated by circulating serine protease inhibitors (e.g., β2-antiplasmin). The catalytic activity of plasmin on the cell surface is directly involved in signal transduction, possibly through activating G-protein coupled PARs. The binding of the kringle domain may not be directly involved in signaling through integrin pathways. Angiostatin effectively blocks plasmin-induced cell migration possibly by competing with plasmin for binding to integrins. Aprotinin, a serine protease inhibitor, also effectively blocks migration. It should be noted that other antiangiogenic agents, RGD-peptide and anti-αvβ3, are effective inhibitors of this process.

of pro-uPA activation by plasmin, by decreasing the apparent Km of uPA to plasmin, and by increasing the Kcat/Km of uPA to plasmin [23]. It is interesting that uPA-knockout mice do not have major thrombotic disorders [24]. This is probably because of the redundant fibrinolytic function by tPA. Indeed, combined uPA and tPA knockout mice show extensive thrombotic disorders very similar to those observed in plasminogen-knockout mice, but these are rarely detected in animals lacking uPA or tPA alone [25]. In contrast to uPA, studies performed in uPAR-knockout mice do not really support a major role of uPAR in fibrinolysis. Fibrin deposits are found within the livers of mice with a combined deficiency in uPAR and tPA, but not in uPAR-knockout mice, indicating a minor role for uPAR in plasminogen activation [25]. The extraordinarily mild consequences of combined uPAR and tPA deficiency raised the question of whether there are other receptors for uPA that might facilitate plasminogen activation [19, 25].

Besides plasminogen activation, uPA has been shown to induce the adhesion and chemotactic movement of myeloid cells [26, 27], to induce cell migration in human epithelial cells [28] and bovine endothelial cells [29], and to promote cell growth [30–32]. Notably these signaling functions of uPA do not require its proteolytic activity. Several studies suggest that uPA has additional, unidentified cell-surface receptor(s) other than uPAR that are involved in signaling events. For

example, blocking of uPA binding to uPAR using a mono-clonal antibody or by depletion of cell surface uPAR with phosphatidylinositol-specific phospholipase C (PIPLC) did not inhibit uPA-induced mitogenic effects in smooth muscle cells [33]. uPA-induced mitogenic effects in melanoma cells are independent of high-affinity binding to uPAR, and this suggests the existence of a low-affinity binding site on this cell type based on the kinetic data [34]. The chemotactic action of uPA on smooth muscle cells depends on its kringle domain, and kinetic evidence indicates that these cells express a lower-affinity kringle receptor distinct from uPAR [35]. The isolated uPA kringle augments vascular smooth muscle cell constriction in vitro [36] and in vivo [37]. Taken together these observations all suggest that cells express uPA-binding proteins (other than uPAR) that mediate signaling from uPA.

We found that uPA binds specifically to integrin αvβ3 on CHO cells depleted of uPAR (Figure 2). The binding of uPA to αvβ3 required the uPA kringle domain (Figure 3). The isolated uPA kringle domain binds specifically to purified, recombinant soluble, and cell surface αvβ3, and other integrins (α4β1 and α9β1), and induces migration of CHO cells in an αvβ3-dependent manner. The binding of the uPA kringle to αvβ3 and uPA kringle-induced αvβ3-dependent cell migration is blocked by angiostatin. We studied whether the binding of uPA to integrin αvβ3 through the kringle

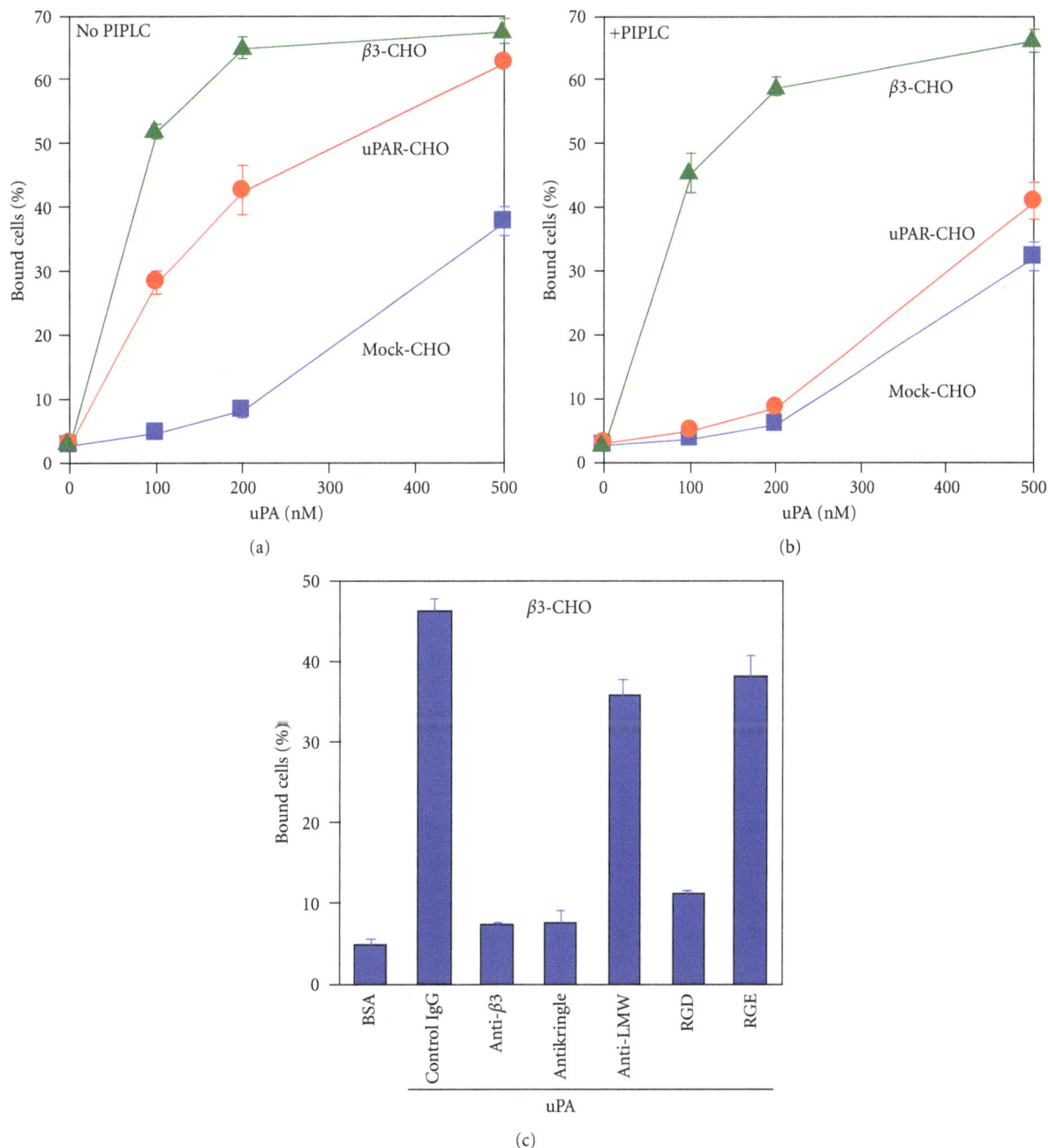

(a)

(b)

(c)

FIGURE 2: uPA binding to the cell surface in an integrin $\alpha v \beta 3$-dependent and uPAR-independent manner [38]. (a) and (b) Depletion of uPAR from the cell surface blocked uPA binding to uPAR-CHO cells, but did not affect uPA binding to $\beta 3$-CHO cells. To deplete GPI-linked uPAR on the cell surface, $\beta 3$-CHO, uPAR-CHO, or control mock-transfected CHO cells were treated with PIPLC. The treatment removed more than 95% of human uPAR from uPAR-CHO cells as determined by flow cytometry with anti-uPAR mAb 3B10 (data not shown). uPA was immobilized to wells of 96-well microtiter plates at the indicated coating concentrations, and incubated with cells without (a) or with (b) pretreatment with PI-PLC. Bound cells were quantified. (c) uPA binding to $\beta 3$-CHO cells is specific to $\alpha v \beta 3$ and the kringle domain. uPA (200 nM coating concentration) was immobilized to wells of 96-well microtiter plates and incubated with $\beta 3$-CHO cells in the presence of mAb 16N7C2 (anti-$\beta 3$), Ab 963 (anti-kringle), mAb UNG-5 (anti-LMW-uPA), or RGD or RGE peptides (100 μM).

domain plays a role in plasminogen activation. On CHO cell depleted of uPAR, uPA enhances plasminogen activation in a kringle and $\alpha v \beta 3$-dependent manner (Figure 4). Endothelial cells bind to and migrate on uPA and uPA kringle in an $\alpha v \beta 3$-dependent manner. These results suggest that uPA binding to integrins through the kringle domain plays an important role in both plasminogen activation and uPA-induced intracellular signaling. The uPA kringle-integrin interaction may represent a novel therapeutic target for cancer, inflammation, and vascular remodeling [38].

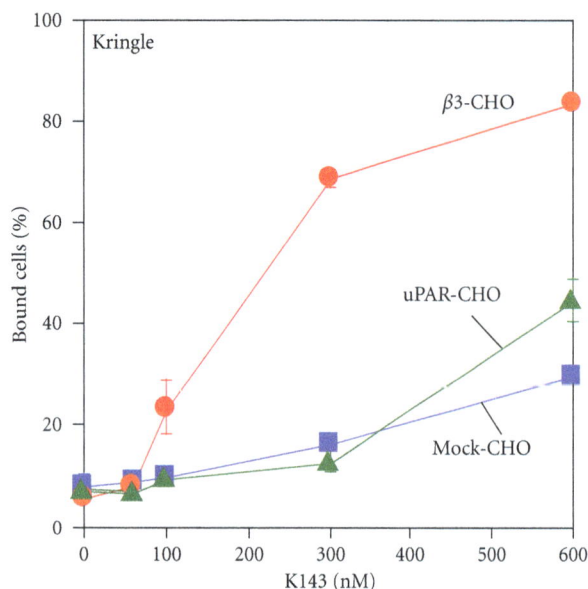

FIGURE 3: The kringle domain of uPA mediates binding to $\alpha v\beta 3$. The uPA kringle domain was immobilized onto wells of 96-well microtiter plates at the indicated coating concentrations and incubated with $\beta 3$-CHO, uPAR-CHO, or mock-CHO cells. The ability of the uPA fragments to support adhesion of these cells was determined [38].

We propose a model, in which the N-terminal GFD of uPA binds to uPAR and the kringle domain of uPA binds to integrins, leading to the uPAR-uPA-integrin ternary complex on the cell surface. It is likely that the ternary complex formation may be involved in uPA signaling and plasminogen activation. The isolated kringle or the isolated GFD domain may suppress uPA signaling or plasminogen activation by suppressing the process. Indeed isolated kringle domain or GFD have been shown to suppress tumorigenesis [39].

3. Another Example of the Role of $\alpha v\beta 3$ in uPA Signaling: uPA Kringle and Integrin $\alpha v\beta 3$ in Neutrophil Activation

It has been reported that antibody to integrin $\alpha v\beta 3$ and RGD peptide suppress the signaling action of uPA in neutrophils, although it is unclear if this include direct uPA-$\alpha v\beta 3$ interaction [40]. The study examined the ability of specific uPA domains to increase cytokine expression in murine and human neutrophils stimulated with lipopolysaccharides (LPS). Whereas the addition of intact uPA to neutrophils cultured with LPS increased mRNA and protein levels of interleukin-1β, macrophage-inflammatory protein-2, and tumor necrosis factor α, deletion of the kringle domain from uPA resulted in loss of these potentiating effects. Addition of purified uPA kringle domain to LPS-stimulated neutrophils increased cytokine expression to a degree comparable with that produced by single-chain uPA. Inclusion of the RGD but not the RGE peptide to neutrophil cultures blocked uPA kringle-induced potentiation of proinflammatory responses,

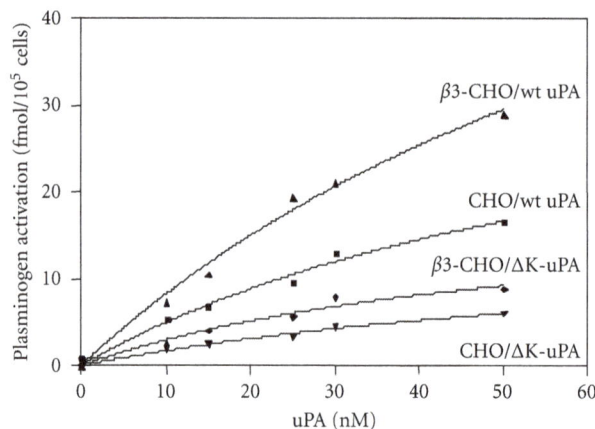

FIGURE 4: Integrin-dependent plasminogen activation on the cell surface. Parental CHO cells and $\beta 3$-CHO cells in wells of 96-well plates were treated with PIPLC to deplete uPAR, and incubated with wt or delta kringle (ΔK) uPA in the cold binding buffer for 1 h at 4°C. The cells were washed with the binding buffer, and plasminogen activation was determined using Glu-plasminogen and SpectrozymePL chromogenic substrate at 37°C. We found that $\beta 3$-CHO cells showed much higher ability to activate plasminogen in a manner dependent on the uPA added. Deletion of the kringle domain (with ΔK-uPA) markedly reduced the plasminogen activation on $\beta 3$-CHO, indicating that $\alpha v\beta 3$ and uPA-dependent plasminogen activation required the kringle domain of uPA. These results suggest that the binding of uPA kringle to integrin $\alpha v\beta 3$ induces plasminogen activation [38].

demonstrating that interactions between the kringle domain and integrins are involved. Antibodies to the αv or $\beta 3$ subunit or to $\alpha v\beta 3$ heterodimer prevented uPA kringle-induced enhancement of expression of proinflammatory cytokines and also of adhesion of neutrophils to the uPA kringle domain. These results demonstrate that the kringle domain of uPA, through interaction with $\alpha v\beta 3$ integrins, potentiates neutrophil activation.

4. A Docking Model of uPAR-uPA Kringle-Integrin Interaction

How does integrin $\alpha v\beta 3$ interact with uPA kringle? This has recently been predicted by docking simulation [41]. They modeled the interaction of uPA on two integrins, $\alpha IIb\beta 3$ in the open configuration and $\alpha v\beta 3$ in the closed configuration. They found that multiple lowest energy solutions point to an interaction of the kringle domain of uPA at the boundary between α and β chains on the surface of the integrins. This region is not far away from peptides that have been previously shown to have a biological role in uPAR/integrins dependent signaling. They demonstrated that in silico docking experiments can be successfully carried out to identify the binding mode of the kringle domain of uPA on the scaffold of integrins in the open and closed conformation. Importantly they found that the binding mode is the same on different integrins and in both configurations. To get a molecular view of the system is a prerequisite to unravel the complex

TABLE 1: Amino acid residues involved in αvβ3-uPA kringle interaction in the docking model. Amino acid residues at the binding interface (within the 6 Angstrom) were selected using Swiss pdb viewer (v. 4.02).

αv	β3	uPA kringle
Ala149, Asp150, Tyr178, Gln214, Ala215, Ile216, Asp218, Asp219, Arg248	Tyr122, Ser123, Met124, Lys125, Asp126, Asp127, Asp179, Met180, Lys181, Thr182, Arg214, Arg216, Asp217, Ala218, Asp251, Ala252, Lys253, Thr311, Glu312, Asn313, Val314, Asn316, Val332, Leu333, Ser334, Met335, Asp336, Ser337	Ser47, Lys48, Thr49, Tyr51, Glu52, Gly53, Asn54, Gly55, His56, Phe57, Tyr58, Arg59, Tyr84, Asp90, Leu92, Gln93, Leu94, Asn104, Pro105, Asp106, Asn107, Arg108, Arg109, Arg110, Glu125

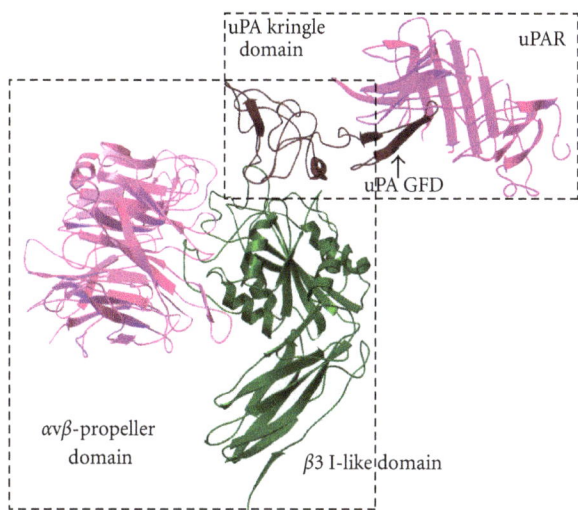

FIGURE 5: A model of integrin, uPA kringle, and uPAR complex. We performed docking simulation of the interaction between uPA kringle (PDB code 2URK) and integrin αvβ3 (PDB code 1L5G) using Autodock3. The simulation predicted the poses in which uPA kringle interacts with αvβ3. The uPA kringle-integrin complex was superposed with the ATF-uPAR complex (PDB code 2I9B).

protein-protein interactions underlying uPA/uPAR/integrin mediated cell motility, adhesion, and proliferation, and to design rational in vitro experiments.

However, in their paper which amino acid residues in uPA kringle are involved in integrin interaction is unclear. Thus, we presented our model here (Figure 5). We performed docking simulation of the interaction between uPA kringle (PDB code 2URK) and integrin αvβ3 (PDB code 1L5G) using Autodock3. The simulation predicted the poses in which uPA kringle interacts with αvβ3 (docking energy −22.3 kcal/mol). The amino acid residues involved in the interaction are shown in Table 1. The uPA kringle-binding site in αvβ3 appears to be common to other known αvβ3 ligands. The uPA kringle-integrin complex was superposed with the ATF-uPAR complex (PDB code 2I9B). Our model predicts that integrin αvβ3 and uPAR can bind to ATF (GFD and kringle) simultaneously without steric hindrance. Obviously, it would be important to identify amino acid residues in uPA kringle that are critical for integrin binding by site-directed mutagenesis. In future studies, using uPA kringles that cannot bind to integrins or uPAR, it would be important to study the role of uPA kringle-integrin interaction in the proinflammatory action of uPA and to establish the role of uPAR in this process.

5. uPAR-Integrin Interaction

Previous studies suggest that uPAR directly binds to integrins [42–44]. How can our hypothesis explain this interaction? Our preliminary docking simulation studies of interaction between uPAR and integrin αvβ3 did not detect high-affinity αvβ3 binding sites in uPAR (not shown). In contrast the docking simulation of interaction between uPA kringle and αvβ3 predicted high affinity binding of αvβ3 to uPA kringle (as shown above). Since uPA binds to uPAR at high-affinity through GFD of uPA, one possibility is that previous studies detected interactions between the uPA-uPAR complex and integrins, in which integrins bind indirectly to uPAR through uPA kringle, but not those between uPAR and integrins. uPA is widely expressed in different cell types and tissues. This hypothesis should be rigorously tested in future studies.

References

[1] Y. Takada, X. Ye, and S. Simon, "The integrins," *Genome Biology*, vol. 8, no. 5, article 215, 2007.

[2] M. S. O'Reilly, L. Holmgren, Y. Shing et al., "Angiostatin: a circulating endothelial cell inhibitor that suppresses angiogenesis and tumor growth," *Cold Spring Harbor Symposia on Quantitative Biology*, vol. 59, pp. 471–482, 1994.

[3] Y. Cao, M. S. O'Reilly, B. Marshall, E. Flynn, R. W. Ji, and J. Folkman, "Expression of angiostatin cDNA in a murine fibrosarcoma suppresses primary tumor growth and produces long-term dormancy of metastases," *Journal of Clinical Investigation*, vol. 101, no. 5, pp. 1055–1063, 1998.

[4] S. Gately, P. Twardowski, M. S. Stack et al., "The mechanism of cancer-mediated conversion of plasminogen to the angiogenesis inhibitor angiostatin," *Proceedings of the National Academy of Sciences of the United States of America*, vol. 94, no. 20, pp. 10868–10872, 1997.

[5] A. J. Lay, X. M. Jiang, O. Kisker et al., "Phosphoglycerate kinase acts in tumour angiogenesis as a disulphide reductase," *Nature*, vol. 408, no. 6814, pp. 869–873, 2000.

[6] Y. Cao, R. W. Ji, D. Davidson et al., "Kringle domains of human angiostatin: characterization of the anti- proliferative activity on endothelial cells," *Journal of Biological Chemistry*, vol. 271, no. 46, pp. 29461–29467, 1996.

[7] T. Tarui, L. A. Miles, and Y. Takada, "Specific interaction of angiostatin with integrin αvβ3 in endothelial cells," *Journal of Biological Chemistry*, vol. 276, no. 43, pp. 39562–39568, 2001.

[8] T. Syrovets, B. Tippler, M. Rieks, and T. Simmet, "Plasmin is a potent and specific chemoattractant for human peripheral monocytes acting via a cyclic guanosine monophosphate-dependent pathway," *Blood*, vol. 89, no. 12, pp. 4574–4583, 1997.

[9] T. Tarui, M. Majumdar, L. A. Miles, W. Ruf, and Y. Takada, "Plasmin-induced migration of endothelial cells: a potential

target for the anti-angiogenic action of angiostatin," *Journal of Biological Chemistry*, vol. 277, no. 37, pp. 33564–33570, 2002.

[10] M. Majumdar, T. Tarui, B. Shi, N. Akakura, W. Ruf, and Y. Takada, "Plasmin-induced migration requires signaling through protease-activated receptor 1 and integrin $\alpha9\beta1$," *Journal of Biological Chemistry*, vol. 279, no. 36, pp. 37528–37534, 2004.

[11] E. Pluskota, D. A. Soloviev, K. Bdeir, D. B. Cines, and E. F. Plow, "Integrin $\alpha M\beta2$ orchestrates and accelerates plasminogen activation and fibrinolysis by neutrophils," *Journal of Biological Chemistry*, vol. 279, no. 17, pp. 18063–18072, 2004.

[12] V. P. Yakubenko, S. P. Yadav, and T. P. Ugarova, "Integrin $\alpha D\beta2$, an adhesion receptor up-regulated on macrophage foam cells, exhibits multiligand binding properties," *Blood*, vol. 107, no. 4, pp. 1643–1650, 2006.

[13] V. K. Lishko, V. V. Novokhatny, V. P. Yakubenko, H. V. Skomorovska-Prokvolit, and T. P. Ugarova, "Characterization of plasminogen as an adhesive ligand for integrins $\alpha M\beta2$ (Mac-1) and $\alpha5\beta1$," *Blood*, vol. 104, no. 3, pp. 719–726, 2004.

[14] H. K. Kim, D. S. Oh, S. B. Lee, J. M. Ha, and A. J. Young, "Antimigratory effect of TK1-2 is mediated in part by interfering with integrin $\alpha2\beta1$," *Molecular Cancer Therapeutics*, vol. 7, no. 7, pp. 2133–2141, 2008.

[15] L. Liu, A. W. Craig, H. D. Meldrum, S. M. Marcovina, B. E. Elliott, and M. L. Koschinsky, "Apolipoprotein(a) stimulates vascular endothelial cell growth and migration and signals through integrin $\alpha V\beta3$," *Biochemical Journal*, vol. 418, no. 2, pp. 325–336, 2009.

[16] H. A. Chapman, "Plasminogen activators, integrins, and the coordinated regulation of cell adhesion and migration," *Current Opinion in Cell Biology*, vol. 9, no. 5, pp. 714–724, 1997.

[17] P. A. Andreasen, L. Kjoller, L. Christensen, and M. J. Duffy, "The urokinase-type plasminogen activator system in cancer metastasis: a review," *International Journal of Cancer*, vol. 72, no. 1, pp. 1–22, 1997.

[18] K. Danø, J. Rømer, B. S. Nielsen et al., "Cancer invasion and tissue remodeling—cooperation of protease systems and cell types," *Acta Pathologica, Microbiologica et Immunologica Scandinavica*, vol. 107, no. 1, pp. 120–127, 1999.

[19] A. Mondino and F. Blasi, "uPA and uPAR in fibrinolysis, immunity and pathology," *Trends in Immunology*, vol. 25, no. 8, pp. 450–455, 2004.

[20] E. Appella, E. A. Robinson, and S. J. Ullrich, "The receptor-binding sequence of urokinase. A biological function for the growth-factor module of proteases," *Journal of Biological Chemistry*, vol. 262, no. 10, pp. 4437–4440, 1987.

[21] K. Bdeir, A. Kuo, B. S. Sachais et al., "The kringle stabilizes urokinase binding to the urokinase receptor," *Blood*, vol. 102, no. 10, pp. 3600–3608, 2003.

[22] D. A. Waltz, R. M. Fujita, X. Yang et al., "Nonproteolytic role for the urokinase receptor in cellular migration in vivo," *American Journal of Respiratory Cell and Molecular Biology*, vol. 22, no. 3, pp. 316–322, 2000.

[23] V. Ellis, C. Pyke, J. Eriksen, H. Solberg, and K. Dano, "The urokinase receptor: involvement in cell surface proteolysis and cancer invasion," *Annals of the New York Academy of Sciences*, vol. 667, pp. 13–31, 1992.

[24] P. Carmeliet, L. Schoonjans, L. Kieckens et al., "Physiological consequences of loss of plasminogen activator gene function in mice," *Nature*, vol. 368, no. 6470, pp. 419–424, 1994.

[25] T. H. Bugge, M. J. Flick, M. J. S. Danton et al., "Urokinase-type plasminogen activator is effective in fibrin clearance in the absence of its receptor or tissue-type plasminogen activator,"

Proceedings of the National Academy of Sciences of the United States of America, vol. 93, no. 12, pp. 5899–5904, 1996.

[26] D. A. Waltz, L. Z. Sailor, and H. A. Chapman, "Cytokines induce urokinase-dependent adhesion of human myeloid cells. A regulatory role for plasminogen activator inhibitors," *Journal of Clinical Investigation*, vol. 91, no. 4, pp. 1541–1552, 1993.

[27] M. R. Gyetko, R. F. Todd, C. C. Wilkinson, and R. G. Sitrin, "The urokinase receptor is required for human monocyte chemotaxis in vitro," *Journal of Clinical Investigation*, vol. 93, no. 4, pp. 1380–1387, 1994.

[28] N. Busso, S. K. Masur, D. Lazega, S. Waxman, and L. Ossowski, "Induction of cell migration by pro-urokinase binding to its receptor: possible mechanism for signal transduction in human epithelial cells," *Journal of Cell Biology*, vol. 126, no. 1, pp. 259–270, 1994.

[29] L. E. Odekon, N. Gilboa, P. Del Vecchio, and P. W. Gudewicz, "Urokinase in conditioned medium from phorbol ester-pretreated endothelial cells promotes polymorphonuclear leukocyte migration," *Circulatory Shock*, vol. 37, no. 2, pp. 169–175, 1992.

[30] S. A. Rabbani, A. P. Mazar, S. M. Bernier et al., "Structural requirements for the growth factor activity of the amino-terminal domain of urokinase," *Journal of Biological Chemistry*, vol. 267, no. 20, pp. 14151–14156, 1992.

[31] J. A. Aguirre Ghiso, K. Kovalski, and L. Ossowski, "Tumor dormancy induced by downregulation of urokinase receptor in human carcinoma involves integrin and MAPK signaling," *Journal of Cell Biology*, vol. 147, no. 1, pp. 89–103, 1999.

[32] K. Fischer, V. Lutz, O. Wilhelm et al., "Urokinase induces proliferation of human ovarian cancer cells: characterization of structural elements required for growth factor function," *FEBS Letters*, vol. 438, no. 1-2, pp. 101–105, 1998.

[33] S. M. Kanse, O. Benzakour, C. Kanthou, C. Kost, H. Roger Lijnen, and K. T. Preissner, "Induction of vascular SMC proliferation by urokinase indicates a novel mechanism of action in vasoproliferative disorders," *Arteriosclerosis, Thrombosis, and Vascular Biology*, vol. 17, no. 11, pp. 2848–2854, 1997.

[34] J. L. Koopman, J. Slomp, A. C. W. de Bart, P. H. A. Quax, and J. H. Verheijent, "Mitogenic effects of urokinase on melanoma cells are independent of high affinity binding to the urokinase receptor," *Journal of Biological Chemistry*, vol. 273, no. 50, pp. 33267–33272, 1998.

[35] S. Mukhina, V. Stepanova, D. Traktouev et al., "The chemotactic action of urokinase on smooth muscle cells is dependent on its kringle domain. Characterization of interactions and contribution to chemotaxis," *Journal of Biological Chemistry*, vol. 275, no. 22, pp. 16450–16458, 2000.

[36] A. Haj-Yehia, T. Nassar, B. S. Sachais et al., "Urokinase-derived peptides regulate vascular smooth muscle contraction in vitro and in vivo," *The FASEB Journal*, vol. 14, no. 10, pp. 1411–1422, 2000.

[37] T. Nassar, A. Haj-Yehia, S. Akkawi et al., "Binding of urokinase to low density lipoprotein-related receptor (LRP) regulates vascular smooth muscle cell contraction," *Journal of Biological Chemistry*, vol. 277, no. 43, pp. 40499–40504, 2002.

[38] T. Tarui, N. Akakura, M. Majumdar et al., "Direct interaction of the kringle domain of urokinase-type plasminogen activator (uPA) and integrin $\alpha v\beta3$ induces signal transduction and enhances plasminogen activation," *Thrombosis and Haemostasis*, vol. 95, no. 3, pp. 524–534, 2006.

[39] A. P. Mazar, "Urokinase plasminogen activator receptor choreographs multiple ligand interactions: implications for tumor

progression and therapy," *Clinical Cancer Research*, vol. 14, no. 18, pp. 5649–5655, 2008.

[40] S. H. Kwak, S. Mitra, K. Bdeir et al., "The kringle domain of urokinase-type plasminogen activator potentiates LPS-induced neutrophil activation through interaction with $\alpha v\beta3$ integrins," *Journal of Leukocyte Biology*, vol. 78, no. 4, pp. 937–945, 2005.

[41] B. Degryse, J. Fernandez-Recio, V. Citro, F. Blasi, and M. V. Cubellis, "In silico docking of urokinase plasminogen activator and integrins," *BMC Bioinformatics*, vol. 9, no. 2, article S8, 2008.

[42] Y. Wei, J. A. Eble, Z. Wang, J. A. Kreidberg, and H. A. Chapman, "Urokinase receptors promote $\beta1$ integrin function through interactions with integrin $\alpha3\beta1$," *Molecular Biology of the Cell*, vol. 12, no. 10, pp. 2975–2986, 2001.

[43] D. I. Simon, N. K. Rao, H. Xu et al., "Mac-1 (CD11b/CD18) and the urokinase receptor (CD87) form a functional unit on monocytic cells," *Blood*, vol. 88, no. 8, pp. 3185–3194, 1996.

[44] Y. Wei, M. Lukashev, D. I. Simon et al., "Regulation of integrin function by the urokinase receptor," *Science*, vol. 273, no. 5281, pp. 1551–1555, 1996.

6

Human L-Ficolin (Ficolin-2) and Its Clinical Significance

David C. Kilpatrick[1] and James D. Chalmers[2]

[1] *Scottish National Blood Transfusion Service, National Science Laboratory, Ellen's Glen Road, Edinburgh EH17 7QT, UK*
[2] *MRC Centre for Inflammation Research, University of Edinburgh, 47 Little France Crescent, Edinburgh EH16 4TJ, UK*

Correspondence should be addressed to David C. Kilpatrick, david.kilpatrick@nhs.net

Academic Editor: Misao Matsushita

Human L-ficolin (P35, ficolin-2) is synthesised in the liver and secreted into the bloodstream where it is one of the major pattern recognition molecules of plasma/serum. Like other ficolins, it consists of a collagen-like tail region linked to a fibrinogen-related globular head; a basic triplet subunit arises via a collagen-like triple helix, and this then forms higher multimers (typically a 12-mer, Mr 400K). Unlike other ficolins, it has a complex set of binding sites arranged within an internal cleft enabling it to recognise a variety of molecular patterns including acetylated sugars and certain 1,3-β-glucans. It is one of the few molecules known to activate the lectin pathway of complement. Recently, some disease association studies (at either the DNA or protein level) have implicated L-ficolin in innate immunity, where it might cooperate with pentraxins and collectins. Emerging lines of evidence point to a role for L-ficolin in respiratory immunity, where its affinity for *Pseudomonas aeruginosa* could be significant.

1. Introduction

1.1. Discovery. Ficolins were first discovered as transforming growth factor β-binding proteins present in porcine uterus, characterised by the possession of both fibrinogen-related and collagen-like domains [1]. However, it was the description of P35 as an opsonic, GlcNAc-specific lectin that first indicated that this family could be involved in innate immunity as pattern recognition molecules [2]. A very similar molecule named L- (for liver) ficolin was independently purified from human plasma on GlcNAc-Sepharose [3]. Both resembled two other previously discovered plasma proteins: EBP-37, which bound elastin [4], and hucolin, which bound a corticosteroid derivative [5].

Despite some minor discrepancies in properties, it was clear that the same protein was being isolated on different affinity matrices, and the term L-ficolin (or ficolin-2) is now used for this protein with a rather catholic taste in ligands. In this, as in several other features, it resembles the intensively studied mannan- (or mannose-) binding lectin [6], but by definition a ficolin has a fibrinogen-like domain combined with a collagen-like domain and is therefore not a collectin. (Collectins combine a collagen-like domain with a C-type lectin domain and resemble ficolins and C1q in tertiary structure.)

1.2. The Ficolin Family. Porcine ficolins consist of two homologous molecules, designated ficolin-α and -β. Although first discovered in uterine tissue, porcine ficolin-α is more abundant in liver and blood where two isoforms, "little ficolin" (Mr~400 000) and "big ficolin" (Mr~800 000) were described [7]. Ficolin-β, with around 80% identity to ficolin-α, was found to be expressed mainly in neutrophils [8].

A similar situation exists in mice. Ficolin A is present in liver and blood plasma, while ficolin B (60% identical) is expressed in bone marrow and spleen and is associated with macrophages [9, 10].

These findings have prompted the generalisation that ficolins can be classified into soluble serum ficolins and cell-bound ficolins whatever the species. This view is supported by a similar dichotomy in the toad, *Xenopus laevis* [11]. The relationships between ficolins in the above species and others have been reviewed in more detail by Matsushita [12] and by Garred et al. 2010 [13].

1.3. Human Ficolins. Unlike pigs or mice, humans have three ficolins, all of which are present in the bloodstream: M-ficolin (monocyte ficolin or ficolin-1); L-ficolin (liver ficolin or ficolin-2); and H-ficolin (Hakata antigen or ficolin-3). M- and L-ficolin have approximately 80% identity in amino acid sequence; H-ficolin has only about 50% identity with

TABLE 1: The human ficolins.

	M-ficolin	L-ficolin	H-ficolin
Molecular size (subunit)	35 K	35 K	34 K
Molecular size (native)	900 K	420 K	610 K
Location	Neutrophils, monocytes > serum	liver; serum	liver; bile; lung; serum
Chemical specificity	Acetylated sugars	acetylated compounds; LPS; 1,3-β-glucans; lipoteichoic acids; elastin; steroids	D-fucose > GlcNAc; polysaccharide from A.viridans
Microbial specificity	E. coli; S. aureus; S. agalactiae	S. aureus; S. pneumoniae; S. typhimurium; E. coli; P. aeruginosa; M. bovis; G. lamblia; T. cruzi; A. fumigatus,	A. viridans; T. cruzi
Complement activation	Yes	Yes	Yes
Opsonic activity	?	Yes	?
Collagenase sensitivity	Yes	Yes	No

the other two. M-ficolin, predominantly found in monocytes and granulocytes, is the homologue of murine ficolin-B and porcine ficolin-β; L-ficolin is the homologue of murine ficolin-A and porcine ficolin-α [12].

The third human ficolin, the Hakata antigen originally identified and defined by autoantibodies present in a small minority of lupus patients, is synthesised in both liver (secreted into bile as well as blood) and lung (and secreted into the bronchi). It is the most abundant plasma ficolin and the most potent at activating complement in vitro [14, 15].

A comparison of the properties of the ficolins is summarised in Table 1. All three have the ability to activate the lectin pathway of complement, an activity known to be shared with just two collectins, mannan-binding lectin (MBL) and CL-L1 [16, 17]. All three ficolins seem to recognise acetylated sugars like GlcNAc to some degree. L-ficolin (like MBL) appears to be a major pattern recognition molecule in human plasma [18]. It has a uniquely complex set of binding sites, potentially conferring the ability to recognise and interact with a wide range of microorganisms [19]. Recently, L-ficolin has been the subject of several disease association studies, providing evidence that L-ficolin complements MBL as an important component of innate immunity in the circulation. The structure, properties and function(s) of L-ficolin form the remainder of this paper.

2. Genetics

The human L-ficolin gene (FCN2) has been localised to chromosome 9 (9q34) [20], like the M-ficolin gene but in contrast to the H-ficolin gene located on chromosome 1. The L-ficolin gene has eight exons (Figure 1). The first exon encodes a signal sequence and the first nine N-terminal residues. Exons 2 and 3 encode a collagen-like region similar to that found in collectins. The fourth exon encodes a link or connecting region. Exons five to eight encode a domain similar in structure to that of the C-terminal portion of the human fibrinogen β and γ chains that is characterized by the conservation of 24 mainly hydrophobic amino acid residues.

The FCN2 gene is undoubtedly polymorphic. Hummelshoj et al. [21] first described 5 polymorphisms in the promotor region and 9 in the structural gene from a Danish population. Compatible results were reported by Herpers et al. [22], describing 10 single nucleotide polymorphisms in 1888 Dutch blood donors. A later study compared five different ethnic groups; some ethnic-specific polymorphisms were noted, but most were found in all populations [23].

Considerable linkage disequilibrium exists between pairs of promoter and structural gene dimorphisms, complicating the investigation of the relationship between allele expression and protein concentration, but such relationships certainly exist. High concentrations appear to be associated with the variant (minority) nucleotide at promoter position−4, while low concentrations are associated with the variant allele at position 6424 on exon 8 [24, 25]. It is perhaps surprising, and certainly confusing, that the latter mutation is also associated with an increased lectin activity (GlcNAc binding). L-ficolin has several independent activities, however, associated with a variety of binding sites (see below), and it is unknown if other activities are affected.

The single nucleotide polymorphisms implicated in influencing protein concentration in one or more studies are listed in Table 2. It is clear however that those mutations have a very modest influence on average L-ficolin values, each being associated with large and overlapping ranges [25]. Although circulating L-ficolin levels appear to be reasonably stable in healthy individuals, it remains to be established if nongenetic influences may have as great or greater an effect than the FCN2 genotype. For example, we found that patients treated for haematological malignancies had significantly lower median serum L-ficolin compared with healthy controls [26]. Moreover, when studying babies with sepsis, we have found large differences in some individuals with time (unpublished).

Most individuals possess one or two of the most common five haplotypes, and Munthe-Fog et al. [25] have helpfully stratified serum L-ficolin concentration according to those haplotypes. From this analysis, it is apparent that FCN2

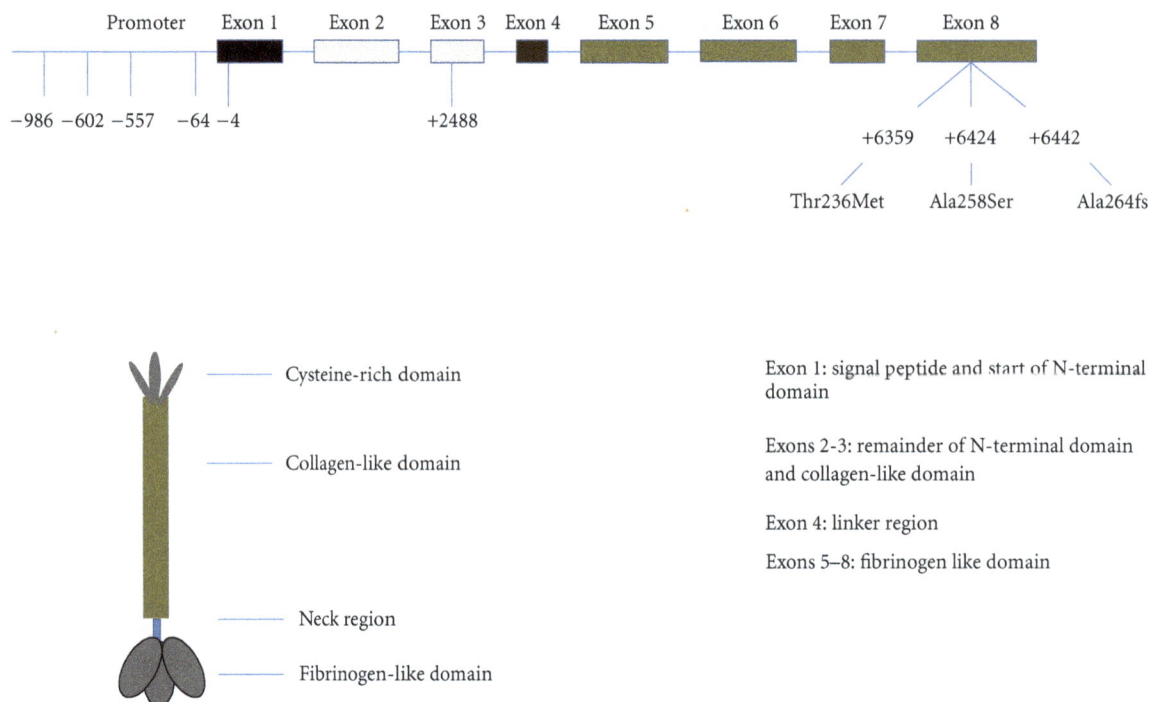

FIGURE 1: The human *FCN2* gene. The positions of the major single nucleotide polymorphisms are shown. The mutation at +6442 leads to a deletion (fs = frameshift mutation).

TABLE 2: Potentially important polymorphisms in the *FCN2* gene.

SNP no.	Region and position	Base substitution	Amino acid substitution
rs3124952	promoter −986	A>G	—
rs3124953	promoter −602	G>A	—
rs3811140	promoter −557	A>G	—
rs28969369	promoter −64	A>C	—
rs17514136	promoter −4	A>G	—
rs17549193	exon 8 +6359	C>T	Thr236Met
rs7851696	exon 8 +6424	G>T	Ala258Ser

genotype has little value in predicting protein concentration in individuals. It has been suggested that the huge difference in MBL concentration in general between groups with different *MBL2* genotypes is woefully inadequate for predicting serum MBL levels in individuals and that using that approach is unreliable for inferring differences in disease cohorts [27]. That limitation is even more true of *FCN2* genotyping, although immunogenetic studies may provide interesting additional information to L-ficolin protein measurement.

3. Structure

The primary structure is composed of 288 amino acids forming a gene product of apparent Mr 35 000 after glycosylation [2]. A short N-terminal region implicated in multimer formation is followed by a series of 19 (Gly-X-Y) repeats forming the collagen-like region or domain. This is attached via a short linking sequence to a large globular domain with a distinctive fold, homologous to the C-terminal domains found in fibrinogen chains. This fibrinogen-like domain occurs in several apparently unrelated proteins, including tenascins, the acetylated sugar-binding tachylectins from a horseshoe crab (*Tachypleus tridentatus*) and the sialic acid-binding lectin from the slug, *Limax flavus* [28].

The combination of fibrinogen-like domain and collagen-like region in L-ficolin (and other ficolins) permits the gene product to form a basic subunit consisting a triple helical tail and a trio of globular heads. This 3-dimensional structure is often likened to a bowl of tulips and resembles the shape of the complement component C1q and the collectin family despite those other molecules not having primary sequence homology with ficolins or with each other. The triplet subunits can then associate to form higher multimers (Figure 2). The major form in plasma is believed to be a tetramer of subunits (12-mer) with an apparent Mr of approximately 400 000 [10, 29].

L-ficolin uniquely possesses a complex set of binding sites constituting a recognition surface that can detect various acetylated structures and neutral sugars in the context of extended polysaccharides. This conclusion is based on studies of its trimeric recombinant recognition domains solved by X-ray crystallography [19, 30]. There is an outer binding site (S1) close to the only calcium binding site. This could be considered the ancestral binding site, as it is homologous to that of the horseshoe crab tachylectin 5A as well as that found in human H- and M-ficolins. Surprisingly, S1 is not responsible for recognition of acetylated sugars unlike its counterpart in tachylectin 5A. Instead, L-ficolin

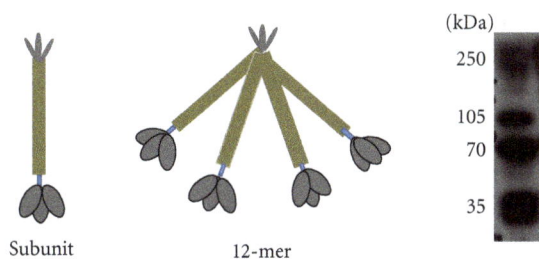

FIGURE 2: L-ficolin structure. The glycosylated gene product of Mr 35 K forms a basic triplet subunit. The subunit can form higher multimers, of which the 12-mer (4×3) is thought to be the most abundant in serum. The inset (right) shows a typical western blot after electrophoresis of L-ficolin under nonreducing conditions.

possesses three inner binding sites (S2, S3, S4) that are located on both sides of a cleft between the upper parts of the protomers. It is S3 that is mainly responsible for binding acetylated structures and also binds 1,3-β-glucans with assistance from the minor site, S4. The innermost S2 has affinity for galactose. The L-ficolin recognition groove with its contiguous subsites (S2–S4) is reminiscent of the peptidoglycan binding proteins of invertebrates [19].

The ficolins, collectins, and complement component C1q all have a similar 3-dimensional shape suited to function as multivalent recognition molecules with increased affinity for ligands achieved by multiple protein oligosaccharide interactions. It is noteworthy that L-ficolin possesses a semi-open structure intermediate between the compact assembly of C1q and the wide open arrangement of MBL which has little interaction between the lectin domains and a buried surface 8% the size of that of C1q [19].

The collagen-like region is responsible for the formation of the helical tails of the molecule and also for functional signalling. In particular, the lysine-57 residue is a key component of the binding site for MBL-associated serine proteases and also for calreticulin, a putative complement/collectin/ficolin receptor on phagocytes [31].

4. Biological Properties

4.1. Biochemical Specificity. L-ficolin was reported to bind to mannan, mannan-Sepharose, and GlcNAc-BSA in a calcium-dependent manner [2], whereas other workers described calcium-independent binding to GlcNAc-Sepharose and elution with GlcNAc [3]. L-ficolin also bound to CNBr-activated (but underivatized) Sepharose blocked with Tris. Since L-ficolin did not bind to mannan coupled to "Emphase" (a polyacrylamide derivative) [32], it seems likely the affinity chromatography was achieved with Sepharose (i.e., polygalactose) not mannan; this interaction was possibly mediated by binding site S2. The inconsistency regarding calcium is to some extent explained by the observation that L-ficolin bound to GlcNAc-Sepharose or CysNAc-Sepharose beads not only in the presence of calcium but also in the absence of calcium at high NaCl concentrations [33]. Given that X-ray crystallography has shown that the ancestral (S1) calcium-associated binding site does not bind galactose or

acetylated sugars, it appears that the other binding sites are influenced indirectly by their ionic environment.

It is clear that L-ficolin binds not just to acetylated sugars but also to nonsaccharide acetylated compounds [33]. Nevertheless, glycan array studies have established that L-ficolin does not bind to most acetylated oligosaccharides found on mammalian glycoconjugates [34]. Rather, L-ficolin has complex recognition requirements and binding probably requires the presence of two or more acetylated sugar groups presented in an appropriate conformation. In another glycan array study, L-ficolin preferentially recognised disulphated N-acetyllactosamine and tri- and tetrasaccharides containing terminal galactose or N-acetylglucosamine and binding was sensitive to the orientation of the bond between N-acetyllactosamine and the adjacent saccharide [35].

L-ficolin is the major 1,3-β-glucan-binding protein in human plasma [18] and can bind to lipoteichoic acid [36]. Potentially, therefore, L-ficolin could bind to a wide variety of fungi and Gram-positive bacteria. L-ficolin recognises and binds to the viral envelope glycoproteins (E1 and E2) of the hepatitis C virus and triggers the lectin pathway of complement by binding to a virally infected human hepatoma cell line [37].

L-ficolin also appears to bind human DNA, suggesting a mechanism for attaching to apoptotic or necrotic autologous cells and thus promoting the removal of dead and moribund cells and tissues [38]. It has already been mentioned that L-ficolin can bind to a protein, elastin [4], and a lipid [5]. Most significantly, it can bind to the pentraxins, C-reactive protein [39], pentraxin 3 (PTX3) [40], and serum amyloid P component (J. D. Chalmers, unpublished).

4.2. Microbial Specificity. L-ficolin has been found to bind to the Gram-negative bacteria, *Salmonella typhimurium* (Ra strain) [2], *Escherichia coli* [41], and *Pseudomonas aeruginosa* [42], as well as the Gram-positive species, (capsulated) *Staphylococcus aureus* and streptococci including the pneumococcus [33, 43, 44]. All interactions were partially sensitive to GlcNAc. L-ficolin binding to the intracellular bacterium *Mycobacterium bovis* has also been reported [45]. Additionally, binding to *Haemophilus influenzae* and *Moraxella catarrhalis* has been observed (J. D. Chalmers, unpublished).

Binding of L-ficolin to the protozoan causing Chagas' disease, *Trypanosoma cruzi*, has been demonstrated [46] and to the intestinal protozoan *Giardia lamblia* [47]. It also binds the opportunistic fungal pathogen, *Aspergillus fumigatus* [40]. This last interaction can be partially inhibited by GlcNAc or Curdlan (a β-1,3-glucose polymer).

The L-ficolin-mediated response to *A. fumigatus* is enhanced by calcium-independent binding to PTX3 [40]. Similar synergy was observed between L-ficolin and C-reactive protein in response to *P. aeruginosa* [42]. These appear to be impressive examples of how L-ficolin can combine with pentraxins to amplify antimicrobial recognition and effector mechanisms.

4.3. Complement Activation. L-ficolin is one of the few molecules known to activate the lectin pathway of complement activation [48]. This arises after forming a complex with MBL-associated serine proteases (MASP)-1, -2, and -3, of which MASP-2 is crucial for complement activation. MASP-2 binding takes place at a site on the collagen-like region [31]. L-ficolin-MASP-2 interaction leads to activation of the latter, enabling it to cleave complement components C2 and C4 in a manner similar to the C1q, r, s complex of the classical pathway initiated by antigen-antibody formation. The roles of MASP-1 and its alternatively spliced gene product MASP-3 are less clear, but evidence is accumulating that MASP-1 may link complement to the coagulation system [49–53] as well as collaborating with MASP-2 in the generation of the C3 convertase [54, 55]. MASP-3 is primarily found complexed to H-ficolin and appears to regulate complement activation mediated by the latter [56].

The other activators of MASP-2 and hence the lectin pathway are the other human ficolins (M- and H-ficolin) and the collectins, MBL and CL-11/CL-K1 [16, 17]. In so far as they have been directly compared, H-ficolin emerged as the most potent complement activator, at least as measured in vitro by a particular C4 deposition assay [15]. L-ficolin and MBL were similar, but had only half the complement activation capacity of H-ficolin.

Unsurprisingly, L-ficolin has been shown to possess opsonic activity by enhancing phagocytosis of *Salmonella* by human neutrophils [2]. It can also promote the phagocytosis of *Pseudomonas aeruginosa* [42] and *Streptococcus pneumoniae* (J. D. Chalmers, unpublished). Although such opsonisation may be mediated by complement activation, it has also been suggested that L-ficolin can opsonise bacteria by binding calreticulin on phagocytes via its collagen-like domain [31].

L-ficolin can cooperate with pentraxins to opsonise bacteria and initiate the lectin pathway, as was mentioned earlier.

Two reports support the view that L-ficolin can opsonise autologous dead or dying cells and cellular debris [38, 57]. Kuraya et al. [57] concluded that L-ficolin binds to apoptotic cells and activates complement via the lectin pathway. Jensen et al. [38], however, observed binding to necrotic but not apoptotic cells, and only at supraphysiological concentrations of $\geq 20\,\mu g/mL$. High concentrations of L-ficolin also promoted the uptake of necrotic Jurkat cells by monocyte-derived macrophages in a phagocytosis assay [38]. Many molecules have been described as binding to apoptotic cells, however, and we have no idea of their relative physiological significance in vivo.

5. L-Ficolin in Health and Disease

5.1. L-Ficolin in Healthy Subjects. In healthy adult individuals, the distribution of serum L-ficolin is perfectly Gaussian, so the mean and median are exactly the same. That value has generally been reported to be between 3 and $4\,\mu g/mL$ [58] although more recently it has been determined at $5.4\,\mu g/mL$ [25]. A complicating consideration is that purified L-ficolin is very labile (unlike in serum); if anything other than a recently isolated preparation is used as a standard, the values obtained from the calibration curve will be higher

than the true values. There is also scope for discrepancy between immunoassayed protein and a measurement based on activity. However, in general, we have found broadly similar values to be obtained with either acetylated BSA or anti-L-ficolin antibody (clone GN4) as solid-phase capture agent combined with the same labelled detection antibody (clone GN5).

Most normal values fall within the range 1000 to 6000 ng/mL, although occasionally much higher values are detected. No value much below 1000 ng/mL has been detected in a healthy adult; therefore, absolute L-ficolin deficiency has not been shown to exist. From genetic studies [21], a homozygous frameshift mutation in exon 8 (rs28357091; Ala264fs; FCN2-D) would be expected to give total deficiency, but this has not yet been found.

Normal values are generally lower in antenatal (cord blood) sera [3, 59–61] and correlate with gestational age, at least until shortly before term [60]. Only one study has investigated serum L-ficolin throughout life [61]. The highest values were found between 1 and 4 years of life (median 11300 ng/mL), then dropped during later childhood (4–16 years, median 8660 ng/mL) before reaching a stable adult level (median 3370) after 16 years of age. Although Sallenbach et al.'s values [61] were in good agreement with our and others' adult data and reasonably close agreement for antenatal data, their elevated values for later childhood were at variance with our finding of similar-to-adult values for such children [62].

Although no absolute deficiency of L-ficolin has yet been discovered, it seems possible that relative deficiency ("insufficiency") defined by low serum L-ficolin ± immunogenetics could contribute to disease susceptibility. Unfortunately, the disease process itself (or treatment) might affect circulating L-ficolin, as has been found for patients with haematological malignancies [26], preeclampsia [63], and sepsis (D. C. Kilpatrick, unpublished).

5.2. General Infections. L-ficolin insufficiency was associated with perinatal infections in preterm Polish babies [60]. In a cohort of over 1800 consecutive deliveries, the rate of perinatal infections in babies with cord blood L-ficolin <1000 ng/mL (the lowest 9%) was twice that of babies with higher concentrations (13.7% versus 7.7%; $P < 0.01$). This relationship was not independent of gestational age and birthweight, but suggests that L-ficolin insufficiency could be one of several factors that contribute to the adverse consequences of prematurity and low birthweight [60].

Schlapback et al. [64], however, found no significant relationship between low cord blood L-ficolin and sepsis in 47 premature infants. This was despite finding significant relationships between low H-ficolin and Gram-positive sepsis and between low MBL and Gram-negative sepsis. Incidentally, that study did confirm correlations between L-ficolin concentration and both gestational age and birthweight. It is possible that study was underpowered for L-ficolin: for Gram-negative sepsis, 47% (7/15) of the patients had low (<1000 ng/mL) L-ficolin compared with 23% (22/94) of infection-free, matched controls ($P < 0.07$) [64].

Uraemic patients have an increased susceptibility to infection, and peritonitis is a common complication in patients on continuous ambulatory peritoneal dialysis (CAPD). The +6359 C>T variant in the *FCN2* gene, causing a Thr ▶ Met alteration and a concomitant decrease in lectin activity, was found to be commoner in CAPD patients with a history of staphylococcal peritonitis compared with CAPD patients without such a history [65]. In the former, exit site *Staphylococcus aureus* was also more prevalent [65].

Another clinical context of interest is that of chemotherapy ± transplantation. MBL has been intensively studied in this context with wildly differing results [66, 67]. In our series of haematological malignancy patients who were rendered severely neutropenic, entirely negative results were obtained for both L-ficolin and H-ficolin [26].

However, the situation appears to be very different in the context of liver transplantation, where the recipient assumes the phenotype of the donor. The variant *FCN2* +6359 allele was associated with a doubling of the bacterial infection risk within the first year following transplant, and the risk was enhanced by coinheritance of *MBL2* variant alleles [68]. Patients with one or more lectin pathway genetic variants and infection had a much increased mortality rate which was highly statistically significant [68]. As well as those bacterial infections causing sepsis and pneumonia, the normal (high lectin activity) *FCN2* +6359 was associated with protection from cytomegalovirus after liver transplantation [69]. Again, the combination of *FCN2* and *MBL2* risk alleles conferred a particularly high degree of susceptibility.

5.3. Respiratory Infections. There is a strong theoretical basis to believe L-ficolin may be important in respiratory infections. As discussed previously, L-ficolin binds to and opsonises a wide variety of important respiratory pathogens (including capsulated *S. pneumoniae*, *S. aureus*, *P. aeruginosa*, *H. influenzae* and others). L-ficolin, like MBL, has been found in the inflamed lung (induced sputum and bronchoalveolar lavage fluid) in concentrations sufficient to cause complement activation (J. D. Chalmers, unpublished).

In perhaps the first disease association study involving L-ficolin, an association between L-ficolin insufficiency and recurrent respiratory infections in children was reported. This relationship was particularly marked for patients with coexisting allergic disorders (mostly rhinitis and/or asthma with high IgE) [70]. This preliminary observation prompted a fresh, prospective study on children aged 1 to 16 years (mean 8.9) to confirm or refute the previous retrospective findings and to distinguish between infection and allergy [62]. L-ficolin insufficiency was indeed significantly associated with asthma and/or allergic rhinitis in the context of recurrent respiratory infections, but not with those allergic disorders in the absence of infection or with recurrent respiratory infections in the absence of allergy [62]. The reason for this relationship is not clear, but it is possible that L-ficolin confers some protection from microorganisms that exacerbate allergic inflammation in the lung.

Another research group examined *FCN2* variants in a birth cohort followed up for 4 years. Analysis based on constructed haplotypes yielded no relationship with recurrent respiratory infection. No serum L-ficolin measurements were made [71].

Similarly negative immunogenetic results were obtained in a study of invasive pneumococcal disease [72], but again no serum L-ficolin data were obtained.

There was a clear association between low serum L-ficolin and idiopathic bronchiectasis [73]. This has since been confirmed in a much larger series, and there is evidence to suggest the basis could lie in protection from colonisation with *Pseudomonas aeruginosa* [74]. Certainly, low serum L-ficolin in those patients identifies a clinical phenotype associated with more severe disease and therefore with poorer quality of life.

5.4. Other Infectious Diseases. The distribution of *FCN2* haplotypes in leprosy patients differed significantly from healthy controls [75]. The authors interpreted their findings as an indication that normal (relatively high) L-ficolin concentrations protect against *M. leprae* infections, but serum L-ficolin was not actually measured.

A comparison of mild and severe *Plasmodium falciparum* malaria revealed that serum L-ficolin concentration is highest during acute severe disease, but this difference was not reflected in the distribution of the *FCN2* haplotypes [76].

In contrast, the distribution of *FCN2* haplotypes was altered in cutaneous leishmaniasis patients when compared with healthy controls [77]. Haplotypes expected to confer normal concentrations of L-ficolin were commoner in the controls, but again serum L-ficolin was not actually measured.

5.5. Pregnancy Disorders. A small proportion of women who had experienced recurrent miscarriage had serum L-ficolin values below the lower limit of normal [59]. These patients were sampled at a single point in time, and a longitudinal study would be required to determine if those low values were stable. It would also be instructive to learn if and how L-ficolin varied during pregnancy in such patients.

According to Wang and coworkers [63], serum L-ficolin increases several-fold during normal pregnancy, although it is not clear whether a direct comparison was made between pregnant and nonpregnant subjects. (Moreover, van de Geijn et al. [78] reported that serum L-ficolin was not influenced by pregnancy.) Certainly, Wang et al. found L-ficolin levels to be significantly lower in preeclamptic pregnancies compared with uncomplicated pregnancies of similar gestational age. Postnatal placentae from preeclamptic pregnancies contained higher concentrations of L-ficolin in lysates, and more intense immunohistochemical staining was observed in syncytiotrophoblast. The co-expression of L-ficolin and Ras protein in preeclamptic syncytiotrophoblast was noted. The authors interpret those observations as evidence of consumption of L-ficolin by apoptotic trophoblast causing depletion in the circulation.

5.6. Miscellaneous Disorders. The distribution of *FCN2* variants was found to be altered in poststreptococcal disease [79]. A haplotype associated with low levels of L-ficolin

TABLE 3: Some disease associations of L-ficolin insufficiency.

Disease	Associated with	Reference
Perinatal infections	Low serum L-ficolin	[60]
Staphylococcal peritonitis	FCN2 +6359 variant allele (FCN2-B)	[65]
Bacterial infections following liver transplantation	FCN2 +6359 variant allele (FCN2-B)	[68]
Childhood infections combined with allergic diseases	Low serum L-ficolin	[62, 70]
Idiopathic bronchiectasis	Low serum L-ficolin	[73, 74]
Preeclampsia	Low serum L-ficolin	[63]
Chronic rheumatic heart disease	Promoter haplotype GGA	[79]

was slightly but significantly more frequent in patients with chronic rheumatic heart disease (CRHD) compared with healthy controls. Conversely, another haplotype was commoner in controls than in either CRHD or rheumatic fever patients. Since L-ficolin can readily bind to *Streptococcus pyogenes* and thereby activate complement [44], it is conceivable that these immunogenetic differences are related to an altered innate response.

A similar *FCN2* investigation in Behcet's disease was essentially negative [80] as was the outcome of serum L-ficolin measurements in sarcoidosis [81]. Nevertheless there was a trend towards lower circulating L-ficolin in sarcoid patients, in contrast to an increase noted for MBL.

Immunohistochemical evidence for L-ficolin involvement in IgA nephropathy has been found [82], but not in renal allograft rejection [83].

The clearest associations of L-ficolin insufficiency with disease susceptibility are summarised in Table 3.

6. Animal Studies

Animal work, though of indirect relevance to clinical situations, often provides valuable supplementary evidence concerning the function of human molecules. For example, native (plasma-derived) and recombinant porcine ficolin-α was found to neutralise porcine reproductive and respiratory virus (a major pathogen of swine) in vitro in a GlcNAc-dependent manner [84]. However, variant alleles of ficolin-α were not associated with common infectious diseases (pneumonia, enteritis, serositis, septicemia) at necropsy, despite significant associations with MBL-A, MBL-C, and surfactant protein A [85]. Of more tenuous significance was the finding that a chimeric molecule combining the lectin domain of MBL and the collagen-like region of L-ficolin had enhanced protective activity towards Ebola virus than either of the parent molecules [86].

Ficolin-A-deficient knockout mice have been established. Survival of these mice after infection with *Streptococcus pneumoniae* was reduced compared with that of wild-type

mice [87], but the phenotype of the knockout mice has not been described in detail.

Human L-ficolin cDNA has been cloned into an expression plasmid and used in a murine model of *Salmonella typhimurium* infection [88]. Administration of L-ficolin in that form protected mice from a potentially lethal challenge with *Salmonella*, with bacterial counts dramatically reduced a week after infection compared with empty-vector-treated controls. The recombinant L-ficolin enhanced monocyte phagocytosis of *Salmonella* in a dose-dependent manner [88].

7. Conclusions

Human L-ficolin is a unique plasma recognition molecule with a broad specificity for microorganisms. It is capable of complementing collectins (such as MBL) and pentraxins in forming a battery of protective molecules constituting the first line of defence. Its functional activities are likely to be mediated through the lectin pathway of complement activation. Although clinical research involving L-ficolin is still in its infancy, evidence is emerging that insufficiency of L-ficolin might increase susceptibility to respiratory infections. In particular, the possibility that L-ficolin is a key factor in protection from *Pseudomonas aeruginosa* warrants further investigation.

References

[1] H. Ichijo, L. Ronnstrand, K. Miyagawa, H. Ohashi, C.-H. Heldin, and K. Miyazono, "Purification of transforming growth factor-β1 binding proteins from porcine uterus membranes," *The Journal of Biological Chemistry*, vol. 266, no. 33, pp. 22459–22464, 1991.

[2] M. Matsushita, Y. Endo, S. Taira et al., "A novel human serum lectin with collagen- and fibrinogen-like domains that functions as an opsonin," *The Journal of Biological Chemistry*, vol. 271, no. 5, pp. 2448–2454, 1996.

[3] Y. Le, S. M. Tan, S. H. Lee, O. L. Kon, and J. Lu, "Purification and binding properties of a human ficolin-like protein," *Journal of Immunological Methods*, vol. 204, no. 1, pp. 43–49, 1997.

[4] S. Harumiya, A. Omori, T. Sugiura, Y. Fukumoto, H. Tachikawa, and D. Fujimoto, "EBP-37, a new elastin-binding protein in human plasma: structural similarity to ficolins, transforming growth factor-β1-binding proteins," *Journal of Biochemistry*, vol. 117, no. 5, pp. 1029–1035, 1995.

[5] P. F. Edgar, "Hucolin, a new corticosteroid-binding protein from human plasma with structural similarities to ficolins, transforming growth factor-β1-binding proteins," *FEBS Letters*, vol. 375, no. 1-2, pp. 159–161, 1995.

[6] D. C. Kilpatrick, "Mannan-binding lectin: clinical significance and applications," *Biochimica et Biophysica Acta*, vol. 1572, no. 2-3, pp. 401–413, 2002.

[7] T. Ohashi and H. P. Erickson, "Two oligomeric forms of plasma ficolin have differential lectin activity," *The Journal of Biological Chemistry*, vol. 272, no. 22, pp. 14220–14226, 1997.

[8] A. S. Brooks, J. Hammermueller, J. P. DeLay, and M. A. Hayes, "Expression and secretion of ficolin β by porcine neutrophils," *Biochimica et Biophysica Acta*, vol. 1624, no. 1–3, pp. 36–45, 2003.

[9] Y. Fujimori, S. Harumiya, Y. Fukumoto et al., "Molecular cloning and characterization of mouse ficolin-a," *Biochemical and Biophysical Research Communications*, vol. 244, no. 3, pp. 796–800, 1998.

[10] T. Ohashi and H. P. Erickson, "Oligomeric structure and tissue distribution of ficolins from mouse, pig and human," *Archives of Biochemistry and Biophysics*, vol. 360, no. 2, pp. 223–232, 1998.

[11] Y. Kakinuma, Y. Endo, M. Takahashi et al., "Molecular cloning and characterization of novel ficolins from Xenopus laevis," *Immunogenetics*, vol. 55, no. 1, pp. 29–37, 2003.

[12] M. Matsushita, "The ficolin family: an overview," in *Collagen-Related Lectins in Innate Immunity*, D. Kilpatrick, Ed., pp. 17–31, Research Signpost, 2007.

[13] P. Garred, C. Honoré, Y. J. Ma et al., "The genetics of ficolins," *Journal of Innate Immunity*, vol. 2, no. 1, pp. 3–16, 2009.

[14] R. Sugimoto, Y. Yae, M. Akaiwa et al., "Cloning and characterization of the Hakata antigen, a member of the ficolin/opsonin p35 lectin family," *The Journal of Biological Chemistry*, vol. 273, no. 33, pp. 20721–20727, 1998.

[15] T. Hummelshoj, L. M. Fog, H. O. Madsen, R. B. Sim, and P. Garred, "Comparative study of the human ficolins reveals unique features of Ficolin-3 (Hakata antigen)," *Molecular Immunology*, vol. 45, no. 6, pp. 1623–1632, 2008.

[16] H. Keshi, T. Sakomoto, T. Kawai et al., "Identificationand characterization of novel human collectin CL-K1," *Microbiology and Immunology*, vol. 50, pp. 1001–1013, 2006.

[17] S. Hansen, L. Selman, N. Palaniyar et al., "Collectin 11 (CL-11, CL K1) is a MASP-1/3-associated plasma collectin with microbial-binding activity," *Journal of Immunology*, vol. 185, no. 10, pp. 6096–6104, 2010.

[18] Y. G. Ma, M. Y. Cho, M. Zhao et al., "Human mannose-binding lectin and L-ficolin function as specific pattern recognition proteins in the lectin activation pathway of complement," *The Journal of Biological Chemistry*, vol. 279, no. 24, pp. 25307–25312, 2004.

[19] V. Garlatti, N. Belloy, L. Martin et al., "Structural insights into the innate immune recognition specificities of L- and H-ficolins," *The EMBO Journal*, vol. 26, no. 2, pp. 623–633, 2007.

[20] Y. Endo, Y. Sato, M. Matsushita, and T. Fujita, "Cloning and characterization of the human lectin P35 gene and its related gene," *Genomics*, vol. 36, no. 3, pp. 515–521, 1996.

[21] T. Hummelshoj, L. Munthe-Fog, H. O. Madsen, T. Fujita, M. Matsushita, and P. Garred, "Polymorphisms in the FCN2 gene determine serum variation and function of Ficolin-2," *Human Molecular Genetics*, vol. 14, no. 12, pp. 1651–1658, 2005.

[22] B. L. Herpers, M. M. Immink, B. A. W. De Jong, H. Van Velzen-Blad, B. M. De Jongh, and E. J. Van Hannen, "Coding and non-coding polymorphisms in the lectin pathway activator L-ficolin gene in 188 Dutch blood bank donors," *Molecular Immunology*, vol. 43, no. 7, pp. 851–855, 2006.

[23] T. Hummelshoj, L. Munthe-Fog, H. O. Madsen, and P. Garred, "Functional SNPs in the human ficolin (FCN) genes reveal distinct geographical patterns," *Molecular Immunology*, vol. 45, no. 9, pp. 2508–2520, 2008.

[24] M. Cedzynski, L. Nuytinck, A. P. M. Atkinson et al., "Extremes of L-ficolin concentration in children with recurrent infections are associated with single nucleotide polymorphisms in the FCN2 gene," *Clinical and Experimental Immunology*, vol. 150, no. 1, pp. 99–104, 2007.

[25] L. Munthe-Fog, T. Hummelshøj, B. E. Hansen et al., "The impact of FCN2 polymorphisms and haplotypes on the Ficolin-2 serum levels," *Scandinavian Journal of Immunology*, vol. 65, no. 4, pp. 383–392, 2007.

[26] D. C. Kilpatrick, L. A. Mclintock, E. K. Allan et al., "No strong relationship between mannan binding lectin or plasma ficolins and chemotherapy-related infections," *Clinical and Experimental Immunology*, vol. 134, no. 2, pp. 279–284, 2003.

[27] A. S. Swierzko, A. Szala, M. Cedzynski et al., "Mannan-binding lectin genotypes and genotype-phenotype relationships in a large cohort of Polish neonates," *Human Immunology*, vol. 70, no. 1, pp. 68–72, 2009.

[28] S. L. MacDonald and D. C. Kilpatrick, "Collagen-related defence proteins as animal lectins," in *Collagen-Related Lectins in Innate Immunity*, D. Kilpatrick, Ed., pp. 1–16, Research Signpost, 2007.

[29] T. Hummelshoj, N. M. Thielens, H. O. Madsen, G. J. Arlaud, R. B. Sim, and P. Garred, "Molecular organization of human Ficolin-2," *Molecular Immunology*, vol. 44, no. 4, pp. 401–411, 2007.

[30] V. Garlatti, L. Martin, M. Lacroix et al., "Structural insights into the recognition properties of human ficolins," *Journal of Innate Immunity*, vol. 2, no. 1, pp. 17–23, 2009.

[31] M. Lacroix, C. Dumestre-Pérard, G. Schoehn et al., "Residue Lys57 in the collagen-like region of human L-ficolin and its counterpart Lys47 in H-ficolin play a key role in the interaction with the mannan-binding lectin-associated serine proteases and the collectin receptor calreticulin," *Journal of Immunology*, vol. 182, no. 1, pp. 456–465, 2009.

[32] D. C. Kilpatrick, "Isolation of human mannan binding lectin, serum amyloid P component and related factors from Cohn Fraction III," *Transfusion Medicine*, vol. 7, no. 4, pp. 289–294, 1997.

[33] A. Krarup, S. Thiel, A. Hansen, T. Fujita, and J. C. Jensenius, "L-ficolin is a pattern recognition molecule specific for acetyl groups," *The Journal of Biological Chemistry*, vol. 279, no. 46, pp. 47513–47519, 2004.

[34] A. Krarup, D. A. Mitchell, and R. B. Sim, "Recognition of acetylated oligosaccharides by human L-ficolin," *Immunology Letters*, vol. 118, no. 2, pp. 152–156, 2008.

[35] E. Gout, V. Garlatti, D. F. Smith et al., "Carbohydrate recognition properties of human ficolins: glycan array screening reveals the sialic acid binding specificity of M-ficolin," *The Journal of Biological Chemistry*, vol. 285, no. 9, pp. 6612–6622, 2010.

[36] N. J. Lynch, S. Roscher, T. Hartung et al., "L-ficolin specifically binds to lipoteichoic acid, a cell wall constituent of gram-positive bacteria, and activates the lectin pathway of complement," *Journal of Immunology*, vol. 172, no. 2, pp. 1198–1202, 2004.

[37] J. Liu, M. A. M. Ali, Y. Shi et al., "Specifically binding of L-ficolin to N-glycans of HCV envelope glycoproteins E1 and E2 leads to complement activation," *Cellular and Molecular Immunology*, vol. 6, no. 4, pp. 235–244, 2009.

[38] M. L. Jensen, C. Honoré, T. Hummelshøj, B. E. Hansen, H. O. Madsen, and P. Garred, "Ficolin-2 recognizes DNA and participates in the clearance of dying host cells," *Molecular Immunology*, vol. 44, no. 5, pp. 856–865, 2007.

[39] P. M. L. Ng, A. Le Saux, C. M. Lee et al., "C-reactive protein collaborates with plasma lectins to boost immune response against bacteria," *The EMBO Journal*, vol. 26, no. 14, pp. 3431–3440, 2007.

[40] Y. J. Ma, A. Doni, T. Hummelshøj et al., "Synergy between ficolin-2 and pentraxin 3 boosts innate immune recognition and complement deposition," *The Journal of Biological Chemistry*, vol. 284, no. 41, pp. 28263–28275, 2009.

[41] J. Lu and Y. Le, "Ficolins and the fibrinogen-like domain," *Immunobiology*, vol. 199, no. 2, pp. 190–199, 1998.

[42] J. Zhang, J. Koh, J. Lu et al., "Local inflammation induces complement crosstalk which amplifies the antimicrobial response," *PLoS Pathogens*, vol. 5, no. 1, Article ID e1000282, 2009.

[43] A. Krarup, U. B. S. Sørensen, M. Matsushita, J. C. Jensenius, and S. Thiel, "Effect of capsulation of opportunistic pathogenic bacteria on binding of the pattern recognition molecules Mannan-binding lectin, L-ficolin, and H-ficolin," *Infection and Immunity*, vol. 73, no. 2, pp. 1052–1060, 2005.

[44] Y. Aoyagi, E. E. Adderson, C. E. Rubens et al., "L-ficolin/mannose-binding lectin-associated serine protease complexes bind to group B streptococci primarily through N-acetylneuraminic acid of capsular polysaccharide and activate the complement pathway," *Infection and Immunity*, vol. 76, no. 1, pp. 179–188, 2008.

[45] M. V. Carroll, N. Lack, E. Sim, A. Krarup, and R. B. Sim, "Multiple routes of complement activation by Mycobacterium bovis BCG," *Molecular Immunology*, vol. 46, no. 16, pp. 3367–3378, 2009.

[46] I. D. S. Cestari, A. Krarup, R. B. Sim, J. M. Inal, and M. I. Ramirez, "Role of early lectin pathway activation in the complement-mediated killing of Trypanosoma cruzi," *Molecular Immunology*, vol. 47, no. 2-3, pp. 426–437, 2009.

[47] I. Evans-Osses, E. A. Ansa-Addo, J. M. Inal, and M. I. Ramirez, "Involvement of lectin pathway activation in the complement killing of Giardia intestinalis," *Biochemical and Biophysical Research Communications*, vol. 395, no. 3, pp. 382–386, 2010.

[48] M. Matsushita, Y. Endo, and T. Fujita, "Cutting edge: complement-activating complex of ficolin and mannose-binding lectin-associated serine protease," *Journal of Immunology*, vol. 164, no. 5, pp. 2281–2284, 2000.

[49] J. S. Presanis, K. Hajela, G. Ambrus, P. Gál, and R. B. Sim, "Differential substrate and inhibitor profiles for human MASP-1 and MASP-2," *Molecular Immunology*, vol. 40, no. 13, pp. 921–929, 2004.

[50] A. Krarup, R. Wallis, J. S. Presanis, P. Gál, and R. B. Sim, "Simultaneous activation of complement and coagulation by MBL-associated serine protease 2," *PLoS One*, vol. 2, no. 7, article e623, 2007.

[51] A. Krarup, K. C. Gulla, P. Gál, K. Hajela, and R. B. Sim, "The action of MBL-associated serine protease 1 (MASP1) on factor XIII and fibrinogen," *Biochimica et Biophysica Acta*, vol. 1784, no. 9, pp. 1294–1300, 2008.

[52] K. C. Gulla, K. Gupta, A. Krarup et al., "Activation of mannan-binding lectin-associated serine proteases leads to generation of a fibrin clot," *Immunology*, vol. 129, no. 4, pp. 482–495, 2010.

[53] K. Takahashi, W. C. Chang, M. Takahashi et al., "Mannose-binding lectin and its associated proteases (MASPs) mediate coagulation and its deficiency is a risk factor in developing complications from infection, including disseminated intravascular coagulation," *Immunobiology*, vol. 216, no. 1-2, pp. 96–102, 2011.

[54] C. B. Chen and R. Wallis, "Two mechanisms for mannose-binding protein modulation of the activity of its associated serine proteases," *The Journal of Biological Chemistry*, vol. 279, no. 25, pp. 26058–26065, 2004.

[55] A. Kocsis, K. A. Kekési, R. Szász et al., "Selective inhibition of the lectin pathway of complement with phage display selected peptides against mannose-binding lectin-associated serine protease (MASP)-1 and-2: significant contribution of MASP-1 to lectin pathway activation," *Journal of Immunology*, vol. 185, no. 7, pp. 4169–4178, 2010.

[56] M. O. Skjoedt, Y. Palarasah, L. Munthe-Fog et al., "MBL-associated serine protease-3 circulates in high serum concentrations predominantly in complex with Ficolin-3 and regulates Ficolin-3 mediated complement activation," *Immunobiology*, vol. 215, no. 11, pp. 921–931, 2010.

[57] M. Kuraya, Z. Ming, X. Liu, M. Matsushita, and T. Fujita, "Specific binding of L-ficolin and H-ficolin to apoptotic cells leads to complement activation," *Immunobiology*, vol. 209, no. 9, pp. 689–697, 2005.

[58] D. C. Kilpatrick, "Clinical significance of mannan binding lectin and L-ficolin," in *Collagen-Related Lectins in Innate Immunity*, D. C. Kilpatrick, Ed., pp. 57–84, Research Signpost, 2007.

[59] D. C. Kilpatrick, T. Fujita, and M. Matsushita, "P35, an opsonic lectin of the ficolin family, in human blood from neonates, normal adults, and recurrent miscarriage patients," *Immunology Letters*, vol. 67, no. 2, pp. 109–112, 1999.

[60] A. S. Swierzko, A. P. M. Atkinson, M. Cedzynski et al., "Two factors of the lectin pathway of complement, l-ficolin and mannan-binding lectin, and their associations with prematurity, low birthweight and infections in a large cohort of Polish neonates," *Molecular Immunology*, vol. 46, no. 4, pp. 551–558, 2009.

[61] S. Sallenbach, S. Thiel, C. Aebi et al., "Serum concentrations of lectin-pathway components, children and in healthy neonates adults: mannan-binding lectin (MBL), M-, L-, and H-ficolin, and MBL-associated serine protease-2 (MASP-2)," *Pediatric Allergy and Immunology*, vol. 22, no. 4, pp. 424–430, 2011.

[62] M. Cedzynski, A. P. M. Atkinson, A. S. Swierzko et al., "L-ficolin (ficolin-2) insufficiency is associated with combined allergic and infectious respiratory disease in children," *Molecular Immunology*, vol. 47, no. 2-3, pp. 415–419, 2009.

[63] C. C. Wang, K. W. Yim, T. C. W. Poon et al., "Innate immune response by ficolin binding in apoptotic placenta is associated with the clinical syndrome of preeclampsia," *Clinical Chemistry*, vol. 53, no. 1, pp. 42–52, 2007.

[64] L. J. Schlapbach, M. Mattmann, S. Thiel et al., "Differential role of the lectin pathway of complement activation in susceptibility to neonatal sepsis," *Clinical Infectious Diseases*, vol. 51, no. 2, pp. 153–162, 2010.

[65] S. C.A. Meijvis, B. L. Herpers, H. Endeman et al., "Mannose-binding lectin (*MBL2*) and ficolin-2 (*FCN2*) polymorphisms in patients on peritoneal dialysis with staphylococcal peritonitis," *Nephrology Dialysis Transplantation*, vol. 26, no. 3, pp. 1042–1045, 2011.

[66] N. J. Klein and D. C. Kilpatrick, "Is there a role for mannan/mannose-binding lectin (MBL) in defence against infection following chemotherapy for cancer?" *Clinical and Experimental Immunology*, vol. 138, no. 2, pp. 202–204, 2004.

[67] D. C. Kilpatrick, "Mannan-binding lectin and stem cell transplantation," in *Hematopoietic Stem Cell Transplantation Research Advances*, K. B. Neuman, Ed., pp. 1–6, Nova Science, 2008.

[68] B. J. F. de Rooij, B. van Hoek, W. R. Ten Hove et al., "Lectin complement pathway gene profile of donor and recipient determine the risk of bacterial infections after orthotopic liver transplantation," *Hepatology*, vol. 52, no. 3, pp. 1100–1110, 2010.

[69] B.-J. F. de Rooij, M. T. van der Beek, B. van Hoek et al., "Combined donor-recipient mannose-binding lectin and ficolin-2 gene polymorphisms predispose to human cytomegalovirus infection after orthotopic liver transplantation," *Journal of Hepatology*, vol. 54, article S236, 2011.

[70] A. P. M. Atkinson, M. Cedzynski, J. Szemraj et al., "L-ficolin in children with recurrent respiratory infections," *Clinical and Experimental Immunology*, vol. 138, no. 3, pp. 517–520, 2004.

[71] J. M. Ruskamp, M. O. Hoekstra, D. S. Postma et al., "Exploring the role of polymorphisms in ficolin genes in respiratory tract infections in children," *Clinical and Experimental Immunology*, vol. 155, no. 3, pp. 433–440, 2009.

[72] S. J. Chapman, F. O. Vannberg, C. C. Khor et al., "Functional polymorphisms in the FCN2 gene are not associated with invasive pneumococcal disease," *Molecular Immunology*, vol. 44, no. 12, pp. 3267–3270, 2007.

[73] D. C. Kilpatrick, J. D. Chalmers, S. L. MacDonald et al., "Stable bronchiectasis is associated with low serum L-ficolin concentrations," *Clinical Respiratory Journal*, vol. 3, no. 1, pp. 29–33, 2009.

[74] J. D. Chalmers, D. C. Kilpatrick, B. McHugh et al., "Single nucleotide polymorphisms in the ficolin-2 gene predispose to *Pseudomonas aeruginosa* infection and disease severity in non-cystic fibrosis bronchiectasis," *Thorax*, vol. 66, supplement 4, pp. A1–A2, 2011.

[75] I. de Messias-Reason, P. G. Kremsner, and J. F. J. Kun, "Functional haplotypes that produce normal ficolin-2 levels protect against clinical leprosy," *Journal of Infectious Diseases*, vol. 199, no. 6, pp. 801–804, 2009.

[76] I. Faik, S. I. Oyedeji, Z. Idris et al., "Ficolin-2 levels and genetic polymorphisms of FCN2 in malaria," *Human Immunology*, vol. 72, no. 1, pp. 74–79, 2011.

[77] A. Assaf, I. Faik, T. Aebischer et al., "Genetic evidence of functional Ficolin-2 haplotype as susceptibility factor in cutaneous leishmaniasis," *PloS One*. In press.

[78] F. E. van de Geijn, A. Roos, Y. A. de Man et al., "Mannose-binding lectin levels during pregnancy: a longitudinal study," *Human Reproduction*, vol. 22, no. 2, pp. 362–371, 2007.

[79] I. J. Messias-Reason, M. D. Schafranski, P. G. Kremsner, and J. F. J. Kun, "Ficolin 2 (FCN2) functional polymorphisms and the risk of rheumatic fever and rheumatic heart disease," *Clinical and Experimental Immunology*, vol. 157, no. 3, pp. 395–399, 2009.

[80] X. Chen, Y. Katoh, K. Nakamura et al., "Single nucleotide polymorphisms of Ficolin 2 gene in Behçet's disease," *Journal of Dermatological Science*, vol. 43, no. 3, pp. 201–205, 2006.

[81] C. B. Svendsen, T. Hummelshøj, L. Munthe-Fog et al., "Ficolins and Mannose-Binding Lectin in Danish patients with sarcoidosis," *Respiratory Medicine*, vol. 102, no. 9, pp. 1237–1242, 2008.

[82] A. Roos, M. P. Rastaldi, N. Calvaresi et al., "Glomerular activation of the lectin pathway of complement in IgA nephropathy is associated with more severe renal disease," *Journal of the American Society of Nephrology*, vol. 17, no. 6, pp. 1724–1734, 2006.

[83] N. Imai, S. Nishi, B. Alchi et al., "Immunohistochemical evidence of activated lectin pathway in kidney allografts with peritubular capillary C4d deposition," *Nephrology Dialysis Transplantation*, vol. 21, no. 9, pp. 2589–2595, 2006.

[84] N. D. Keirstead, C. Lee, D. Yoo, A. S. Brooks, and M. A. Hayes, "Porcine plasma ficolin binds and reduces infectivity of porcine reproductive and respiratory syndrome virus (PRRSV) in vitro," *Antiviral Research*, vol. 77, no. 1, pp. 28–38, 2008.

[85] N. D. Keirstead, M. A. Hayes, G. E. Vandervoort, A. S. Brooks, E. J. Squires, and B. N. Lillie, "Single nucleotide polymorphisms in collagenous lectins and other innate immune genes in pigs with common infectious diseases," *Veterinary Immunology and Immunopathology*, vol. 142, no. 1-2, pp. 1–13, 2011.

[86] I. C. Michelow, M. Dong, B. A. Mungall et al., "A novel L-ficolin/mannose-binding lectin chimeric molecule with enhanced activity against Ebola virus," *The Journal of Biological Chemistry*, vol. 285, no. 32, pp. 24729–24739, 2010.

[87] Y. Endo, M. Matsushita, and T. Fujita, "The role of ficolins in the lectin pathway of innate immunity," *International Journal of Biochemistry and Cell Biology*, vol. 43, no. 5, pp. 705–712, 2011.

[88] Y. Ma, F. Luo, T. Xiang et al., "Effects of L-ficolin on host resistance, gamma interferon production and phagocytosis against Salmonella infection," *Molecular Immunology*, vol. 44, p. 211, 2011.

7

So Many Plasminogen Receptors: Why?

Edward F. Plow, Loic Doeuvre, and Riku Das

Department of Molecular Cardiology, Joseph J. Jacobs Center for Thrombosis and Vascular Biology, Cleveland Clinic, 9500 Euclid Avenue, NB50, Cleveland, OH 44195, USA

Correspondence should be addressed to Edward F. Plow, plowe@ccf.org

Academic Editor: Lindsey A. Miles

Plasminogen and plasmin tether to cell surfaces through ubiquitously expressed and structurally quite dissimilar family of proteins, as well as some nonproteins, that are collectively referred to as plasminogen receptors. Of the more than one dozen plasminogen receptors that have been identified, many have been shown to facilitate plasminogen activation to plasmin and to protect bound plasmin from inactivation by inhibitors. The generation of such localized and sustained protease activity is utilized to facilitate numerous cellular responses, including responses that depend on cellular migration. However, many cells express multiple plasminogen receptors and numerous plasminogen receptors are expressed on many different cell types. Furthermore, several different plasminogen receptors can be used to support the same cellular response, such as inflammatory cell migration. Here, we discuss the perplexing issue: why are there so many different Plg-Rs?

1. Introduction

Plasminogen receptors (Plg-Rs) are a broadly distributed and heterogeneous group of cell surface proteins that share a common feature, the ability to interact with plasminogen (Plg) and plasmin. The list in Table 1, not necessarily all inclusive, identifies 12 different Plg-Rs. Many of these Plg-Rs are expressed by many different cell types, and many are present on the same cell type. Indeed, the number of Plg binding sites on any particular cell type can be extraordinarily high (range from 10^5 to 10^7 Plg binding sites per cell). The similarities among these Plg-Rs are very limited and appear to rest only on their ability to be expressed at cell surfaces where they can display their Plg and Plm binding function. Nevertheless, this binding function allows many different Plg-Rs to orchestrate diverse biological responses including fibrinolysis, inflammation, wound healing, and angiogenesis. The question then arises as to why there are so many Plg-Rs and whether there is a plausible explanation for this extensive functional redundancy? This paper will consider these basic questions. As a forewarning, we do not purport to provide clear answers to these questions but hopefully our speculations will be challenging and stimulating.

2. So Many Plg-Rs: Do Different Plg-Rs Bind Plg Differently?

Almost all of the Plg-Rs listed in Table 1 engage the lysine binding sites (LBS) of Plg and Plm by virtue of a C-terminal lysine or by presenting an internal amino acid residue in a context that mimics a C-terminal lysine. As a consequence of a common mechanism of engagement, Plg-Rs are projected to enhance Plg activation by either urokinase plasminogen activator (uPA) or tissue plasminogen activator (tPA), to enhance the catalytic activity of plasmin and to protect bound plasmin from inactivation by plasmin inhibitors [1–3]. Indeed, several Plg-Rs have been reported to have one or more of these functional attributes [4–6]. Also, with a similar mechanism of binding, the affinities of the various Plg-Rs for Plg should be similar. The context of the LBS binding residue within a Plg-R might be influenced by adjacent amino acids or local conformation and thereby influence the affinity of specific subset of Plg-Rs for Plg. However, even for Plg-Rs that utilize an internal residue rather than a C-terminal lysine to engage Plg, affinities for the ligand appear to be similar ($\sim 1 \mu$M), [7]. One potential exception to this assertion could be the annexin A2/p11

TABLE 1: Plg-Rs on various cell types.

Plg-Rs	Cell types	C-terminal lysine	Major cellular localization	Secretory pathways
(1) Annexin A2	Endothelial cells, monocytoid lineage	Absent*	Cytosol and or nucleus	Translocation depends on p11 and phosphorylation; activity of L-type like Ca^{2+} channels and intracellular Ca^{2+}; associates with plasma membrane via phosphatidylserine.
(2) Actin	Endothelial cells, carcinoma, catecholaminergic cells, PC-3, HT1080	Absent	Cytoskeleton	Not known
(3) Amphoterin	Neuronal cells	Absent	Cytoplasmic and extracellular	Not known
(4) $\alpha V\beta_3$	Endothelial cells	Absent	Integral membrane protein	Classical endoplasmic reticulum and Golgi pathway
(5) $\alpha M\beta_2$	Neutrophils, monocytes, macrophages	Absent	Integral membrane protein	Classical endoplasmic reticulum and Golgi pathway
(6) $\alpha IIb\beta_3$	Platelets, RA synovial fibroblasts	Absent	Integral membrane protein	Classical endoplasmic reticulum and Golgi pathway
(7) Cytokeratin 8	Hepatocellular, breast carcinoma	Present	Cytoskeleton	Not known
(8) α-Enolase	Monocytes, neutrophils, carcinoma, lymphoid, myoblast neurons	Present	Cytosol	L-type-like Ca^{2+} channel and intracellular Ca^{2+}
(9) Histone 2B	Neutrophils, monocytoid cells, endothelial cells	Present	Nucleus	L type like Ca^{2+} channel and intracellular Ca^{2+}. Associates with plasma membrane via phosphatidylserine and heparin sulfate
(10) P11	Endothelial cells, HT1080 cells	Present	Cytosol and or nucleus	L-type-like Ca^{2+} channel and intracellular Ca^{2+}. Associates with multiple plasma membrane binding partners, including annexin 2
(11) Plg-RKT	Monocytes, macrophages, neuronal cells	Present	Integral membrane protein	Classical endoplasmic reticulum and Golgi pathway
(12) TATA-binding protein-interacting protein	Monocytoid cells	Present	Nucleus	Not known

* requires cleavage to bind Plg [8].

heterotetramer, where the proximity of multiple Plg binding sites within a single molecular species could enhance affinity substantially. To support this possibility or other reports of higher-affinity Plg-Rs, variability in ligand preparations used (e.g., presence of Lys-Plg in Glu-Plg preparation) must be controlled. Furthermore, since ligand availability seems not to be limiting (Plg is present at high concentrations), differences in apparent affinity may have less impact than anticipated.

3. So Many Plg-Rs: Do Different Cell Types Use Different Plg-Rs?

Not all Plg-Rs are expressed on all cell types. As an example of a Plg-R with a restricted cellular distribution, integrin $\alpha M\beta2$ is a Plg-R [7] and its expression is confined to leukocytes. However, leukocytes express many other Plg-Rs, including annexin A2/p11, which has long been promulgated as the major Plg-R on endothelial cells (ECs). Indeed, inactivation of either the annexin A2 or p11 genes does affect EC-dependent responses, including angiogenesis, tumorogenesis, fibrinolysis, and inflammation [9–12]. However, ECs do express other Plg-Rs. As an illustrative example, histone H2B, a high-abundance Plg-R on monocytoid cells, is also readily detected on the surface of HUVEC (Figure 1). In Figure 1, H2B was detected on the surface of HUVEC by a cell-surface biotinylation approach [13] in which the cells were surface labeled with biotin, lysed, and the biotinylated proteins were isolated on streptavidin beads and then identified by western blotting with specific antibodies (see Figure 1 and its legend for details). H2B was labeled with biotin, whereas p65, a control intracellular protein, was not even though both H2B and p65 were readily detected in the whole cell lysates of HUVEC. H2B associates with the surface of monocytoid cells by binding to phosphatidylserine (PS) [14]. Annexin V, another PS binding protein displaces the H2B from the surface of monocytoid cells [14] and also chases biotin labeled H2B from the surface of HUVEC (see

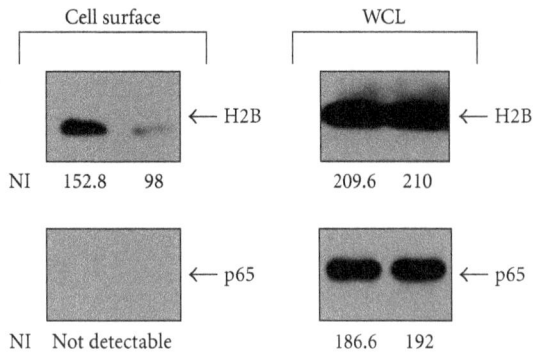

FIGURE 1: H2B exposure on the surface of endothelial cells. Human umbilical vein endothelial cells (HUVECs) were either untreated or treated with annexin V (250 nM) for 48 hr. Cells were surface labeled with biotin, and the biotinylated proteins were isolated using streptavidin-conjugated beads. H2B and p65 (a transcription factor with a cytosolic and nuclear localization) that were bound and eluted from the streptavidin beads were detected by western blotting with a rabbit anti-H2B or rabbit anti-p65. The absence of biotinylated p65 serves as a control for surface labeling of H2B. Band intensities of the western blots were analyzed using Kodak ID 3.6 software, and net intensity (NI) of each band is indicated below each lane. In each set of two lanes, the right-hand lane is in the presence of annexin V and the left-hand lane in its absence. (WCL: whole cell lysates).

Figure 1). Biotinylation also labeled H2B on the surface of microvascular endothelial cells as well as on large-vessel endothelial cells (not shown). As an independent approach, we confirmed the presence of H2B on the surface of HUVECs by flow cytometry. Also, α-enolase, the first identified Plg-R, has been implicated in the binding of Plg to microparticles released from ECs [15]. Thus, in addition to annexin A2/p11, other Plg-Rs have been detected on endothelial cells. Hence, the notion of the preeminence of a specific Plg-R on a particular cell type does not seem tenable.

The compartmentalization of specific Plg-Rs to select locations on the cell surface could provide a mechanism to distinguish the function of one Plg-R from another. Several integrins serve as Plg-Rs (Table 1) and integrins do localize to the leading edge of migrating cells [16], and uPAR also localizes to the leading edge of migrating cells [17]. Furthermore, annexin 2 has been localized to the leading edge of migrating retinal glial cells and malignant glioma cell [18, 19]. Thus, an advantageous microenvironment may be created in which one Plg-R is particularly proficient in Plg activation. However, recent data have suggested that cell-surface-bound Plg can be efficiently activated or even more efficiently activated by uPA bound to another cell than that on the same cell [20]. The boost in efficiency of Plg activation gained by localization on a single cell may be offset by the restricted diffusion or orientation of the Plg activator on the cell surface. Thus, localization of certain Plg-Rs to a specific microdomain on the cell surface and the functional advantage of such localization remain a possibility. We did note a uniform distribution of several Plg-Rs, as well as bound Plg, on monocytoid cells by confocal microscopy although changes in distribution

under stimulated conditions were not tested [13]. A common mechanism dependent on L-type like calcium channels has been implicated in translocation of several Plg-Rs to the surface of monocytoid cells [21], but the mechanisms by which these Plg-R tether to the cell surface are distinct [14]. Hence, Plg-Rs could compartmentalize on the cell surface.

4. So Many Plg-Rs: Are Plg-Rs Differentially Regulated on Cells?

It is well established that Plg binding to cells can be markedly modulated; changes in Plg binding capacity of specific cell types can increase 3- to 20-fold in response to specific stimuli. Cellular events and responses that can induce such changes include oncogenic transformation (breast and adenocarcinoma cancer) [22, 23] differentiation (monocytes, adipocytes) [21, 24], agonist stimulation, (leukocytes, endothelial cells, platelets) [7, 25–27], adhesion (monocytoid cells) [28], and apoptosis (monocytoid cells) [14]. In addition, Plg binding can be enhanced by proteolysis of existing cell surface proteins to generate new C-terminal lysines [29, 30]. This latter mechanism for exposing new Plg-Rs can be triggered by plasmin itself and depends on the availability of uPA on the cell surface [31]. Thus, even though a cell type can express multiple Plg-Rs, a subset of Plg-Rs may be differentially upregulated and utilized to mediate a specific cellular response.

The data in Figure 2 provides an illustrative example of how different Plg-Rs maybe utilized by the same cell in responding to different stimuli. THP-1 monocytoid cells were either stimulated to undergo differentiation using vitamin D3 + IFNγ or apoptosis using camptothecin. Consistent with our prior report [13, 14], the cells respond to these stimuli by markedly upregulating their Plg binding capacity. In association with differentiation, Plg binding increased by 3.3-fold. Of the Plg-Rs analyzed by FACS, enolase, annexin2, p11, and H2B, surface expression increased most markedly for H2B (4.7-fold) in response to differentiation. In response to apoptosis induced by camptothecin, Plg binding increased by 10-fold. While surface localization of H2B did increase significantly (4.6-fold), much more striking was the 20-fold upregulation of p11 in the camptothecin-treated THP-1 cells. This pattern of enhanced p11 expression was also observed in U937 monocytoid cells treated with camptothecin, where 5.8-fold increase of Plg binding was associated with 6.3-fold increase in p11 expression. Of note, these increases in p11 expression on apoptotic cells were not paralleled by substantial increases of the annexinA2 subunit. In the camptothecin-treated THP-1 cells, surface expression of the annexinA2 subunit increased by 2.8-fold and for U397 cells, the increase was 2.3-fold. As explanations for this disproportional upregulation of p11, the subpopulation of annexinA2 molecules that escort p11 to the cell surface may not react with the antibody used in this analysis, or the anti-p11 may selectively penetrate apoptotic cells, which are known to be leaky [32]. A more interesting possibility is that a portion of the p11 that becomes surface expressed is in a free form or is associated with other binding partners.

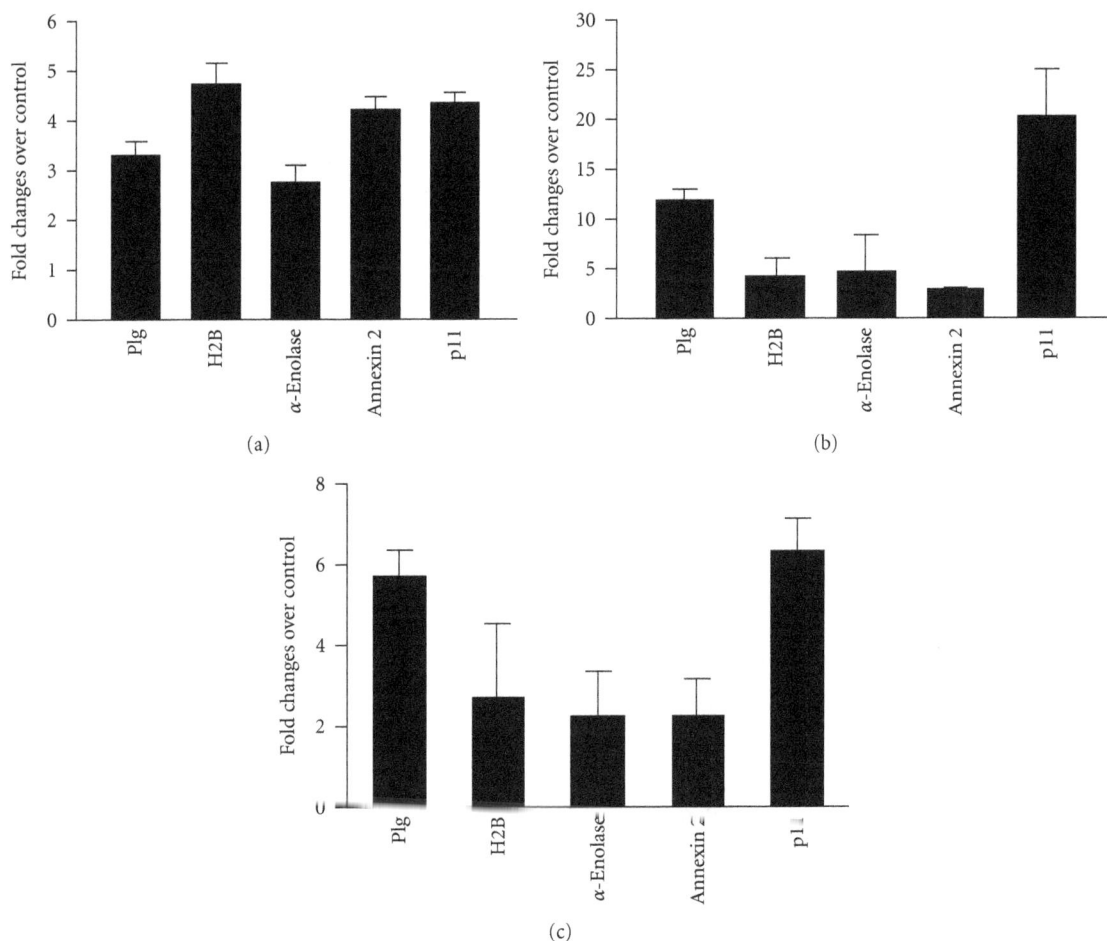

FIGURE 2: THP-1 (a), (b) and U937 (c) cells were either differentiated with IFNγ + VD3 for 48 h (a) or induced to undergo apoptosis with camptothecin for 24 h (b), (c). Cells are labeled with Alexa-488-Plg or anti-Plg-Rs antibodies against H2B, α-enolase, annexin A2, and p11 followed by Alexa-488-ant-rabbit IgG (c) and analyzed by FACS. Early apoptotic populations are used to analyze the data. Data are means ± SD of two to three independent experiments and presented as the fold change relative to untreated THP-1 or U937 cells.

Besides annexin2, other plasma membrane proteins, NaV1.8 sodium channel, TASK1 potassium channel, TRPV5/TRPV6 channels, and cathepsin B [33] have been shown to interact with p11, could assist in its transport to the cell surface, and may still further extend the repertoire of Plg-Rs expressed by monocytoid cells.

In vivo data also support the proposition that different Plg-Rs mediate the response of the same cell type to different stimuli. In a thioglycollate-induced peritonitis model, an antibody to H2B that blocks Plg binding inhibited macrophage recruitment by ~50% while an antibody to α-enolase that also blocks Plg binding to its target produced less than 25% inhibition of macrophage recruitment [13]. In contrast, in an LPS-induced lung inflammation model, Plg binding to α-enolase overexpressing U937 cells produced a substantial enhancement of macrophage migration [34].

5. So, Why So Many Plg-Rs?

While the utilization of different Plg-Rs to orchestrate different cellular responses is supported by data cited above, blocking of several different Plg-Rs has been shown to markedly suppress what appears to be the same inflammatory response thioglycollate=induced peritonitis. The contribution of H2B (45% [13]), p11 (53% [12]), and Plg-RKT (58% [35]), either with antibodies or gene inactivation, exceeds 100%. Such extensive inhibition becomes even more incomprehensible since macrophage recruitment is decreased by only 65% in Plg$^{-/-}$ mice compared to wild-type littermates [36]. At least four explanations can be considered to explain such observations. First, these various Plg-Rs may exert an effect on macrophage recruitment unrelated to Plg. The effect of blockade of individual Plg-Rs in a Plg$^{-/-}$ background could be used to identify such functions. Second, a threshold of bound Plg must be attained in order for Plg to facilitate cell migration. No single Plg-R may harness sufficient Plg to reach this threshold, and, hence, cooperation among several Plg-Rs is required. Third, while many different Plg-Rs enhance Plg activation, the intracellular signaling responses that they elicit may be distinct. Cellular recruitment is a complex response requiring activation of many different intracellular signaling pathways. Different Plg-Rs may trigger distinct signaling events, and these pathways may need to cooperate to yield efficient migration. Blunting the signaling response elicited by occupancy of any one Plg-R may lead to suppressed signaling and diminished migration. Fourth, recruitment into the peritoneum is a temporally extended and multi-step response, and different Plg-Rs may come into play at different times and stages during the response. Hence, difference Plg-Rs may be utilized to achieve specific steps in the recruitment cascade.

6. Summary and Concluding Remarks

In this brief discussion, we have raised the question as to why there are so many Plg-Rs. With so many different receptors frequently, with many of them expressed on the same cell type, it is difficult to envision how the cell would prioritize its utilization among these multiple Plg-R. Affinity differences between Plg-Rs for their Plg and plasmin ligands could distinguish one receptor from another, but this can only be tested by direct comparisons among Plg-Rs. Utilization of specific Plg-Rs to mediate tissue specific or cell specific responses can also be envisioned, but such analyses again mandate comparative studies. In fact, in each of the explanations suggested above, to account for the profound role of many different Plg-Rs in what is globally visualized as a single cellular response, macrophage recruitment into the peritoneum, comparative studies are again needed. The goal of such comparative studies is not to prove that one particular Plg-R is *better* than another, but rather to help dissect the ways in which Plg orchestrates cell migration and other cellular responses in vivo.

Acknowledgments

The authors thank Sidney Jones from Plow Lab for assisting in endothelial experiments. This work was supported by NIH grant HL17964 (E. F. Plow) and an American Heart Association Scientist Development Grant 11SDG7390041(R. Das).

References

[1] E. F. Plow, T. Herren, A. Redlitz, L. A. Miles, and J. L. Hoover-Plow, "The cell biology of the plasminogen system," *FASEB Journal*, vol. 9, no. 10, pp. 939–945, 1995.

[2] R. Das, E. Pluskota, and E. F. Plow, "Plasminogen and its receptors as regulators of cardiovascular inflammatory responses," *Trends in Cardiovascular Medicine*, vol. 20, no. 4, pp. 120–124, 2010.

[3] L. A. Miles, S. B. Hawley, N. Baik, N. M. Andronicos, F. J. Castellino, and R. J. Parmer, "Plasminogen receptors: the sine qua non of cell surface plasminogen activation," *Frontiers in Bioscience*, vol. 10, no. 2, pp. 1754–1762, 2005.

[4] G. M. Cesarman, C. A. Guevara, and K. A. Hajjar, "An endothelial cell receptor for plasminogen/tissue plasminogen activator (t-PA). II. Annexin II-mediated enhancement of t-PA-dependent plasminogen activation," *Journal of Biological Chemistry*, vol. 269, no. 33, pp. 21198–21203, 1994.

[5] A. Redlitz, B. J. Fowler, E. F. Plow, and L. A. Miles, "The role of an enolase-related molecule in plasminogen binding to cells," *European Journal of Biochemistry*, vol. 227, no. 1-2, pp. 407–415, 1995.

[6] M. Kwon, T. J. MacLeod, Y. Zhang, and D. M. Waisman, "S100A10, annexin A2, and annexin A2 heterotetramer as candidate plasminogen receptors," *Frontiers in Bioscience*, vol. 10, no. 1, pp. 300–325, 2005.

[7] E. Pluskota, D. A. Soloviev, K. Bdeir, D. B. Cines, and E. F. Plow, "Integrin $\alpha M\beta 2$ orchestrates and accelerates plasminogen activation and fibrinolysis by neutrophils," *Journal of Biological Chemistry*, vol. 279, no. 17, pp. 18063–18072, 2004.

[8] K. A. Hajjar, A. T. Jacovina, and J. Chacko, "An endothelial cell receptor for plasminogen/tissue plasminogen activator. I. Identity with annexin II," *Journal of Biological Chemistry*, vol. 269, no. 33, pp. 21191–21197, 1994.

[9] K. D. Phipps, A. P. Surette, P. A. O'Connell, and D. M. Waisman, "Plasminogen receptor S100A10 is essential for the migration of tumor-promoting macrophages into tumor sites," *Cancer Research*, vol. 71, no. 21, pp. 6676–6683, 2011.

[10] Q. Ling, A. T. Jacovina, A. Deora et al., "Annexin II regulates fibrin homeostasis and neoangiogenesis in vivo," *Journal of Clinical Investigation*, vol. 113, no. 1, pp. 38–48, 2004.

[11] A. P. Surette, P. A. Madureira, K. D. Phipps, V. A. Miller, P. Svenningsson, and D. M. Waisman, "Regulation of fibrinolysis by S100A10 in vivo," *Blood*, vol. 118, no. 11, pp. 3172–3181, 2011.

[12] P. A. O'Connell, A. P. Surette, R. S. Liwski, P. Svenningsson, and D. M. Waisman, "S100A10 regulates plasminogen-dependent macrophage invasion," *Blood*, vol. 116, no. 7, pp. 1136–1146, 2010.

[13] R. Das, T. Burke, and E. F. Plow, "Histone H2B as a functionally important plasminogen receptor on macrophages," *Blood*, vol. 110, no. 10, pp. 3763–3772, 2007.

[14] R. Das and E. F. Plow, "Phosphatidylserine as an anchor for plasminogen and its plasminogen receptor, Histone H2B, to the macrophage surface," *Journal of Thrombosis and Haemostasis*, vol. 9, no. 2, pp. 339–349, 2011.

[15] R. Lacroix, F. Sabatier, A. Mialhe et al., "Activation of plasminogen into plasmin at the surface of endothelial microparticles: a mechanism that modulates angiogenic properties of endothelial progenitor cells in vitro," *Blood*, vol. 110, no. 7, pp. 2432–2439, 2007.

[16] J. D. Hood and D. A. Cheresh, "Role of integrins in cell invasion and migration," *Nature Reviews Cancer*, vol. 2, no. 2, pp. 91–100, 2002.

[17] H. W. Smith and C. J. Marshall, "Regulation of cell signalling by uPAR," *Nature Reviews Molecular Cell Biology*, vol. 11, no. 1, pp. 23–36, 2010.

[18] M. J. Hayes, D. Shao, M. Bailly, and S. E. Moss, "Regulation of actin dynamics by annexin 2," *EMBO Journal*, vol. 25, no. 9, pp. 1816–1826, 2006.

[19] M. E. Beckner, X. Chen, J. An, B. W. Day, and I. F. Pollack, "Proteomic characterization of harvested pseudopodia with differential gel electrophoresis and specific antibodies," *Laboratory Investigation*, vol. 85, no. 3, pp. 316–327, 2005.

[20] T. Dejouvencel, L. Doeuvre, R. Lacroix et al., "Fibrinolytic cross-talk: a new mechanism for plasmin formation," *Blood*, vol. 115, no. 10, pp. 2048–2056, 2010.

[21] R. Das, T. Burke, D. R. Van Wagoner, and E. F. Plow, "L-type calcium channel blockers exert an antiinflammatory effect by suppressing expression of plasminogen receptors on macrophages," *Circulation Research*, vol. 105, no. 2, pp. 167–175, 2009.

[22] M. Ranson, N. M. Andronicos, M. J. O'Mullane, and M. S. Baker, "Increased plasminogen binding is associated with

metastatic breast cancer cells: differential expression of plasminogen binding proteins," *British Journal of Cancer*, vol. 77, no. 10, pp. 1586–1597, 1998.

[23] P. Burtin, G. Chavanel, J. Andre-Bougaran, and A. Gentile, "The plasmin system in human adenocarcinomas and their metastases. A comparative immunofluorescence study," *International Journal of Cancer*, vol. 39, no. 2, pp. 170–178, 1987.

[24] J. Hoover-Plow and L. Yuen, "Plasminogen binding is increased with adipocyte differentiation," *Biochemical and Biophysical Research Communications*, vol. 284, no. 2, pp. 389–394, 2001.

[25] E. A. Peterson, M. R. Sutherland, M. E. Nesheim, and E. L. G. Pryzdial, "Thrombin induces endothelial cell-surface exposure of the plasminogen receptor annexin 2," *Journal of Cell Science*, vol. 116, no. 12, pp. 2399–2408, 2003.

[26] D. Ratel, S. Mihoubi, E. Beaulieu et al., "VEGF increases the fibrinolytic activity of endothelial cells within fibrin matrices: involvement of VEGFR-2, tissue type plasminogen activator and matrix metalloproteinases," *Thrombosis Research*, vol. 121, no. 2, pp. 203–212, 2007.

[27] L. A. Miles and E. F. Plow, "Binding and activation of plasminogen on the platelet surface," *Journal of Biological Chemistry*, vol. 260, no. 7, pp. 4303–4311, 1985.

[28] H.-S. Kim, E. F. Plow, and L. A. Miles, "Regulation of plasminogen receptor expression on monocytoid THP- 1 cells by adherence to extracellular matrix proteins," *Circulation*, vol. 86, I-140, 1992.

[29] M. Camacho, M. C. Fondaneche, and P. Burtin, "Limited proteolysis of tumor cells increases their plasmin-binding ability," *FEBS Letters*, vol. 245, no. 1-2, pp. 21–24, 1989.

[30] M. Gonzalez-Gronow, S. Stack, and S. V. Pizzo, "Plasmin binding to the plasminogen receptor enhances catalytic efficiency and activates the receptor for subsequent ligand binding," *Archives of Biochemistry and Biophysics*, vol. 286, no. 2, pp. 625–628, 1991.

[31] G. E. Stillfried, D. N. Saunders, and M. Ranson, "Plasminogen binding and activation at the breast cancer cell surface: the integral role of urokinase activity," *Breast Cancer Research*, vol. 9, no. 1, article no. R14, 2007.

[32] M. J. O'Mullane and M. S. Baker, "Elevated plasminogen receptor expression occurs as a degradative phase event in cellular apoptosis," *Immunology and Cell Biology*, vol. 77, no. 3, pp. 249–255, 1999.

[33] P. Svenningsson and P. Greengard, "p11 (S100A10)—an inducible adaptor protein that modulates neuronal functions," *Current Opinion in Pharmacology*, vol. 7, no. 1, pp. 27–32, 2007.

[34] M. Wygrecka, L. M. Marsh, R. E. Morty et al., "Enolase-1 promotes plasminogen-mediated recruitment of monocytes to the acutely inflamed lung," *Blood*, vol. 113, no. 22, pp. 5588–5598, 2009.

[35] S. Lighvani, N. Baik, J. E. Diggs, S. Khaldoyanidi, R. J. Parmer, and L. A. Miles, "Regulation of macrophage migration by a novel plasminogen receptor Plg-R $_{KT}$," *Blood*, vol. 118, no. 20, pp. 5622–5630, 2011.

[36] V. A. Ploplis, E. L. French, P. Carmeliet, D. Collen, and E. F. Plow, "Plasminogen deficiency differentially affects recruitment of inflammatory cell populations in mice," *Blood*, vol. 91, no. 6, pp. 2005–2009, 1998.

α-Enolase, a Multifunctional Protein: Its Role on Pathophysiological Situations

Àngels Díaz-Ramos, Anna Roig-Borrellas, Ana García-Melero, and Roser López-Alemany

Biological Clues of the Invasive and Metastatic Phenotype Research Group, (IDIBELL) Institut d'Investigacions Biomèdiques de Bellvitge, L'Hospitalet de Llobregat, 08908 Barcelona, Spain

Correspondence should be addressed to Roser López-Alemany, rlopez@idibell.cat

Academic Editor: Lindsey A. Miles

α-Enolase is a key glycolytic enzyme in the cytoplasm of prokaryotic and eukaryotic cells and is considered a multifunctional protein. α-enolase is expressed on the surface of several cell types, where it acts as a plasminogen receptor, concentrating proteolytic plasmin activity on the cell surface. In addition to glycolytic enzyme and plasminogen receptor functions, α-Enolase appears to have other cellular functions and subcellular localizations that are distinct from its well-established function in glycolysis. Furthermore, differential expression of α-enolase has been related to several pathologies, such as cancer, Alzheimer's disease, and rheumatoid arthritis, among others. We have identified α-enolase as a plasminogen receptor in several cell types. In particular, we have analyzed its role in myogenesis, as an example of extracellular remodelling process. We have shown that α-enolase is expressed on the cell surface of differentiating myocytes, and that inhibitors of α-enolase/plasminogen binding block myogenic fusion *in vitro* and skeletal muscle regeneration in mice. α-Enolase could be considered as a marker of pathological stress in a high number of diseases, performing several of its multiple functions, mainly as plasminogen receptor. This paper is focused on the multiple roles of the α-enolase/plasminogen axis, related to several pathologies.

1. Introduction

Enolase, also known as phosphopyruvate hydratase, was discovered in 1934 by Lohman and Mayerhof. It is one of the most abundantly expressed cytosolic proteins in many organisms. It is a key glycolytic enzyme that catalyzes the dehydratation of 2-phosphoglycerate to phosphoenolpyruvate, in the last steps of the catabolic glycolytic pathway [1] (Figure 1). It is a metalloenzyme that requires the metal ion magnesium (Mg^{2+}) to be catalytically active. Enolase is found from archaebacteria to mammals, and its sequence is highly conserved [2]. In vertebrates, the enzyme occurs as three isoforms: α-enolase (*Eno1*) is found in almost all human tissues, whereas β-enolase (*Eno3*) is predominantly found in muscle tissues, and γ-enolase (*Eno2*) is only found in neuron and neuroendocrine tissues [3]. The three enolase isoforms share high-sequence identity and kinetic properties [4–6]. Enzymatically active enolase which exists in a dimeric

(homo- or heterodimers) form is composed of two subunits facing each other in an antiparallel fashion [6, 7]. The crystal structure of enolase from yeast and human has been determined and catalytic mechanisms have been proposed [8–10].

Although it is expressed in most of the cells, the gene that encodes enolase is not considered a housekeeping gene since its expression varies according to the pathophysiological, metabolic, or developmental conditions of cells [11]. α-Enolase mRNA translation which is primarily under developmental control is significantly upregulated during cellular growth and practically undetectable during quiescent phases [12, 13].

Recent accumulation of evidence revealed that, in addition to its innate glycolytic function, α-enolase plays an important role in several biological and pathophysiological processes: by using an alternative stop codon, the α-enolase mRNA can be translated into a 37 kDa protein which lacks

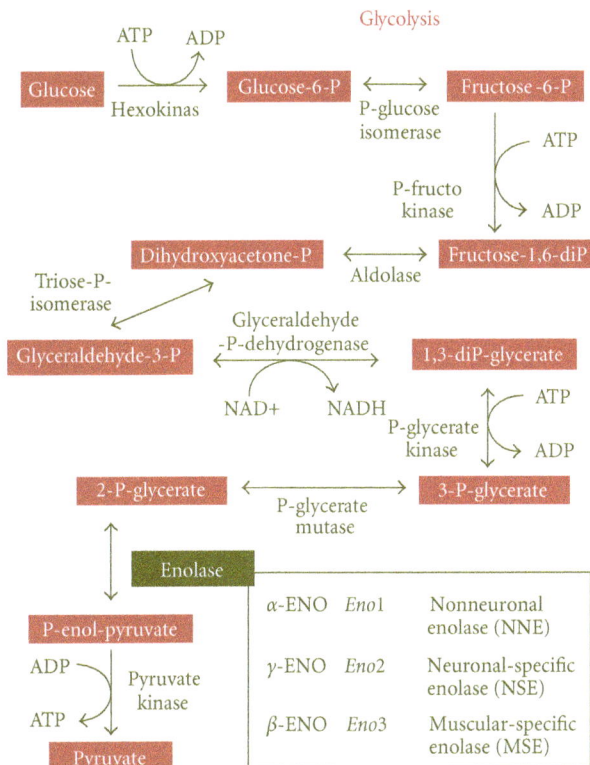

FIGURE 1: Summary of glycolytic metabolic pathway. Metabolic chain reactions of glycolysis, the central pathway for the catabolism of carbohydrates that takes place in the cytoplasm of almost all prokaryotic and eukaryotic cells. The insert shows different enolase isoenzymes in vertebrates.

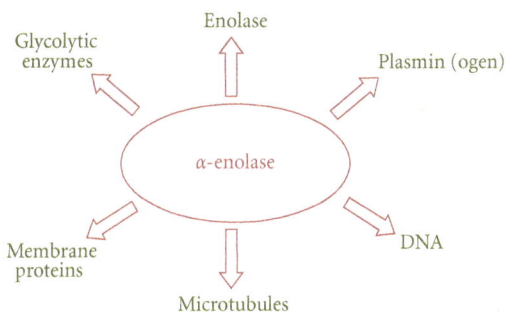

FIGURE 2: Interaction of α-enolase with other nuclear, cytoplasmic, or membrane molecules. α-Enolase can directly interact with other enolase isoforms (α, β, and γ) to form enzymatically active dimers, other glycolytic enzymes as pyruvate kinase, phosphoglycerate mutase and aldolase. It can also bind to microtubules network proteins, as F-actin and tubulin, and it is localized on the cell surface, interacting with other membrane proteins, where it binds to plasminogen and plasmin.

the first 96 amino acid residues. This protein, named c-myc promoter-binding protein 1 (MBP-1) is localized in the nucleus and can bind to the c-myc P2 promoter and negatively regulates transcription of the protooncogene [14]. α-Enolase has been detected on the surface of hematopoietic cells such as monocytes, T cells and B cells, neuronal cells,

and endothelial cells as a strong plasminogen receptor, modulating pericellular fibrinolytic activity. The expression of α-enolase on the surface of a variety of eukaryotic cells has been found to be dependent on the pathophysiological conditions of these cells [15–19].

α-Enolase has also been described as a neurotrophic factor [20], a heat-shock protein (HSP48) [21], and a hypoxic stress protein [22]. Furthermore, α-enolase is part of the crystallin lens of vertebrates [23], binds to fragments of F-actin and tubulin [24], and has been detected associated to centrosomes in HeLa cells [25]. α-Enolase also binds with high affinity to other glycolytic enzymes: pyruvate kinase, phosphoglycerate mutase, which are adjacent to enolase in the glycolytic pathway, and to aldolase, which is known to associate with cytoskeletal proteins [26] (Figure 2).

It has also been suggested that upregulation of α-enolase contributes to hypoxia tolerance through nonglycolytic mechanisms [27]. Increased expression of α-enolase has been reported to correlate with progression of tumors, neuroblastoma, and lung cancer, and enolase has been considered to be a potential diagnostic markers for many tumors [28–32].

Thus, α-enolase appears to be a "moonlighting protein," one of a growing list of proteins that are recognized as identical gene products exhibiting multiple functions at distinct cellular sites through "gene sharing" [33, 34]. This paper is focused on the multiple roles of the α-enolase/plasminogen axis, related to several pathologies.

2. The Plasminogen Activation System

In multicellular organisms, extracellular proteolysis is important to many biological processes involving a dynamic rearrangement of cell-cell and cell-matrix interactions, being the plasminogen activation (PA) system among the most important extracellular proteases. The PA system comprises an inactive proenzyme, plasminogen, and ubiquitous in body fluid, that can be converted into the active enzyme, plasmin, by two physiological activators (PAs): tissue-type plasminogen activator (tPA) and urokinase-type plasminogen activator (uPA). Inhibition of the plasminogen system occurs at the level of the PA, by specific inhibitors (PAI-1 and PAI-2), or at the level of plasmin, by α2-antiplasmin (reviewed in [35]). The PA/plasmin system is a key regulator in extracellular matrix (ECM) remodeling directly by its ability to degrade ECM components, such as laminin or fibronectin, and indirectly via activation of matrix metalloproteinases (MMPs), which will degrade collagen(s) subsequently. Furthermore, plasmin is able to activate latent growth factors, such as transforming growth factor β (TGFβ) and basic fibroblast growth factor (bFGF) (reviewed in [35]).

Work from numerous groups has clearly demonstrated that the localization of plasminogen and its activators uPA and tPA on the cell surface, though association to specific cell membrane receptors, provides a mechanism for cells to harness and regulate the activities of these proteases [36, 37]. Binding sites for plasminogen, tPA, and uPA have been identified on a variety of cell types, including monocytes, fibroblasts, and endothelial cells [38, 39]. uPA is recruited

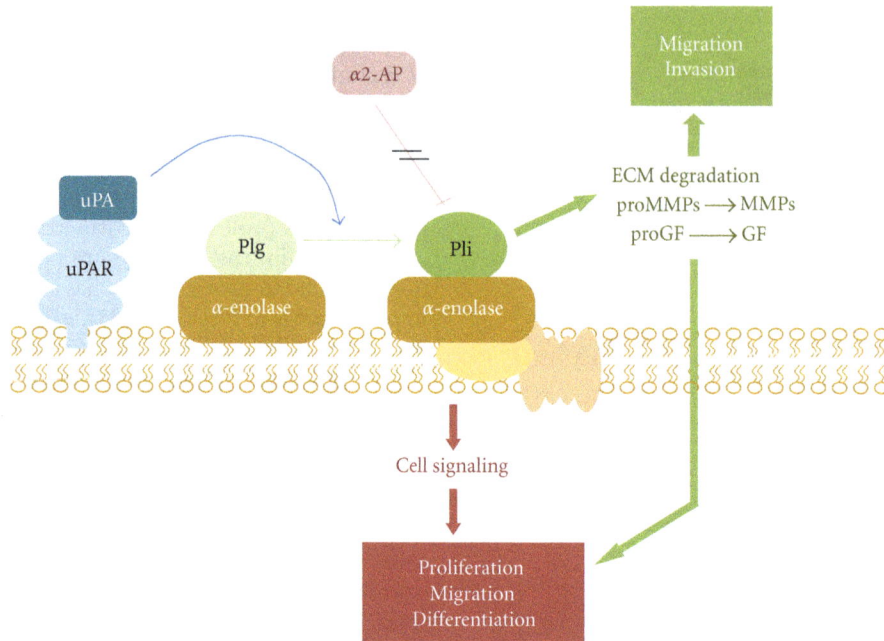

FIGURE 3: Schematic overview represents α-enolase/plasminogen interaction on the cell surface. α-Enolase enhances plasminogen activation on the cell surface, concentrates plasmin proteolytic activity on the pericellular area and protects plasmin from its inhibitor α2-antiplasmin. Once activated, plasmin can degrade most of the components of the extracellular matrix, directly or indirectly by activating metalloproteases. It is also capable to activate prohormones of progrowing factors. Abbreviations: Plg, plasminogen; Pli, plasmin, α2-AP, α2-antiplasmin; uPA, urokinase-type plasminogen activator, uPAR, urokinase-type plasminogen activator; ECM, extracellular matrix; MMPs, metalloproteases; GF, growing factors.

to the cell membrane immediately after its secretion via a specific uPA receptor (uPAR, CD87), expressed on the cell surface, that localize extracellular proteolysis and induces cell migration, cell adhesion, and proliferation (reviewed in [40, 41]).

Described binding sites for plasminogen include α-enolase [18, 42], annexin A2 [43], p11 [44], histone H2B [45, 46], actin [47], gp330 [48], cytokeratin 8 [49], histidine-proline rich glycoprotein [50], glyceraldehide-3-phosphate dehydrogenase [51] gangliosides [18], and Plg-R_{TK} [52]. α-Enolase and most of these proteins have C-terminal lysines predominantly responsible for plasminogen binding/activation [53]. Notably, most of these proteins have other described functions than plasminogen receptors, and lack a transmembrane domain, Plg-R_{TK} being an exception, as it is a transmembrane receptor [52].

3. α-Enolase as a Plasminogen Receptor

We and others have previously identified α-enolase as a plasminogen receptor on the surfaces of several diverse cell types including carcinoma cells [42], monocytoid cells [15, 18], leukocytic cell lines [54], rat neuronal cells [16], and pathogenic streptococci [1].

On the cell surface, interaction of plasminogen with α-enolase enhances its activation by PAs, concentrates protease activity pericellularly [55–57], and protects plasmin from inhibition by α2-antiplasmin [18, 58] (Figure 3). In order to examine the role of α-enolase in the pericellular

generation of plasmin activity, we produced a monoclonal antibody, MAb11G1, that specifically blocked plasminogen binding to purified α-enolase [54]. MAb11G1 allowed us to demonstrate that α-enolase occupancy by plasminogen on leukocytoid cells and on peripheral blood neutrophils is required for pericellular plasminogen activation and plasmin generation [54].

Considering the extraordinarily high number of plasminogen binding sites/cells that have been described in different cell types, no single surface protein can account for all plasminogen binding sites, suggesting that different receptors coexist on the cell surface [18]. Evidence from monocytoid cells suggested that α-enolase was only one of several plasminogen receptors and its contribution to plasmin activation was only modest [18, 54]. Posterior studies have emphasized the role of annexin A2 and histone H2B as plasminogen receptors in the same cells [46], suggesting a minor contribution of α-enolase as plasminogen receptor. In more recent studies, the role of α-enolase has been resurrected, showing a central role for α-enolase in monocyte recruitment in inflammatory lung disease [59]. These results imply that different plasminogen receptors could be targeted to regulate inflammatory cell recruitment in a temporal-specific manner.

The α-enolase-plasminogen interaction is mediated by binding of plasminogen kringle domains to the C-terminal residues of α-enolase (K_{434}) [15, 18]. Furthermore, interaction of plasminogen lysine binding sites with α-enolase depends upon recognition of C-terminal lysines K_{420}, K_{422}, and K$_{434}$, suggesting that amino acid residues upstream and/or secondary structure may be responsible for the high affinity of α-enolase for plasminogen [15, 18]. Another putative plasminogen-binding motif has been proposed in view of its crystal structure at position, $_{250}$FFRSGKY$_{256}$, that remains exposed when α-enolase forms a dimer, necessary for its glycolytic activity [10]. Human α-enolase structure has been determined and it has been found that it exhibits specific surface properties that are distinct from those of other enolases despite high-sequence similarity. These differences in structure explain its various activities, including plasmin(ogen) and DNA binding [10].

The mechanism by which α-enolase, that lacks a signal sequence, is associated with the cell membrane remains unknown. Some authors have speculated that a hydrophobic domain within α-enolase might serve as an internal signal sequence [60], while others suggest that posttranslational acetylation [61] or phosphorylation [62] may control membrane association. Nevertheless, α-enolase forms part of a growing list of proteins that lack signal sequences, but are transported to the cell surface by a yet unknown mechanism.

4. α-Enolase in Myogenesis and Muscle Regeneration

Proteolysis associated with the cell surface is a usual mechanism in several physiological processes involving tissue remodeling. Myogenesis is an example of tissue remodeling in which massive extracellular matrix degradation takes place. Components of the PA system play important, yet distinct roles in muscle regeneration after injury. Using genetically modified mice for uPA and plasminogen, we and others have shown that loss of uPA-mediated plasmin activity blunts muscle repair in vivo [63–66]. In contrast, a negative role for PAI-1 in muscle regeneration was suggested [65]. The PA system has also been shown to have an increasingly important role in muscular dystrophies. For example, greater expression of uPA has been found in mdx muscle, the mouse model for Duchenne muscle dystrophy (DMD). Conversely, genetic loss of uPA exacerbated dystrophy and reduced muscle function in mdx mice [66]. Satellite cells derived from human DMD patients produce more uPAR and PAI-1 and less uPA than normal satellite cells [67]. uPA and plasmin appear to be required for infiltration of macrophages into the damaged or dystrophic muscle in mdx mice. However, an interesting observation underpinning these results was that genetic loss of the uPAR in mdx mice failed to exacerbate muscular dystrophy, suggesting that uPA exerts its proteolytic effects independently of its cell surface receptor uPAR [66].

β-enolase is considered the specific muscular enolase isoform, it is expressed in proliferating adult myoblasts as well as in differentiated myotubes [68]. It is upregulated in muscle during embryogenic development and it is considered an early marker of myogenesis [69]. The increase of the β-isoform is accompanied by a decrease of the α and γ isoform; the γ-isoform is completely absent in the adult muscle, but the expression of the α-isoform is maintained in the adult muscle and in muscular cells [70, 71]. Furthermore, we have described that α-enolase is upregulated in murine myoblasts C2C12 differentiation in vitro and in muscle regeneration in vivo [72], thus raising the question of whether plasminogen receptors may also function in myogenesis and skeletal regeneration as a mechanism for regulating plasmin activity.

We have investigated the role of α-enolase plasminogen receptor in muscle regeneration after injury, a process involving extensive cell infiltration and ECM remodeling. Injured wild-type mice and dystrophic mdx mice were treated with inhibitors of α-enolase/plasminogen binding: MAb11G1 (an inhibitory monoclonal antibody against α-enolase) and ε-aminocaproic acid (EACA, a lysine analogue). These treatments had negative impacts on muscle repair by impairing adequate inflammatory cell infiltration and promoting extracellular matrix deposition, which resulted in persistent degeneration. Furthermore, satellite cell-derived myoblasts (i.e., MPCs) expressed α-enolase on the cell surface, and this expression was upregulated during myogenic differentiation, correlating with an increase of plasminogen binding to the cell surface. We found that both MAb11G1 and EACA treatments impaired satellite cell-derived myoblasts functions in vitro in agreement with blunted growth of new myofibers in vivo (Diaz-Ramos et al., unpublished results).

Loss of uPAR in vivo did not affect the degeneration/regeneration process; in addition, cultured myoblasts from uPAR-deficient mice showed efficient myoblast differentiation and fusion [66, 73], indicating that uPAR is dispensable for efficient muscle repair. This reinforces the idea that α-enolase is the main functional plasminogen receptor during muscle tissue remodeling. Altogether, these results demonstrate the novel requirement of α-enolase for restoring homeostasis of injured muscle tissue, by concentrating plasmin activity on the cell surface of inflammatory and myogenic cells.

5. The α-Enolase Expression in Injured Cardiac Muscle

The actuation of the PA system in tissue healing after a cardiac failure, driving the degradation the ECM and scar tissue after an ischemic injury and allowing the inflammatory cell invasion, has been extensively demonstrated.

The regulation of α-enolase in cardiac tissue as regulator of glucose metabolism has been analyzed by several authors. A decrease of α-enolase expression in the aging heart of old male monkeys has been described, paralleling left ventricular dysfunction, and could be involved in the mechanism for the cardiomyopathy of aging [74]. α-Enolase expression has been identified as a strongly induced factor in response to ischemic hypoxia and reoxygenation in rat hearts subjected to ischemia-reperfusion [75]. Furthermore, α-Enolase improved the contractility of cardiomyocytes impaired by ischemic hypoxia [76]. α-Enolase has also been proposed as a marker for early diagnosis for acute myocardial infarction [77].

On the other hand, recent evidences indicates an involvement of proteinases, including the PAs and MMPs systems, in the process of extracellular matrix degradation and cell migration during cardiac wound healing [78]. In a recent study, Heymans et al. demonstrated that uPA-deficient mice showed impaired infarct healing and were completely protected against cardiac rupture after induction of a myocardial infarction [79]. Wound healing after infarct was abolished in plasminogen-deficient mice, indicating that the plasminogen system is required for the repair process of the heart after infarction. In the absence of plasminogen, inflammatory cells did not migrate into the infarcted myocardium, necrotic cardiomyocytes were not removed and there was no formation of granulation tissue and fibrous tissue [80]. Furthermore, PAI-1, which has been shown to be expressed in mammalian cardiomyocytes [81], has been implicated in the process of the cardiac remodeling by inhibiting activation of MMPs as well as plasmin generation. A dramatic induction of PAI-1 in a mouse model of infarct has been described [82]. Experiments using mice deficient in PAI-1 suggest that increased expression of cardiac PAI-1 may contribute to the development of fibrous change after acute myocardial infarction (AMI). *In vivo* studies also showed that PAI-1 expression was induced in hearts under pathological conditions as ventricular hypertrophy [83].

All these results demonstrate that the PA system plays a role in ECM remodeling after a cardiac injury and allows inflammatory cell invasion. Furthermore, it can also play a role in cardiomyocyte survival. Cardiomyocytes, which are terminally differentiated cells, cannot proliferate, even when they are damaged; the damage can lead to cell death in the case of serious diseases such as acute myocardial infarction and myocarditis [84]. Recent studies have identified myocyte apoptosis in the failing human heart [85, 86]. Plasminogen could also drive cardiomyocyte apoptosis, because plasmin induces cell detachment and apoptosis of smooth muscle cells through its binding to the cell surface, although the receptor responsable for plasminogen binding has not yet been identified [87].

Knowing that the PA system has been associated with cardiac remodeling, and that α-enolase is upregulated in cardiac infarction, it is tempting to speculate that α-enolase could act as plasminogen receptor, regulating PA activity on cardiac cells. Previous results from our laboratory have shown that plasmin activity is concentrated on the cell surface of cardiac fibroblasts in a lysine-dependent manner, and this binding capacity is increased by hypoxic conditions. Furthermore, plasminogen binding drives the activation of fibroblasts to myofibroblasts, the main cells responsible of tissue remodeling after a cardiac injury (Garcia-Melero et al., unpublished results).

6. α-Enolase/Plasmin Role in Apoptosis

It has been described that plasminogen binding to the cell surface and its further activation to plasmin induces cell detachment and apoptosis in smooth muscle cells, neurons and vascular myofibroblasts [88–90], although the molecular responsible for plasminogen interaction with the cell surface has not been identified.

Externalization of glycolytic enzymes is a common and early aspect of cell death in different cell types triggered to die with different suicidal stimuli [91]. Apoptotic cells are recognized by phagocytes and trigger an active immuno-suppressive response. The lack of inflammation associated normally with the clearance of apoptotic cells has been linked to inflammatory and autoimmune disease as systemic lupus erythematosis and rheumatic diseases [92–95]. Regarding apoptotic cell surface proteins, a new concept has been defined, SUPER, referring to Surface-exposed (during apoptotic cell death), Ubiquitously expressed, Protease sensitive, Evolutionary-conserved, and Resident normally in viable cells (SUPER), to emphasize defining properties of apoptotic determinants for recognition and immune modulation. Ucker et al. have recently demonstrated that almost all members of the glycolytic pathway are enriched among apoptotic cell membranes, with α-enolase being the more abundant enzyme in the cell membrane, and considered the most paradigmatic SUPER protein [91]. In the cell membrane of apoptotic cells, α-enolase has lost its glycolytic activity, but it acts as plasminogen receptor, coinciding with the description of the association of plasminogen binding with apoptotic cell death [96]. In contrast to α-enolase, other molecular plasminogen receptors as annexin A2 [97] or H2B [46], were not preferentially enriched on the apoptotic cell surface.

7. α-Enolase in Cancer

Several reports have shown an upregulation of α-enolase in several types of cancer [98–100]. The role of α-enolase as a plasminogen receptor on cancer cells has been extensively documented, where it acts as a key protein, promoting cellular metabolism in anaerobic conditions, and driving tumor invasion through plasminogen activation and ECM degradation (reviewed in [101]).

Recently, an analysis of disease-specific gene network identified desmin, interleukin 8, and α-enolase as central elements for colon cancer tumorogenesis [102]. Knockdown of α-enolase expression in different tumor cell lines caused a dramatic increase in their sensitivity to microtubule targeted drugs (e.g., taxanes and vincristine), probably due to α-enolase-tubulin interactions [103], suggesting a role for α-enolase in modulating the microtubule network. Downregulation of α-enolase gene product deceased invasiveness of the follicular thyroid carcinoma cell lines [104]. α-Enolase overexpression has been associated with head and neck cancer cells, and this increase associated not only with cancer progression but also with poor clinical outcomes. Furthermore, exogenous α-enolase expression promoted cell proliferation, migration, invasion, and tumorogenesis [105].

During tumor formation and expansion, tumor cells must increase glucose metabolism [106]. Hypoxia is common feature of solid tumors. Consistent with this, overexpression of glycolytic genes has been found in a myriad of human cancers [107]. In tumor cells, α-enolase is upregulated and supports anaerobic proliferation (Warburg effect), and it is expressed on the cell surface, where it promotes cancer invasion. Thus, it seems that α-enolase is playing a pleitropic role on cancer cell progression. Furthermore, it has been demonstrated that α-enolase is upregulated in pancreatic ductal adenocarcinoma, where it is subjected to a array of posttranslational modifications, namely acetylation, methylation, and phosphorylation [108]. Both, α-enolase expression and posttranslational modifications could be of diagnostic and prognostic value in cancer (reviewed in [101]).

8. Posttranslational Modifications of α-Enolase

Posttranslational protein modifications, such as phosphorylation, acetylation, and methylation are common and important mechanisms of acute and reversible regulation of protein function in mammalian cells, and largely control cellular signaling events that orchestrate biological functions. Several posttranslational modifications have been described for α-enolase. α-Enolase phosphorylation has been associated with pancreatic cancer, and induces specific autoantibody production in pancreatic ductal adenocarcinoma patients with diagnostic value [109]. Lysine acetylated α-enolase has been detected in mouse brain [110]. Nitration of tyrosine residues in α-enolase has been detected in diabetic rat hearts, contributing to the impaired glycolytic activity in diabetic cardiomyopathy [111]. Phosphorylated α-enolase has been detected in gastrocnemious muscle, and phosphorylation decreased with age [112]. Furthermore, carbonylation of α-enolase has been detected on human myoblasts under oxidative stress [113].

It remains to be determined how the posttranslational modifications of α-enolase can affect its catalytic activity, localization of the cell, protein stability, and the ability to dimerize or form a complex with other molecules. Investigations of these modifications patterns in different pathologies will provide insights into its important role in pathophysiological processes.

9. α-Enolase in Rheumatoid Arthritis

The overexpression of α-enolase has also been found associated with chronic autoimmune diseases like rheumatoid arthritis [19, 114], systemic sclerosis [115], and primary nephropathies [116]. Autoantibodies to α-enolase, are present in the sera of patients with very early rheumatoid arthritis and have potential diagnostic and prognostic value [117]. Recently, citrullinated proteins have been considered the main autoantigen of rheumatoid arhritis. Citrullination, also termed deimination, is a modification of arginine side chains catalyzed by peptidylarginine deaminase. This posttranscriptional modification has the potential to alter the structure, antigenicity, and function of proteins. α-Enolase is abundantly expressed in the sinovial membrane, and antibodies against citrullinated α-enolase were specific for rheumatoid arthritis. Citrullination changes the conformation of α-enolase and interferes with the noncovalent interaction involved in the formation of the enolase dimer, then results in an altered glycolytic activity and plasminogen binding. It is likely that citrullination of cell-surface α-enolase abrogates its plasminogen binding and activating function and contributes to the decreased fibrinolysis observed in rheumatoid arthritis [118]. Curiously, other glycolytic enzymes such as glucose phosphate isomerase and aldolase also promote rheumatoid arthritis autoimmunity by acting as autoantigens [119].

10. α-Enolase in Alzheimer's Disease

Although γ-enolase is the specific neuronal enolase isoform, α-isoform is also present in neurological tissues. Plasmin formation enhanced by α-enolase has been proposed to enhance neuritogenesis [16, 120]. Furthermore, cathepsin X cleavage of C-terminal lysine of α-enolase impaired survival and neuritogenesis of neuronal cells [121]. α-Enolase has been reported as a strong plasminogen receptor within the brain; it is known to be upregulated in the Alzheimer's disease brain and has been proposed as a promising therapeutic target for this disease (reviewed in [122]). Glucose hypometabolism and upregulation of glycolytic enzymes is a predominant feature in Alzheimer's disease [123], but accumulating results suggest that α-enolase may have other functions that just metabolic processing of glucose: plasminogen bound to α-enolase stimulates plasmin activation of mitogen-activated protein kinase (MAPK)/extracellular-signal regulated kinase 1/2 (ERK1/2) prosurvival factor and also can drive plasmin degradation of amyloid-β (Aβ) protein, the main component of amyloid plaques. Thus, α-enolase might play a neuroprotective role through its multiple functions (reviewed in [122]).

Recently, several posttranslational modifications to α-enolase have been found in Alzheimer's disease. Elevated levels of glycosylated-α-enolase [124], oxidized [123], or glutathionylated [125] have been found related to Alzheimer's disease. These modifications would render enolase catalyticaly inactive, related to the metabolic deficit associated to Alzheimer's disease. The effect of these modifications in other multiple functions of α-enolase is a subject of ongoing experiments, but it is possible that α-enolase modifications alter not only glucose metabolism, but also its role as plasminogen receptor, controlling neuronal survival and Aβ degradation.

11. Plasmin and Intracellular Signaling

Other than its role in concentrating proteolytic activity on the cell surface, several recent studies have shown that plasmin is able to activate several intracellular signaling pathways, that led to the activation of several transcription factors, in a cell surface binding dependent way. In most of the cases, the molecular mechanism responsible remains unknown: it could be due to the proteolytic activation of a second factor or due to direct binding of plasmin(ogen) to a specific receptor. Several pieces of work show that the plasmin proteolytic activity is essential for the induction of an intracellular response, as in monocytes, where plasmin bound to the cell surface proteolytically activates annexin A2 and stimulates MMP-1 production through the activation of ERK and p38 pathways [126]. The phosphorylation of Janus Kinase 1 (JAK1)/Tyrosin Kinase 2 (TYK2) that drives to the activation to the transcription factors AP-1 and Nuclear Factor κB (NFκB), and the expression of several cytokines: interleukin-1α and -1β (IL-1α and IL-1β), tissue factor (TF), and the Tumoral Necrosis Factor-α (TNF-α), are a consequence of plasmin interaction with the cell surface [127–129]. Plasmin promotes p38 and p44/42 MAPK activation and fibroblast proliferation through Protease Activated Receptor-1 (PAR-1) [130, 131]. Other authors have described that plasminogen and plasmin regulate the gene transcription of genes as *c-fos, erg-1,* and *Eno1* in mononucleated blood cells and fibroblasts, by activating the MEK/ERK pathways [132, 133].

In most of the cases, the receptor responsible for this cellular response remains to be identified. Most of the protein candidates for plasminogen receptors are small proteins that lack a transmembrane domain and are not able to induce directly an intracellular response. Some work suggests an association between the plasminogen receptor and other membrane proteins, that could serve as molecular collaborators to induce the activation of intracellular signaling pathways. Several proteins have been identified as such molecular collaborators. For instance, plasmin can activate PAR-1 in fibroblasts, by the phosphorylation of Erk [130]; plasminogen and plasmin activate the expression of several genes in fibroblasts and monocytes through G-Protein Coupled Receptors, (GPCR) [132, 133]; some integrins such as $\alpha9\beta1$ integrin in Chinese Hamster Ovary (CHO) cells [134] and $\alpha v\beta3$ integrin, in vascular endothelial cells [135], participate actively in plasmin-induced cell migration.

In none of these cases, the plasmin receptor associated with these proteins have been identified. Some work have identified annexin A2 as the receptor that concentrates plasmin activity to the cell surface and drives a subsequent intracellular response [127–129]. Other authors have described a collaboration between α-enolase and GPCR in fibroblasts and mononucleated blood cells [132, 133]. Plasmin induces smooth muscle cell proliferation through extracellular transactivation of the epidermal growth factor receptor (EGFR) by a MMP-mediated, heparin binding—epidermal growth factor (HB-EGF-) dependent process [136]. Future studies will be necessary to determine the molecular mechanism of the plasminogen receptor on several cell types and the putative proteins associated with it.

We have shown that plasmin activity is able to activate MAPK/ERK and phosphatidyl-inositol 3-kinase (PI3K)/Akt pathways in C2C12 murine myoblast cell lines and in primary cultures of muscle precursor cells, and that intracellular activation depends on plasmin activity, but also on plasmin(ogen) binding to the cell surface in a lysine binding sites dependent way (Roig-Borrellas et al., unpublished results), although the receptor responsible and the molecular mechanism remains to be elucidated.

12. Concluding Remarks

Recently, a proteomic meta-analysis of 169 published articles, including differently expressed 4700 proteins, based on 2-dimensional electrophoresis analysis of human, mouse, and rat tissues, identified α-enolase as the first protein differentially expressed in mice and the second in human pathologies, regardless of the tissue used and experiment performed [137], suggesting that α-enolase could be part of a group of universal cellular sensors that respond to multiple different stimuli. Thus, α-enolase could be considered as a marker of pathological stress in a high number of diseases. The importance of α-enolase as plasminogen receptor has been determined in several pathologies such as cancer, skeletal myogenesis, Alzheimer's disease, and rheumatoid arthritis, among others. α-Enolase upregulation has also been described in a myriad of other pathologies, as inflammatory bowel disease [138, 139], autoimmune hepatitis [140], or membranous glomerulonephritis [141], not discussed in this paper, although its role on concentrating plasmin activity on the cell surface has not always been stablished. It will not be surprising that in many of these pathologies, α-enolase could exert one of its multiple functions, mainly as a plasminogen receptor, focalizing plasmin activity on the cell membrane and promoting ECM degradation/remodeling, but also activating intracellular survival pathways and controlling survival/apoptosis of cells.

Further studies of posttranslational modifications of α-enolase and its implications on α-enolase subcellular distribution and function, especially interaction with other proteins will be necessary. Also, the role of α-enolase as activator of intracellular signaling pathways, probably in collaboration with other membrane proteins, will serve to elucidate the multiples roles of this functionally complex protein.

Unexpectedly, other glycolytic enzymes have been described as having other nonglycolytic functions in transcriptional regulation (hexokinase-2, HK; lactate dehydrogenase A, LDH; glyceraldehydes-3-phosphate dehydrogenase, GAPDH), stimulation of cell motility (glucose-6-phosphate isomerase), and regulation of apoptosis (glucokinase, HK and GAPDH), indicating that they are more complicated, multifunctional proteins rather than simply components of the glycolytic pathway (reviewed in [142]).

Some of the more interesting and challenging issues, regarding α-enolase multifunction that need to be addressed are (i) the mechanism of its export to the cell surface, (ii) the role of α-enolase as an inductor of intracellular signaling pathways, and (iii) the role of posttranslational modifications of α-enolase and implications on its subcellular distribution and function. Investigations of these subjects in different human pathologies will provide insights into its important role on pathophysiological processes and it would make this protein an interesting drug target for different diseases.

Acknowledgments

This study was supported by the Ministerio de Ciencia e Innovación, Spain (SAF2004-04717, SAF2007-63596) and Association Française contre les Myopathies, France (AFM 9869). A. Díaz-Ramos was supported by an AFM predoctoral fellowship. A. Roig-Borrellas was supported by a F.P.U. Fellowship. A. García-Melero was supported by a Marató-TV3 Foundation fellow. A. Roig-Borrellas is a registered student in the PhD Doctorate Program in Biochemistry, Molecular Biology and Biomedicine of the Universitat Autònoma de Barcelona (UAB), Spain.

References

[1] V. Pancholi, "Multifunctional α-enolase: its role in diseases," *Cellular and Molecular Life Sciences*, vol. 58, no. 7, pp. 902–920, 2001.

[2] M. Piast, I. Kustrzeba-Wójcicka, M. Matusiewicz, and T. Banaś, "Molecular evolution of enolase," *Acta Biochimica Polonica*, vol. 52, no. 2, pp. 507–513, 2005.

[3] P. J. Marangos, A. M. Parma, and F. K. Goodwin, "Functional properties of neuronal and glial isoenzymes of brain enolase," *Journal of Neurochemistry*, vol. 31, no. 3, pp. 727–732, 1978.

[4] S. Feo, D. Oliva, G. Barbieri, W. Xu, M. Fried, and A. Giallongo, "The gene for the muscle-specific enolase is on the short arm of human chromosome 17," *Genomics*, vol. 6, no. 1, pp. 192–194, 1990.

[5] A. Giallongo, S. Feo, R. Moore, C. M. Croce, and L. C. Showe, "Molecular cloning and nucleotide sequence of a full-length cDNA for human α enolase," *Proceedings of the National Academy of Sciences of the United States of America*, vol. 83, no. 18, pp. 6741–6745, 1986.

[6] L. Fletcher, C. C. Rider, and C. B. Taylor, "Enolase isoenzymes. III. Chromatographic and immunological characteristics of rat brain enolase," *Biochimica et Biophysica Acta*, vol. 452, no. 1, pp. 245–252, 1976.

[7] K. Kato, Y. Okagawa, F. Suzuki, A. Shimizu, K. Mokuno, and Y. Takahashi, "Immunoassay of human muscle enolase subunit in serum: a novel marker antigen for muscle diseases," *Clinica Chimica Acta*, vol. 131, no. 1-2, pp. 75–85, 1983.

[8] L. Lebioda and B. Stec, "Crystal structure of enolase indicates that enolase and pyruvate kinase evolved from a common ancestor," *Nature*, vol. 333, no. 6174, pp. 683–686, 1988.

[9] E. Zhang, J. M. Brewer, W. Minor, L. A. Carreira, and L. Lebioda, "Mechanism of enolase: the crystal structure of asymmetric dimer enolase- 2-phospho-D-glycerate/enolase-phosphoenolpyruvate at 2.0 A resolution," *Biochemistry*, vol. 36, no. 41, pp. 12526–12534, 1997.

[10] H. J. Kang, S. K. Jung, S. J. Kim, and S. J. Chung, "Structure of human α-enolase (hENO1), a multifunctional glycolytic enzyme," *Acta Crystallographica D*, vol. 64, no. 6, pp. 651–657, 2008.

[11] L. McAlister and M. J. Holland, "Targeted deletion of a yeast enolase structural gene. Identification and isolation of yeast enolase isozymes," *Journal of Biological Chemistry*, vol. 257, no. 12, pp. 7181–7188, 1982.

[12] J. P. Holland, L. Labieniec, C. Swimmer, and M. J. Holland, "Homologous nucleotide sequences at the 5' termini of messenger RNAs synthesized from the yeast enolase and glyceraldehyde-3-phosphate dehydrogenase gene families. The primary structure of a third yeast glyceraldehyde-3-phosphate dehydrogenase gene," *Journal of Biological Chemistry*, vol. 258, no. 8, pp. 5291–5299, 1983.

[13] A. Giallongo, D. Oliva, L. Calì, G. Barba, G. Barbieri, and S. Feo, "Structure of the human gene for alpha-enolase," *European Journal of Biochemistry*, vol. 190, no. 3, pp. 567–573, 1990.

[14] S. Feo, D. Arcuri, E. Piddini, R. Passantino, and A. Giallongo, "ENO1 gene product binds to the c-myc promoter and acts as a transcriptional repressor: relationship with Myc promoter-binding protein 1 (MBP-1)," *FEBS Letters*, vol. 473, no. 1, pp. 47–52, 2000.

[15] L. A. Miles, C. M. Dahlberg, J. Plescia, J. Felez, K. Kato, and E. F. Plow, "Role of cell-surface lysines in plasminogen binding to cells: identification of α-enolase as a candidate plasminogen receptor," *Biochemistry*, vol. 30, no. 6, pp. 1682–1691, 1991.

[16] K. Nakajima, M. Hamanoue, N. Takemoto, T. Hattori, K. Kato, and S. Kohsaka, "Plasminogen binds specifically to α-enolase on rat neuronal plasma membrane," *Journal of Neurochemistry*, vol. 63, no. 6, pp. 2048–2057, 1994.

[17] A. K. Dudani, C. Cummings, S. Hashemi, and P. R. Ganz, "Isolation of a novel 45 kDa plasminogen receptor from human endothelial cells," *Thrombosis Research*, vol. 69, no. 2, pp. 185–196, 1993.

[18] A. Redlitz, B. J. Fowler, E. F. Plow, and L. A. Miles, "The role of an enolase-related molecule in plasminogen binding to cells," *European Journal of Biochemistry*, vol. 227, no. 1-2, pp. 407–415, 1995.

[19] P. A. Fontan, V. Pancholi, M. M. Nociari, and V. A. Fischetti, "Antibodies to streptococcal surface enolase react with human α-enolase: implications in poststreptococcal sequelae," *Journal of Infectious Diseases*, vol. 182, no. 6, pp. 1712–1721, 2000.

[20] N. Takei, J. Kondo, K. Nagaike, K. Ohsawa, K. Kato, and S. Kohsaka, "Neuronal survival factor from bovine brain is identical to neuron-specific enolase," *Journal of Neurochemistry*, vol. 57, no. 4, pp. 1178–1184, 1991.

[21] H. Iida and I. Yahara, "Yeast heat-shock protein of M(r) 48,000 is an isoprotein of enolase," *Nature*, vol. 315, no. 6021, pp. 688–690, 1985.

[22] R. M. Aaronson, K. K. Graven, M. Tucci, R. J. McDonald, and H. W. Farber, "Non-neuronal enolase is an endothelial hypoxic stress protein," *Journal of Biological Chemistry*, vol. 270, no. 46, pp. 27752–27757, 1995.

[23] R. L. Mathur, M. C. Reddy, S. Yee, R. Imbesi, B. Groth-Vasselli, and P. N. Farnsworth, "Investigation of lens glycolytic enzymes: species distribution and interaction with supramolecular order," *Experimental Eye Research*, vol. 54, no. 2, pp. 253–260, 1992.

[24] J. L. Walsh, T. J. Keith, and H. R. Knull, "Glycolytic enzyme interactions with tubulin and microtubules," *Biochimica et Biophysica Acta*, vol. 999, no. 1, pp. 64–70, 1989.

[25] S. A. Johnstone, D. M. Waisman, and J. B. Rattner, "Enolase is present at the centrosome of HeLa cells," *Experimental Cell Research*, vol. 202, no. 2, pp. 458–463, 1992.

[26] T. Merkulova, M. Lucas, C. Jabet et al., "Biochemical characterization of the mouse muscle-specific enolase: developmental changes in electrophoretic variants and selective binding to other proteins," *Biochemical Journal*, vol. 323, no. 3, pp. 791–800, 1997.

[27] A. Subramanian and D. M. Miller, "Structural analysis of α-enolase: mapping the functional domains involved in down-regulation of the c-myc protooncogene," *Journal of Biological Chemistry*, vol. 275, no. 8, pp. 5958–5965, 2000.

[28] B. Eriksson, K. Öberg, and M. Stridsberg, "Tumor markers in neuroendocrine tumors," *Digestion*, vol. 62, supplement 1, pp. 33–38, 2000.

[29] J. Niklinski and M. Furman, "Clinical tumour markers in lung cancer," *European Journal of Cancer Prevention*, vol. 4, no. 2, pp. 129–138, 1995.

[30] E. H. Cooper, "Neuron-specific enolase," *International Journal of Biological Markers*, vol. 9, no. 4, pp. 205–210, 1994.

[31] J. A. Ledermann, "Serum neurone-specific enolase and other neuroendocrine markers in lung cancer," *European Journal of Cancer A*, vol. 30, no. 5, pp. 574–576, 1994.

[32] M. Takashima, Y. Kuramitsu, Y. Yokoyama et al., "Overexpression of alpha enolase in hepatitis C virus-related hepatocellular carcinoma: association with tumor progression as determined by proteomic analysis," *Proteomics*, vol. 5, no. 6, pp. 1686–1692, 2005.

[33] C. J. Jeffery, "Moonlighting proteins," *Trends in Biochemical Sciences*, vol. 24, no. 1, pp. 8–11, 1999.

[34] J. Piatigorsky, "Multifunctional lens crystallins and corneal enzymes: more than meets the eye," *Annals of the New York Academy of Sciences*, vol. 842, pp. 7–15, 1998.

[35] J. P. Irigoyen, P. Muñoz-Cánoves, L. Montero, M. Koziczak, and Y. Nagamine, "The plasminogen activator system: biology and regulation," *Cellular and Molecular Life Sciences*, vol. 56, no. 1-2, pp. 104–132, 1999.

[36] L. A. Miles and E. F. Plow, "Binding and activation of plasminogen on the platelet surface," *Journal of Biological Chemistry*, vol. 260, no. 7, pp. 4303–4311, 1985.

[37] K. A. Hajjar, P. C. Harpel, E. A. Jaffe, and R. L. Nachman, "Binding of plasminogen to cultured human endothelial cells," *Journal of Biological Chemistry*, vol. 261, no. 25, pp. 11656–11662, 1986.

[38] E. F. Plow, D. E. Freaney, J. Plescia, and L. A. Miles, "The plasminogen system and cell surfaces: evidence for plasminogen and urokinase receptors on the same cell type," *Journal of Cell Biology*, vol. 103, no. 6, pp. 2411–2420, 1986.

[39] L. A. Miles and E. F. Plow, "Receptor mediated binding of the fibrinolytic components, plasminogen and urokinase, to peripheral blood cells," *Thrombosis and Haemostasis*, vol. 58, no. 3, pp. 936–942, 1987.

[40] F. Blasi and P. Carmeliet, "uPAR: a versatile signalling orchestrator," *Nature Reviews Molecular Cell Biology*, vol. 3, no. 12, pp. 932–943, 2002.

[41] L. Ossowski and J. A. Aguirre-Ghiso, "Urokinase receptor and integrin partnership: coordination of signaling for cell adhesion, migration and growth," *Current Opinion in Cell Biology*, vol. 12, no. 5, pp. 613–620, 2000.

[42] R. Lopez-Alemany, P. Correc, L. Camoin, and P. Burtin, "Purification of the plasmin receptor from human carcinoma cells and comparison to α-enolase," *Thrombosis Research*, vol. 75, no. 4, pp. 371–381, 1994.

[43] K. A. Hajjar, A. T. Jacovina, and J. Chacko, "An endothelial cell receptor for plasminogen/tissue plasminogen activator. I. Identity with annexin II," *Journal of Biological Chemistry*, vol. 269, no. 33, pp. 21191–21197, 1994.

[44] G. Kassam, B. H. Le, K. S. Choi et al., "The p11 subunit of the annexin II tetramer plays a key role in the stimulation of t-PA-dependent plasminogen activation," *Biochemistry*, vol. 37, no. 48, pp. 16958–16966, 1998.

[45] T. Herren, T. A. Burke, R. Das, and E. F. Plow, "Identification of histone H2B as a regulated plasminogen receptor," *Biochemistry*, vol. 45, no. 31, pp. 9463–9474, 2006.

[46] R. Das, T. Burke, and E. F. Plow, "Histone H2B as a functionally important plasminogen receptor on macrophages," *Blood*, vol. 110, no. 10, pp. 3763–3772, 2007.

[47] A. K. Dudani and P. R. Ganz, "Endothelial cell surface actin serves as a binding site for plasminogen, tissue plasminogen activator and lipoprotein(a)," *British Journal of Haematology*, vol. 95, no. 1, pp. 168–178, 1996.

[48] J. J. Kanalas and S. P. Makker, "Identification of the rat Heymann nephritis autoantigen (GP330) as a receptor site for plasminogen," *Journal of Biological Chemistry*, vol. 266, no. 17, pp. 10825–10829, 1991.

[49] T. A. Hembrough, L. Li, and S. L. Gonias, "Cell-surface cytokeratin 8 is the major plasminogen receptor on breast cancer cells and is required for the accelerated activation of cell-associated plasminogen by tissue-type plasminogen activator," *Journal of Biological Chemistry*, vol. 271, no. 41, pp. 25684–25691, 1996.

[50] D. B. Borza and W. T. Morgan, "Acceleration of plasminogen activation by tissue plasminogen activator on surface-bound histidine-proline-rich glycoprotein," *Journal of Biological Chemistry*, vol. 272, no. 9, pp. 5718–5726, 1997.

[51] S. B. Winram and R. Lottenberg, "The plasmin-binding protein Plr of group A streptococci is identified as glyceraldehyde-3-phosphate dehydrogenase," *Microbiology*, vol. 142, no. 8, pp. 2311–2320, 1996.

[52] N. M. Andronicos, E. I. Chen, N. Baik et al., "Proteomics-based discovery of a novel, structurally unique, and developmentally regulated plasminogen receptor, Plg-RKT, a major regulator of cell surface plasminogen activation," *Blood*, vol. 115, no. 7, pp. 1319–1330, 2010.

[53] J. Felez, "Plasminogen binding to cell surfaces," *Fibrinolysis and Proteolysis*, vol. 12, no. 4, pp. 183–189, 1998.

[54] R. López-Alemany, C. Longstaff, S. Hawley et al., "Inhibition of cell surface mediated plasminogen activation by a monoclonal antibody against α-enolase," *American Journal of Hematology*, vol. 72, no. 4, pp. 234–242, 2003.

[55] C. Longstaff, R. E. Merton, P. Fabregas, and J. Felez, "Characterization of cell-associated plasminogen activation catalyzed by urokinase-type plasminogen activator, but independent of urokinase receptor (uPAR, CD87)," *Blood*, vol. 93, no. 11, pp. 3839–3846, 1999.

[56] V. Sinniger, R. E. Mertont, P. Fabregas, J. Felez, and C. Longstaff, "Regulation of tissue plasminogen activator activity by cells: domains responsible for binding and mechanism of stimulation," *Journal of Biological Chemistry*, vol. 274, no. 18, pp. 12414–12422, 1999.

[57] J. Félez, L. A. Miles, P. Fábregas, M. Jardí, E. F. Plow, and R. H. Lijnen, "Characterization of cellular binding sites and interactive regions within reactants required for enhancement of plasminogen activation by tPA on the surface of leukocytic cells," *Thrombosis and Haemostasis*, vol. 76, no. 4, pp. 577–584, 1996.

[58] L. A. Miles, G. M. Fless, A. M. Scanu et al., "Interaction of Lp(a) with plasminogen binding sites on cells," *Thrombosis and Haemostasis*, vol. 73, no. 3, pp. 458–465, 1995.

[59] M. Wygrecka, L. M. Marsh, R. E. Morty et al., "Enolase-1 promotes plasminogen-mediated recruitment of monocytes to the acutely inflamed lung," *Blood*, vol. 113, no. 22, pp. 5588–5598, 2009.

[60] G. von Heijne, P. Liljestrom, P. Mikus, H. Andersson, and T. Ny, "The efficiency of the uncleaved secretion signal in the plasminogen activator inhibitor type 2 protein can be enhanced by point mutations that increase its hydrophobicity," *Journal of Biological Chemistry*, vol. 266, no. 23, pp. 15240–15243, 1991.

[61] L. A. Bottalico, N. C. Kendrick, A. Keller, Y. Li, and I. Tabas, "Cholesteryl ester loading of mouse peritoneal macrophages is associated with changes in the expression or modification of specific cellular proteins, including increase in an α-enolase isoform," *Arteriosclerosis and Thrombosis*, vol. 13, no. 2, pp. 264–275, 1993.

[62] J. A. Cooper, F. S. Esch, S. S. Taylor, and T. Hunter, "Phosphorylation sites in enolase and lactate dehydrogenase utilized

by tyrosine protein kinase in vivo and in vitro," *Journal of Biological Chemistry*, vol. 259, no. 12, pp. 7835–7841, 1984.

[63] F. Lluís, J. Roma, M. Suelves et al., "Urokinase-dependent plasminogen activation is required for efficient skeletal muscle regeneration in vivo," *Blood*, vol. 97, no. 6, pp. 1703–1711, 2001.

[64] M. Suelves, R. López-Alemany, F. Lluís et al., "Plasmin activity is required for myogenesis in vitro and skeletal muscle regeneration in vivo," *Blood*, vol. 99, no. 8, pp. 2835–2844, 2002.

[65] M. Suelves, B. Vidal, V. Ruiz et al., "The plasminogen activation system in skeletal muscle regeneration: antagonistic roles of Urokinase-type Plasminogen Activator (UPA) and its inhibitor (PAI-1)," *Frontiers in Bioscience*, vol. 10, no. 3, pp. 2978–2985, 2005.

[66] M. Suelves, B. Vidal, A. L. Serrano et al., "uPA deficiency exacerbates muscular dystrophy in MDX mice," *Journal of Cell Biology*, vol. 178, no. 6, pp. 1039–1051, 2007.

[67] G. Fibbi, E. Barletta, G. Dini et al., "Cell invasion is affected by differential expression of urokinase plasminogen activator/urokinase plasminogen activator receptor system in muscle satellite cells from normal and dystrophic patients," *Laboratory Investigation*, vol. 81, no. 1, pp. 27–39, 2001.

[68] J. M. Taylor, J. D. Davies, and C. A. Peterson, "Regulation of the myoblast-specific expression of the human β-enolase gene," *Journal of Biological Chemistry*, vol. 270, no. 6, pp. 2535–2540, 1995.

[69] F. Fougerousse, F. Edom-Vovard, T. Merkulova et al., "The muscle-specific enolase is an early marker of human myogenesis," *Journal of Muscle Research and Cell Motility*, vol. 22, no. 6, pp. 535–544, 2001.

[70] A. Keller, J. Peltzer, G. Carpentier et al., "Interactions of enolase isoforms with tubulin and microtubules during myogenesis," *Biochimica et Biophysica Acta*, vol. 1770, no. 6, pp. 919–926, 2007.

[71] T. Merkulova, M. Dehaupas, M. C. Nevers, C. Créminon, H. Alameddine, and A. Keller, "Differential modulation of α, β and γ enolase isoforms in regenerating mouse skeletal muscle," *European Journal of Biochemistry*, vol. 267, no. 12, pp. 3735–3743, 2000.

[72] R. López-Alemany, M. Suelves, and P. Muñoz-Cánoves, "Plasmin generation dependent on α-enolase-type plasminogen receptor is required for myogenesis," *Thrombosis and Haemostasis*, vol. 90, no. 4, pp. 724–733, 2003.

[73] S. C. Bryer and T. J. Koh, "The urokinase-type plasminogen activator receptor is not required for skeletal muscle inflammation or regeneration," *American Journal of Physiology*, vol. 293, no. 3, pp. R1152–R1158, 2007.

[74] L. Yan, H. Ge, H. Li et al., "Gender-specific proteomic alterations in glycolytic and mitochondrial pathways in aging monkey hearts," *Journal of Molecular and Cellular Cardiology*, vol. 37, no. 5, pp. 921–929, 2004.

[75] L. A. Zhu, N. Y. Fang, P. J. Gao, X. Jin, and H. Y. Wang, "Differential expression of α-enolase in the normal and pathological cardiac growth," *Experimental and Molecular Pathology*, vol. 87, no. 1, pp. 27–31, 2009.

[76] Y. Mizukami, A. Iwamatsu, T. Aki et al., "ERK1/2 regulates intracellular ATP levels through α-enolase expression in cardiomyocytes exposed to ischemic hypoxia and reoxygenation," *Journal of Biological Chemistry*, vol. 279, no. 48, pp. 50120–50131, 2004.

[77] J. Mair, "Progress in myocardial damage detection: new biochemical markers for clinicians," *Critical Reviews in Clinical Laboratory Sciences*, vol. 34, no. 1, pp. 1–66, 1997.

[78] H. E. Kim, S. S. Dalal, E. Young, M. J. Legato, M. L. Weisfeldt, and J. D'Armiento, "Disruption of the myocardial extracellular matrix leads to cardiac dysfunction," *Journal of Clinical Investigation*, vol. 106, no. 7, pp. 857–866, 2000.

[79] S. Heymans, A. Luttun, D. Nuyens et al., "Inhibition of plasminogen activators or matrix metalloproteinases prevents cardiac rupture but impairs therapeutic angiogenesis and causes cardiac failure," *Nature Medicine*, vol. 5, no. 10, pp. 1135–1142, 1999.

[80] E. Creemers, J. Cleutjens, J. Smits et al., "Disruption of the plasminogen gene in mice abolishes wound healing after myocardial infarction," *American Journal of Pathology*, vol. 156, no. 6, pp. 1865–1873, 2000.

[81] K. Macfelda, T. W. Weiss, C. Kaun et al., "Plasminogen activator inhibitor 1 expression is regulated by the inflammatory mediators interleukin-1α, tumor necrosis factor-α, transforming growth factor-β and oncostatin M in human cardiac myocytes," *Journal of Molecular and Cellular Cardiology*, vol. 34, no. 12, pp. 1681–1691, 2002.

[82] K. Takeshita, M. Hayashi, S. Iino et al., "Increased expression of plasminogen activator inhibitor-1 in cardiomyocytes contributes to cardiac fibrosis after myocardial infarction," *American Journal of Pathology*, vol. 164, no. 2, pp. 449–456, 2004.

[83] S. M. Carroll, L. E. Nimmo, P. S. Knoepfler, F. C. White, and C. M. Bloor, "Gene expression in a swine model of right ventricular hypertrophy: intercellular adhesion molecule, vascular endothelial growth factor and plasminogen activators are upregulated during pressure overload," *Journal of Molecular and Cellular Cardiology*, vol. 27, no. 7, pp. 1427–1441, 1995.

[84] K. B. S. Pasumarthi and L. J. Field, "Cardiomyocyte cell cycle regulation," *Circulation Research*, vol. 90, no. 10, pp. 1044–1054, 2002.

[85] G. Olivetti, R. Abbi, F. Quaini et al., "Apoptosis in the failing human heart," *The New England Journal of Medicine*, vol. 336, no. 16, pp. 1131–1141, 1997.

[86] S. Guerra, A. Leri, X. Wang et al., "Myocyte death in the failing human heart is gender dependent," *Circulation Research*, vol. 85, no. 9, pp. 856–866, 1999.

[87] B. Ho-Tin-Noé, G. Rojas, R. Vranckx, H. R. Lijnen, and E. Anglés-Cano, "Functional hierarchy of plasminogen kringles 1 and 4 in fibrinolysis and plasmin-induced cell detachment and apoptosis," *FEBS Journal*, vol. 272, no. 13, pp. 3387–3400, 2005.

[88] O. Meilhac, B. Ho-Tin-Noé, X. Houard, M. Philippe, J. B. Michel, and E. Anglés-Cano, "Pericellular plasmin induces smooth muscle cell anoikis," *The FASEB Journal*, vol. 17, no. 10, pp. 1301–1303, 2003.

[89] B. Ho-Tin-Noé, H. Enslen, L. Doeuvre, J. M. Corsi, H. R. Lijnen, and E. Anglés-Cano, "Role of plasminogen activation in neuronal organization and survival," *Molecular and Cellular Neuroscience*, vol. 42, no. 4, pp. 288–295, 2009.

[90] N. Kochtebane, C. Choqueux, S. Passefort et al., "Plasmin induces apoptosis of aortic valvular myofibroblasts," *Journal of Pathology*, vol. 221, no. 1, pp. 37–48, 2010.

[91] D. S. Ucker, M. R. Jain, G. Pattabiraman, K. Palasiewicz, R. B. Birge, and H. Li, "Externalized glycolytic enzymes are novel, conserved, and early biomarkers of apoptosis," *Journal of Biological Chemistry*, vol. 287, pp. 10625–10343, 2012.

[92] M. Herrmann, R. E. Voll, O. M. Zoller, M. Hagenhofer, B. B. Ponner, and J. R. Kalden, "Impaired phagocytosis of

apoptotic cell material by monocyte-derived macrophages from patients with systemic lupus erythematosus," *Arthritis and Rheumatism*, vol. 41, pp. 1241–1250, 1998.

[93] P. L. Cohen, R. Caricchio, V. Abraham et al., "Delayed apoptotic cell clearance and lupus-like autoimmunity in mice lacking the c-mer membrane tyrosine kinase," *Journal of Experimental Medicine*, vol. 196, no. 1, pp. 135–140, 2002.

[94] C. V. Rothlin, S. Ghosh, E. I. Zuniga, M. B. A. Oldstone, and G. Lemke, "TAM receptors are pleiotropic inhibitors of the innate immune response," *Cell*, vol. 131, no. 6, pp. 1124–1136, 2007.

[95] E. B. Thorp, "Mechanisms of failed apoptotic cell clearance by phagocyte subsets in cardiovascular disease," *Apoptosis*, vol. 15, no. 9, pp. 1124–1136, 2010.

[96] M. J. O'Mullane and M. S. Baker, "Loss of cell viability dramatically elevates cell surface plasminogen binding and activation," *Experimental Cell Research*, vol. 242, no. 1, pp. 153–164, 1998.

[97] G. M. Cesarman, C. A. Guevara, and K. A. Hajjar, "An endothelial cell receptor for plasminogen/tissue plasminogen activator (t-PA). II. Annexin II-mediated enhancement of t-PA-dependent plasminogen activation," *Journal of Biological Chemistry*, vol. 269, no. 33, pp. 21198–21203, 1994.

[98] G. C. Chang, K. J. Liu, C. L. Hsieh et al., "Identification of α-enolase as an autoantigen in lung cancer: its overexpression is associated with clinical outcomes," *Clinical Cancer Research*, vol. 12, no. 19, pp. 5746–5754, 2006.

[99] C. López-Pedrera, J. M. Villalba, E. Siendones et al., "Proteomic analysis of acute myeloid leukemia: identification of potential early biomarkers and therapeutic targets," *Proteomics*, vol. 6, supplement 1, pp. S293–S299, 2006.

[100] M. Katayama, H. Nakano, A. Ishiuchi et al., "Protein pattern difference in the colon cancer cell lines examined by two-dimensional differential in-gel electrophoresis and mass spectrometry," *Surgery Today*, vol. 36, no. 12, pp. 1085–1093, 2006.

[101] M. Capello, S. Ferri-Borgogno, P. Cappello, and F. Novelli, "α-enolase: a promising therapeutic and diagnostic tumor target," *FEBS Journal*, vol. 278, no. 7, pp. 1064–1074, 2011.

[102] W. Jiang, X. Li, S. Rao et al., "Constructing disease-specific gene networks using pair-wise relevance metric: application to colon cancer identifies interleukin 8, desmin and enolase 1 as the central elements," *BMC Systems Biology*, vol. 2, article 72, 2008.

[103] E. Georges, A. M. Bonneau, and P. Prinos, "RNAi-mediated knockdown of alpha-enolase increases the sensitivity of tumor cells to antitubulin chemotherapeutics," *International Journal of Biochemistry and Molecular Biology*, vol. 2, pp. 303–308, 2011.

[104] B. Trojanowicz, A. Winkler, K. Hammje et al., "Retinoic acid-mediated down-regulation of ENO1/MBP-1 gene products caused decreased invasiveness of the follicular thyroid carcinoma cell lines," *Journal of Molecular Endocrinology*, vol. 42, no. 3, pp. 249–260, 2009.

[105] S. T. Tsai, I. H. Chien, W. H. Shen et al., "ENO1, a potential prognostic head and neck cancer marker, promotes transformation partly via chemokine CCL20 induction," *European Journal of Cancer*, vol. 46, no. 9, pp. 1712–1723, 2010.

[106] S. Jin, R. S. DiPaola, R. Mathew, and E. White, "Metabolic catastrophe as a means to cancer cell death," *Journal of Cell Science*, vol. 120, no. 3, pp. 379–383, 2007.

[107] Y. Abiko, M. Nishimura, K. Kusano et al., "Expression of MIP-3α/CCL20, a macrophage inflammatory protein in oral squamous cell carcinoma," *Archives of Oral Biology*, vol. 48, no. 2, pp. 171–175, 2003.

[108] W. Zhou, M. Capello, C. Fredolini et al., "Mass spectrometry analysis of the post-translational modifications of r-enolase from pancreatic ductal adenocarcinoma cells," *Journal of Proteome Research*, vol. 9, no. 6, pp. 2929–2936, 2010.

[109] B. Tomaino, P. Cappello, M. Capello et al., "Circulating autoantibodies to phosphorylated α-enolase are a hallmark of pancreatic cancer," *Journal of Proteome Research*, vol. 10, no. 1, pp. 105–112, 2011.

[110] H. Iwabata, M. Yoshida, and Y. Komatsu, "Proteomic analysis of organ-specific post-translational lysine-acetylation and -methylation in mice by use of anti-acetyllysine and -methyllysine mouse monoclonal antibodies," *Proteomics*, vol. 5, no. 18, pp. 4653–4664, 2005.

[111] N. Lu, Y. Zhang, H. Li, and Z. Gao, "Oxidative and nitrative modifications of α-enolase in cardiac proteins from diabetic rats," *Free Radical Biology and Medicine*, vol. 48, no. 7, pp. 873–881, 2010.

[112] J. Gannon, L. Staunton, K. O'Connell, P. Doran, and K. Ohlendieck, "Phosphoproteomic analysis of aged skeletal muscle," *International Journal of Molecular Medicine*, vol. 22, no. 1, pp. 33–42, 2008.

[113] M. A. Baraibar, J. Hyzewicz, A. Rogowska-Wrzesinska et al., "Oxidative stress-induced proteome alterations target different cellular pathways in human myoblasts," *Free Radical Biology and Medicine*, vol. 51, pp. 1522–1532, 2011.

[114] A. Kinloch, V. Tatzer, R. Wait et al., "Identification of citrullinated alpha-enolase as a candidate autoantigen in rheumatoid arthritis," *Arthritis Research & Therapy*, vol. 7, no. 6, pp. R1421–1429, 2005.

[115] F. Pratesi, S. Moscato, A. Sabbatini, D. Chimenti, S. Bombardieri, and P. Migliorini, "Autoantibodies specific for α-enolase in systemic autoimmune disorders," *Journal of Rheumatology*, vol. 27, no. 1, pp. 109–115, 2000.

[116] K. Wakui, M. Tanemura, K. Suzumori et al., "Clinical applications of two-color telomeric fluorescence in situ hybridization for prenatal diagnosis: identification of chromosomal translocation in five families with recurrent miscarriages or a child with multiple congenital anomalies," *Journal of Human Genetics*, vol. 44, no. 2, pp. 85–90, 1999.

[117] V. Saulot, O. Vittecoq, R. Charlionet et al., "Presence of autoantibodies to the glycolytic enzyme α-enolase in sera from patients with early rheumatoid arthritis," *Arthritis and Rheumatism*, vol. 46, no. 5, pp. 1196–1201, 2002.

[118] N. Wegner, K. Lundberg, A. Kinloch et al., "Autoimmunity to specific citrullinated proteins gives the first clues to the etiology of rheumatoid arthritis," *Immunological Reviews*, vol. 233, no. 1, pp. 34–54, 2010.

[119] X. Chang and C. Wei, "Glycolysis and rheumatoid arthritis," *International Journal of Rheumatic Diseases*, vol. 14, no. 3, pp. 217–222, 2011.

[120] A. T. Jacovina, F. Zhong, E. Khazanova, E. Lev, A. B. Deora, and K. A. Hajjar, "Neuritogenesis and the nerve growth factor-induced differentiation of PC-12 cells requires annexin II-mediated plasmin generation," *Journal of Biological Chemistry*, vol. 276, no. 52, pp. 49350–49358, 2001.

[121] N. Obermajer, B. Doljak, P. Jamnik, U. P. Fonović, and J. Kos, "Cathepsin X cleaves the C-terminal dipeptide of alpha- and gamma-enolase and impairs survival and neuritogenesis of neuronal cells," *International Journal of Biochemistry and Cell Biology*, vol. 41, no. 8-9, pp. 1685–1696, 2009.

[122] D. A. Butterfield and M. L. B. Lange, "Multifunctional roles of enolase in Alzheimer's disease brain: beyond altered glucose metabolism," *Journal of Neurochemistry*, vol. 111, no. 4, pp. 915–933, 2009.

[123] A. Castegna, M. Aksenov, V. Thongboonkerd et al., "Proteomic identification of oxidatively modified proteins in Alzheimer's disease brain. Part II: dihydropyrimidinase-related protein 2, α-enolase and heat shock cognate 71," *Journal of Neurochemistry*, vol. 82, no. 6, pp. 1524–1532, 2002.

[124] J. B. Owen, F. D. Domenico, R. Suitana et al., "Proteomics-determined differences in the concanavalin-A-fractionated proteome of hippocampus and inferior parietal lobule in subjects with alzheimer's disease and mild cognitive impairment: implications for progression of AD," *Journal of Proteome Research*, vol. 8, no. 2, pp. 471–482, 2009.

[125] S. F. Newman, R. Sultana, M. Perluigi et al., "An increase in S-glutathionylated proteins in the Alzheimer's disease inferior parietal lobule, a proteomics approach," *Journal of Neuroscience Research*, vol. 85, no. 7, pp. 1506–1514, 2007.

[126] Y. Zhang, Z. H. Zhou, T. H. Bugge, and L. M. Wahl, "Urokinase-type plasminogen activator stimulation of monocyte matrix metalloproteinase-1 production is mediated by plasmin-dependent signaling through annexin A2 and inhibited by inactive plasmin," *Journal of Immunology*, vol. 179, no. 5, pp. 3297–3304, 2007.

[127] T. Syrovets, M. Jendrach, A. Rohwedder, A. Schüle, and T. Simmet, "Plasmin-induced expression of cytokines and tissue factor in human monocytes involves AP-1 and IKKβ-mediated NF-κB activation," *Blood*, vol. 97, no. 12, pp. 3941–3950, 2001.

[128] L. Burysek, T. Syrovets, and T. Simmet, "The serine protease plasmin triggers expression of MCP-1 and CD40 in human primary monocytes via activation of p38 MAPK and Janus kinase (JAK)/STAT signaling pathways," *Journal of Biological Chemistry*, vol. 277, no. 36, pp. 33509–33517, 2002.

[129] Q. Li, Y. Laumonnier, T. Syrovets, and T. Simmet, "Plasmin triggers cytokine induction in human monocyte-derived macrophages," *Arteriosclerosis, Thrombosis, and Vascular Biology*, vol. 27, no. 6, pp. 1383–1389, 2007.

[130] U. R. Pendurthi, M. Ngyuen, P. Andrade-Gordon, L. C. Petersen, and L. V. M. Rao, "Plasmin induces *Cyr61* gene expression in fibroblasts via protease-activated receptor-1 and p44/42 mitogen-activated protein kinase-dependent signaling pathway," *Arteriosclerosis, Thrombosis, and Vascular Biology*, vol. 22, no. 9, pp. 1421–1426, 2002.

[131] S. K. Mandal, L. V. M. Rao, T. T. T. Tran, and U. R. Pendurthi, "A novel mechanism of plasmin-induced mitogenesis in fibroblasts," *Journal of Thrombosis and Haemostasis*, vol. 3, no. 1, pp. 163–169, 2005.

[132] L. P. Sousa, B. M. Silva, B. S. A. F. Brasil et al., "Plasminogen/plasmin regulates α-enolase expression through the MEK/ERK pathway," *Biochemical and Biophysical Research Communications*, vol. 337, no. 4, pp. 1065–1071, 2005.

[133] L. P. De Sousa, B. S. A. F. Brasil, B. M. Silva et al., "Plasminogen/plasmin regulates c-fos and egr-1 expression via the MEK/ERK pathway," *Biochemical and Biophysical Research Communications*, vol. 329, no. 1, pp. 237–245, 2005.

[134] M. Majumdar, T. Tarui, B. Shi, N. Akakura, W. Ruf, and Y. Takada, "Plasmin-induced migration requires signaling through protease-activated receptor 1 and integrin α9β1," *Journal of Biological Chemistry*, vol. 279, no. 36, pp. 37528–37534, 2004.

[135] T. Tarui, M. Majumdar, L. A. Miles, W. Ruf, and Y. Takada, "Plasmin-induced migration of endothelial cells: a potential target for the anti-angiogenic action of angiostatin," *Journal of Biological Chemistry*, vol. 277, no. 37, pp. 33564–33570, 2002.

[136] E. Roztocil, S. M. Nicholl, I. I. Galaria, and M. G. Davies, "Plasmin-induced smooth muscle cell proliferation requires epidermal growth factor activation through an extracellular pathway," *Surgery*, vol. 138, no. 2, pp. 180–186, 2005.

[137] J. Petrak, R. Ivanek, O. Toman et al., "Déjà vu in proteomics. A hit parade of repeatedly identified differentially expressed proteins," *Proteomics*, vol. 8, no. 9, pp. 1744–1749, 2008.

[138] C. Roozendaal, M. H. Zhao, G. Horst et al., "Catalase and α-enolase: two novel granulocyte autoantigens in inflammatory bowel disease (IBD)," *Clinical and Experimental Immunology*, vol. 112, no. 1, pp. 10–16, 1998.

[139] N. Vermeulen, I. Arijs, S. Joossens et al., "Anti-α-enolase antibodies in patients with inflammatory bowel disease," *Clinical Chemistry*, vol. 54, no. 3, pp. 534–541, 2008.

[140] E. Ballot, A. Bruneel, V. Labas, and C. Johanet, "Identification of rat targets of anti-soluble liver antigen autoantibodies by serologic proteome analysis," *Clinical Chemistry*, vol. 49, no. 4, pp. 634–643, 2003.

[141] M. Bruschi, M. L. Carnevali, C. Murtas et al., "Direct characterization of target podocyte antigens and auto-antibodies in human membranous glomerulonephritis: alfa-enolase and borderline antigens," *Journal of Proteomics*, vol. 74, no. 10, pp. 2008–2017, 2011.

[142] J. W. Kim and C. V. Dang, "Multifaceted roles of glycolytic enzymes," *Trends in Biochemical Sciences*, vol. 30, no. 3, pp. 142–150, 2005.

9

Clinical Value of CD24 Expression in Retinoblastoma

Jia Li,[1] Changqing Li,[2] Hongfeng Yuan,[1] and Fang Gong[3]

[1] Department of Ophthalmology, Research Institute of Surgery and Daping Hospital, Third Military Medical University,
  Chongqing, 400042, China
[2] Department of Pharmacy, Chongqing Red Cross Hospital, Chongqing 400020, China
[3] Department of Neonatology, Yong Chuan Affiliated Hospital, Chongqing Medical University, Chongqing 402160, China

Correspondence should be addressed to Hongfeng Yuan, yhf871@yahoo.com.cn and Fang Gong, gong_fyc@163.com

Academic Editor: Xin-Yuan Guan

Background. The expression of CD24 has been detected in a wide variety of human malignancies. Downregulation of CD24 inhibits proliferation and induces apoptosis in tumor cells, whereas its upregulation increases tumor growth and metastasis. However, no data on CD24 protein levels in retinoblastoma are available, and the mechanism of CD24 involvement in retinoblastoma progress has not been elucidated. The aim of this study was to explore the expression profile of CD24 in the retinoblastoma tumor samples and to correlate with clinicopathological parameters. *Methods.* Immunohistochemistry was performed for CD24 on the archival paraffin sections of retinoblastoma and correlated with clinicopathological features. Western blotting was performed to confirm immunoreactivity results. *Results.* CD24 immunoreactivity was observed in 72.0% (36/50) of the retinoblastoma specimens. Among the 35 low-risk tumors, CD24 was expressed in 62.9% (22/35) tumors and among the 15 high-risk tumors, CD24 was expressed in 93.3% (14/15) tumors. High-risk tumors showed significantly increased expression of CD24 compared to tumors with low-risk ($P < 0.05$). *Conclusions.* This is the first correlation between CD24 expression and histopathology in human retinoblastoma. Our study showed increased expression of CD24 in high risk tumors compared to low risk tumors. Further functional studies are required to explore the role of CD24 in retinoblastoma.

1. Introduction

Retinoblastoma is the most common primary intraocular malignancy in children [1]. It is more frequent in some developing areas, such as Latin America, Africa, China, and India [2]. Factors that play a role in tumor invasion and metastasis include early genetic events such as increased copy number of chromosome 6p and 1q, and late events such as high levels of telomerase activity, loss of chromosome 1p, and p53 inactivation [3]. Retinoblastoma is diagnosed late, usually when extraocular dissemination has occurred and the prognosis is poor. Primary chemoreduction is used for intraocular retinoblastoma and systemic chemotherapy is used following enucleation in patients with optic nerve and deep choroidal invasion, orbital extension, and metastatic disease. New treatment modalities, such as subconjunctival injection, selective ophthalmic artery injection, and vitreous injection, are being investigated and have achieved favorable results [4]. Although many modalities are used, almost half

of eyes with retinoblastoma have to be enucleated. Thus, there is an urgent need to further study the biology and molecular mechanisms of retinoblastoma and identify the specific proteins that cause tumor progression and predict prognosis in order to improve the therapeutic outcome of patients with retinoblastoma.

CD24 is a small, heavily glycosylated, phosphatidyli-nositol-anchored ucin-like cell surface protein [5]. Several studies have identified CD24 as an alternative ligand for P-selectin, an adhesion molecule which is expressed by activated endothelial cells and platelets [6]. Functionally, it is considered to play a critical role in the metastasis of tumor cells through P-selectin. Many tumor cell lines can bind to platelets via P-selectin, and CD24 expression might enhance the metastatic potential of tumor cells [7]. Recently, several studies have investigated CD24 overexpression in a wide variety of human malignancies lung cancer, chorio-carcinoma, cholangiocarcinoma, glioma, pancreatic cancer, prostatic cancers, renal cell carcinoma, and ovarian and breast cancers [8–15]. This increased expression is usually tied with a more aggressive course of the disease. CD24 positivity was significantly related to younger patient age, higher pT stages, and higher PSA relapse rate in prostatic adenocarcinomas [10]. Wei et al. have shown that CD24 is overexpressed in pancreatic cancer cell lines in comparison to nine normal pancreatic cell lines [11]. In hepatocellular carcinomas, CD24 expression was also correlated with serum levels of HBs-Ag, elevated serum AFP levels, and p53 mutations [12]. Lee et al. noted strong CD24 staining in renal cell carcinoma irrespective of the histological tumor type [13]. A recent study by Smith et al. reported CD24 overexpression in bladder cancer compared to normal urothelium; however, no correlation with tumor stage and grade was noted. Additionally, bladder cancer patients with strong CD24 immunoreactivity tended to have a shorter disease-free survival [14]. Moreover, it has been demonstrated that the down-regulation of CD24 inhibits proliferation and induces apoptosis in tumor cells, whereas its up-regulation increases tumor growth and metastasis. Although there have been substantial advances in the understanding of the basic biology and pathogenesis of retinoblastoma, there is no data on CD24 protein levels in retinoblastoma, and the mechanism of CD24 involvement in retinoblastoma progress has not been elucidated. The purpose of this study was to investigate if CD24 is present in retinoblastoma and to evaluate the correlation between CD24 expression and the severity of the malignancy of this tumor.

2. Materials and Methods

2.1. Patients and Tissue Samples. This study was approved by the Research Ethics Committee of DaPing Hospital, Research Institute of Surgery Third Military Medical University, Chongqing, China. Written informed consent was obtained from all of the patients. All specimens were handled and made anonymous according to the ethical and legal standards.

Fifty tumors were available from 50 eyes for the present study. Among them, there were tumors from 35 males and 15 females. The age ranged from 1 month to 8 years. (median = 22 months). There were 38 unilateral retinoblastomas and 12 bilateral retinoblastomas. Ten fresh tumors were used for studying the expression of CD24.

2.2. Histopathological Information. There were 35 tumors with no massive invasion of choroid or optic nerve (lower risk to extraocular relapse) and 15 tumors with invasion of choroid/optic nerve/orbit (higher risk to extraocular relapse). Among the 15 high-risk tumors, 2 tumors had diffuse choroidal invasion, 2 tumors with postlaminar optic nerve invasion. There are 5 tumors with both diffuse choroidal and post-laminar ON invasion, 2 tumors with both diffuse choroidal and prelaminar ON invasion, 2 tumors with focal choroidal and post-laminar ON invasion, and 1 tumor with post-laminar optic nerve invasion with surgical end involved. There was 1 tumor with orbital invasion. There were 31 tumors with poorly differentiated cells, 12 tumors

with well-differentiated and 7 tumors with moderately differentiated cells.

2.3. Immunohistochemistry Assay. Formalin-fixed, paraffin-embedded, sectioned tissues (5 μm thick) were immunostained using the Labelled Streptavidin Biotin 2 System (BioGenex; San Ramon, CA, USA). Following peroxidase blocking with 0.3% H_2O_2/methanol for 30 min, specimens were blocked with phosphate-buffered saline (PBS) containing 5% normal horse serum (Vector Laboratories Inc., Burlingame, CA, USA). All incubations with mouse monoclonal antibody CD24 (Neomarkers, Clone 24C02, Fremont, CA, USA) at 1:100 dilution were carried out overnight at 4°C. Then the specimens were briefly washed in PBS and incubated at room temperature with the anti-mouse antibody and avidin-biotin peroxidase (Vector Laboratories Inc., Burlingame, CA, USA). The specimens were then washed in PBS and color-developed by diaminobenzidine solution (Dako Corporation, Carpinteria, CA, USA). After washing with water, specimens were counterstained with Meyer's hematoxylin (Sigma Chemical Co., St Louis, MO, USA). For CD24-negative controls, immunohistochemistry was performed using normal mouse immunoglobin with a dilution of 1:100 (0.4 μg) and normal goat immunoglobin with a dilution of 1:100 (8 μg/mL).

Assessment of immunohistochemical staining was evaluated by two independent pathologists. The CD24-positive cells showed immunoreactivity in the cytoplasm of tumor cells. Randomly 10 vital tumor fields were scanned for protein expression and percentage of positive tumor cells was noted for each field. Then finally the average expression was calculated from the 10 values for the entire slide. Depending on the percentage of positive cells, 4 categories were established: 0, no positive cells; 1+, positive cells in less than one-third (faint); 2+, positive cells in 33~67% (heterogeneous); 3+, positive cells in more than two-thirds (positive) of total tumor cell population.

2.4. Western Blot Analysis. Retinoblastoma tissues were homogenized in lysis buffer (PBS, 1% nonidet P-40 (NP-40), 0.5% sodium deoxycholate, 0.1% sodium dodecyl sulfate (SDS), 100 μg/mL aprotinin, 100 μg/mL phenylmethylsulfonyl fluoride (PMSF), sodium orthovanadate) at 4°C throughout all procedures, and sonicated for 70 s, then added to 300 μg PMSF per gram of tissue and incubated on ice for 30 min, followed by centrifugation at 15,000 rpm for 20 min at 4°C. The protein content was determined according to Bradford's method (Bradford, 1976), with bovine serum albumin used as a standard. Equal amounts of protein were separated electrophoretically on 7.5% SDS-polyacrylamide gels and transferred onto polyvinylidene difluoride membranes (Roche; Basel, Switzerland). The membranes were probed with monoclonal mouse anti-human CD24 (SC-7034, 1:500, Santa Cruz Biotechnology). The expression level of CD24 was determined by incubating the membranes with horseradish peroxidase-conjugated anti-mouse immunoglobulin G (1:3000 dilution) and enhanced chemiluminescence reagent (Pierce; Minneapolis, MN, USA), according to the manufacturers'

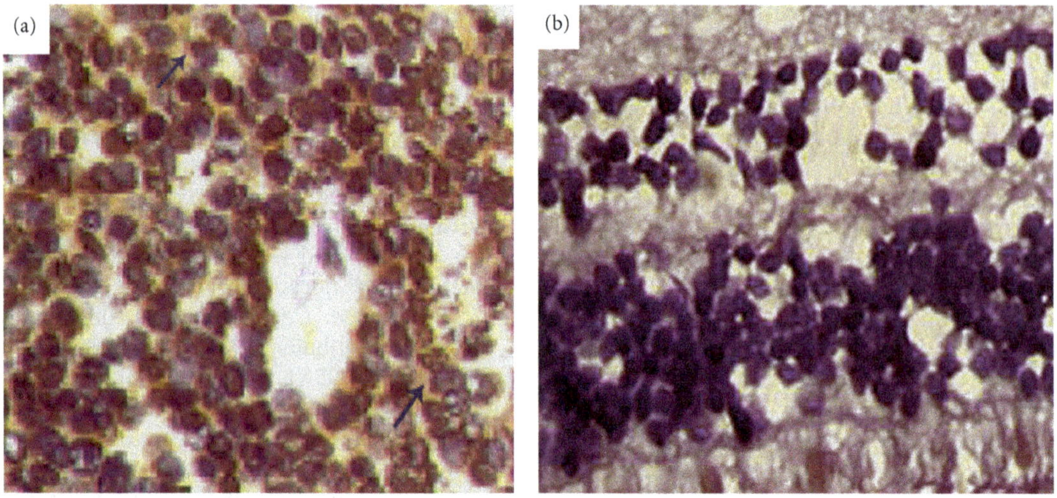

FIGURE 1: Immunohistochemical staining for CD24 in human retinoblastoma tissues (original magnification ×200). (a) CD24 staining is negative in non-neoplastic retina; (b) CD24-positive expression was found in the cytoplasm of retinoblastoma tissues.

suggested protocols. Detection of positivity was by using ECL system (Amersham Pharmacia) or the Supersignal West Femto Maximum Sensitivity Substrate (Pierce).

2.5. Statistical Analysis. All computations were carried out using the software of SPSS version13.0 for Windows (SPSS Inc, IL, USA). Data were expressed as means ± standard deviation (SD). Mann-Whitney U test was used to determine association of immunoreactivity of CD24 with tumor invasion and differentiation. For statistical purposes, moderately differentiated and well-differentiated tumors were grouped together and were compared with poorly differentiated tumors. Differences were considered statistically significant when P was less than 0.05.

3. Results

3.1. Immunohistochemical Staining of CD24 Protein in Human Retinoblastoma Tissues. Immunohistochemistry analysis showed no positive CD24 protein expression in the nonneoplastic retina (Figure 1(a)). CD24 immunoreactivity was observed in the cytoplasm of retinoblastoma specimens with varying levels of percentage tumor staining and intensities (Figure 1(b)). CD24 immunoreactivity was observed in 72.0% (36/50) of the retinoblastoma specimens.

We observed that tumors with invasion showed significantly higher expression of CD24 compared to tumors without invasion ($P < 0.05$). The immunoreactivity of CD24 in high-risk and low-risk tumors is shown in Table 1. The raw data of the individual immunoscores for CD24 in the high-risk and low-risk tumors are available in Tables 2 and 3, respectively. Among the 35 low-risk tumors, CD24 was expressed in 62.9% (22/35) tumors and among the 15 high-risk tumors, CD24 was expressed in 93.3% (14/15) tumors. High-risk tumors showed significantly increased expression of CD24 compared to tumors with low risk ($P < 0.05$).

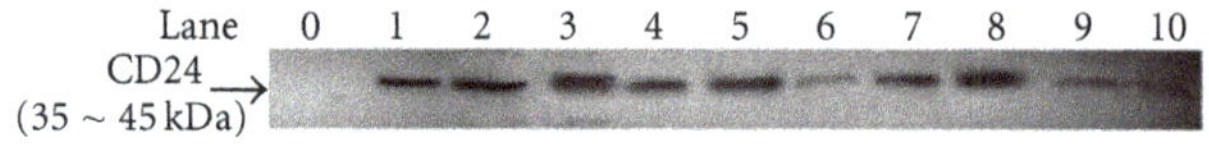

FIGURE 2: Western blot analysis for CD24 protein expression in retinoblastoma tissues, with non-neoplastic retina as control. The picture shows that CD24 was negative in non-neoplastic retina (lane 0), strongly expressed in 5 (83.3%, lanes 1~5) and faintly expressed in 1 (16.7%, lanes 6) high-risk tumors, whereas it was strongly expressed in 2 (50.0%, lanes 7~8) and faintly expressed in 2 (50.0%, lanes 9~10) low-risk tumors.

3.2. Western Blot Analysis on Expression Levels of CD24 Protein in Human Retinoblastoma Tissues. Consistent with the results of immunohistochemical staining, western blot analysis also showed no CD24 protein in the non-neoplastic retina (lane 0), but positive expression in the retinoblastoma tissues which showed a single band of approximately 35~45 kDa for CD24 protein in all the tumors (Figure 2). Of the 10 tumors studied, 6 tumors were of high-risk and 4 tumors were of low-risk. CD24 was strongly expressed in 5 (83.3%, lanes 1~5) and faintly expressed in 1 (16.7%, lanes 6) high-risk tumors, whereas it was strongly expressed in 2 (50.0%, lanes 7~8) and faintly expressed in 2 (50.0%, lanes 9~10) low-risk tumors, also suggesting high-risk tumors showed significantly increased expression of CD24 compared to tumors with low-risk ($P < 0.05$).

4. Discussion

To the best of our knowledge, this is the first study to demonstrate the correlation between CD24 and retinoblastoma. There are three main findings in the present study. At first, there was no CD24 expression in non-neoplastic retina; secondly, positive CD24 protein expression was observed in more than half of the retinoblastoma tissues by both

TABLE 1: The expression of CD24 protein in high-risk and low-risk retinoblastoma.

Group	No. of cases	CD24 immunohistochemical staining			
		Positive	Heterogeneous	Faint	Negative
Overall cohort	50	19	11	6	14
High-risk tumors	15	9	4	1	1
Low-risk tumors	35	10	7	5	13
P value			<0.05		

TABLE 2: The expression of CD24 protein and clinicopathological features of retinoblastoma tissues with high risk of extraocular relapse.

Case no.	Age/sex	Clinicopathological features	CD24 positivity (%)
1	3 mon/M	OS: PD, Post Lam ON Inv	30
2	5/M	OD: PD, rectus orbital invasion	80
3	3/M	OS: PD, Diff Ch Inv	40
4	7/M	OS: PD, Post Lam ON Inv	70
5	2/M	OS: PD, Diff Ch, Post Lam ON Inv	10
6	20 mon/F	OS: PD, Diff Ch, Post Lam ON Inv	30
7	5/M	OS: PD, Diff Ch, Pre Lam ON inv	60
8	3/M	OS: PD, Diff Ch, Post Lam ON Inv	0
9	1/F	OD: PD, Post Lam ON Inv with SE involved	70
10	2/F	OS: PD, Diff Ch, Pre Lam ON Inv	50
11	1/F	OD: PD, Diff Ch Inv	60
12	4/M	OS: PD, Diff Ch Inv	80
13	3/M	OD: PD, Focal Ch, Post Lam ON Inv	90
14	2/F	OS: PD, Diff Ch, Post Lam ON Inv	50
15	2/M	OD: PD, Focal Ch, Post Lam ON Inv	70

Note: mon: months; M: male; F: female; OD: right eye; OS: left eye; PD: poorly differentiated; MD: moderately differentiated; WD: well differentiated; UL: unilateral disease; BL: bilateral disease; Diff Ch Inv: diffused choroidal invasion of tumor; Focal Ch Inv: focal invasion of tumor cells into choroids; Pre Lam ON Inv: pre-laminar invasion of optic nerve; Post Lam ON Inv: post-laminar invasion of the optic nerve.

immunohistochemistry and western blot analysis; and thirdly the level of CD24 was significantly higher in the retinoblastoma tissues with high risk of extraocular relapse than those with low risk of extraocular relapse. These results suggest that CD24 protein may be involved in the pathogenesis of retinoblastoma, and the level of CD24 protein expression may be associated with the severity of retinoblastoma.

CD24 was identified 30 years ago [16]. Since then, it has been extensively used as a marker for the differentiation of hematopoietic and neuronal cells, in addition to tissue and tumor stem cells [17]. CD24 has been reported to be involved in the control of cell proliferation, apoptosis, and cell adhesion. CD24 gene encodes a sialoglycoprotein that is expressed mainly on mature granulocytes, premature lymphocytes, and epithelial and neural cells. The encoded CD24 protein is anchored to the cell surface by a glycosylphosphatidylinositol and acts as a ligand for P-selectin on activated endothelial cells and platelets [18]. In lymphocytes, CD24 can modulate the capacities of early T and B lymphoid progenitor cells to proliferate and survive. Antibody-mediated cross-linking of CD24 induces apoptosis

in a process involving the B-cell receptor and mitogen-activated protein kinases [19]. CD24 expression is only limited to several cell types in physiological condition. However, a large variety of tumors express CD24. For example, CD24 is broadly overexpressed on esophageal squamous cell carcinoma, small cell lung cancer, hepatocellular carcinoma, cholangiocarcinoma, pancreatic adenocarcinoma, urothelial carcinoma, prostate carcinomas, ovarian cancer, breast cancer, B-cell lymphomas, erythroleukemia, gliomas, and primary neuroendocrine carcinomas [8–15, 20]. In most cancer types, CD24 expression is significantly associated with a more aggressive course of the disease, indicating CD24 may play a role in the progression of cancer. In the pancreatic cancers, Ikenaga et al. demonstrated that higher tumor stage, nodal metastasis, and higher-grade tumors were more frequent in the CD24-positive group compared with the CD24-negative group. CD24 expression was also associated with shorter survival in univariate analysis [21]. In addition, Lee et al. reported that the non-small-cell lung carcinoma patients with CD24-high tumors tended to have a higher risk of disease progression and cancer-related death. Multivariate analysis proved CD24-high expression as

TABLE 3: The expression of CD24 protein and clinicopathological features of retinoblastoma tissues with low risk of extraocular relapse.

Case no.	Age/sex	Clinicopathological features	CD24 positivity (%)
1	22 mon/M	OD: WD	0
2	3/M	OS: WD	10
3	3 mon/F	OS: PD, Pre Lam ON Inv	30
4	8/F	OD: PD, Pre Lam ON Inv	60
5	1 mon/F	OD: PD, Pre Lam ON Inv	60
6	5/F	OD: MD, Focal Ch, Pre Lam ON Inv	90
7	4/F	OD: MD, Focal Ch Inv	50
8	2/F	OD: WD, Pre Lam ON Inv	70
9	3/F	OD: WD, Focal Ch, Pre Lam ON Inv	50
10	2/M	OS: PD, Focal Ch Inv	80
11	3/M	OD: WD, Focal Ch Inv	60
12	3/M	OS: PD	0
13	5/M	OS: PD	0
14	1/F	OD: WD	50
15	3/F	OS: PD	0
16	4 mon/M	OS: PD	0
17	2 mon/M	OS: WD	0
18	2/M	OD: PD	40
19	2/M	OD: MD	10
20	16 mon/M	OS: PD, Pre Lam ON Inv	70
21	6 mon/M	OS: PD	30
22	2/M	OD: WD	60
23	20 mon/M	OD: WD	0
24	23 mon/M	OS: PD	50
25	3 mon/M	OD: PD	0
26	1/M	OS: WD	0
27	5 mon/M	OD: WD	30
28	1 mon/M	OS: WD	20
29	3/M	OS: PD	0
30	11 mon/M	OD: MD	10
31	5/M	OS: MD	0
32	2/F	OS: PD	20
33	1/M	OS: MD	0
34	4 mon/M	OD: MD	0
35	5/M	OS: PD, Pre Lam ON Inv	60

Note: mon: months; M: male; F: female; OD: right eye; OS: left eye; PD: poorly differentiated; MD: moderately differentiated; WD: well differentiated; UL: unilateral disease; BL: bilateral disease; Focal Ch Inv: focal invasion of tumor cells into choroids; Pre Lam ON Inv: pre-laminar invasion of optic nerve.

independent prognostic factors of disease progression and cancer-related death [22]. Moreover, Sano group indicated that CD24 expression was associated with lymph node metastasis status, pathologic stage, number of nodal metastases, lymphatic invasion status, and venous invasion status of patients with esophageal squamous cell carcinoma. They also concluded that the overexpression of CD24 was a novel independent prognostic marker for identifying patients with poor prognosis after curative resection of esophageal squamous cell carcinoma [23]. In agreement with these previous reports, our study demonstrated an overexpression of CD24 in human retinoblastoma tissues, which also correlated significantly to the disease progression. Non-neoplastic

retina failed to show any detectable immunoreactivity for CD24.

The significance and clinical implications of CD24 expression in retinoblastoma tissues are yet to be elucidated. In this study, the expression levels of CD24 protein in patients with high risk of extraocular relapse were significantly higher than in patients with lower risk of extraocular relapse. These results suggest tumor tissue CD24 may act as a biomarker for prediction of the severity of retinoblastoma and the prognosis of the patients.

In conclusion, this preliminary study has demonstrated for the first time that more than half of retinoblastoma express the CD24 protein. Tissue levels of CD24 protein are increased in high-risk tumors compared to low-risk tumors. Further functional studies are required to explore the role of CD24 in retinoblastoma.

Conflict of Interests

The authors declare no conflict of interests.

References

[1] K. Mallikarjuna, C. S. Sundaram, Y. Sharma et al., "Comparative proteomic analysis of differentially expressed proteins in primary retinoblastoma tumors," *Proteomics—Clinical Applications*, vol. 4, no. 4, pp. 449–463, 2010.

[2] S. Vandhana, P. R. Deepa, U. Jayanthi, J. Biswas, and S. Krishnakumar, "Clinico-pathological correlations of fatty acid synthase expression in retinoblastoma: an indian cohort study," *Experimental and Molecular Pathology*, vol. 90, no. 1, pp. 29–37, 2011.

[3] P. Indovina, A. Acquaviva, G. De Falco et al., "Downregulation and aberrant promoter methylation of *p16ink4a*: a possible novel heritable susceptibility marker to retinoblastoma," *Journal of Cellular Physiology*, vol. 223, no. 1, pp. 143–150, 2010.

[4] S. Kase, J. G. Parikh, and N. Rao, "Expression of α-crystallin in retinoblastoma," *Archives of Ophthalmology*, vol. 127, no. 2, pp. 187–192, 2009.

[5] C. Bleckmann, H. Geyer, A. Lieberoth et al., "O-glycosylation pattern of CD24 from mouse brain," *Biological Chemistry*, vol. 390, no. 7, pp. 627–645, 2009.

[6] C. Bleckmann, H. Geyer, V. Reinhold et al., "Glycomic analysis of N-linked carbohydrate epitopes from CD24 of mouse brain," *Journal of Proteome Research*, vol. 8, no. 2, pp. 567–582, 2009.

[7] N. Cremers, M. A. Deugnier, and J. Sleeman, "Loss of CD24 expression promotes ductal branching in the murine mammary gland," *Cellular and Molecular Life Sciences*, vol. 67, no. 13, pp. 2311–2322, 2010.

[8] B. Liu, Y. Zhang, M. Liao et al., "Clinicopathologic and prognostic significance of CD24 in gallbladder carcinoma," *Pathology and Oncology Research*, vol. 17, no. 1, pp. 45–50, 2011.

[9] M. O. Riener, A. Vogetseder, B. C. Pestalozzi et al., "Cell adhesion molecules P-cadherin and CD24 are markers for carcinoma and dysplasia in the biliary tract," *Human Pathology*, vol. 41, no. 11, pp. 1558–1565, 2010.

[10] J. H. Lee, S. H. Kim, E. S. Lee, and Y. S. Kim, "CD24 overexpression in cancer development and progression: a meta-analysis," *Oncology Reports*, vol. 22, no. 5, pp. 1149–1156, 2009.

[11] H. J. Wei, T. Yin, Z. Zhu, P. F. Shi, Y. Tian, and C. Y. Wang, "Expression of CD44, CD24 and esa in pancreatic adenocarcinoma cell lines varies with local microenvironment," *Hepatobiliary and Pancreatic Diseases International*, vol. 10, no. 4, pp. 428–434, 2011.

[12] X. R. Yang, Y. Xu, B. Yu et al., "CD24 is a novel predictor for poor prognosis of hepatocellular carcinoma after surgery," *Clinical Cancer Research*, vol. 15, no. 17, pp. 5518–5527, 2009.

[13] H. J. Lee, D. I. Kim, C. Kwak, J. H. Ku, and K. C. Moon, "Expression of CD24 in clear cell renal cell carcinoma and its prognostic significance," *Urology*, vol. 72, no. 3, pp. 603–607, 2008.

[14] S. C. Smith and D. Theodorescu, "The ral gtpase pathway in metastatic bladder cancer: key mediator and therapeutic target," *Urologic Oncology*, vol. 27, no. 1, pp. 42–47, 2009.

[15] S. Bektas, B. Bahadir, B. H. Ucan, and S. O. Ozdamar, "CD24 and galectin-1 expressions in gastric adenocarcinoma and clinicopathologic significance," *Pathology and Oncology Research*, vol. 16, no. 4, pp. 569–577, 2010.

[16] Z. Wang, Q. Shi, Z. Wang et al., "Clinicopathologic correlation of cancer stem cell markers CD44, CD24, VEGF and HIF-1α in ductal carcinoma in situ and invasive ductal carcinoma of breast: an immunohistochemistry-based pilot study," *Pathology Research and Practice*, vol. 207, no. 8, pp. 505–513, 2011.

[17] J. B. Overdevest, S. Thomas, G. Kristiansen, D. E. Hansel, S. C. Smith, and D. Theodorescu, "CD24 offers a therapeutic target for control of bladder cancer metastasis based on a requirement for lung colonization," *Cancer Research*, vol. 71, no. 11, pp. 3802–3811, 2011.

[18] K. Taniuchi, I. Nishimori, and M. A. Hollingsworth, "Intracellular CD24 inhibits cell invasion by posttranscriptional regulation of bart through interaction with g3bp," *Cancer Research*, vol. 71, no. 3, pp. 895–905, 2011.

[19] G. Kristiansen, E. MacHado, N. Bretz et al., "Molecular and clinical dissection of CD24 antibody specificity by a comprehensive comparative analysis," *Laboratory Investigation*, vol. 90, no. 7, pp. 1102–1116, 2010.

[20] X. Fang, P. Zheng, J. Tang, and Y. Liu, "CD24: from A to Z," *Cellular and Molecular Immunology*, vol. 7, no. 2, pp. 100–103, 2010.

[21] N. Ikenaga, K. Ohuchida, K. Mizumoto et al., "Characterization of CD24 expression in intraductal papillary mucinous neoplasms and ductal carcinoma of the pancreas," *Human Pathology*, vol. 41, no. 10, pp. 1466–1474, 2010.

[22] H. J. Lee, G. Choe, S. Jheon, S. W. Sung, C. T. Lee, and J. H. Chung, "CD24, a novel cancer biomarker, predicting disease-free survival of non-small cell lung carcinomas: a retrospective study of prognostic factor analysis from the viewpoint of forthcoming (seventh) new tnm classification," *Journal of Thoracic Oncology*, vol. 5, no. 5, pp. 649–657, 2010.

[23] A. Sano, H. Kato, S. Sakurai et al., "CD24 expression is a novel prognostic factor in esophageal squamous cell carcinoma," *Annals of Surgical Oncology*, vol. 16, no. 2, pp. 506–514, 2009.

10

Stable Plastid Transformation for High-Level Recombinant Protein Expression: Promises and Challenges

Meili Gao,[1] Yongfei Li,[2] Xiaochang Xue,[3] Xianfeng Wang,[4] and Jiangang Long[5]

[1] Department of Biological Science and Engineering, Key Laboratory of Biomedical Information Engineering of the Ministry of Education, School of Life Science and Technology, Xi'an Jiaotong University, Xi'an, Shaanxi 710049, China
[2] School of Materials and Chemical Engineering, Xi'an Technological University, Xi'an, Shaanxi 710032, China
[3] State Key Laboratory of Cancer Biology, Department of Biopharmaceutics School of Pharmacy, Fourth Military Medical University, Xi'an, Shaanxi 710032, China
[4] Department of Anesthesiology, Wake Forest School of Medicine, Winston-Salem, NC 27157, USA
[5] Institute of Mitochondrial Biology and Medicine, Key Laboratory of Biomedical Information Engineering of the Ministry of Education, School of Life Science and Technology, Xi'an Jiaotong University, Xi'an, Shaanxi 710049, China

Correspondence should be addressed to Jiangang Long, jglong@126.com

Academic Editor: Elvira Gonzalez De Mejia

Plants are a promising expression system for the production of recombinant proteins. However, low protein productivity remains a major obstacle that limits extensive commercialization of whole plant and plant cell bioproduction platform. Plastid genetic engineering offers several advantages, including high levels of transgenic expression, transgenic containment via maternal inheritance, and multigene expression in a single transformation event. In recent years, the development of optimized expression strategies has given a huge boost to the exploitation of plastids in molecular farming. The driving forces behind the high expression level of plastid bioreactors include codon optimization, promoters and UTRs, genotypic modifications, endogenous enhancer and regulatory elements, posttranslational modification, and proteolysis. Exciting progress of the high expression level has been made with the plastid-based production of two particularly important classes of pharmaceuticals: vaccine antigens, therapeutic proteins, and antibiotics and enzymes. Approaches to overcome and solve the associated challenges of this culture system that include low transformation frequencies, the formation of inclusion bodies, and purification of recombinant proteins will also be discussed.

1. Introduction

The demand for recombinant proteins such as biopharmaceutical proteins and industrial enzymes is expected to rise dramatically in the near future. However, the current capacity and cost of production for most recombinant proteins limits their availability [1]. Therefore, the strong global demand for low-cost and high-yield recombinant proteins is the impetus driving molecular farming, particularly in developing nations [2]. Commercial production of such recombinant proteins has traditionally relied on bacterial fermentation or mammalian cell-based production. However, limitations including cost, scalability, safety, and

protein authenticity with these expression systems have prompted research into alternative platforms [3, 4].

Recently, plant-based systems potentially provide a low-cost alternative for the production of recombinant proteins. Strategies for plant transformation contain stable nuclear transformation, stable plastid transformation, plant cell-suspension, and transient expression systems [5]. Plant cell suspension cultures have several advantages including the capacity for shorter life cycles, independence from environmental effects such as climate, soil quality, season, day length and weather, the lack of biosafety issues such as gene flow via pollen, and the possibility of bacterial contamination from the plant growth environment [6]. But, the yield and quality of recombinant proteins in plant cell culture-based expression systems need to be further improved. In addition, the transient expression systems, which are perhaps the fastest and the most convenient production platform for plant molecular farming, are mainly used for quick validation of expression constructs [7].

Production of recombinant proteins in transgenic plants was initially based on integration of a target gene into the nuclear genome and later included transformation of the chloroplast genome [1]. Stable nuclear transformation leads to the expression of the transgene after integration with the host genome. This transformation confers stably inheritable traits that were not present in the untransformed host plant [5]. Plant-based systems combine advantages of both production systems: as higher eukaryotes, plants synthesis complex multimeric proteins with posttranslational modifications closely resembling mammalian modifications. In addition, production in plants eliminates the risk of product contamination by human pathogens possibly hidden in mammalian cell lines or in their complex organic production media [8]. However, except for few recombinant proteins, most often very low expression levels of foreign proteins (less than 1% of the total soluble protein, TSP) were observed in nuclear transgenic plants. Also, gene silencing can occur in nuclear transformation, which results in lower expression of recombinant proteins [9]. The impinging problems of nuclear transformation associated with position effects due to random gene integration, and safety due to environmental dissemination of genes by pollen has hampered its expediency for commercialization [10]. For commercial exploitation of the therapeutic proteins and vaccine antigens, high and reliable levels of expression are required, which could be achieved by alternative approaches [9].

Plastid transformation provides a valuable alternative to nuclear transformation because it combines numerous advantages, especially high expression levels that the nuclear transformation lacks. This review focuses on stable plastid transformation in plant. Here in, we give main advantages on plastid information, factors for high-yield production, the expression level of recombinant proteins in plastid, the challenges directions in the development and commercialization of recombinant proteins in plastid expression system are discussed.

2. Advantages of Plastid Expression Systems

Plant cells contain three genomes: a large one in the nucleus and two smaller ones in the mitochondria and plastids. Plastids are a group of organelles that include the sites of photosynthesis of chloroplasts, as well as several other differentiation forms, including the carotenoid-accumulating chromoplasts in flowers and fruits, and the starch-storing amyloplasts in roots and tubers. As semiautonomous organelles, each cell contains a large number of plastids, ~100 chloroplasts per cell and each chloroplast contains about 100 genomes. Therefore, plastid transformation permits the introduction of thousands of copies of transgenes

per plant cell. It dramatically enhances the protein production in the cell [11, 12]. Though both plastid transformation and nuclear transformation are stable recombinant protein expression systems in plants, the protein expression level is far higher in the former transformation than that of the latter transformation.

The issue of transgene containment and prevention of its escape into the environment and into wild-type plant populations is becoming increasingly relevant due to the exponential growth of the use of genetically modified plants in agriculture [13, 14]. Generally, nuclear transgenes can be transmitted by pollen and thus require additional genetic modifications to ensure transgene containment, such as engineering of male sterility [15]. However, chloroplast genomes defy the laws of Mendelian inheritance in that they are maternally inherited in most species and the pollen does not contain chloroplasts. The chloroplast expression system has a natural biocontainment of transgene flow by out-crossing. In this regard, transplastomic plants are much safer than plants with nuclear transgenes. Therefore, the plastid expression system is an environmentally friendly approach and is allaying public concerns [5, 9].

In addition, transgene integration into the plastome is based on two homologous recombination events between the targeting regions of the transformation vector and the wild-type ptDNA (plastid genome or plastome) [16]. Chloroplast transformation eliminates the concerns of position effect, frequently observed in nuclear transgenic lines [17]. Recently, it has been demonstrated that the extent of similarity between the plastidial sequences involved in homologous recombination is important to ensure high transformation efficiency [18, 19]. Hence, the lack of transgene silencing has been observed in chloroplast transformation accompanied with higher expression levels than in nuclear transgenic plants. For example, no gene silencing has been observed in spite of high translation levels, up to 46.1% TSP. It has been observed that there is also no gene silencing when transcripts accumulated 169-fold and 150-fold higher in transgenic chloroplasts than nuclear transgenics [20, 21].

Besides previous advantages, it is possible to express several genes using a single promoter which makes the plastid transformation approach a highly attractive method [22]. Several heterologous operons have been expressed in transgenic chloroplasts, and polycistrons are translated without processing into monocistrons [23]. Factually, most plastid genes are arranged in operons, which are transcribed as polycistronic mRNAs. The processing mechanisms for translation regulation in chloroplast genes mainly include posttranscriptional RNA processing and intercistronic processing. Posttranscriptional RNA processing of primary transcripts represents an important control which relies more on RNA stability than on transcriptional regulation of chloroplast gene expression. RNA stability is mainly influenced by the presence of 5′-UTRs, nucleus-encoded factors and 3′-UTRs. Intercistronic processing (i.e., RNA cutting) refers to many primary polycistronic transcripts in plastids undergoing posttranscriptional cleavage into monocistronic or oligocistronic units. This process enhances translation of chloroplast operons such as psbB, petD, and pet clusters in maize. Additionally, different species may experience various processing mechanisms for the same gene cluster. On the other hand, some polycistronic mRNAs in plastids are translatable and do not undergo posttranscriptional cleavage into monocistronic units. These polycistronic mRNAs often have canonical SD sequences upstream of each individual cistron. Simultaneously, some polycistronic transcripts are not translatable that endonucleolytic processing can be a prerequisite for protein biosynthesis [23, 24]. Chaperones, which present in chloroplasts, facilitate chloroplasts to show the correct folding and assembling complex mammalian proteins [25]. This was demonstrated through the formation of functional protein such as interferon alpha and gamma as well as cholera toxin-B subunit (CTB) in transgenic chloroplasts [9]. Further, chloroplasts can also be a good place to store the biosynthetic products that could otherwise be harmful when accumulated in cytosol [26]. As described above, CTB was toxic when expressed and accumulated in the cytosol in very small quantities. However, CTB was accumulated in large quantities and it had no toxic effect in chloroplast [27]. Trehalose is used as a preservative in the pharmaceutical industry. Similarly, trehalose was toxic when accumulated in cytosol but was nontoxic when compartmentalized within chloroplasts [20].

3. Factors for High-Yield Production in Chloroplast Expression Systems

Chloroplast offers an alternative stable expression system to nuclear transformation. Highly polyploid nature of the plastid genome will allow the integration of thousands of copies of transgenes per cell. This will result production of very high levels of proteins in the transgenic plants produced by plastid transformation [22]. The regulation of recombinant protein expression is a complex system. It includes the interacting elements and the extent of interdependence between different factors, which is not completely understood. Some factors, such as promoters, UTR sequences, codon optimization, post-translational modification, construction of transcriptional fusions, protease activity, as well as accumulation of toxic recombinant protein in chloroplast have been concentrated. Recent progress and development on these factors affecting recombinant protein levels are reviewed here.

3.1. Codon Optimization. The genetic code is made up of many redundant triplets that encode for the same amino acid. This implies many alternative nucleic acid sequences can encode a protein. Since the rules for deciphering a DNA sequence to determine the amino acid sequence of the encoded protein were established over 40 years ago, the genomes of different organisms, and the different genomes of single organisms, employ codon biases as mechanisms for optimizing and regulating protein expression are well established [28]. Optimizing the codon usage of most heterologous genes further reflects that codon biases can

increase their expression efficiency by increasing their translation rates, and may decrease their susceptibility to gene silencing [29].

The expression level of recombinant protein was very low when recombinant genes were directly taken from other systems and were not optimized for expression in *Chlamydomonas reinhardtii* chloroplasts [30]. As described above, codon bias is the most important determinant of protein expression in prokaryotic genomes [31], and adjustment of codons in transgenes is necessary for high level (i.e., commercially viable) expression [32]. This was further evidenced by a green fluorescent protein (gfp) reporter gene in accordance with such codon bias. The codon-optimized gfp and nonoptimized native gfp genes were transformed into chloroplasts, both driven by the same promoter and UTR [33]. The codon-optimized gfp gene resulted in an 80-fold increase in GFP accumulation compared with the nonoptimized version. Transformation of a codon-optimized human antibody gene [34] and codon-optimized luciferase gene [32] in *C. reinhardtii* chloroplasts also confirmed that codon bias plays a significant role in protein accumulation. Thus, increased protein production in these strains highlights the necessity for codon optimization of any gene for which high levels of protein production are desired. Further, in a recent study, a hepatitis C virus core polypeptide expressed in chloroplasts, the results suggested that the codonoptimised gene increased monocistronic core mRNA levels by at least 2-fold and core polypeptides by over 5-fold, relative to the native viral gene [35].

Recently, the codon adaptation index (CAI) is used as a quantitative tool to predict heterologous gene expression levels based on their codon usage. As the chloroplast, mitochondrial, and nuclear genomes may exhibit different codon biases, genome-specific CAI values should be used for optimal translational [36]. Codon optimization is an effective and necessary step in gene sequence optimization, and one relatively simple to address with recent advances in DNA synthesis technology. Analysis of codon usage in chloroplast genes by e-CAI evaluation software (http://genomes.urv.es/CAIcal/E-CAI/) showed that translational selection does indeed operate for a majority of genes in the chloroplast. However, codon optimization is not the only factor to be considered for selection a desired gene for high level expression of recombinant protein in plastid transformation. Further, the work in [37] describes an "optimal" gene which the codon choices do not limit expression is a desired need in the future study.

3.2. Promoters and UTRs. The gene expression level in plastids is predominantly determined by promoter and 5′-UTR elements. Promoters contain the sequences which are required for RNA polymerase binding to start transcription and regulation of transcription. As the strength and expression profile of the key regulatory element "promoter", it plays an important role in driving the transcription to achieve high level of transcription. Hence, in order to obtain high-level protein accumulation from expression of the transgene, the first requirement is a strong promoter to ensure high levels of mRNA. Chloroplast-specific promoters have also been utilized for targeting the foreign protein expression into chloroplasts [12]. For example, the 16S ribosomal RNA promoter (Prrn) like psbA and atpA gene promoters are commonly used for chloroplast transformation. Several molecules including CTB, LTB, protective antigen, insulin, or vaccines have been produced in chloroplasts using either Prrn or psbA promoter [27, 31, 38–41]. These promoters drive the high level of recombinant protein expression in plastid transformation. Other promoters may be found from Dow AgroSciences LLC which has secured exclusive rights to Chloroplast Transformation Technology (CTT) from Chlorogen, Inc. (http://www.dowagro.com/search.aspx?q=promoter).

Stability of the transgenic mRNA is ensured by the 5′-UTR and 3′-UTR sequences flanking the transgenes. The 5′-UTR is very important for translation initiation and plays a critical role in determining the translational efficiency. This was also evidenced by a series of studies in plastid transformation. For example, transcriptional efficiency was shown to be regulated by both chloroplast gene promoters as well as sequences contained within the 5′-UTR [42]. A variety of studies have revealed that translational efficiency is a rate limiting step for chloroplast gene expression [43] and have shown that the 5′-UTRs of plastid mRNAs contain key elements for translational regulation [44]. Additionally, sequences found within the 5′-UTR are also important for optimal levels of transcription, although the nature of these internal enhancer sequences has yet to be defined [42]. The 3′-UTR plays an important role in gene expression as it contains message for transcript polyadenylation that directly affects mRNA stability [45]. The most commonly used 5′-UTR and 3′-UTR are psbA/TpsbA [46–50]. However, in a recent study, a gfp reporter gene and the 5′- and 3′-UTRs from five different chloroplast genes were used to construct a series of chimeric genes in the chloroplast genome. The results showed that the highest levels of recombinant protein expression were obtained using either the atpA or psbD 5′-UTRs, while the nature of the 3′-UTR invariably had little effect on reporter protein accumulation [51]. Hence, more 5′- and 3′-UTRs and the sequences within the coding region on the expression of recombinant protein in plastid should be further studied.

3.3. Genotypic Modifications.

A synthesized gene is generally modified from the natural version because natural genes are often poorly expressed in heterologous hosts, even when the expression system is related to the organism from which the gene originated [37]. Transgenes are inserted in the chloroplast genome by homologous recombination, which implies that each transformant obtained should be identical if using a single integration vector. Identical recombinant protein expression profiles for each transformant are therefore expected [36]. However, Surzycki et al. [31] reported the protein yields varying from 0.88% to 20.9% total cell protein (TCP) in transgenic lines obtained from a single biolistic transformation in chloroplast. This variation may be due to genotypic modifications resulting from the transformation process. The low expression levels of transgenic proteins may depend more on these modifications than on the selection of promoters, UTRs, or insertion sites.

3.4. Endogenous Enhancer and Regulatory Elements.

As previously reported, for posttranscriptional regulation in chloroplast, the light plays a vital role in the translation of many chloroplast mRNAs [43, 52]. The highest level of light induction is for the psbA mRNA which encodes D1, a core protein of photosystem II [53]. Additionally, the psbA 5′-UTR is capable of conferring light-regulated translation to recombinant mRNAs [34, 51]. This offers the potential to regulate recombinant protein expression, which might be necessary to express proteins that are not well tolerated by the chloroplast. Further, psbA-driven heterologous protein expression is increased when the endogenous psbA gene is deleted [32]. This may be due to either the removal of autoattenuation from D1 protein feedback [54] or to reduced competition with endogenous psbA for limited transcription or translation factors. In fact, this is a process that we refer to as control by epistasy of synthesis (CES process) and occurs during chloroplast protein biogenesis in C. reinhardtii [55]. The synthesis of a CES subunit is markedly reduced in the absence of its assembly partners, which involves negative feedback or feedback inhibition [56, 57]. Therefore, in C. reinhardtii chloroplast studies, gene products, including cytochrome f, photosystem I (PSI), regulate the translation of their own mRNA through feedback inhibition [54, 57]. Nevertheless, the low expression levels of heterologous genes were observed in microalgal chloroplasts but not in tobacco chloroplast expression systems, in which this inhibition is not observed. In addition, placing the chloroplast transgene under the transcriptional regulation of an inducible factor is for inducible expression of recombinant proteins in higher plant chloroplasts [58, 59]. Further, it has been evidenced that a large set of nucleus-encoded factors act mostly at posttranscriptional steps of chloroplast gene expression. Among these proteins, the Nac2 protein of C. reinhardtiiis, is specifically required for the stable accumulation of the psbD mRNA encoding the D2 reaction center polypeptide of PSII [60].

Generally, an intrinsic helicase which exists in ribosomes has the ability to allow translation through even very strong hairpins and to preclude many structures from limiting the translation rate [61]. Hence, an actively translated message can be densely packed with ribosomes, unwinding structure as they move along. The rates of ribosome binding and clearing of the ribosome-binding site (RBS) after initial elongation (approx. 13–20 codons) play an important role in translational initiation. Slow translation through the initial leader may reduce or eliminate any benefits of a strong RBS sequence. Gene design strategies often seek to minimize mRNA structure. Structures that involve or otherwise occlude the RBS and/or start codon in genes expressed in prokaryotes can impair expression, presumably by interfering with ribosomal binding and translational initiation [62, 63]. An algorithm, which designs prokaryotic RBSs to achieve desired rates for initiation of translation considering the structure of the mRNA and the affinity of the RBS for the ribosome, has been recently developed (https://github.com/hsalis/ribosome-binding-site-calculator).

The fusion of recombinant products to native proteins has also resulted in an increase of protein yield. It has been reported that the endogenous rubisco LSU protein was fused to a recombinant luciferase through a cleavable domain in algal chloroplasts. This resulted in a 33-fold increase in luciferase expression compared to luciferase expressed alone. The results suggest that recombinant protein accumulation can be enhanced by fusion with a native protein. Also, the liberation of recombinant proteins from the native ones would simplify product purification and increase the yield of recombinant protein [64].

3.5. Posttranslational Modification and Proteolysis. Chloroplasts lack the machinery to perform complex posttranslational modifications, like glycosylation, on proteins. So, proteins not requiring posttranslational modifications can be expressed in chloroplast expression systems. Many proteins do require posttranslational modifications not performed in the prokaryotic plastids. As we know, nonglycosylated antibodies have greatly reduced activation of complement and somewhat reduced Fc-mediated binding in activation of antibody-dependent cell-mediated cytotoxicity (ADCC). In addition, complement fixation and ADCC activation are not required for antigen binding. Moreover, therapeutic antibodies that function to sequester molecules or block binding sites do not require the activation of ADCC and strive to avoid activation of complement. Therefore, chloroplast-expressed nonglycosylated antibodies might actually be superior to glycosylated antibodies for some therapeutic uses [65]. In some cases, for instance, for the production of the therapeutic protein human alpha1-anti-trypsin (A1AT) as well as *Plasmodium falciparum* surface protein 25 (Pfs25) and 28 (Pfs28), the absence of glycosylation could be considered an advantage. Human alpha1-antitrypsin (A1AT), a major therapeutic protein, is that mature A1AT is a glycosylated protein containing three N-linked carbohydrate sidechains. Though the glycosylation is important for the half-life of A1AT in the plasma, it is not required for the binding to elastase. As a consequence, the production of an active unglycosylated version in plants can be envisaged. Nadai and his colleagues have produced A1AT, in genetically engineered tobacco plastids. These chloroplast-made therapeutic proteins are fully active and bind to porcine pancreatic elastase [66]. Malaria transmission blocking vaccine candidates, *Plasmodium falciparum* surface protein 25 (Pfs25) and 28 (Pfs28), are structurally complex aglycosylated outer membrane proteins that contain four tandem epidermal growth factor-like (EGF) domains, each with several disulfide bonds. The chloroplast of green algae can fold complex proteins and make disulfide bonds, but lacks the machinery for glycosylation. Thus, these proteins have been produced in chloroplast of green algae which are structurally similar to the native proteins and antibodies raised to these recombinant proteins recognize Pfs25 and Pfs28 from *P. falciparum* [67].

Since the first evidence of disulfide-bond formation of human somatotropin has been expressed in tobacco chloroplasts [41], many recombinant proteins which contain the expected disulfide bonds have been expressed in chloroplasts of both higher plants and *C. reinhardtii* [48]. Consequently, chloroplasts could be an excellent system for the expression of proteins that require structural disulfide bonds. The chloroplast proteins responsible for disulfide-bond formation could be the same proteins used to transduce the light activation signals and used to regulate chloroplast translation, as one of these redox-dependent proteins is a protein disulfide isomerase [65]. As discussed above, Pfs25 and Pfs28 expressed in chloroplast of gree alage further evidenced that chloroplast could fold complex proteins and make disulfide bonds and form functional proteins [67].

As we know, the proteases of prokaryotic origin play critical roles in chloroplast development and maintenance [68]. Clearly, except the chloroplast processing peptidases which cleave transit peptides, processive proteases such as the serine ATP-dependent Clp protease, the ATP-dependent metalloprotease FtsH and the serine-dependent DegP protease, degrade abnormal soluble and membrane-bound proteins, unassembled proteins and the D1 protein of PSII, as well as misfolded and periplasmic proteins, respectively. The identity of the cross-reacting protein of ATP-dependent Lon protease still needs to be confirmed [69, 70]. Factually, chloroplasts of both higher plants and algae contain proteases commonly found in bacteria; Clp, Deg, and FtsH proteases are all found in the nuclear genome of *C. reinhardtii* with at least one ortholog of each targeted to the chloroplast [69, 71]. However, these are a minor proportion of the proteases normally encountered in the cytoplasm of eukaryotic cells. Therefore, the chloroplast could potentially be a more sheltered environment for the production of proteins that are particularly susceptible to proteolysis [65]. The level of foreign protein accumulation results from a balance between rates of protein synthesis and degradation, the latter of which is increasingly found to impact recombinant product yields. Proteolytic enzymes, which are essential for endogenous protein processing, may lead to the degradation of foreign proteins after synthesis, or interfere with their correct assembly and posttranslational modification. Proteolysis may also lead to inconsistent results and to difficulties in downstream processing or purification due to degraded or nonfunctional protein fragments [36]. Hence, several strategies are available to minimize proteolytic degradation of foreign proteins in plants. For example, the coexpression of protease inhibitors has proven useful in increasing recombinant protein yields in plants, without affecting normal growth and development [36].

4. High-Level Expression of Recombinant Proteins in Plastids

Attainment of high expression level of foreign proteins in plastids is a major breakthrough, which makes this system ideal for large-scale production of recombinant protein [9]. As previously reported, recombinant proteins expressed from chloroplast genomes normally make up 5–25% of TSP [72, 73]. Some chloroplast transgenes have even been reported to accumulate to levels of 72% TSP [18], which is over to recombinant protein expression levels in bacteria. In contrast, the recombinant protein expression levels in nuclear do not exceed 1-2% TSP in most cases. For example, the expression level of human interferon-gamma (hINF-γ) in chloroplast was 100 times higher than those of hINF-γ in nuclear [74]. In another study, a phage-derived protein antibiotic, PlyGBS, expressed from a chloroplast transgene accumulated to extremely high levels, exceeding 70% TSP [75]. Thus, if plants are intended to be used as bioreactors for large-scale production of recombinant proteins, chloroplast, rather than nuclear, genome should be targeted for genetic modification [76].

4.1. Vaccine Antigens. A vaccine is an antigenic preparation used to establish immunity against a disease and the main aim of the vaccination is to eradicate infectious diseases. The development and use of vaccines represent one of the greatest achievements in medical history. Today, numerous life-threatening diseases can be prevented efficiently by immunization, and one of them, smallpox, has even been eradicated [77]. The transplastomic plants have emerged as an attractive production system for vaccines which includes the often attainable high antigen yields, the low production costs once stable lines established, and the potential of producing orally applicable (as a result of high expression levels). Several vaccine candidates have been produced successfully via plastid transformation in the past few years [9]. This demonstrates that transplastomic plants, as a second-generation expression system, have great potential to fill gaps in conventional production platforms. A salient feature of plastids is that they combine characteristics of prokaryotic and eukaryotic expression systems [78]. Vaccines produced in plastid vaccines were proved to be fully functional and able to elicit the appropriate immune responses in experimental animals and to protect against toxin or pathogen challenges. Several recent reviews [16, 78–80] describe the production of vaccine antigens in transplastomic plants. Herein, we will concentrate on the relative high expression level of recombinant antigens produced in plastid.

CTB is responsible for inducing both mucosal and serum immunity. Since the cholera toxin is internalized by the receptors present on mucosal lining, the CTB was one of the early toxins selected for testing the concept of edible vaccines. Further, CTB being a bacterial protein is not glycosylated in native form. Hence, its feasibility for developing vaccine antigen has been examined by expressing the gene in plants both by transformation into chloroplastic and nuclear genome [81]. CTB subunit, the first vaccine expressed in tobacco chloroplasts, showed the expression level as high as 4.1% TSP [27]. Conversely, the tobacco leaves expressed CTB protein at 0.02% of TSP [82]. Mishra et al. [83] fused CTB with ubiquitin at N-terminal end increases the level further to 0.91% of TSP. In some cases, CTB fusions with target antigens have been used as a potent mucosal immunogen

and adjuvant because of its high binding affinity for the GM1-ganglioside receptor in mucosal epithelium. Then, it was shown to be functional by the GM1 binding assay [27, 84]. Recently, two CTB fusion proteins were expressed in tobacco and lettuce chloroplasts. Fusion proteins, containing CTB and the antigens AMA1 and MSP1 of malaria, were reported to be completely protected against cholera toxin (CT) challenge upon oral immunization. The expression level was 13.17% TSP and 10.11% TSP, respectively [85]. In another study, the fusion protein of CTB with fibronectin-binding domain (D2) of *Staphylococcus aureus*, a bacterium responsible for skin infections and bacteraemia, which may lead to life-threatening secondary infections such as endocarditis, showed the expression level as high as 23% TSP. The results revealed the induction of specific mucosal and systemic immune responses of fusion protein of the CTB-D2 in mice sera and faeces. The pathogen load in the spleen and the intestine of treated mice was significantly reduced in treated mice. Algae-based vaccination protected 80% of animals against lethal doses of *S. aureus*. Importantly, the alga vaccine was stable for more than 1.5 years at room temperature [86]. As described above, the fusion protein of CTB with proinsulin (CTB-pins, diabetes type 1) was achieved overexpression up to 72% of TSP in plastids [40].

The expression in most plastid transformation studies is higher than the expected threshold of 1% TSP which is considered as threshold to allow commercial or economical production [80]. Many other reports exist about high expression of vaccine antigens in plastids. For instance, Tregoning et al. [87] reported fairly high expression of a vaccine candidate TetC in tobacco plastids, accumulating up to 18–27% of TSP. Mucosal immunization of mice with the plastid-produced TetC induced protective levels of TetC antibodies [88]. Further, Tregoning and his colleagues reported that a single intranasal (i.n.) vaccination was as efficient as oral delivery, inducing high levels of activated CD4+ T cells and antitoxin antibody [87]. The high expression level of vaccine proteins such as vaccine virus envelope protein (A27L, Smallpox), human papillomavirus (L1), anthrax protective antigen (*pagA*), and HIV (p24-Nef) was up to 18% [89], 20–26% [90], 29% [18], 40% [91], respectively. Also, the immune responses of chloroplast made of these proteins were reported in the related studies. The chloroplast-made A27L protein was recognized by serum from a patient recently infected with a zoonotic OPV. Other characteristics included the ability to form oligomerize and the stability over a wide range of pH values. Hence, A27L protein could be a distinct advantage for the induction of intestinal secretory immunoglobulin A (IgA) following oral immunization [89]. The chloroplast-derived HPV16 L1 protein displayed conformation-specific epitopes and the ability of assembled into virus-like particles, highly immunogenic in mice after intraperitoneal injection, and neutralizing antibodies [90]. High antibody titers, especially IgG1 titers and IgG2a titers, were produced in plastid *pagA* gene expression protein treated mice. The antibodies from various groups were efficient in neutralizing the lethal toxin *in vitro*. When mice were challenged with *B. anthracis*, mice immunized with the protein imparted 60% and 40%

protection upon intraperitoneal and oral immunizations [10]. The largest cleavage product of HIV major gene-*gag*, p24, forms the conical core of HIV-1 viral particles and is the target of T-cell immune responses in both primary and chronically infected patients. The p24-Nef fusion and the p17/p24 protein effectively boosted T cell and humoral responses in mice [91, 92]. Therefore, chloroplast-derived vaccines have very promising and competitive potential for commercialization [80].

4.2. Therapeutic Proteins and Antibiotics. Recombinant therapeutic proteins are useful for treatment of various conditions such as genetic diseases that result in the production of an insufficient quantity or quality of a particular protein [79]. Plastid transgenic plant strategies have been used to demonstrate the production of many valuable human proteins, including insulin-like growth factor-1 (IGF-1), cardiotrophin-1, aprotinin, alphal-antitrypsin, and thioredoxin 1 [16]. Most therapeutic proteins also have competitive potential for commercialization. Among these proteins, the native and the optimized synthetic IGF-1 genes were expressed in *E. coli* and in transplastomic plants [93]. Expression of both genes was obtained only in transgenic tobacco plant line where the expression of IGF-1 reached up to 32% TSP using plastid light regulatory elements and under continuous illumination [79].

Both the rapid spread of antibiotic resistance and the stagnating discovery of new antibiotics have created an urgent need for alternative antimicrobials for clinical use. Recently, the use of the components of the phage that are needed to kill the bacterium would be a much-preferred strategy. These components, which possess highly efficient hydrolytic enzymes, are referred to as endolysins [78]. At the same time, plastid transformation has also been used as a novel strategy for large-scale, cost-effective production of next-generation antibiotics for topical and systemic treatments [75, 94, 95]. Three recent studies have explored and tested the potential of using chloroplasts as cheap factories for high-level production of endolysin-type protein antibiotics in tobacco chloroplasts: PlyGBS, Pal, and Cpl-1. Expression levels of the Pal, Cpl-1, and synthetic plyGBS gene in chloroplasts were 30%, 10%, and 70%, respectively [75, 95, 96]. In addition, two other potent disulphide-bonded antimicrobial peptides, protegrin-1 (PG1) and retrocyclin-101 (RC101), have also been investigated in transgenic tobacco plastids [94]. The expression levels of two peptides were as high as 26% and 38%, respectively. The antibacterial activity or antiviral infection of these antibiotic proteins was proved efficiently in plastid. Further details can be seen in recent two reviews by Bock and Warzecha [78] and Scotti et al. [16].

4.3. Enzymes and Others. Plastid transformation has been explored for the expression of various enzymes of biological and pharmaceutical importance. The gene for thermostable xylanase was expressed in the chloroplasts of tobacco plants [74]. Xylanase accumulated in the cells to approximately 6% TSP. Zymography assay demonstrated that the estimated activity was $421\,U\,mg^{-1}$ in crude TSP [97]. Recently, GH10 xylanase Xyl10B from *Thermotoga maritima* was expressed in tobacco chloroplasts and accumulated to high level of 13% TSP [98]. Enzymes such as endoglucanase from *Thermobifida fusca* and the endo-β-1,4-glucanase E1 catalytic domain of *Acidothermus celluloyticus*, which involved in biofuel production, were expressed in tobacco chloroplast and showed the expression level as high as 10.7% TSP [99] and 12% TSP [100], respectively. Endoglucanase Cel9A from *Thermobifida fusca* showed the expression level as high as 40% TSP [101]. Many other enzymes recently produced in plant chloroplast were described by Scotti et al. [16].

Also, chloroplasts have been used as bioreactors for production of biomaterial and amino acids. Normally, p-hydroxybenzoic acid (pHBA) is produced in small quantities in all plants. However, stable integration of the ubiC gene into the tobacco chloroplast resulted in hyperexpression of the enzyme and accumulation of this polymer up to 25% of dry weight [102, 103]. As we know, synthetic plastics are difficult to dispose off and continually accumulating nondegradable wastes have become a significant source of environmental pollution. Hence, biodegradable plastics seem to be a viable alternative to synthetic plastics [12]. As an alternative for efficient and inexpensive biomaterial production system, plants have also been engineered to produce polyhydroxybutyrates (PHBs) in the various plant cell compartments [1, 12]. Regenerated plants which an ecdysone analogue-based system was recently employed for induced gene expression in nuclear, contained up to 1-2% PHB (dry weight) in leaves after 6–8 weeks of induction [104]. Conversely, for the economic feasibility of transgenic plants-derived biodegradable plastics, accumulation of at least 15% of the tissue dry weight is required [105]. The biodegradable PHBs were expressed in plastids to increase the expression level. The expression levels ranging up to 40% of dry weight have been obtained [106]. Further, there is a need to aim at higher accumulation without any side effects on plants [12].

5. Current Problems and Future Prospects

Transgene expression from the plant's plastid genome has unique attractions to biotechnologists, including the plastids' potential to accumulate foreign proteins to extraordinarily high levels and the increased biosafety provided by the maternal mode of plastid inheritance, which greatly reduces unwanted transgene transmission via pollen. However, recent data [107] indicate that no transplastomic plant products have been licensed for this purpose as yet; even among all potential transgenic plant products, so far, only two have completed the regulatory processes for licensing: a recombinant single chain antibody to the hepatitis B surface antigen [108], and a Newcastle disease virus vaccine [109]. In addition, recombinant monoclonal antibodies for treatment of non-Hodgkin's lymphoma produced from a plant viral vector [110] were approved by the Food and Drug Administration for manufacturing, but the potential risks in their production and use, and the significant investment of plant-based technology discouraged further progress [107]. This enormous potential notwithstanding, plastid transformation is still limited in its applications for the following issues to be resolved.

5.1. Transformation Frequency and Transformation Vector.

As we know, plastid transformation was limited by low transformation frequencies in potato and other crops. Hence, a breakthrough in chloroplast genetic engineering of agronomically important specie is a highly desirable goal. An approach, which used a modified regeneration procedure and novel vectors containing potato flanking sequence for transgene integration by homologous recombination, achieved efficiency up to one shoot every bombardment in potato transformation in large single-copy region of the plastome [111]. As vector delivery was performed by the biolistic approach, such efficiency corresponds to 15–18 fold improvement, and it is comparable to that usually achieved with tobacco. This represents a significant advancement toward the implementation of the plastid transformation technology.

Although chloroplast transformation technology has advanced significantly in the past two decades, the available plastid transformation vectors still lack several of the important functional features found in binary vectors [76]. First, only a limited number of genes of interest (GOIs) can be cloned into a single chloroplast transformation vector due to intrinsic multiple cloning site (MCS) limitations. Second, unlike binary vectors, which produce T-DNA capable of integrating into any nuclear genome sequence, chloroplast transformation vectors require a certain degree of homology with the plastid genome and, thus, may not be suitable for plant species with insufficiently conserved transgene integration sites [103]. To achieve high efficiency of transplastomic plant production, therefore, replacement of the homologous recombination sequences for each particular species/group of species through the time-consuming and laborious process of vector reconstruction may be required. Alternatively, the target plant itself can be converted into a "universal" recipient by integrating into its plastid genome-specific recombination sites, such as those for the phiC31 phage integrase, albeit also through lengthy experimentation [112].

5.2. Regulation of Gene Expression.

The high expression of recombinant proteins within plastid-engineered systems offers a cost-effective solution for using plants as a bioreactor. In the expression of rHSA in tobacco chloroplasts, the yield of rHSA was increased 500-fold when compared with the expression level in nuclear transformation, reaching 11.1% of TSP [113]. However, this resulted in the formation of inclusion bodies, which needed a further renaturation or refolding process resulting in low recovery of rHSA after purification [50]. Although much of the forgoing discussion has implicitly assumed that maximizing the rate of translational elongation is unequivocally desirable, this is not entirely accurate. It has been suggested that too rapid translation may not allow for efficient "self" or chaperone-aided folding. So, slower codons or codon runs, perhaps at protein domain boundaries, were strategically placed. This could maximize folding efficiency while maintaining a high overall translation rate [114]. In addition, constitutive expression of pharmaceutical proteins or unique metabolic pathways from the plastid genome can result in mutant phenotypes and/or severe growth retardation of transplastomic plants due to metabolite toxicities, interference with photosynthesis, or disturbance of the plastid endomembrane system. Recently, the approach based on an engineered riboswitch is established. This approach acts as a translational regulator of transgene expression in transformed plastids in response to the application of the ligand theophylline. The theophylline riboswitch offers a "plastid-only" solution to inducible gene expression from the chloroplast genome that does not require additional (nuclear or plastid) transgenes and thus should be widely applicable [115].

5.3. Downstream Processing.

The main reason for the high cost of pharmaceutical protein production is purification of recombinant proteins. Also, this technology has not resulted in any product commercialization because problems in the protein purification still need to be solved as discussed later. Recently, a novel protein purification method is described carefully that do not require the use of expensive column chromatography [9]. Factually, this method is based on the inverse transition temperature $\langle T_t \rangle$ of the polymers of elastin's repeating VPGXG or GVGVP sequences. The inverse temperature transition property exhibited the phenomenon as temperature rises, the polymer collapses from an extended chain to a β-spiral structure with three VPGVG or GVGVP units per turn [9, 116]. Elastin as well as Elastin-like polypeptide (ELP) is well solvated and is highly soluble in aqueous solution below T_t. When the solution is heated and the T_t is reached, elastin become insoluble and form large micron-size aggregates that are visible to the naked eye [117]. Also, the environmental sensitivity and reversible solubility of ELPs are retained when an ELP is fused at the gene level with other proteins, and the activity of the ELP fusion protein is retained after cycling through the inverse phase transition [118]. This characteristic transition allows the recombinant ELP fusion protein to be isolated from the cell lysate by repeated steps of aggregation, centrifugation, and resolubilization without chromatography. Based on the previous description, a nonchromatographic method for protein purification was termed as inverse transition cycling (ITC). In ITC, a target protein or peptide is fused to the ELP at the gene level, expressed in E. coli or another expression system. After expression, the cells are lysed, and the cell debris is removed by centrifugation. The ELP fusion protein is then separated from soluble contaminants by triggering the phase transition of the ELP fusion protein [117, 119–121]. As for crystal structural and biochemical characterization, using proteases or self-cleaving ELP tags has been devised for purification of the tag-free recombinant proteins [120, 121]. So, ITC is both cost and time efficient because this purification method eliminates chromatography and scales up of this purification method is easy because it is not limited by resin capacity. To date, ITC has been used to purify many protein ELP fusions, including cytokines, antibodies, and spider silk proteins from transgenic plants [120, 122]. Furthermore, this method will be used to purify protein ELP fusions in chloroplast expression system and to promote chloroplast transformation product commercialization.

Information about the stability of a protein can lead to a more thorough understanding of the mode of action as well as the effects of exposure to various conditions on the transgenic protein [123]. Protein stability represents more common causes of the lack of foreign protein accumulation in transgenic chloroplasts. Unfortunately, the rules governing protein stability in plastids and the identity of sequence motifs and/or structural motifs within the protein that confer stability or susceptibility to degradation are still a mystery. Unraveling the determinants of protein stability in plastids should potentially provide ways of stabilizing, otherwise, unstable recombinant proteins and, therefore, would be of enormous value to plastid biotechnology [78]. Recent studies have reported higher stability of the fused recombinant protein in chloroplast transformation. For example, thioredoxins (Trxs) are small ubiquitous disulphide proteins widely known to enhance expression and solubility of recombinant proteins in microbial expression systems. Sanz-Barrio and his colleagues reported that Trxs-HSA fusions markedly increased the final yield of human serum albumin (HAS, up to 26% of total protein) by higher HAS stability of the fused protein [124]. As discussed previously, RC101 and PG1, two important antimicrobial peptides, accumulate high expression levels by fusion with GFP to confer stability [94]. Additionally, disulphide bond formation is crucial for the biological activity of many therapeutic proteins. Alkaline phosphatase, whose activity and stability strictly depend on the correct formation of two intramolecular disulphide bonds, was expressed in tobacco chloroplast with the efficient formation of disulphide bonds [125]. Moreover, it is also envisaged that protein stability will change over time even with refrigeration [126]. The protein that the transgene encodes should be characterized to determine stability to pH, temperature, and chemical or biochemical agents involved downstream processing [123]. These need to be further studied on stability of recombinant protein in chloroplast transformation in the future.

As previous description, except proteins not requiring posttranslational modifications and nonglycosylated antibodies, it is not suitable to produce glycosylated recombinant proteins because chloroplasts are derived from ancient bacteria that are unable to do protein glycosylation [127], limiting the number of different proteins that can be produced using this system. In addition, plants are presently incapable of authentic human N-glycosylation to produce N-glycans that are essential for stability, folding, and biological activity of most therapeutic proteins. Thus, several glycoengineering strategies for the production of N-glycosylation in plants have emerged, including glycoprotein subcellular targeting, the inhibition of plant-specific glycosyltranferases, or the addition of human-specific glycosyltransferases [128, 129]. For example, based on the chloroplast proteome assay, Villarejo and his colleagues reported a chloroplast-located protein which encoded a α-carbonic anhydrase (α-CA). The respective cDNA was denoted CAH1, and CAH1 was enriched in intact chloroplasts and the stroma fraction. CAH1 protein is not only taken up into the ER but is also glycosylated prior to being targeted to the chloroplast [130]. O-glycosylation is one of the most complex regulated posttranslational modifications of proteins and also one of the least understood [131]. Mucin-type (GalNAc-type) O-glycosylation is found in eumetazoan cells, but absent in plants and yeast. Recently, stably engineered mammalian-type O-glycosylation was established in transgenic plants, demonstrating that plants may serve as host cells for production of recombinant O-glycoproteins [132]. A large single-chain antibody against herpes simplex virus glycoprotein D was expressed and assembled correctly to form fully functional dimers by disulfide bond formation [34]. In addition, other posttranslational modifications, such as lipidation, can be achieved, as for the outer surface lipoprotein A of *Borrelia burgdorferi* expressed from the tobacco plastid genome [133]. These posttranslation modifications still poorly explored open interesting possibilities for plastid-based biotechnology and needed to further study in the future.

5.4. Plant Species and Oral Delivery. Although it has been achieved the development of improved selection/regeneration protocols and/or transformation vectors containing homologous flanking sequences in other plant species such as tomato, potato, eggplant, lettuce, and soybeans [18, 19, 134–137], routine chloroplast transformation is only possible in tobacco, which is inedible and highly regulated, being rich in toxic alkaloids. Hence, the extension of plastid transformation technology to other crops, especially those belonging to monocots, is still limited. In addition, to bring the plant to homoplastomy, where only transgenic genome-copies remain, further two or three rounds of regeneration on selective media are typically required [138, 139]. As with nuclear transgenics, it takes at least a year to generate production lines and to scale up. More experiments should be undertaken to move this technology toward practical utilization.

Some vaccine antigens and therapeutic proteins have only been expressed in tobacco. But tobacco is not edible and the addictiveness of nicotine also makes it unsuitable for oral delivery of therapeutic proteins [79]. Recently, lettuce (*Lactuca sativa* L.) chloroplast transformation has been developed. This system has been optimized and several therapeutic proteins have been expressed [140]. The level of expression in lettuce chloroplasts is similar to that in tobacco chloroplasts, and lettuce can be transformed as rapidly as tobacco. Thus, transformation in lettuce opens the door to practical oral delivery of chloroplast-expressed proteins. Chloroplast-derived therapeutic proteins, delivered orally via plant cells, are protected from degradation in the stomach, presumably because of bioencapsulation of the antigen by the plant cell wall [16]. To facilitate translocation of vaccine antigens or therapeutics from the gut lumen into the circulatory system, target proteins have been fused to the CTB transmucosal carrier protein, which can bind to the epithelial receptor GM1 [141]. This approach has been widely applied to many orally delivered antigens, both for stable nuclear transgenics and for transplastomic approaches [93].

5.5. Production Cost and Purification Cost. Recombinant proteins may be expressed in bacteria, fungi and yeast, insect cells, mammalian cells, animals, or plants. The cost of the resulting product, especially downstream processing or purification cost has been described as the following. (1) The generation of recombinant proteins in bacterial system is faster and easier and thus allows for the easy progression to large-scale manufacturing [129]. Currently, the most widely used recombinant protein production systems are bacterial systems. However, the limitation of the presence of endogenously produced endotoxins and pathogens in *E. coli* is difficult and, therefore, costly to remove from target preparations and creates additional complexities [1]. (2) Foreign proteins production in yeasts and fungi offers the cost effectiveness and scaleup benefits of *E. coli* combined with the advantages of eukaryotic expression. Moreover, protein is expressed into the culture supernatant allowing faster and easier purification; as well, the purified protein contains less contaminating endotoxins as compared to the bacterially expressed counterparts [129]. However, they have a number of technical issues, such as the loss of plasmid and dramatic decrease in protein yield during large-scale production, which should be needed to resolve [1]. (3) Though insect cells have a number of advantages, several disadvantages exist. It has been shown that internal cleavage, at arginine- or lysine-rich sequences, is extremely inefficient and leads to improperly processed proteins. Furthermore, glycosylation capability is limited to only high mannosetype [1]. (4) There are currently many established mammalian cell lines for the production of proteins. It should be noted, however, that the development of large-scale expression techniques is time consuming and requires high initial financial investment. Also, this system requires a nutrient media that is more complex than that of bacteria, fungi, or plants. So, the cost of the resulting product is quite substantial [1, 142]. (5) Transgenic technology has allowed for recombinant protein production in living animals such as rabbits, goats, pigs, and cows. Though the transformation of recombinant proteins in mammary glands has been showing a great promise, the production of recombinant protein in blood requires the use of high cost and complicated procedures. Moreover, the process of producing transgenic animals is labor, time, and cost intensive [1, 143]. (6) As for foreign proteins expressed in plants, it may be transiently transformed in nuclear, stably transformed in nuclear and chloroplast or plastid. More details are described as follows: (a) When recombinant proteins were generated by transient expression using RNA virus vectors or agrobacterium-mediated transient expression through Agroinfection, higher yields can generally be obtained. Recombinant proteins including antigen products have to be extracted and purified. Downstream processing of recombinant products is very expensive, amounting to more

than 80% of the total expense [7, 144]. (b) Product yields of recombinant proteins are usually <1% of total soluble proteins in nuclear transformation in plant. Hence, efficient protein purification from transgenic plants is a major challenge, and high impurity content in the feed streams. These drive up the costs of downstream processing [129, 144, 145]. However, if therapeutic proteins and vaccines are delivered orally, then edible transgenic plant offers the possibility of eliminating the need for expensive downstream protein purification and processing, especially in seed-based production [9, 144, 146]. (c) As described above, expression of a target gene from the chloroplast genome generally provides higher yields. The high expression of recombinant proteins within plastid-engineered systems offers a cost-effective solution for using plants as a bioreactor. As with nuclear transgenics, the cost for purification of therapeutic proteins and vaccines can be eliminated if they are orally delivered or minimized by the use of novel purification strategies [1, 147].

Recently, purification cost and production cost of bioreactor engineering for recombinant protein were assayed in bacterial, yeast, insect cells, mammalian cells, and plant cells. The low purification cost has been shown in plant cells, the medium purification cost has been shown in insect cells, and the high purification cost has been indicated in the three other cell lines. The results of the production cost of recombinant protein were different from that of the purification cost. The low production cost has been existed in bacterial ($20–100/g), in yeast ($20–100/g), in plant cells ($50–100/g), respectively. The medium production cost has been exhibited in insect cells ($50–200/g) and the high production cost exhibited in mammalian cells ($1000–10000/g) [148]. In addition, according to the comparison with other expression systems, the overcost of plant-based production platforms is as low as that of bacteria and yeast, and the purification cost is as high as that of bacteria and yeast [149].

5.6. Other Problems. Chloroplast genetic engineering is considered a "plant safe" strategy. However, few reports indicated that pleiotropic effects of vaccine antigens exhibited some detrimental effects, such as male sterility, yellow leaves, and stunted growth of transplastomic plants, especially in the expression of phaA gene in chloroplasts [80]. Except the limitations described above, there are many confrontations that lie in issues that still need to be addressed and solved: these apply mainly for evaluation of the transplastomic plants, the efficacy of plastid derived vaccines, the regulatory issues development of plastid transformation system for edible plant species and marker excision from transplastomic plants. Many of these challenges are discussed in detail in the previous and recent reviews [80, 107, 150].

Based on the previous description, edible transgenic plant tissue will eliminate the downstream processing or purification cost. Production of recombinant proteins in plant is no need to maintain the cold chain as the plant parts expressing the vaccine or plant extracts can be stored and transported at room temperature [9]. Nevertheless, due to the perishable nature of the fruits and vegetables they require immediate processing to avoid postharvest losses (20–25%) [151]. So, nonthermal processing of fruit and vegetable has been revealed as a useful tool to extend their shelf-life and quality as well as to preserve their nutritional and functional characteristics [152]. In addition, most batches of feed or food derived from genetically modified (GM) plants for the safety study should be kept as cold as possible, depending on the storage time needed [153]. When tobacco and lettuce chloroplasts were transformed with the CTB fused with human proinsulin, old tobacco leaves and old lettuce leaves accumulated proinsulin up to 47% of total leaf protein (TLP) and 53% TLP, respectively. Even in senescent and dried lettuce leaves, accumulation was so stable that up to ~40% proinsulin in TLP was observed. This may promote to facilitate their processing and storage in the field [154]. Typically, intake of a protein is estimated by considering actual expression levels in consumed tissues (i.e., fruit or grain versus leaves) and by considering a comprehensive evaluation of food consumption practices of the population [123]. The exposure levels are based on the concentrations of the transgenic protein likely to be encountered in the human diet. According to the OECD Guideline, a single-dose and repeated-dose toxicologies need further studied to predict human exposure [123]. Based on the previous information, except tobacco, recombinant proteins were mainly explored in a number of plant species, including a few vegetables species. Therefore, the shelf life and dose of edible transgenic plants, especially in plastid transformation, are rarely studied. Further research will be necessary for exploring food or feed demands on edible plants.

6. Conclusions

The plastid transformation offered a good platform of foreign gene expression in high plants. In the joint efforts of researchers all over the world, plastid transformation in plants has made considerable progress. To date, more than 50 different transgenes have been stably integrated and expressed via the plant plastid genome to confer important agronomic traits, as well as to produce industrially valuable biomaterials and therapeutic proteins. The ability to engineer chloroplast as an alternative site for the expression of foreign genes, proteins, reactions, and products has gained prominence relatively recently. Considering the recent scientific and technological developments in plastid transformation technology, such as the marker gene elimination systems, the possibility to induce gene expression, the development of novel purification method, and the selection of novel regulatory sequences for expression in chloroplasts, it can be predicted that in the next future the plastid transformation approach will be applied to a larger set of species and for a wider range of purposes.

Acknowledgments

This study was funded by National Natural Science Foundation of China (30700720) and New Century Excellent Talents in University and the National Natural Science Foundation of China (31070740).

References

[1] V. Mett, C. E. Farrance, B. J. Green, and V. Yusibov, "Plants as biofactories," *Biologicals*, vol. 36, no. 6, pp. 354–358, 2008.

[2] A. Pelosi, R. Shepherd, and A. M. Walmsley, "Delivery of plant-made vaccines and therapeutics," *Biotechnology Advances*, vol. 30, no. 2, pp. 440–480, 2012.

[3] R. Boehm, "Bioproduction of therapeutic proteins in the 21st century and the role of plants and plant cells as production platforms," *Annals of the New York Academy of Sciences*, vol. 1102, pp. 121–134, 2007.

[4] S. J. Streatfield, "Approaches to achieve high-level heterologous protein production in plants," *Plant Biotechnology Journal*, vol. 5, no. 1, pp. 2–15, 2007.

[5] O. O. Obembe, J. O. Popoola, S. Leelavathi, and S. V. Reddy, "Advances in plant molecular farming," *Biotechnology Advances*, vol. 29, no. 2, pp. 210–222, 2011.

[6] Q. Y. Sun, L. W. Ding, G. P. Lomonossoff et al., "Improved expression and purification of recombinant human serum albumin from transgenic tobacco suspension culture," *Journal of Biotechnology*, vol. 155, no. 2, pp. 164–172, 2011.

[7] E. P. Rybicki, "Plant-made vaccines for humans and animals," *Plant Biotechnology Journal*, vol. 8, no. 5, pp. 620–637, 2010.

[8] E. L. Decker and R. Reski, "Moss bioreactors producing improved biopharmaceuticals," *Current Opinion in Biotechnology*, vol. 18, no. 5, pp. 393–398, 2007.

[9] S. Chebolu and H. Daniell, "Chloroplast-derived vaccine antigens and biopharmaceuticals: expression, folding, assembly and functionality," *Current Topics in Microbiology and Immunology*, vol. 332, pp. 33–54, 2010.

[10] J. Gorantala, S. Grover, D. Goel et al., "A plant based protective antigen [PA(dIV)] vaccine expressed in chloroplasts demonstrates protective immunity in mice against anthrax," *Vaccine*, vol. 29, no. 27, pp. 4521–4533, 2011.

[11] P. Maliga, "Plastid transformation in higher plants," *Annual Review of Plant Biology*, vol. 55, pp. 289–313, 2004.

[12] A. K. Sharma and M. K. Sharma, "Plants as bioreactors: recent developments and emerging opportunities," *Biotechnology Advances*, vol. 27, no. 6, pp. 811–832, 2009.

[13] H. Daniell, S. Chebolu, S. Kumar, M. Singleton, and R. Falconer, "Chloroplast-derived vaccine antigens and other therapeutic proteins," *Vaccine*, vol. 23, no. 15, pp. 1779–1783, 2005.

[14] J. J. Grevich and H. Daniell, "Chloroplast genetic engineering: recent advances and future perspectives," *Critical Reviews in Plant Sciences*, vol. 24, no. 2, pp. 83–107, 2005.

[15] H. J. Cho, S. Kim, M. Kim, and B. D. Kim, "Production of transgenic male sterile tobacco plants with the cDNA encoding a ribosome inactivating protein in Dianthus sinensis L.," *Molecules and Cells*, vol. 11, no. 3, pp. 326–333, 2001.

[16] N. Scotti, M. Rigano M, and T. Cardi, "Production of foreign proteins using plastid transformation," *Biotechnology Advances*, vol. 30, no. 2, pp. 387–397, 2012.

[17] H. Daniell, "Molecular strategies for gene containment in transgenic crops," *Nature Biotechnology*, vol. 20, no. 6, pp. 581–586, 2002.

[18] T. Ruhlman, D. Verma, N. Samson, and H. Daniell, "The role of heterologous chloroplast sequence elements in transgene integration and expression," *Plant Physiology*, vol. 152, no. 4, pp. 2088–2104, 2010.

[19] V. T. Valkov, D. Gargano, C. Manna et al., "High efficiency plastid transformation in potato and regulation of transgene expression in leaves and tubers by alternative 5′ and 3′ regulatory sequences," *Transgenic Research*, vol. 20, no. 1, pp. 137–151, 2011.

[20] S. B. Lee, H. B. Kwon, S. J. Kwon et al., "Accumulation of trehalose within transgenic chloroplasts confers drought tolerance," *Molecular Breeding*, vol. 11, no. 1, pp. 1–13, 2003.

[21] A. Dhingra, A. R. Portis, and H. Daniell, "Enhanced translation of a chloroplast-expressed RbcS gene restores small subunit levels and photosynthesis in nuclear RbcS antisense plants," *Proceedings of the National Academy of Sciences of the United States of America*, vol. 101, no. 16, pp. 6315–6320, 2004.

[22] K. Chowdhury and O. Bagasra, "An edible vaccine for malaria using transgenic tomatoes of varying sizes, shapes and colors to carry different antigens," *Medical Hypotheses*, vol. 68, no. 1, pp. 22–30, 2006.

[23] T. Quesada-Vargas, O. N. Ruiz, and H. Daniell, "Characterization of heterologous multigene operons in transgenic chloroplasts. Transcription, processing, and translation," *Plant Physiology*, vol. 138, no. 3, pp. 1746–1762, 2005.

[24] O. Drechsel and R. Bock, "Selection of Shine-Dalgarno sequences in plastids," *Nucleic Acids Research*, vol. 39, no. 4, pp. 1427–1438, 2011.

[25] H. Daniell, O. Carmona-Sanchez, and B. Burns, "Chloroplast derived antibodies, biopharmaceuticals and edible vaccines," in *Molecular Farming*, R. Fischer and S. Schillberg, Eds., pp. 113–133, Wiley-Verlag VCH, Weinheim, Germany, 2004.

[26] L. Bogorad, "Engineering chloroplasts: an alternative site for foreign genes, proteins, reactions and products," *Trends in Biotechnology*, vol. 18, no. 6, pp. 257–263, 2000.

[27] H. Daniell, S. B. Lee, T. Panchal, and P. O. Wiebe, "Expression of the native cholera toxin B subunit gene and assembly as functional oligomers in transgenic tobacco chloroplasts," *Journal of Molecular Biology*, vol. 311, no. 5, pp. 1001–1009, 2001.

[28] C. Gustafsson, S. Govindarajan, and J. Minshull, "Codon bias and heterologous protein expression," *Trends in Biotechnology*, vol. 22, no. 7, pp. 346–353, 2004.

[29] M. Heitzer, A. Eckert, M. Fuhrmann, and C. Griesbeck, "Influence of codon bias on the expression of foreign genes in microalgae," *Advances in Experimental Medicine and Biology*, vol. 616, pp. 46–53, 2007.

[30] K. Ishikura, Y. Takaoka, K. Kato, M. Sekine, K. Yoshida, and A. Shinmyo, "Expression of a foreign gene in *Chlamydomonas reinhardtii* chloroplast," *Journal of Bioscience and Bioengineering*, vol. 87, no. 3, pp. 307–314, 1999.

[31] R. Surzycki, K. Greenham, K. Kitayama et al., "Factors effecting expression of vaccines in microalgae," *Biologicals*, vol. 37, no. 3, pp. 133–138, 2009.

[32] S. P. Mayfield and J. Schultz, "Development of a luciferase reporter gene, luxCt, for *Chlamydomonas reinhardtii* chloroplast," *Plant Journal*, vol. 37, no. 3, pp. 449–458, 2004.

[33] S. Franklin, B. Ngo, E. Efuet, and S. P. Mayfield, "Development of a GFP reporter gene for *Chlamydomonas reinhardtii* chloroplast," *Plant Journal*, vol. 30, no. 6, pp. 733–744, 2002.

[34] S. P. Mayfield, S. E. Franklin, and R. A. Lerner, "Expression and assembly of a fully active antibody in algae," *Proceedings of the National Academy of Sciences of the United States of America*, vol. 100, no. 2, pp. 438–442, 2003.

[35] P. Madesis, M. Osathanunkul, U. Georgopoulou et al., "A hepatitis C virus core polypeptide expressed in chloroplasts detects anti-core antibodies in infected human sera," *Journal of Biotechnology*, vol. 145, no. 4, pp. 377–386, 2010.

[36] G. Potvin and Z. Zhang, "Strategies for high-level recombinant protein expression in transgenic microalgae: a review," *Biotechnology Advances*, vol. 28, no. 6, pp. 910–918, 2010.

[37] M. Welch, A. Villalobos, C. Gustafsson, and J. Minshull, "You're one in a googol: optimizing genes for protein expression," *Journal of the Royal Society Interface*, vol. 6, no. 4, pp. S467–S476, 2009.

[38] T. J. Kang, S. C. Han, M. Y. Kim, Y. S. Kim, and M. S. Yang, "Expression of non-toxic mutant of Escherichia coli heat-labile enterotoxin in tobacco chloroplasts," *Protein Expression and Purification*, vol. 38, no. 1, pp. 123–128, 2004.

[39] V. Koya, M. Moayeri, S. H. Leppla, and H. Daniell, "Plant-based vaccine: mice immunized with chloroplast-derived anthrax protective antigen survive anthrax lethal toxin challenge," *Infection and Immunity*, vol. 73, no. 12, pp. 8266–8274, 2005.

[40] T. Ruhlman, R. Ahangari, A. Devine, M. Samsam, and H. Daniell, "Expression of cholera toxin B-proinsulin fusion protein in lettuce and tobacco chloroplasts-oral administration protects against development of insulitis in non-obese diabetic mice," *Plant Biotechnology Journal*, vol. 5, no. 4, pp. 495–510, 2007.

[41] J. M. Staub, B. Garcia, J. Graves et al., "High-yield production of a human therapeutic protein in tobacco chloroplasts," *Nature Biotechnology*, vol. 18, no. 3, pp. 333–338, 2000.

[42] U. Klein, M. L. Salvador, and L. Bogorad, "Activity of the Chlamydomonas chloroplast rbcL gene promoter is enhanced by a remote sequence element," *Proceedings of the National Academy of Sciences of the United States of America*, vol. 91, no. 23, pp. 10819–10823, 1994.

[43] S. Eberhard, D. Drapier, and F. A. Wollman, "Searching limiting steps in the expression of chloroplast-encoded proteins: relations between gene copy number, transcription, transcript abundance and translation rate in the chloroplast of *Chlamydomonas reinhardtii*," *Plant Journal*, vol. 31, no. 2, pp. 149–160, 2002.

[44] J. Nickelsen, "Chloroplast RNA-binding proteins," *Current Genetics*, vol. 43, no. 6, pp. 392–399, 2003.

[45] M. T. Chan and S. M. Yu, "The 3′ untranslated region of a rice α-amylase gene functions as a sugar-dependent mRNA stability determinant," *Proceedings of the National Academy of Sciences of the United States of America*, vol. 95, no. 11, pp. 6543–6547, 1998.

[46] O. V. Zoubenko, L. A. Allison, Z. Svab, and P. Maliga, "Efficient targeting of foreign genes into the tobacco plastid genome," *Nucleic Acids Research*, vol. 22, no. 19, pp. 3819–3824, 1994.

[47] J. Watson, V. Koya, S. H. Leppla, and H. Daniell, "Expression of Bacillus anthracis protective antigen in transgenic chloroplasts of tobacco, a non-food/feed crop," *Vaccine*, vol. 22, no. 31-32, pp. 4374–4384, 2004.

[48] H. Daniell, S. Kumar, and N. Dufourmantel, "Breakthrough in chloroplast genetic engineering of agronomically important crops," *Trends in Biotechnology*, vol. 23, no. 5, pp. 238–245, 2005.

[49] C. Kittiwongwattana, K. Lutz, M. Clark, and P. Maliga, "Plastid marker gene excision by the phiC31 phage site-specific recombinase," *Plant Molecular Biology*, vol. 64, no. 1-2, pp. 137–143, 2007.

[50] A. Fernández-San Millán, A. Mingo-Castel, M. Miller, and H. Daniell, "A chloroplast transgenic approach to hyper-express and purify human serum albumin, a protein highly

susceptible to proteolytic degradation," *Plant Biotechnology Journal*, vol. 1, no. 2, pp. 71–79, 2003.

[51] D. Barnes, S. Franklin, J. Schultz et al., "Contribution of 5′- and 3′-untranslated regions of plastid mRNAs to the expression of *Chlamydomonas reinhardtii* chloroplast genes," *Molecular Genetics and Genomics*, vol. 274, no. 6, pp. 625–636, 2005.

[52] D. L. Herrin and J. Nickelsen, "Chloroplast RNA processing and stability," *Photosynthesis Research*, vol. 82, no. 3, pp. 301–314, 2004.

[53] P. Malnoe, S. P. Mayfield, and J. D. Rochaix, "Comparative analysis of the biogenesis of photosystem II in the wild-type and y-1 mutant of *Chlamydomonas reinhardtii*," *Journal of Cell Biology*, vol. 106, no. 3, pp. 609–616, 1988.

[54] L. Minai, K. Wostrikoff, F. A. Wollman, and Y. Choquet, "Chloroplast biogenesis of photosystem II cores involves a series of assembly-controlled steps that regulate translation," *Plant Cell*, vol. 18, no. 1, pp. 159–175, 2006.

[55] Y. Choquet, D. B. Stern, K. Wostrikoff, R. Kuras, J. Girard-Bascou, and F. A. Wollman, "Translation of cytochrome f is autoregulated through the 5′ untranslated region of petA mRNA in Chlamydomonas chloroplasts," *Proceedings of the National Academy of Sciences of the United States of America*, vol. 95, no. 8, pp. 4380–4385, 1998.

[56] K. Wostrikoff, Y. Choquet, F. A. Wollman, and J. Girard-Bascou, "TCA1, a single nuclear-encoded translational activator specific for petA mRNA in *Chlamydomonas reinhardtii* chloroplast," *Genetics*, vol. 159, no. 1, pp. 119–132, 2001.

[57] K. Wostrikoff, J. Girard-Bascou, F. A. Wollman, and Y. Choquet, "Biogenesis of PSI involves a cascade of translational autoregulation in the chloroplast of Chlamydomonas," *EMBO Journal*, vol. 23, no. 13, pp. 2696–2705, 2004.

[58] L. Buhot, E. Horvàth, P. Medgyesy, and S. Lerbs-Mache, "Hybrid transcription system for controlled plastid transgene expression," *Plant Journal*, vol. 46, no. 4, pp. 700–707, 2006.

[59] A. Lössl, K. Bohmert, H. Harloff, C. Eibl, S. Mühlbauer, and H. U. Koop, "Inducible trans-activation of plastid transgenes: expression of the R. eutropha phb operon in transplastomic tobacco," *Plant and Cell Physiology*, vol. 46, no. 9, pp. 1462–1471, 2005.

[60] R. Surzycki, L. Cournac, G. Peltier, and J. D. Rochaix, "Potential for hydrogen production with inducible chloroplast gene expression in Chlamydomonas," *Proceedings of the National Academy of Sciences of the United States of America*, vol. 104, no. 44, pp. 17548–17553, 2007.

[61] S. Takyar, R. P. Hickerson, and H. F. Noller, "mRNA helicase activity of the ribosome," *Cell*, vol. 120, no. 1, pp. 49–58, 2005.

[62] K. E. Griswold, N. A. Mahmood, B. L. Iverson, and G. Georgiou, "Effects of codon usage versus putative 5′-mRNA structure on the expression of Fusarium solani cutinase in the Escherichia coli cytoplasm," *Protein Expression and Purification*, vol. 27, no. 1, pp. 134–142, 2003.

[63] S. M. Studer and S. Joseph, "Unfolding of mRNA Secondary Structure by the Bacterial Translation Initiation Complex," *Molecular Cell*, vol. 22, no. 1, pp. 105–115, 2006.

[64] M. Muto, R. E. Henry, and S. P. Mayfield, "Accumulation and processing of a recombinant protein designed as a cleavable fusion to the endogenous Rubisco LSU protein in Chlamydomonas chloroplast," *BMC Biotechnology*, vol. 9, p. 45, 2009.

[65] S. P. Mayfield, A. L. Manuell, S. Chen et al., "*Chlamydomonas reinhardtii* chloroplasts as protein factories," *Current Opinion in Biotechnology*, vol. 18, no. 2, pp. 126–133, 2007.

[66] M. Nadai, J. Bally, M. Vitel et al., "High-level expression of active human alpha1-antitrypsin in transgenic tobacco chloroplasts," *Transgenic Research*, vol. 18, no. 2, pp. 173–183, 2009.

[67] J. A. Gregory, F. Li, L. M. Tomosada et al., "Algae-produced Pfs25 elicits antibodies that inhibit Malaria transmission," *PLoS ONE*, vol. 7, no. 5, Article ID e37179, 2012.

[68] D. Zhang, Y. Kato, L. Zhang et al., "The FtsH protease heterocomplex in Arabidopsis: dispensability of Type-B protease activity for proper chloroplast development," *Plant Cell*, vol. 22, no. 11, pp. 3710–3725, 2010.

[69] Z. Adam, "Chloroplast proteases: possible regulators of gene expression?" *Biochimie*, vol. 82, no. 6-7, pp. 647–654, 2000.

[70] W. Sakamoto, "Protein degradation machineries in plastids," *Annual Review of Plant Biology*, vol. 57, pp. 599–621, 2006.

[71] R. Bock, "Transgenic plastids in basic research and plant biotechnology," *Journal of Molecular Biology*, vol. 312, no. 3, pp. 425–438, 2001.

[72] H. Daniell, "Production of biopharmaceuticals and vaccines in plants via the chloroplast genome.," *Biotechnology journal*, vol. 1, no. 10, pp. 1071–1079, 2006.

[73] P. Maliga, "Progress towards commercialization of plastid transformation technology," *Trends in Biotechnology*, vol. 21, no. 1, pp. 20–28, 2003.

[74] S. Leelavathi, N. Gupta, S. Maiti, A. Ghosh, and V. S. Reddy, "Overproduction of an alkali- and thermo-stable xylanase in tobacco chloroplasts and efficient recovery of the enzyme," *Molecular Breeding*, vol. 11, no. 1, pp. 59–67, 2003.

[75] M. Oey, M. Lohse, B. Kreikemeyer, and R. Bock, "Exhaustion of the chloroplast protein synthesis capacity by massive expression of a highly stable protein antibiotic," *Plant Journal*, vol. 57, no. 3, pp. 436–445, 2009.

[76] B. Meyers, A. Zaltsman, B. Lacroix, S. V. Kozlovsky, and A. Krichevsky, "Nuclear and plastid genetic engineering of plants: comparison of opportunities and challenges," *Biotechnology Advances*, vol. 28, no. 6, pp. 747–756, 2010.

[77] J. B. Ulmer, U. Valley, and R. Rappuoli, "Vaccine manufacturing: challenges and solutions," *Nature Biotechnology*, vol. 24, no. 11, pp. 1377–1383, 2006.

[78] R. Bock and H. Warzecha, "Solar-powered factories for new vaccines and antibiotics," *Trends in Biotechnology*, vol. 28, no. 5, pp. 246–252, 2010.

[79] H. Daniell, N. D. Singh, H. Mason, and S. J. Streatfield, "Plant-made vaccine antigens and biopharmaceuticals," *Trends in Plant Science*, vol. 14, no. 12, pp. 669–679, 2009.

[80] A. G. Lössl and M. T. Waheed, "Chloroplast-derived vaccines against human diseases: achievements, challenges and scopes," *Plant Biotechnology Journal*, vol. 9, no. 5, pp. 527–539, 2011.

[81] S. Tiwari, P. C. Verma, P. K. Singh, and R. Tuli, "Plants as bioreactors for the production of vaccine antigens," *Biotechnology Advances*, vol. 27, no. 4, pp. 449–467, 2009.

[82] D. Jani, N. K. Singh, S. Bhattacharya et al., "Studies on the immunogenic potential of plant-expressed cholera toxin B subunit," *Plant Cell Reports*, vol. 22, no. 7, pp. 471–477, 2004.

[83] S. Mishra, D. K. Yadav, and R. Tuli, "Ubiquitin fusion enhances cholera toxin B subunit expression in transgenic plants and the plant-expressed protein binds GM1 receptors more efficiently," *Journal of Biotechnology*, vol. 127, no. 1, pp. 95–108, 2006.

[84] A. Molina, S. Hervás-Stubbs, H. Daniell, A. M. Mingo-Castel, and J. Veramendi, "High-yield expression of a viral peptide animal vaccine in transgenic tobacco chloroplasts," *Plant Biotechnology Journal*, vol. 2, no. 2, pp. 141–153, 2004.

[85] A. Davoodi-Semiromi, M. Schreiber, S. Nalapalli et al., "Chloroplast-derived vaccine antigens confer dual immunity against cholera and malaria by oral or injectable delivery," *Plant Biotechnology Journal*, vol. 8, no. 2, pp. 223–242, 2010.

[86] I. A. J. Dreesen, G. C. E. Hamri, and M. Fussenegger, "Heat-stable oral alga-based vaccine protects mice from *Staphylococcus aureus* infection," *Journal of Biotechnology*, vol. 145, no. 3, pp. 273–280, 2010.

[87] J. S. Tregoning, S. Clare, F. Bowe et al., "Protection against tetanus toxin using a plant-based vaccine," *European Journal of Immunology*, vol. 35, no. 4, pp. 1320–1326, 2005.

[88] J. S. Tregoning, P. Nixon, H. Kuroda et al., "Expression of tetanus toxin Fragment C in tobacco chloroplasts," *Nucleic Acids Research*, vol. 31, no. 4, pp. 1174–1179, 2003.

[89] M. M. Rigano, C. Manna, A. Giulini et al., "Transgenic chloroplasts are efficient sites for high-yield production of the vaccinia virus envelope protein A27L in plant cells," *Plant Biotechnology Journal*, vol. 7, no. 6, pp. 577–591, 2009.

[90] A. Fernández-San Millán, S. M. Ortigosa, S. Hervás-Stubbs et al., "Human papillomavirus L1 protein expressed in tobacco chloroplasts self-assembles into virus-like particles that are highly immunogenic," *Plant Biotechnology Journal*, vol. 6, no. 5, pp. 427–441, 2008.

[91] F. Zhou, J. A. Badillo-Corona, D. Karcher et al., "High-level expression of human immunodeficiency virus antigens from the tobacco and tomato plastid genomes," *Plant Biotechnology Journal*, vol. 6, no. 9, pp. 897–913, 2008.

[92] A. Meyers, E. Chakauya, E. Shephard et al., "Expression of HIV-1 antigens in plants as potential subunit vaccines," *BMC Biotechnology*, vol. 8, p. 53, 2008.

[93] H. Daniell, G. Ruiz, B. Denes, L. Sandberg, and W. Langridge, "Optimization of codon composition and regulatory elements for expression of human insulin like growth factor-1 in transgenic chloroplasts and evaluation of structural identity and function," *BMC Biotechnology*, vol. 9, p. 33, 2009.

[94] S. B. Lee, B. Li, S. Jin, and H. Daniell, "Expression and characterization of antimicrobial peptides Retrocyclin-101 and Protegrin-1 in chloroplasts to control viral and bacterial infections," *Plant Biotechnology Journal*, vol. 9, no. 1, pp. 100–115, 2011.

[95] M. Oey, M. Lohse, L. B. Scharff, B. Kreikemeyer, and R. Bock, "Plastid production of protein antibiotics against pneumonia via a new strategy for high-level expression of antimicrobial proteins," *Proceedings of the National Academy of Sciences of the United States of America*, vol. 106, no. 16, pp. 6579–6584, 2009.

[96] Z. Adam, "Protein stability and degradation in plastids," *Topics in Current Genetics*, vol. 19, pp. 315–338, 2007.

[97] D. Verma, A. Kanagaraj, S. Jin, N. D. Singh, P. E. Kolattukudy, and H. Daniell, "Chloroplast-derived enzyme cocktails hydrolyse lignocellulosic biomass and release fermentable sugars," *Plant Biotechnology Journal*, vol. 8, no. 3, pp. 332–350, 2010.

[98] J. Y. Kim, M. Kavas, W. M. Fouad, G. Nong, J. F. Preston, and F. Altpeter, "Production of hyperthermostable GH10 xylanase Xyl10B from *Thermotoga maritima* in transplastomic plants enables complete hydrolysis of methylglucuronoxylan to fermentable sugars for biofuel production," *Plant Molecular Biology*, vol. 76, no. 3–5, pp. 357–369, 2011.

[99] B. N. Gray, B. A. Ahner, and M. R. Hanson, "High-level bacterial cellulase accumulation in chloroplast-transformed tobacco mediated by downstream box fusions," *Biotechnology and Bioengineering*, vol. 102, no. 4, pp. 1045–1054, 2009.

[100] T. Ziegelhoffer, J. A. Raasch, and S. Austin-Phillips, "Expression of Acidothermus cellulolyticus E1 endo-β-1,4-glucanase catalytic domain in transplastomic tobacco," *Plant Biotechnology Journal*, vol. 7, no. 6, pp. 527–536, 2009.

[101] K. Petersen and R. Bock, "High-level expression of a suite of thermostable cell wall-degrading enzymes from the chloroplast genome," *Plant Molecular Biology*, vol. 76, no. 3–5, pp. 311–321, 2011.

[102] P. V. Viitanen, A. L. Devine, M. S. Khan, D. L. Deuel, D. E. Van Dyk, and H. Daniell, "Metabolic engineering of the chloroplast genome using the Echerichia coli ubiC gene reveals that chorismate is a readily abundant plant precursor for p-hydroxybenzoic acid biosynthesis," *Plant Physiology*, vol. 136, no. 4, pp. 4048–4060, 2004.

[103] D. Verma and H. Daniell, "Chloroplast vector systems for biotechnology applications," *Plant Physiology*, vol. 145, no. 4, pp. 1129–1143, 2007.

[104] D. A. Dalton, C. Ma, S. Shrestha, P. Kitin, and S. H. Strauss, "Trade-offs between biomass growth and inducible biosynthesis of polyhydroxybutyrate in transgenic poplar," *Plant Biotechnology Journal*, vol. 9, no. 7, pp. 759–767, 2011.

[105] J. Scheller and U. Conrad, "Plant-based material, protein and biodegradable plastic," *Current Opinion in Plant Biology*, vol. 8, no. 2, pp. 188–196, 2005.

[106] K. Bohmert, I. Balbo, J. Kopka et al., "Transgenic Arabidopsis plants can accumulate polyhydroxybutyrate to up to 4% of their fresh weight," *Planta*, vol. 211, no. 6, pp. 841–845, 2000.

[107] E. P. Rybicki, "Plant-produced vaccines: promise and reality," *Drug Discovery Today*, vol. 14, no. 1-2, pp. 16–24, 2009.

[108] M. Pujol, N. I. Ramírez, M. Ayala et al., "An integral approach towards a practical application for a plant-made monoclonal antibody in vaccine purification," *Vaccine*, vol. 23, no. 15, pp. 1833–1837, 2005.

[109] United States Department of Agriculture (USDA) Animal and Plant Health Inspection Service, http://www.aphis.usda.gov/newsroom/content/2006/01/ndvaccine.shtml, 2006.

[110] A. A. McCormick, S. Reddy, S. J. Reinl et al., "Plant-produced idiotype vaccines for the treatment of non-Hodgkin's lymphoma: safety and immunogenicity in a phase I clinical study," *Proceedings of the National Academy of Sciences of the United States of America*, vol. 105, no. 29, pp. 10131–10136, 2008.

[111] N. Scotti, V. T. Valkov, and T. Cardi, "Improvement of plastid transformation efficiently in potato by using vectors with homologous flanking sequences," *GM Crops and Food*, vol. 2, no. 2, pp. 89–91, 2011.

[112] K. A. Lutz, S. Corneille, A. K. Azhagiri, Z. Svab, and P. Maliga, "A novel approach to plastid transformation utilizes the phiC31 phage integrase," *Plant Journal*, vol. 37, no. 6, pp. 906–913, 2004.

[113] I. Farran, J. J. Sánchez-Serrano, J. F. Medina, J. Prieto, and A. M. Mingo-Castel, "Targeted expression of human serum albumin to potato tubers," *Transgenic Research*, vol. 11, no. 4, pp. 337–346, 2002.

[114] E. Angov, C. J. Hillier, R. L. Kincaid, and J. A. Lyon, "Heterologous protein expression is enhanced by harmonizing the codon usage frequencies of the target gene with those of the expression host," *PLoS ONE*, vol. 3, no. 5, Article ID e2189, 2008.

[115] A. Verhounig, D. Karcher, and R. Bock, "Inducible gene expression from the plastid genome by a synthetic riboswitch," *Proceedings of the National Academy of Sciences of the United States of America*, vol. 107, no. 14, pp. 6204–6209, 2010.

[116] H. Reiersen, A. R. Clarke, and A. R. Rees, "Short elastin-like peptides exhibit the same temperature-induced structural transitions as elastin polymers: implications for protein engineering," *Journal of Molecular Biology*, vol. 283, no. 1, pp. 255–264, 1998.

[117] D. E. Meyer and A. Chilkoti, "Genetically encoded synthesis of protein-based polymers with precisely specified molecular weight and sequence by recursive directional ligation: examples from the the elastin-like polypeptide system," *Biomacromolecules*, vol. 3, no. 2, pp. 357–367, 2002.

[118] K. Trabbic-Carlson, L. Liu, B. Kim, and A. Chilkoti, "Expression and purification of recombinant proteins from Escherichia coli: comparison of an elastin-like polypeptide fusion with an oligohistidine fusion," *Protein Science*, vol. 13, no. 12, pp. 3274–3284, 2004.

[119] X. Ge, K. Trabbic-Carlson, A. Chilkoti, and C. D. M. Filipe, "Purification of an elastin-like fusion protein by microfiltration," *Biotechnology and Bioengineering*, vol. 95, no. 3, pp. 424–432, 2006.

[120] W. Hassouneh, T. Christensen, and A. Chilkoti, "Elastin-like polypeptides as a purification tag for recombinant proteins," *Current Protocols in Protein Science*, no. 61, pp. 6.11.1–6.11.16, 2010.

[121] D. Lan, G. Huang, H. Shao et al., "An improved nonchromatographic method for the purification of recombinant proteins using elastin-like polypeptide-tagged proteases," *Analytical Biochemistry*, vol. 415, no. 2, pp. 200–202, 2011.

[122] A. J. Conley, J. J. Joensuu, R. Menassa, and J. E. Brandle, "Induction of protein body formation in plant leaves by elastin-like polypeptide fusions," *BMC Biology*, vol. 7, p. 48, 2009.

[123] B. Delaney, J. D. Astwood, H. Cunny et al., "Evaluation of protein safety in the context of agricultural biotechnology," *Food and Chemical Toxicology*, vol. 46, no. 2, pp. S71–S97, 2008.

[124] R. Sanz-Barrio, A. F. S. Millán, P. Corral-Martínez, J. M. Seguí-Simarro, and I. Farran, "Tobacco plastidial thioredoxins as modulators of recombinant protein production in transgenic chloroplasts," *Plant Biotechnology Journal*, vol. 9, no. 6, pp. 639–650, 2011.

[125] J. Bally, E. Paget, M. Droux, C. Job, D. Job, and M. Dubald, "Both the stroma and thylakoid lumen of tobacco chloroplasts are competent for the formation of disulphide bonds in recombinant proteins," *Plant Biotechnology Journal*, vol. 6, no. 1, pp. 46–61, 2008.

[126] M. E. Horn, S. L. Woodard, and J. A. Howard, "Plant molecular farming: systems and products," *Plant Cell Reports*, vol. 22, no. 10, pp. 711–720, 2004.

[127] V. Yusibov and S. Rabindran, "Recent progress in the development of plant-derived vaccines," *Expert Review of Vaccines*, vol. 7, no. 8, pp. 1173–1183, 2008.

[128] M. H. Ahn, Y. K. Choo, K. Ko et al., "Glyco-engineering of biotherapeutic proteins in plants," *Molecules and Cells*, vol. 25, no. 4, pp. 494–503, 2008.

[129] N. E. Weisser and J. C. Hall, "Applications of single-chain variable fragment antibodies in therapeutics and diagnostics," *Biotechnology Advances*, vol. 27, no. 4, pp. 502–520, 2009.

[130] A. Villarejo, S. Burén, S. Larsson et al., "Evidence for a protein transported through the secretory pathway en route to the higher plant chloroplast," *Nature Cell Biology*, vol. 7, no. 12, pp. 1124–1131, 2005.

[131] E. P. Bennett, U. Mandel, H. Clausen, T. A. Gerken, T. A. Fritz, and L. A. Tabak, "Control of mucintype O-glycosylation-a classification of the polypeptide GalNAc-transferase gene family," *Glycobiology*, vol. 22, no. 6, pp. 736–756, 2012.

[132] Z. Yang, E. P. Bennett, B. Jørgensen, D. P. Drew, E. Arigi, and U. Mandel, "Towards stable genetic engineering of human O-glycosylation in plants," *Plant Physiology Preview,˙* vol. 160, no. 1, pp. 450–463, 2012.

[133] K. Glenz, B. Bouchon, T. Stehle, R. Wallich, M. M. Simon, and H. Warzecha, "Production of a recombinant bacterial lipoprotein in higher plant chloroplasts," *Nature Biotechnology*, vol. 24, no. 1, pp. 76–77, 2006.

[134] R. Bock, "Plastid biotechnology: prospects for herbicide and insect resistance, metabolic engineering and molecular farming," *Current Opinion in Biotechnology*, vol. 18, no. 2, pp. 100–106, 2007.

[135] A. K. Singh, S. S. Verma, and K. C. Bansal, "Plastid transformation in eggplant (*Solanum melongena* L.)," *Transgenic Research*, vol. 19, no. 1, pp. 113–119, 2010.

[136] S. Ruf, M. Hermann, I. J. Berger, H. Carrer, and R. Bock, "Stable genetic transformation of tomato plastids and expression of a foreign protein in fruit," *Nature Biotechnology*, vol. 19, no. 9, pp. 870–875, 2001.

[137] N. Dufourmantel, B. Pelissier, F. Garçon, G. Peltier, J. M. Ferullo, and G. Tissot, "Generation of fertile transplastomic soybean," *Plant Molecular Biology*, vol. 55, no. 4, pp. 479–489, 2004.

[138] K. A. Lutz, Z. Svab, and P. Maliga, "Construction of marker-free transplastomic tobacco using the Cre-loxP site-specific recombination system," *Nature Protocols*, vol. 1, no. 2, pp. 900–910, 2006.

[139] D. Verma, N. P. Samson, V. Koya, and H. Daniell, "A protocol for expression of foreign genes in chloroplasts," *Nature Protocols*, vol. 3, no. 4, pp. 739–758, 2008.

[140] A. Davoodi-Semiromi, N. Samson, and H. Daniell, "The green vaccine: a global strategy to combat infectious and outoimmune diseases," *Human Vaccines*, vol. 5, no. 7, pp. 488–493, 2009.

[141] A. Limaye, V. Koya, M. Samsam, and H. Daniell, "Receptor-mediated oral delivery of a bioencapsulated green fluorescent protein expressed in transgenic chloroplasts into the mouse circulatory system," *The FASEB Journal*, vol. 20, no. 7, pp. 959–961, 2006.

[142] T. G. Warner, "Enhancing therapeutic glycoprotein production in Chinese hamster ovary cells by metabolic engineering endogenous gene control with antisense DNA and gene targeting," *Glycobiology*, vol. 9, no. 9, pp. 841–850, 1999.

[143] K. Mahmoud, "Recombinant protein production: strategic technology and a vital research tool," *Research Journal of Cell and Molecular Biology*, vol. 1, no. 1, pp. 9–22, 2007.

[144] F. Takaiwa, "Seed-based oral vaccines as allergen-specific immunotherapies," *Human Vaccines*, vol. 7, no. 3, pp. 357–366, 2011.

[145] R. M. Twyman, E. Stoger, S. Schillberg, P. Christou, and R. Fischer, "Molecular farming in plants: host systems and expression technology," *Trends in Biotechnology*, vol. 21, no. 12, pp. 570–578, 2003.

[146] J. Boothe, C. Nykiforuk, Y. Shen et al., "Seed-based expression systems for plant molecular farming," *Plant Biotechnology Journal*, vol. 8, no. 5, pp. 588–606, 2010.

[147] H. H. Wang, W. B. Yin, and Z. M. Hu, "Advances in chloroplast engineering," *Journal of Genetics and Genomics*, vol. 36, no. 7, pp. 387–398, 2009.

[148] T. K. Huang and K. A. McDonald, "Bioreactor engineering for recombinant protein production in plant cell suspension cultures," *Biochemical Engineering Journal*, vol. 45, no. 3, pp. 168–184, 2009.

[149] J. Xu, X. Ge, and M. C. Dolan, "Towards high-yield production of pharmaceutical proteins with plant cell suspension cultures," *Biotechnology Advances*, vol. 29, no. 3, pp. 278–299, 2011.

[150] L. Faye, A. Boulaflous, M. Benchabane, V. Gomord, and D. Michaud, "Protein modifications in the plant secretory pathway: current status and practical implications in molecular pharming," *Vaccine*, vol. 23, no. 15, pp. 1770–1778, 2005.

[151] R. L. Bhardwaj and S. Pandey, "Juice blends-a way of utilization of under-utilized fruits, vegetables, and spices: a review," *Critical Reviews in Food Science and Nutrition*, vol. 51, no. 6, pp. 563–570, 2011.

[152] C. Sánchez-Moreno, B. De Ancos, L. Plaza, P. Elez-Martinez, and M. P. Cano, "Nutritional approaches and health-related properties of plant foods processed by high pressure and pulsed electric fields," *Critical Reviews in Food Science and Nutrition*, vol. 49, no. 6, pp. 552–576, 2009.

[153] EFSA GMO Panel, "Safety and nutritional assessment of GM plants and derived food and feed: the role of animal feeding trials," *Food and Chemical Toxicology*, vol. 46, supplement 1, pp. S2–S70, 2008.

[154] D. Boyhan and H. Daniell, "Low-cost production of proinsulin in tobacco and lettuce chloroplasts for injectable or oral delivery of functional insulin and C-peptide," *Plant Biotechnology Journal*, vol. 9, no. 5, pp. 585–598, 2011

11

Antilithiasic and Hypolipidaemic Effects of *Raphanus sativus* L. *var. niger* on Mice Fed with a Lithogenic Diet

Ibrahim Guillermo Castro-Torres,[1] **Elia Brosla Naranjo-Rodríguez,**[2]
Miguel Ángel Domínguez-Ortíz,[3] **Janeth Gallegos-Estudillo,**[3]
and Margarita Virginia Saavedra-Vélez[1]

[1] *Facultad de Química Farmacéutica Biológica, Universidad Veracruzana, Xalapa, Veracruz, Mexico*
[2] *Laboratorio de Neurofarmacología, Departamento de Farmacia, Facultad de Química,*
  Universidad Nacional Autónoma de México, México, DF, Mexico
[3] *Laboratorio de Productos Naturales, Instituto de Ciencias Básicas, Universidad Veracruzana, Xalapa, Veracruz, Mexico*

Correspondence should be addressed to Elia Brosla Naranjo-Rodríguez, eliab@unam.mx

Academic Editor: Ayman El-Kadi

In Mexico, *Raphanus sativus* L. *var. niger* (black radish) has uses for the treatment of gallstones and for decreasing lipids serum levels. We evaluate the effect of juice squeezed from black radish root in cholesterol gallstones and serum lipids of mice. The toxicity of juice was analyzed according to the OECD guidelines. We used female C57BL/6 mice fed with a lithogenic diet. We performed histopathological studies of gallbladder and liver, and measured concentrations of cholesterol, HDL cholesterol and triglycerides. The juice can be considered bioactive and non-toxic; the lithogenic diet significantly induced cholesterol gallstones; increased cholesterol and triglycerides levels, and decreased HDL levels; gallbladder wall thickness increased markedly, showing epithelial hyperplasia and increased liver weight. After treatment with juice for 6 days, cholesterol gallstones were eradicated significantly in the gallbladder of mice; cholesterol and triglycerides levels decreased too, and there was also an increase in levels of HDL ($P < 0.05$). Gallbladder tissue continued to show epithelial hyperplasia and granulocyte infiltration; liver tissue showed vacuolar degeneration. The juice of black radish root has properties for treatment of cholesterol gallstones and for decreasing serum lipids levels; therefore, we confirm in a preclinical study the utility that people give it in traditional medicine.

1. Introduction

Gallstones represent an important problem of public health; its prevalence is about 10% in middle-age persons and 20% in aged persons [1]. Cholesterol is the main component of the majority of gallstones [2]. This disease is strongly associated with atherosclerosis and metabolic syndrome [3], and its pharmacological treatment is limited, being cholecystectomy, an invasive surgical treatment, the only treatment for symptomatic gallstones [4]. Cholesterol gallstones formation is a complex process mediated by genetic and environmental factors [5, 6]. Many proteins (ATP-binding cassette (ABC)) are implicated in its formation, mainly the biliary lipids transporters ABCB4, ABCB11, ABCG5, ABCG8, ABCC7,

and Niemann-Pick C1L1 protein, and these transporters are regulated in the liver by several transcription factors, including nuclear receptors farnesoid X receptor and liver X receptors [7–9]. Other important factors for gallstones formation are high levels of serum lipids (mainly cholesterol and triglycerides) [10]. New studies should focus on the more important proteins involved in biliary secretion and intestinal absorption of cholesterol, because those are key sites of cholesterol transport, on which new drugs might inhibit the formation of cholesterol gallstones; ezetimibe is a drug that showed potential benefit in cholesterol gallstone prevention, and ezetimibe inhibits the expression of Niemann-Pick protein in the small intestine, thereby decreasing the intestinal absorption of cholesterol [11].

There are pharmacological treatments, but there are also treatments based on traditional medicine; in Mexico, knowledge about the medicinal properties of plants is the basis for their use as home remedies. Black radish (*Raphanus sativus* L. *var niger*) is a plant belonging to the Brassicaceae family which contains a high concentration of glucosinolates [12]; experimental studies demonstrated that the aqueous extract and juice of this root possess pharmacological properties, mostly antioxidant, against urinary stones and for detoxifying enzyme activity [13–15]. In Mexican traditional medicine, black radish root has ethnopharmacological uses for the treatment of pigment and cholesterol gallstones and for decreasing serum lipids levels; juice squeezed from the root of black radish significantly decreases ($P < 0.05$) serum lipids levels in C57BL/6 mice fed with a lithogenic diet [16]; however, the active metabolites of black radish responsible for its therapeutic effects are unknown. The principal aim of this work was to evaluate the effects of the juice of black radish root in a biological model of cholesterol gallstones, in order to establish the scientific basis that explains its ethnopharmacological uses.

2. Material and Methods

2.1. Chemicals. Reagent grade cholesterol ($C_{27}H_{45}OH$) was purchased in Hyrel of Mexico; reagent grade cholic acid ≥98% ($C_{24}H_{40}O_5$) purchased in Sigma-Aldrich; reagent grade β-cyclodextrin ($C_{42}H_{70}O_{35}$) H_2O 7.0 mol/mol purchased in Sigma Aldrich; ursodeoxycholic acid ($C_{24}H_{40}O_4$), Ursofalk 250 mg; sodium pentobarbital, Pfizer.

2.2. Plant Material. Fresh tubercles of black radish were collected in Veracruz State, Mexico, in 2011 and were used all experiments. The plant was classified taxonomically for the Instituto de Investigaciones Biológicas in Universidad Veracruzana. The voucher specimen of the plant (CIB 14655) is deposited at the herbarium for reference. The juice of black radish root was prepared in Laboratorio de Neurofarmacología in Facultad de Química. A fresh tubercle (557.7 g) was divided in six parts (60 g), and the juice was squeezed with extraction equipment Moulinex; from each part approximately 10 mL of juice were obtained.

2.3. HPLC Analysis. 500 mL of juice were lyophilized, and the dry material produced 4.5 g. The sample was treated using a previously described method [14]; however, we did not quantify glucosinolates of black radish root. Measurements were done on HPLC equipment (Agilent Technologies 1200 Series Binary SL), RP-C18 column (Zorbax Bonus 100 × 2.1 mm id, 3.5 μm). The mobile phase was 10 : 90 MeOH : H_2O with a flow rate of 2.0 mL/min.

2.4. Acute Toxicity Test. In the toxicity assay, we used a female mouse with the same characteristics of the mice of the main study (female C57BL/6 mice), according to the Organization of Economic Co-operation and Development (OECD) guideline for testing of chemicals. We used a maximum of 7 animals with the following doses: control, vehicle (water), 175, 550, 1750, 2000, and 5000 mg/kg (0.1 mL/10 g body weight). Juice of black radish was lyophilized and was dissolved in purified water. The treatments were administered intragastrically. Neither food nor water was given up to 4 h after the treatment. Body weight, body weight change, signs of toxicity, behaviors, and mortality were observed during the first 4 hours. At the end of the experiment period, the sacrifice of female mice was performed after anesthesia (sodium pentobarbital) in order to avoid animal pain or stress.

2.5. Experimental Animals and Treatments. Thirty-six adult female C57BL/6 mice over 7–9 weeks of age and weighing 18–22 g were purchased in Harlan Laboratories of Mexico and were used in all experiments. Animal care and procedures were conducted according to the guidelines approved by Norma Oficial Mexicana (NOM-062-ZOO-1999) and were subjected to experimental protocols approved by the Programa Institucional para el Cuidado y Uso de Animales de Laboratorio, Facultad de Química, UNAM. All animals were housed in plastic cages in a temperature-controlled room with 12 : 12-h light-dark cycles and provided *ad libitum* access to food and water. Mice had free access to commercial rodent food (Purina Rat Chow). The animals were divided into 6 groups ($n = 6$), and treatments were administered intragastrically. One group was fed a normal diet (ND); five groups were also fed an experimental lithogenic diet containing 2% cholesterol and 0.5% cholic acid (10 g cholesterol/kg and 5 g cholic acid/kg) [17]. The vehicle group received purified water 0.1 mL/10 g (VEH). After feeding with lithogenic diet for 34 days, four groups were treated against cholesterol gallstones for 6 days: one group with ursodeoxycholic acid 0.5% in beta-cyclodextrin 2%, 0.1 mL/10 g (UDCA) and the other groups with juice of black radish root (JBR) diluted one hundredfold (1 : 100), tenfold (1 : 10) in purified water (1 : 100) and juice concentrate. All administrations were at doses of 0.1 mL/10 g of juice. During treatment against cholesterol gallstones (6 days), the VEH + LD group continued receiving lithogenic diet (40 days).

2.6. Gallbladder and Gallstones Analysis. Gallstones were usually visible through the gallbladder. To collect gallstones, the tip of the gallbladder was cut, the gallbladder was squeezed gently with forceps to remove stones, and the inside was washed with an isotonic saline solution to remove any gallstones adhering to the walls; the gallbladder mucosa was observed also. This process was carried out under a stereoscope and a dissecting microscope. The number of cholesterol gallstones in mice was counted.

2.7. Measurements of Total Cholesterol, HDL Cholesterol and Triglycerides. After being treated, all mice were weighed and anesthetized and had blood drawn from the retroorbital vein after an overnight fast. Sacrifice was performed after anesthesia (sodium pentobarbital) in order to avoid animal pain or stress. Serum was separated by centrifugation for further analysis. Determination of biochemical parameters was carried out using a photocolorimeter Dayton Randox.

2.8. Histological Evaluation of Gallbladder and Liver. Gallbladder and liver specimens were prepared and fixed in 4% neutral buffered formaldehyde embedded in paraffin, and 5 μm thin slices were cut and stained with haematoxylin-eosin. All histological sections were reviewed by a pathologist.

2.9. Statistical Analysis. Statistics were done using Sigma Stat Program. Statistical analyses were performed one-way ANOVA with Student-Newman-Keuls *post hoc*. Each point in the table and figure represents the mean ± standard deviation SD.

3. Results

3.1. HPLC Analysis. Our HPLC chromatogram did not prove the presence of glucosinolates; these results are probably due to how the sample was processed. The peaks on the HPLC chromatogram showed components with a retention time of less than 5, as can be seen in Figure 1. The results indicate that these compounds present in the juice of black radish root could be partly responsible for the therapeutic effects in mice.

3.2. Acute Toxicity Test. The results of toxicity test after testing the JBR are shown in Table 1. Single intragastric administration of JBR at all doses did not show visible signs of toxicity, abnormal behaviors, and mortality. Neither body weight was significantly changed relative to that of the control group during the first 4 hours. The juice can be considered bioactive and nontoxic (LD_{50} = 0; no toxicity).

3.3. Analysis of Gallstones. After 34 and 40 days, the lithogenic diet significantly induced the formation of gallstones in female mice and increased liver weight (Figure 2). In Figure 2(a) gallstones were usually visible and moved freely through the gallbladder. In Figure 2(b) an increase in liver weight in mice fed with a lithogenic diet was observed. Cholesterol gallstones were present in female mice as spherical aggregates (white/yellow) with diameters ranging between 0.4 and 0.8 mm; the gallbladders of the mice fed with a lithogenic diet were markedly expanded by the accumulation of bile fluid. The number of gallstones was compared between different experimental groups. In Figure 2(c) two groups of mice showed no cholesterol gallstones after treatment with the JBR concentrate and JBR 1/10. In gallbladders of the group treated with JBR 1/100 there were showen residues of cholesterol gallstones, while in the groups treated with UDCA 0.5% and LD, the cholesterol gallstones persisted within the gallbladder.

3.4. Cholesterol, HDL Cholesterol, and Triglycerides. The results from serum lipids analysis are shown in Table 2 and Figure 3. After 40 days, the mice of group LD + VEH had increased cholesterol and triglycerides levels; levels of HDL cholesterol decreased. The results from different experimental groups were compared with LD + VEH group. In Figure 3(a) after treatment with JBR, cholesterol levels

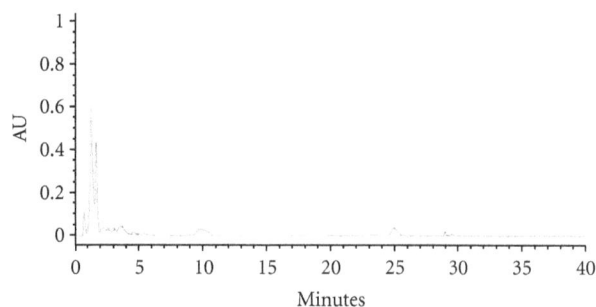

FIGURE 1: HPLC chromatogram of juice of black radish root.

TABLE 1: Toxicity test according to OECD guidelines.

Doses (mg/kg) of JBR	Mortality %	Signs of toxicity	Signs of toxicity %	Survival %
Female control	0	0/1	0	100
Vehicle (water)	0	0/1	0	100
175	0	0/1	0	100
550	0	0/1	0	100
1750	0	0/1	0	100
2000	0	0/1	0	100
5000	0	0/1	0	100

LD_{50} = 0; no toxicity.

decreased in mice, compared with mice fed with a lithogenic diet ($P < 0.05$). Although in two groups (UDCA 0.5% and JBR 1 : 10) gallstones did not remove the, those groups did show a decreased plasma cholesterol levels. In Figure 3(b) results of triglyceride levels are important, because concentration of JBR was crucial. Groups treated with JBR 1 : 10, and concentrated JBR had decreased levels of triglycerides to normal ranges; as the results presented in the graph of the group ND show, there was a significant difference. JBR 1 : 00 and UDCA 0.5% groups showed triglyceride levels above the normal range, significant difference in comparison with the LD + VEH group. In Figure 3(c) levels of HDL cholesterol decreased significantly in LD + VEH group, but in groups treated with BRJ and UDCA 0.5%, the levels of HDL cholesterol increased.

3.5. Microscopy Studies of Gallbladder Mucosa. All tissue of gallbladders was analyzed in optical microscope with 40x magnification. In Figure 4(a) there were no significant changes in the gallbladder mucosa in ND group; however, other experimental groups had significant changes when compared with the ND group. The groups with cholesterol gallstones (VEH and UDCA 0.5%) showed accumulation of biliary sludge, and its mucosas were the most damaged; these changes may be due to movement of gallstones and biliary sludge accumulation. In Figure 4(b) white arrows indicate cholesterol gallstones in VEH group, after fed for 40 days with a lithogenic diet. In Figure 4(c) Mucosa of group treated with UDCA 0.5%. In Figure 4(d) the groups treated with JBR showed less damage in mucosa of gallbladder.

(a)   (b)

FIGURE 2: Cholesterol gallstones and liver in female mice.

TABLE 2: Black radish juice modifies some components in serum of mice.

Group	Biochemical parameter (mg/dL)		
	Cholesterol	HDL cholesterol	Triglycerides
ND	82 ± 2	67 ± 1.5	92.3 ± 0.8
LD + VEH	103.6 ± 3.5	51 ± 1.4	101.5 ± 1.2
LD + UDCA 0.5%	91 ± 2.4	60.5 ± 2	98.3 ± 1.3
LD + JBR 1:100	91.5 ± 2.2*	62.8 ± 0.7*	96.1 ± 0.1*
LD + JBR 1:10	80.6 ± 1.5*	62.8 ± 0.7*	93.8 ± 0.7*
LD + JBR	79.8 ± 1.8*	62.8 ± 1.1*	93 ± 0.6*

ND: normal diet; LD: lithogenic diet; VEH: vehicle; UDCA: ursodeoxycholic acid; JBR: juice of black radish root. Values are means ± SD using 6 mice in each experimental group. *Values indicate significant differences between groups treated with JBR ($P < 0.05$) versus LD + VEH group.

3.6. Histopathology Study. Histopathology results were compared between different experimental groups, and all tissue specimens were stained with hematoxilin eosin (Figure 5). There were no apparent histological changes in ND group. In Figure 5(a) Black arrows indicate the epithelial hyperplasia, and the white arrow indicates granulocyte infiltration in gallbladder of C57BL/6 mice after being fed for 40 days with a lithogenic diet (40x magnification). In Figure 5(b) the gallbladder of mice treated with JBR also presented inflammatory infiltration and epithelial hyperplasia, despite eradication of cholesterol gallstones (10x magnification). In Figure 5(c) all liver tissue specimens showed perivascular infiltration, vacuolar degeneration, and dilated central and portal veins (10x magnification).

4. Discussion

Our general aim was to evaluate the effectiveness of black radish juice against cholesterol gallstone disease and serum lipids (cholesterol, HDL cholesterol and triglycerides) in mice. The black radish contains high concentrations of a glucosinolates, but the exact mechanism of the biologically active compounds in black radish is not clear yet. Regarding possible active metabolites, our investigation proved that intact glucosinolates disappeared during processing and/or storage, for there were no peaks of glucosinolates on the HPLC chromatogram, as can be seen in Figure 1. There are two studies about HPLC analysis of black radish: in 1998 Lugasi and coworkers reported an HPLC chromatogram

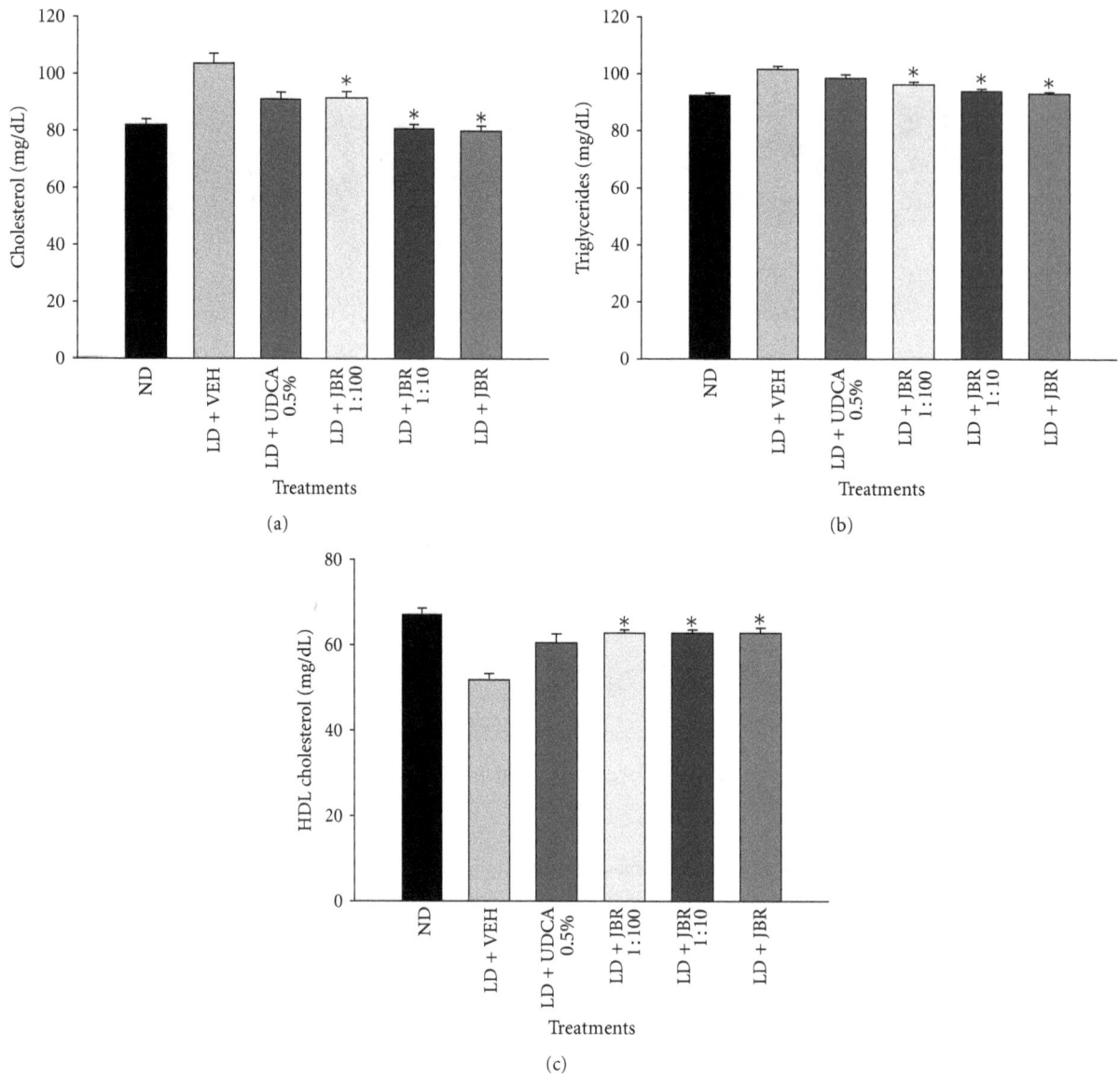

(a)

(b)

(c)

FIGURE 3: Effect of juice in serum lipids.

of juice squeezed of black radish, which did not detect presence of glucosinolates, only the peak corresponding to the internal standard (benzyl glucosinolates) [18]. Hanlon and coworkers (2007) reported an HPLC chromatogram of desulfoglucosinolates in an aqueous extract of black radish [14]. Both authors performed a quantification of secondary metabolites in black radish by HPLC. In our analysis we did not quantify secondary metabolites. In the HPLC chromatogram of Hanlon and coworkers, degradation products of glucoraphasatin were the most apparent; also observed were degradation products of glucoraphanin; these results are very important, because glucoraphanin and its degradation product, sulforaphane, have been shown to have a potential hypocholesterolemic effect in animal models [19]. It is likely that compounds related to glucoraphanin are responsible for lowering cholesterol and triglycerides in the serum of mice in our study although we must emphasize

that the hypocholesterolaemic effects do not necessarily cause the dissolution of cholesterol gallstones in mice. A hypolipidaemic effect is directly related to the prevention of gallstones, rather than to treatment, but it is important to know that the reduction of plasma cholesterol is a direct consequence of the decrease in intestinal absorption although we did not evaluate this. When absorption is decreased, biliary cholesterol secretion can be decreased, thereby reducing the saturation of this lipid in the vesicle, which favors the activity of bile salts on cholesterol gallstones, since they are the major solutes of bile. An important aspect was that we did not evaluate the elimination of cholesterol in feces, as neutral sterol, because, by demonstrating the reduction of plasma cholesterol, there must have been be a route of elimination, or accumulation. The toxicity test was undertaken prior to the therapeutic evaluation of the juice of black radish root; there is currently a tendency to

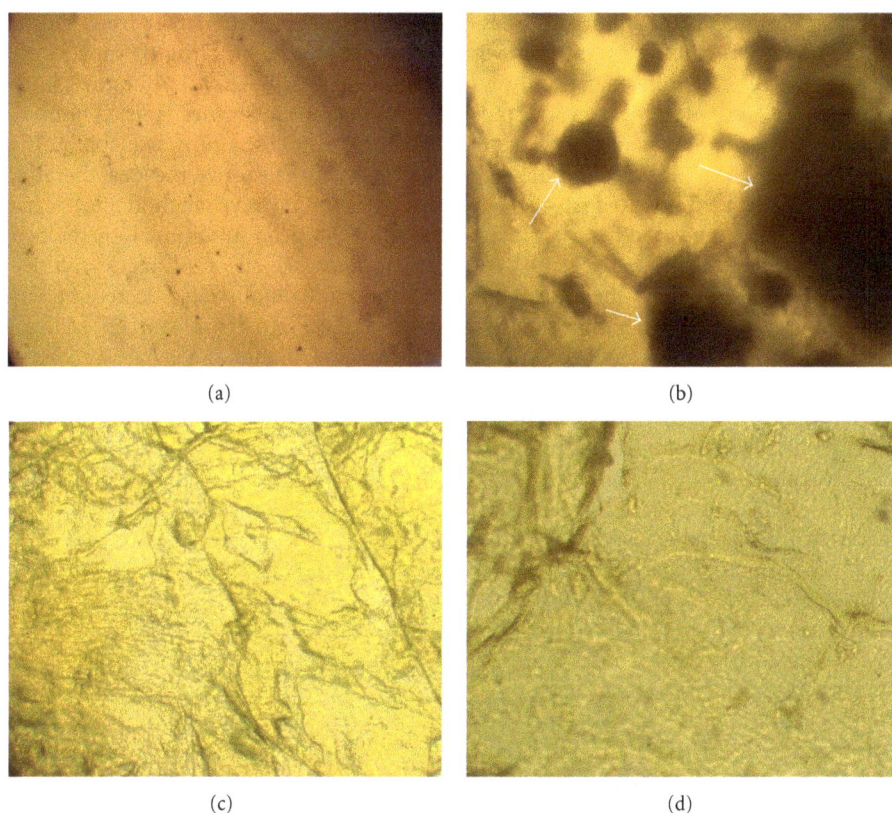

FIGURE 4: Microscopic analysis in mucosa of gallbladder.

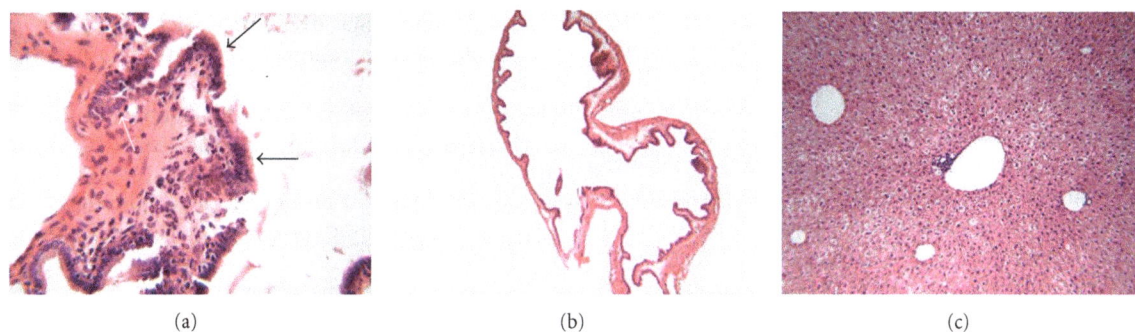

FIGURE 5: Histopathological analysis of gallbladder and liver.

call not using laboratory animals in toxicological tests, due to the high cost and the suffering imposed on the animals. We used the toxicity test according to OECD guidelines for minimizing the number of animals required to estimate the acute oral toxicity of juice black radish. In addition to the estimation of LD_{50} and confidence intervals, the test allows the observation of signs of toxicity. According to OECD guideline and Lipnick et al. [20], substances that explain the median lethal dose (LD_{50}) higher than 5 g/kg body weight by oral route can be considered practically nontoxic. Thus, it can be concluded that juice black radish is absent of the acute oral toxicity. Although the exact mechanism of the biologically active compounds in black radish is not clear yet, a beneficial effect of the drug was evident against cholesterol

gallstones and both cholesterol and triglyceride levels. Our study is new and represents part of the verification of the uses that are given to the plant in Mexican traditional medicine. Our mice developed cholesterol gallstones after 34 days of being fed with a lithogenic diet; Ebihara and Kiriyama (1985) reported similar observations in male ICR mice also using a high cholesterol and cholic acid diet [21]. We did not determine the concentrations of cholesterol, bile salts, and phospholipids in the hepatic bile, as is usually done in studies to inhibit the formation of cholesterol gallstones, because we did not do a preventative treatment; on the contrary, we administered a treatment to destroy gallstones when they were already formed. The juice of black radish root diluted tenfold and concentrated juice decreased cholesterol and triglycerides levels and also removed cholesterol gallstones in mice. The juice diluted one hundredfold did not remove all gallstones in mice; the effect of concentration was determinant. In another study related to this work, the juice of black radish root exhibited significant antioxidant properties in rats fed with a diet rich in lipids (20% sunflower oil, 2% cholesterol, 0.5% cholic acid in normal chow); the juice was diluted tenfold with water and the animals drank it instead of water for 9 days [11]. These results are related to cholesterol gallstones formation; the rats were fed a diet rich in lipids, but rats have no gallbladder in which gallstones can be formed or accumulated; therefore, this diet is a lithogenic diet for mice, because formation of gallstones was only produced by simultaneous feeding of cholesterol, cholic acid, and fats. This might lead us to think that it also produced antioxidant effects in our mice. The experimental group treated with ursodeoxycholic acid had decreased levels of cholesterol and triglycerides; however, in this group gallstones was not removed. This is logical because treatment with ursodeoxycholic acid requires a long time to dissolve cholesterol gallstones, and we only administered ursodeoxycholic acid for 6 days. Ursodeoxycholic acid has been recommended as the first-line pharmacological therapy in a subgroup of symptomatic patients with small, radiolucent cholesterol gallstones and its longterm administration has been shown to promote the dissolution of cholesterol gallstones [22]. The effect to decreasing lipid levels could also be enhanced by the administration of beta-cyclodextrin in ursodeoxycholic acid, because it is a cyclic oligosaccharide that binds cholesterol and bile acids *in vitro*, and it has been previously shown to be an effective plasma cholesterol-lowering agent in hamsters and domestic pigs and also produces antilithiasic effects in hamsters [23]. The lithogenic diet produced granulocyte infiltration in mice; according to data reported in the literature, the presence of gallstones in gallbladder produces neutrophilic infiltration and reactive epithelial changes (increased nucleus-to-cytoplasm ratio, mitoses, and prominent nucleoli) that can easily be confused with neoplasia [24]. After the administration of treatments against the effects of the lithogenic diet, the gallbladders of mice continued to show granulocyte infiltration and hyperplasia in the epithelium of gallbladder. Although changes in the gallbladder mucosa were different between treatment groups and the vehicle group, the histopathological changes were similar.

We consider this paper as an important contribution because we showed antilithiasic effect in mice, which is different from demonstrating an antilithogenic effect that prevents the formation of gallstones. There are currently few studies about molecular mechanisms that destroy gallstones once they are formed; therefore, it is important to start studying which processes could lead to create antilithiasic effects. Our work did not determine the molecular mechanism that destroys gallstones, but it is a background for continuing the experimental studies about the action mechanism of phytochemical components in black radish. It is understandable that medical science is more interested in the prevention of cholesterol gallstones disease, because as there are many risk factors for developing it; but there are many people who suffer from gallstones now, and they require a treatment. Drug therapy for cholesterol gallstone disease plays a limited role, but novel interesting information about the molecular mechanisms responsible for the formation of gallstones is now available [25]. Unfortunately, for symptomatic gallstones, the ideal therapy is cholecystectomy, which is an invasive method [26]. Thanks to the knowledge of Mexican traditional medicine, we demonstrated an antilithiasic effect of black radish juice in mice; in addition we demonstrated decreasing serum cholesterol and triglycerides, two important lipids involved in cholesterol gallstones formation.

5. Conclusion

Juice squeezed from black radish root has properties against cholesterol gallstones and for decreasing serum lipids levels. Considering that it is a non-toxic natural product, more studies should be made to find the action mechanism of secondary metabolites.

Conflict of Interests

The authors declare that they have no conflicts of interests.

Acknowledgments

The experimental work of the authors was supported by the Laboratorio de Neurofarmacología and Programa de Apoyo a Licenciatura (PAL, clave: 6300-06), in the Universidad Nacional Autónoma de México (UNAM). The authors appreciate the cooperation of the Instituto de Investigaciones Biológicas in the Universidad Veracruzana for taxonomic classification of black radish. They also appreciate the support of Espacio Común de Educación Superior (ECOES), UNAM, and the Programa de Movilidad de la Universidad Veracruzana (PROMUV). They thank Dr. Atonatiu E. Gómez-Martínez, Jefe del Bioterio de la Facultad de Química, UNAM.

References

[1] P. Portincasa, A. Moschetta, and G. Palasciano, "Cholesterol gallstone disease," *Lancet*, vol. 368, no. 9531, pp. 230–239, 2006.

[2] C. E. Ruhl and J. E. Everhart, "Gallstone disease is associated with increased mortality in the United States," *Gastroenterology*, vol. 140, no. 2, pp. 508–516, 2011.

[3] N. Méndez-Sánchez, J. Bahena-Aponte, N. C. Chávez-Tapia et al., "Strong association between gallstones and cardiovascular disease," *American Journal of Gastroenterology*, vol. 100, no. 4, pp. 827–830, 2005.

[4] A. Di Ciaula, D. Q. H. Wang, H. H. Wang, L. Bonfrate, and P. Portincasa, "Targets for current pharmacologic therapy in cholesterol gallstone disease," *Gastroenterology Clinics of North America*, vol. 39, no. 2, pp. 245–264, 2010.

[5] K. J. Maurer, M. C. Carey, and J. G. Fox, "Roles of infection, inflammation, and the immune system in cholesterol gallstone formation," *Gastroenterology*, vol. 136, no. 2, pp. 425–440, 2009.

[6] C. S. Stokes, M. Krawczyk, and F. Lammert, "Gallstones: environment, lifestyle and genes," *Digestive Diseases*, vol. 29, no. 2, pp. 191–201, 2011.

[7] K. J. Van Erpecum, "Pathogenesis of cholesterol and pigment gallstones: an update," *Clinics and Research in Hepatology and Gastroenterology*, vol. 35, no. 4, pp. 281–287, 2011.

[8] M. C. Vázquez, A. Rigotti, and S. Zanlungo, "Molecular mechanisms underlying the link between nuclear receptor function and cholesterol gallstone formation," *Journal of Lipids*, vol. 2012, Article ID 547643, 7 pages, 2012.

[9] I. G. Castro-Torres, "Cholesterol gallstones formation: new scientific advances," *Revista GEN (Gastroenterología Nacional)*, vol. 66, no. 1, pp. 57–62, 2012.

[10] A. H. M. Smelt, "Triglycerides and gallstone formation," *Clinica Chimica Acta*, vol. 411, no. 21-22, pp. 1625–1631, 2010.

[11] O. de Bari, B.A. Neuschwander-Tetri, M. Liu M et al., "Ezetimibe: its novel effects on the prevention and the treatment of cholesterol gallstones and nonalcoholic fatty liver disease," *Journal of Lipids*, vol. 2012, Article ID 302847, 16 pages, 2012.

[12] A. Lugasi, A. Blázovics, K. Hagymási, I. Kocsis, and Á. Kéry, "Antioxidant effect of squeezed juice from black radish (*Raphanus sativus* L. *var niger*) in alimentary hyperlipidaemia in rats," *Phytotherapy Research*, vol. 19, no. 7, pp. 587–591, 2005.

[13] R. Vargas, R. M. Perez, S. Perez, M. A. Zavala, and C. Perez, "Antiurolithiatic activity of *Raphanus sativus* aqueous extract on rats," *Journal of Ethnopharmacology*, vol. 68, no. 1–3, pp. 335–338, 1999.

[14] P. R. Hanlon, M. G. Robbins, L. D. Hammon, and D. M. Barnes, "Aqueous extract from the vegetative portion of Spanish black radish (*Raphanus sativus* L. *var. niger*) induces detoxification enzyme expression in HepG2 cells," *Journal of Functional Foods*, vol. 1, no. 4, pp. 356–365, 2009.

[15] P. R. Hanlon, D. M. Webber, and D. M. Barnes, "Aqueous extract from Spanish black radish (*Raphanus sativus* L. *var. niger*) induces detoxification enzymes in the HepG2 human hepatoma cell line," *Journal of Agricultural and Food Chemistry*, vol. 55, no. 16, pp. 6439–6446, 2007.

[16] I. G. Castro-Torres and E. B. Naranjo-Rodríguez, "Treatment with black radish (*Raphanus Sativus* L. *var niger*) reduces serum lipids levels in mice fed with a lithogenic diet," *Revista Latinoamericana de Quimica*, vol. 39, no. 3, p. 273, 2012.

[17] E. Reihnér and D. Ståhlberg, "Lithogenic diet and gallstone formation in mice: integrated response of activities of regulatory enzymes in hepatic cholesterol metabolism," *British Journal of Nutrition*, vol. 76, no. 5, pp. 765–772, 1996.

[18] A. Lugasi, A. Blázovics, K. Hagymási et al., "Antioxidant and free radical scavenging properties of squeezed juice from black radish (*Raphanus sativus* L. *var niger*) root," *Phytotherapy Research*, vol. 12, pp. 502–506, 1998.

[19] L. N. Rodríguez-Cantú, J. A. Gutiérrez-Uribe, J. Arriola-Vucovich, R. I. Díaz-De La Garza, J. W. Fahey, and S. O. Serna-Saldivar, "Broccoli (*brassica oleracea* var. *italica*) sprouts and extracts rich in glucosinolates and isothiocyanates affect cholesterol metabolism and genes involved in lipid homeostasis in hamsters," *Journal of Agricultural and Food Chemistry*, vol. 59, no. 4, pp. 1095–1103, 2011.

[20] R. L. Lipnick, J. A. Cotruvo, R. N. Hill et al., "Comparison of the up-and-down, conventional LD50, and fixed-dose acute toxicity procedures," *Food and Chemical Toxicology*, vol. 33, no. 3, pp. 223–231, 1995.

[21] K. Ebihara and S. Kiriyama, "Prevention of cholesterol gallstones by a water-soluble dietary fiber, konjac mannan, in mice," *Nutrition Reports International*, vol. 32, no. 1, pp. 223–229, 1985.

[22] P. Portincasa, A. Di Ciaula, H. H. Wang, A. Moschetta, and D. Q. H. Wang, "Medicinal treatments of cholesterol gallstones: old, current and new perspectives," *Current Medicinal Chemistry*, vol. 16, no. 12, pp. 1531–1542, 2009.

[23] N. Boehler, M. Riottot, J. Férézou et al., "Antilithiasic effect of β-cyclodextrin in LPN hamster: comparison with cholestyramine," *Journal of Lipid Research*, vol. 40, no. 4, pp. 726–734, 1999.

[24] D. E. Hansel, A. Maitra, and P. Argani, "Pathology of the gallbladder: a concise review," *Current Diagnostic Pathology*, vol. 10, no. 4, pp. 304–317, 2004.

[25] D. Q. Wang, D. E. Cohen, and M. C. Carey, "Biliary lipids and cholesterol gallstone disease," *Journal of Lipid Research*, vol. 50, supplement, pp. S406–S411, 2009.

[26] V. I. Reshetnyak, "Concept of the pathogenesis and treatment of cholelithiasis," *World Journal of Hepatology*, vol. 4, no. 2, pp. 18–34, 2012.

12

Interaction of Human Dopa Decarboxylase with L-Dopa: Spectroscopic and Kinetic Studies as a Function of pH

Riccardo Montioli, Barbara Cellini, Mirco Dindo, Elisa Oppici, and Carla Borri Voltattorni

Section of Biological Chemistry, Department of Life Sciences and Reproduction, University of Verona, Strada Le Grazie 8, 37134 Verona, Italy

Correspondence should be addressed to Carla Borri Voltattorni; carla.borrivoltattorni@univr.it

Academic Editor: Alessandro Paiardini

Human Dopa decarboxylase (hDDC), a pyridoxal $5'$-phosphate (PLP) enzyme, displays maxima at 420 and 335 nm and emits fluorescence at 384 and 504 nm upon excitation at 335 nm and at 504 nm when excited at 420 nm. Absorbance and fluorescence titrations of hDDC-bound coenzyme identify a single pK_{spec} of ~7.2. This pK_{spec} could not represent the ionization of a functional group on the Schiff base but that of an enzymic residue governing the equilibrium between the low- and the high-pH forms of the internal aldimine. During the reaction of hDDC with L-Dopa, monitored by stopped-flow spectrophotometry, a 420 nm band attributed to the $4'$-N-protonated external aldimine first appears, and its decrease parallels the emergence of a 390 nm peak, assigned to the $4'$-N-unprotonated external aldimine. The pH profile of the spectral change at 390 nm displays a pK of 6.4, a value similar to that (~6.3) observed in both k_{cat} and k_{cat}/K_m profiles. This suggests that this pK represents the $ESH^+ \rightarrow ES$ catalytic step. The assignment of the pKs of 7.9 and 8.3 observed on the basic side of k_{cat} and the PLP binding affinity profiles, respectively, is also analyzed and discussed.

1. Introduction

Dopa decarboxylase (DDC; EC 4.1.1.28) is a pyridoxal $5'$-phosphate- (PLP-) dependent enzyme which catalyzes the irreversible decarboxylation of L-Dopa and L-5-hydroxy-tryptophan (5-HTP), thus producing the neurotransmitters dopamine and serotonin. The enzyme accepts other catechol-or indole-related L-amino acids and has been therefore also named L-aromatic amino acid decarboxylase (AADC). Like other PLP enzymes [1–7], DDC is of clinical interest since it is involved either in Parkinson's disease, a degenerative disorder of the central nervous system resulting from the death of dopamine-generating cells in the *substantia nigra*, or in AADC deficiency, a rare inherited neurometabolic disease due to mutations on the *AADC* gene leading to deficit of catecholamines and serotonin in the central nervous system and periphery [8]. Thus, the elucidation of the structural and functional features of the enzyme is relevant for the development of treatment strategies for both disorders. In this regard, a structure-based design search aimed at identifying inhibitors of peripheral DDC more selective than those currently administered to patients with Parkinson's disease has been recently reported [9]. Moreover, the molecular basis of AADC deficiency analyzed by comparing the characteristics of normal human DDC (hDDC) with those of some pathogenic variants in their recombinant purified form has allowed not only to unravel their molecular defects but also to suggest appropriate therapeutic treatments for patients bearing the examined mutations [10, 11].

Progress in understanding the structure/function relationships operating in DDC has been obtained by means of kinetic, spectroscopic, and structural studies on the pig kidney and rat liver enzymes, either in the naturally occurring [12–17] or in the recombinant purified [18–26] forms, as well as, more recently, on recombinant purified hDDC [9–11, 27]. The crystal structure of pig kidney holo DDC alone and in complex with carbidopa (a substrate analog endowed with a hydrazinic group) has been solved at 2.6 and 2.5 Å resolutions, respectively [28]. The overall structure of the protein is a tightly associated dimer in which the active site is buried in the central part. Each monomer is composed of a large domain and a C-terminal domain, typical of the aspartate aminotransferase family (Fold Type I), as well as an N-terminal domain characteristic of Group II decarboxylases. Several other features of DDC are evident in these structures: (i) the way in which PLP is anchored to the enzyme involving His302 and His192, two highly conserved residues in α-decarboxylases [29]; (ii) how the inhibitor binds; and (iii) which amino acid residues might be involved in the catalytic activity. Unexpectedly, the crystal structure of hDDC in the apo form reveals that it exists in an open conformation in which the dimer subunits move 20 Å apart and the two active sites become solvent exposed. Moreover, by varying the PLP concentration in the crystals of the open DDC, two more structures have been solved, thus allowing to identify the structural determinants of the conformational change occurring upon PLP binding [27].

Although DDC enzymes share similar absorption spectra, that is, absorption maxima at 420 and 335 nm, pig kidney and rat liver enzymes display different PLP emission fluorescent properties, possibly due to the presence, even if to a different degree, of a species absorbing at 335 nm and emitting fluorescence at 390 nm in their apoenzyme forms [15, 24, 25]. These findings, together with the fact that the coenzyme absorbing bands show a modest pH dependence, do not allow to unequivocally assign the 335 nm absorbing band to a form of the internal aldimine and to understand how the equilibrium between the 420 and 335 nm species is governed. Moreover, although previous spectroscopic analyses of the reaction of both pig kidney and rat liver DDC with L-Dopa provided evidence for the appearance of two intermediates absorbing at 420 and 385 nm, the assignment of these species is conflicting. These absorbance bands have been attributed to 1-N-protonated-4'-N-protonated and 1-N-protonated-4'-N-unprotonated Schiff bases shown by the low- and high-pH forms of the external aldimine [13, 17]. On the other hand, other authors suggested that these species were formed during the course of the decarboxylation reaction, being the 420 nm and the 385 nm either the adsorption complex and the external aldimine with L-Dopa, respectively [25, 26], or two different external aldimines [30].

The present study presents a detailed investigation of the pH dependence of the bound coenzyme absorbance and fluorescence features and of the steady-state kinetic parameters of hDDC. Additionally, changes of the absorbance bands of hDDC upon L-Dopa binding as a function of pH have been monitored by rapid scanning stopped-flow experiments. Taken together, the results allow us to identify three observable ionizations in hDDC and to propose their involvement in the structures of the bound coenzyme and in the intermediates along the decarboxylation pathway.

2. Material and Methods

2.1. Chemicals. L-Dopa, PLP, 2,4,6-trinitrobenzene-1-sulfonic acid, isopropyl-β-D-thiogalactopyranoside, and protease inhibitor cocktail were purchased from Sigma-Aldrich. Bis-Tris-propane at a final concentration of 50 mM was used

over the pH range 6–8.8. The other chemicals were of the highest purity available.

2.2. Enzyme Preparation. The conditions used for expression and purification of hDDC were as previously described [11]. The apo form was prepared as previously reported [11].

2.3. Binding Affinity of hDDC for PLP. The equilibrium dissociation constant for PLP, $K_{D(PLP)}$, was determined by measuring the quenching of the intrinsic fluorescence of the apoenzyme (0.15 μM) in the presence of PLP at a concentration range of 0.01–20 μM. The experiments were carried out in 50 mM Bis-Tris-propane in the pH range 6–8.8. The $K_{D(PLP)}$ values were obtained by fitting the data to the following equation:

$$Y = Y_{MAX} \frac{[E]_t + [PLP]_t + K_{D(PLP)} - \sqrt{([E]_t + [PLP]_t + K_{D(PLP)})^2 - 4[E]_t[PLP]_t}}{2[E]_t},$$

$$(1)$$

where $[E]_t$ and $[PLP]_t$ represent the total concentrations of hDDC dimer and PLP, respectively, Y refers to the intrinsic fluorescence quenching changes at a PLP concentration, $[PLP]$, and Y_{max} refers to the aforementioned changes when all enzyme molecules are complexed with the coenzyme.

2.4. Enzyme Assay. The decarboxylase activity toward L-Dopa was measured by the spectrophotometric assay described by Sherald et al. [31], and it was modified by Charteris and John [32]. Measurements were performed in the presence of 10 μM PLP. Data of enzymatic activity as a function of substrate concentration were fitted to Michaelis-Menten equation.

2.5. Spectral Measurements. Absorption spectra were made with a Jasco V-550 spectrophotometer at a protein concentration of 10 μM. Fluorescence spectra were taken with an FP750 Jasco spectrofluorometer using 5 nm excitation and emission bandwidths at a protein concentration of 10 μM. The enzyme solution was drawn through a 0.2 μm filter to reduce light scattering from the small amount of precipitate. Spectra of blanks, that is, samples containing all components except hDDC, were taken immediately before the measurements of samples containing protein.

2.6. pH Studies. Absorbance and fluorescence data were fitted to (2) and (3):

$$A = \frac{A_1 - A_2}{1 + 10^{pH - pK_{spec}}} + A_2,$$

$$(2)$$

$$A = \frac{A_1 - A_2}{1 + 10^{pK_{spec} - pH}} + A_2,$$

$$(3)$$

where A_1 and A_2 are the higher and the lower absorbance limits at a particular wavelength, respectively.

The log $K_{D(PLP)}$, log k_{cat}, log k_{cat}/K_m, and values for hDDC versus pH were fitted to the following appropriate equations:

$$\log Y = \log \frac{Y_L + Y_H\left(H/K_1\right)}{1 + H/K_1}, \qquad (4)$$

$$\log Y = \log \frac{C}{1 + H/K_A + K_B/H}, \qquad (5)$$

$$\log Y = \log \frac{C}{1 + H/K_A}, \qquad (6)$$

where K_A and K_B represent the ionization constants for enzyme or reactant functional groups, Y is the value of the parameter observed as a function of pH, C is the pH-independent value of Y, H is the hydrogen ion concentration, and Y_L and Y_H are constant values at low and high pH, respectively.

2.7. Pre-Steady-State Kinetic Analysis by UV-Vis Stopped-Flow Spectroscopy. hDDC was mixed using a Biologic SFM300 mixer with an equal volume of 2 mM L-Dopa in 50 mM Bis-Tris-propane at pH ranging from 6.0 to 8.0. Coenzyme absorbance changes were monitored with a TC-100 (path length of 1 cm) quartz cell coupled to a Biokine PMS-60 instrument. The dead time was 3.8 ms at a flow velocity of 8 mL/sec. Absorbance scans (800) from 300 to 600 nm were collected on a logarithmic time scale with a J&M Tidas 16256 diode array detector (Molecular kinetics, Pullman, WA). Data were analyzed using either SPECFIT (Spectrum Software, Claix, France) or Biokine 4.01 (Biologic, Claix, France) to determine the observed rate constants. Single-wavelength time courses were fit to an equation of the following general form:

$$A_t = A_\infty \pm \sum A_i \exp^{(-k_{obs}t)}, \qquad (7)$$

where A_t is the absorbance at time t, A_i is the amplitude of each phase, k_{obs} is the observed rate constant for each phase, and A_∞ is the final absorbance.

3. Results

3.1. pH Dependence of the Internal Aldimine and Coenzyme Binding Affinity. Apo hDDC completely lacks absorbance in the visible region, while the holo form is characterized by absorbance bands at 420 and 335 nm associated with positive dichroic signals at the same wavelengths [11]. The absorbance spectrum of hDDC in the holo form changes as a function of pH over the range 6–8.5 (Figure 1): the 335 nm band increases with pH, while the 420 nm band decreases. The spectra do not show a clear isosbestic point, thus suggesting the presence of multiple species. When we fitted the data to curves with one, two, or three ionizations, we found that they fit best to a model with one ionization ((2) and (3)): the pK_{spec} values obtained were 7.2 ± 0.1 and 7.3 ± 0.1 for the absorbance at 420 and 335 nm, respectively (Figure 1, inset) (Table 1).

Excitation of hDDC at 420 nm results in an emission band at 504 nm, whose intensity decreases as pH increases below

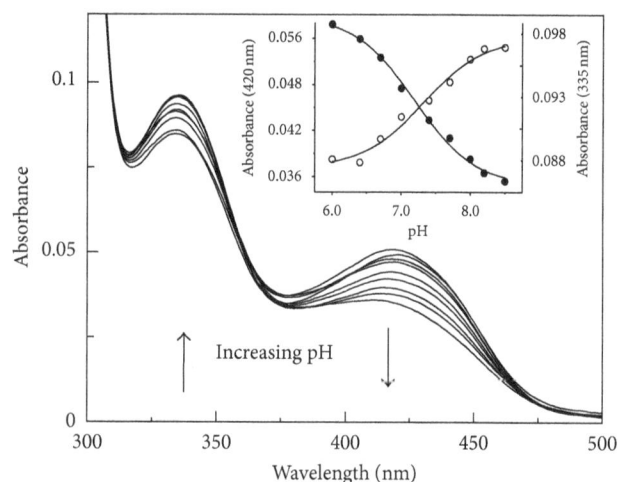

FIGURE 1: pH dependence of the visible spectra of hDDC. Absorbance spectra of 10 μM hDDC were acquired in 50 mM Bis-Tris-propane at pH 6.0, 6.4, 7.0, 7.4, 7.7, 8.0, 8.2, and 8.5. The inset shows the pH dependence of the absorbance at 420 ($\bullet$) and 335 nm ($\circ$). The solid lines represent the theoretical curves according to (2) and (3).

TABLE 1: Summary of pK_a values for hDDC.

Parameter	pH-independent value	pK_1	pK_2
k_{cat}	$5.1 \pm 0.4\,\mathrm{s}^{-1}$	6.3 ± 0.1	7.9 ± 0.1
k_{cat}/K_m	$174 \pm 3\,\mathrm{mM}^{-1}\mathrm{s}^{-1}$	6.1 ± 0.1	
Absorbance at 420 nm pH titration			7.2 ± 0.1
Absorbance at 335 nm pH titration			7.3 ± 0.1
Emission fluorescence pH titration (exc. 335 nm)			
emis.$_{384\,nm}$			7.3 ± 0.1
emis.$_{504\,nm}$			7.2 ± 0.1
Emission fluorescence pH titration (exc. 420 nm)			
emis.$_{504\,nm}$			7.2 ± 0.1
Amplitude ext. ald. 390 nm		6.4 ± 0.3	
$K_{D(PLP)}$			8.3 ± 0.2

a single pK of 7.2 ± 0.1 (Figure 2(a) and inset). Moreover, hDDC emits at 384 and 504 nm upon excitation at 335 nm. Emission fluorescence intensity at 384 nm increases with increasing pH, while that at 504 nm decreases (Figure 2(b)). As shown in the inset of Figure 2(b), the pH profile for the 384 nm emission intensity increases above a single pK of 7.3 ± 0.1, while that at 504 nm decreases below a single pK of 7.1 ± 0.1 (Table 1). When emission was observed at 384 nm, the excitation spectrum exhibits a maximum at 337 nm, whereas at 500 nm, the excitation spectrum displays maxima at 341 and 410 nm.

The $K_{D(PLP)}$ value for hDDC at pH 7.4 is 0.2 μM, and it increases as the pH increases. The pH dependence of $K_{D(PLP)}$

(a)

(b)

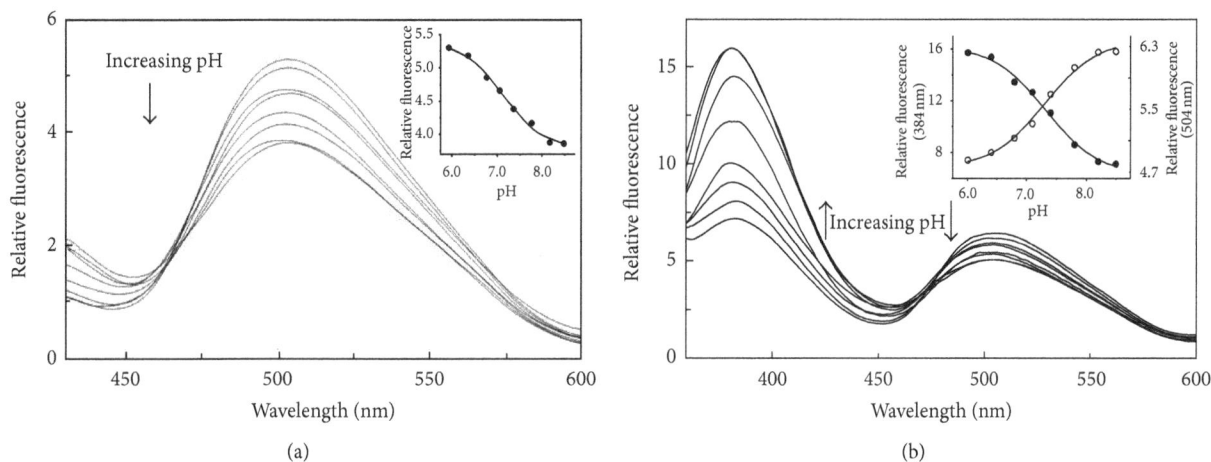

FIGURE 2: pH dependence of the internal aldimine emission fluorescence of hDDC. Emission fluorescence spectra of 10 μM hDDC in 50 mM Bis-Tris-propane at pH 6.0, 6.4, 6.8, 7.1, 7.4, 7.8, 8.2, and 8.5 upon excitation at 420 nm (a) and 335 nm (b). The inset of (a) shows the pH dependence of the emission intensity at 504 nm ($\bullet$), while that of (b) shows the pH dependence of the emission intensity at 504 nm ($\circ$) and 384 nm ($\bullet$). The solid lines represent the theoretical curves according to (2) and (3).

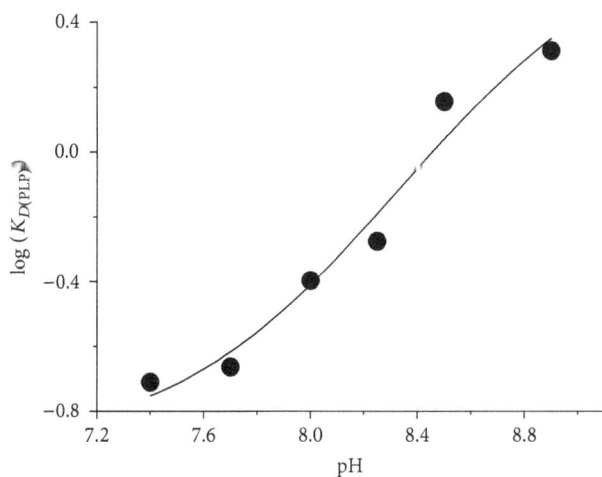

FIGURE 3: pH dependence of $K_{D(PLP)}$ of hDDC. $K_{D(PLP)}$ values were determined in 50 mM Bis-Tris-propane at the indicated pH as described under "Section 2". The points shown are the experimental values (expressed in μM), while the curve is from fit to the data using (4).

fits well (4), giving a $pK_{D(PLP)}$ value of 8.3 ± 0.2 (Figure 3, Table 1).

3.2. pH Dependence of Kinetic Parameters for hDDC.

The pH dependence of the kinetic parameters for hDDC toward L-Dopa was determined, and the results are shown in Figures 4(a) and 4(b). The variation with pH of log k_{cat} gives rise to a bell-shaped profile: fitting the data to (5) and (6) yields pK values of 6.3 ± 0.1 and 7.9 ± 0.1; Log k_{cat}/K_m decreases below a pK of 6.1 ± 0.1 (Table 1).

3.3. pH Dependence of External Aldimine with L-Dopa.

To obtain information about the identity of intermediates in the reaction of hDDC with L-Dopa we carried out rapid-kinetic spectroscopic studies over the pH range 6–8. Upon mixing the enzyme with L-Dopa at a saturating concentration, a biphasic spectral change was observed. In the first phase, a rapid increase in the absorbance at 420 nm and a decrease in the 335 nm band were observed within 50 ms, followed by a second phase, in which the absorbance at 420 nm decreases and concomitantly a new absorbance band appears at 390, occurring with a rate of 31 s^{-1} (Figure 5). The amplitude of the absorbance changes at 390 nm increases with pH above a single pK of 6.4 ± 0.3 (Figure 6) (Table 1). It should be also noted that at pH higher than 6.4 the appearance of the 390 nm band is accompanied by that of a shoulder absorbing at ~440 nm. Considering that the enzyme-dopamine complex absorbs at ~400 nm and that only a modest amount of dopamine (20 μM) is formed at the end of the second phase, the shoulder cannot be attributed to the enzyme-dopamine complex. A likely attribution might be a quinonoid species, which, according to Metzler et al. [33], could absorb at wavelengths lower than 500 nm.

In order to establish if the intermediate absorbing at 420 nm represents a Michaelis complex or an external aldimine, its rate of formation as a function of L-Dopa concentration was measured. We decided to carry out these measurements at 15°C and at pH 6.0, that is, under experimental conditions in which the decarboxylation reaction is slow enough so that its kinetics can be measured. As shown in Figure 7, upon addition of L-Dopa to hDDC, an increase in the 420 nm band with a concomitant decrease in the 335 nm signal can be seen. The apparent first-order rate constant of the appearance of the 420 nm band, k_{obs}, shows a hyperbolic dependence on L-Dopa concentration in the range 0.08–1 mM (inset of Figure 7). The k_{obs} data were fitted to the following equation:

$$k_{obs} = k_{+2} \frac{[\text{L-Dopa}]}{K_1 + [\text{L-dopa}]} + k_{-2}, \qquad (8)$$

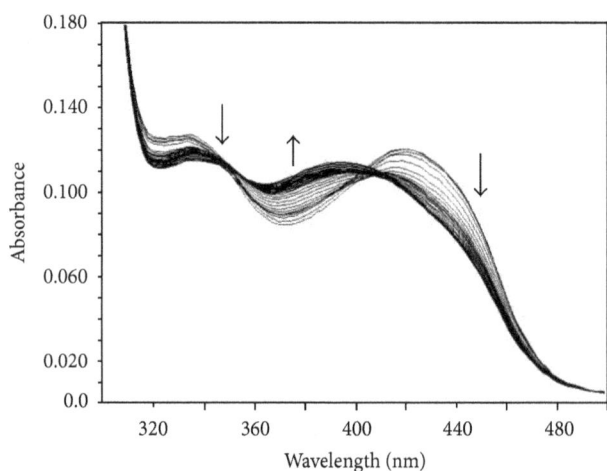

(a)

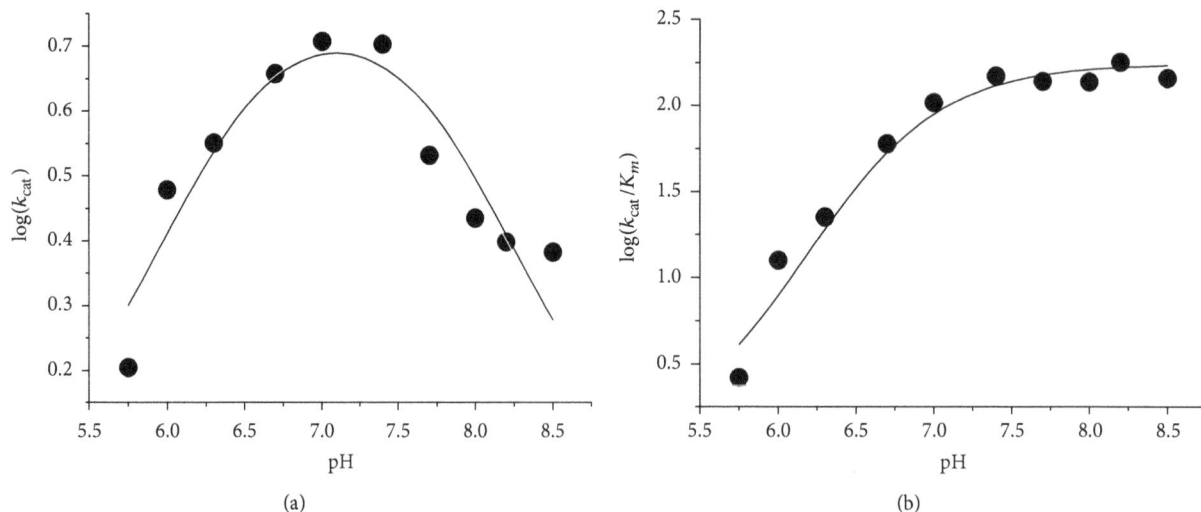

(b)

FIGURE 4: pH dependence of the kinetic parameters for the decarboxylase activity of hDDC toward L-Dopa. (a) Log k_{cat} profile and (b) log k_{cat}/K_m profile. The assays were performed at 25°C in 50 mM Bis-Tris-propane at the indicated pH using 50 nM enzyme concentration in the presence of 10 μM exogenous PLP. The points shown are the experimental values, while the curves are from fits to the data using (5) for log k_{cat} and (6) for log k_{cat}/K_m.

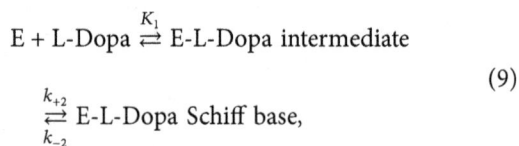

FIGURE 5: Time-resolved spectra for the reaction of hDDC with L-Dopa. Rapid scanning stopped-flow spectra obtained upon reaction of hDDC (20 μM) with L-Dopa (2 mM) in Tris-Bis-propane, pH 7.4, at 25°C. Spectra were taken between 1 and 100 ms at 1 ms intervals and between 101 and 200 ms at 10 ms intervals.

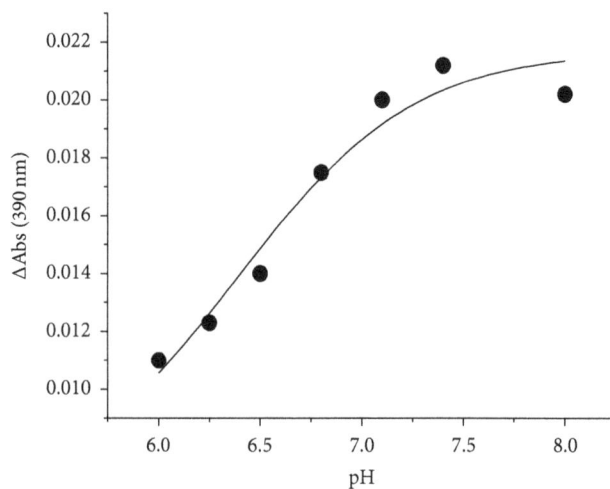

FIGURE 6: pH dependence of the amplitude of the change of the 390 nm absorbance band in the reaction of hDDC with L-Dopa. Amplitude of the 390 nm absorbance band monitored by rapid scanning stopped-flow spectra upon reaction of hDDC (20 μM) and L-Dopa (2 mM) in Tris-Bis-propane, at 25°C at the indicated pH. The points shown are the experimental values, while the curve is from fit to the data using (3).

which describes the following two-step binding model assuming that the first step is rapid:

$$E + \text{L-Dopa} \overset{K_1}{\rightleftharpoons} \text{E-L-Dopa intermediate}$$

$$\overset{k_{+2}}{\underset{k_{-2}}{\rightleftharpoons}} \text{E-L-Dopa Schiff base,} \tag{9}$$

where K_1 is the dissociation constant for the intermediate (Michaelis complex or geminal diamine) formed prior to the formation of the Schiff base species, and k_{+2} and k_{-2} are first-order rate constants for an interconversion between the intermediate and the Schiff base. Estimated values of k_{+2} and

K_1 based on the data in the inset of Figure 7 are 124 $\pm$ 3 s^{-1} and 0.32 $\pm$ 0.02 mM, respectively, while the k_{-2} value is nearly zero. The K_1 value is consistent with the K_m value at pH 6.0 (0.23 $\pm$ 0.03 mM) measured under steady-state conditions, thus strongly suggesting that the intermediate absorbing at 420 nm represents a 1-N-4'-N-protonated external aldimine. It follows that the rate of formation of the Schiff base at 25°C can be estimated to be around 250 s^{-1} using the empirical rule of a 2-fold reduction for a 10°C reduction in temperature. This value is considerably higher (~80-fold) than the k_{cat}

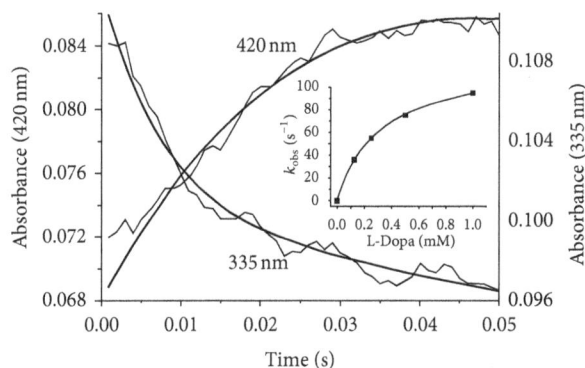

FIGURE 7: Single-wavelength stopped-flow measurements of the reaction of hDDC with L-Dopa at pH 6.0 and 15°C. Reaction of hDDC (20 μM) with L-Dopa (1 mM) was carried out at 15°C in 50 mM Bis-Tris-propane (pH 6.0). Time courses at 420 and 335 nm are reported. The solid lines are from fits to (7). The inset shows the dependence of k_{obs} for the increase in the intensity of the 420 nm band as a function of the final concentrations of L-Dopa after mixing. The points shown are the experimental values, while the curve is from fit to the data using (8).

value at 25°C at pH 6.0 (3 s^{-1}), thus indicating that one of the catalytic steps after the external Schiff base formation, including the decarboxylation step, is rate determining in the entire catalytic reaction [34]. All together, these data indicate the consecutive formation of two external aldimines, one absorbing at 420 nm and the other at 390 nm, whose structures are presented in Figure 8.

4. Discussion

The pH dependence of catalysis and spectral features has been studied in details for only a few PLP-dependent enzymes [35–39]. This allowed to elucidate how ionizations control their activities. Up to date, such analyses have been hampered for DDC for the following reasons: (i) a large portion of the coenzyme covalently bound is present in both pig kidney and rat liver enzymes in an inactive form and shows absorbance and PLP emission fluorescence similar to those of the corresponding holoenzymes and (ii) both pig kidney and rat liver DDC enzymes show little absorbance changes with altering pH.

Unlike the apo form of pig kidney and rat liver DDC, the apo hDDC does not display any absorbance in the 330 nm region and does not exhibit PLP emission fluorescence. Thus, we decided to perform a detailed investigation of the pH dependence of the spectroscopic properties of hDDC in its internal and external aldimine forms, of the PLP binding affinity as well as of the steady-state kinetic parameters. The following discussion attempts to assign the pK values observed in these analyses.

The titration of hDDC-bound coenzyme absorbance and fluorescence over the pH range 6–8.5 is consistent with a deprotonation event with a pKa value of ~7.3, which could be the result of the deprotonation of the internal aldimine. However, there is no red shift in the 335 nm band at pH values

higher than the apparent pK, as would be expected for the unprotonated aldimine absorbing at 360 nm. Structures which could account for an increase in the 330 nm region at high pH have been postulated to arise either from the formation of an adduct upon addition of a deprotonated nucleophilic or a hydroxyl group to the imine double bond or from the conversion from the ketoenamine to the enolimine tautomer. The attribution of the 335 nm absorbance band to a substituted aldamine is in contrast with the following data: (i) when hDDC is treated with NH$_2$OH, the absorbance bands at 420 and 335 nm are completely lost, and the resultant apoenzyme lacks the PLP emission fluorescence, (ii) binding of L-Dopa causes the disappearance of the 335 absorbance band, (iii) upon excitation at 335 nm, the 384 nm fluorescence emission remains at low pH where the substituted aldamine would be destabilized by protonation, and (iv) the fluorescence excitation spectrum at the emission wavelength of 384 nm shows that the absorbance band that gives the excited state is seen at 341 nm, which is apparently longer than the wavelength generally observed for substituted aldamine structures, 330–335 nm. Thus, it should be taken in consideration the possibility that the 335 nm band could be attributed to the enolimine tautomer of the Schiff base. Both 384 and 504 nm emission maxima are seen upon excitation of hDDC at 335 nm, and acid promotes the 504 nm emission at the expense of the 384 nm emission. Honikel and Madsen have shown that the enolimine can emit either solely at ~400 or ~500 nm, or at a combination of both wavelengths, depending on the polarity and acidity of the solvent. Fluorescence emission at ~400 nm versus 500 nm is determined by a competition between (i) proton transfer from the enolimine structure at the excited state to the aldimine nitrogen of the ketoenamine in the singlet excited state and (ii) radiative decay of the excited to the ground state [40]. On the basis of all these considerations, we can conclude that the 384 nm fluorescence emission of hDDC results from the excited state of the enolimine tautomer of the Schiff base before it has tautomerized to the ketoenamine excited state. A similar explanation has been proposed for the pH-dependent spectral changes observed for dialkylglycine decarboxylase [37], serine glyoxylate aminotransferase [38], histidine decarboxylase [41], and glutamate decarboxylase [42]. Examination of the absorbance titration curves indicates that at pH values much lower than the apparent pK (~7.2) some 335 nm is still present and that at pH values higher than pK some 420 nm absorbing species remains. Thus, one might expect from our results that the pyridine nitrogen in hDDC is not fully protonated, as it is usually assumed for PLP enzymes having an aspartate residue interacting with the pyridine nitrogen. Accordingly, the model depicted in Scheme 1 is proposed: the N-protonated (I) and N-unprotonated (II) ketoenamine forms absorb at 420 nm, while the N-protonated (III) and N-unprotonated (IV) enolimine tautomers absorb at 335 nm. At pH values less than pK I and III will be present, while II and IV represent the forms at pH much higher than the pK. Ring nitrogen protonation state governs the two equilibria between I and II absorbing at 420 nm as well as between III and IV absorbing at 335 nm, that is, between species spectroscopically indistinguishable. Thus, the most likely

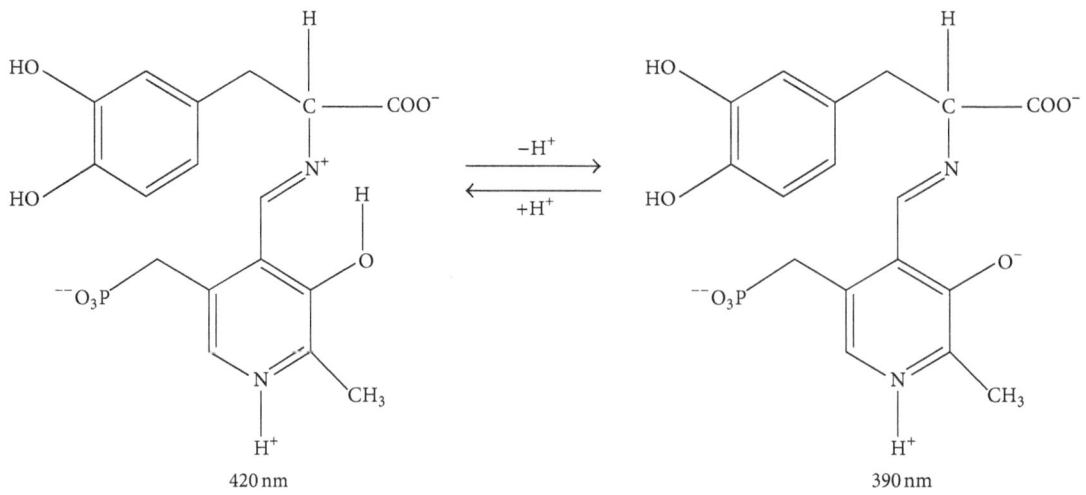

FIGURE 8: Structures of the coenzyme-L-Dopa complexes absorbing at 420 nm and 390 nm.

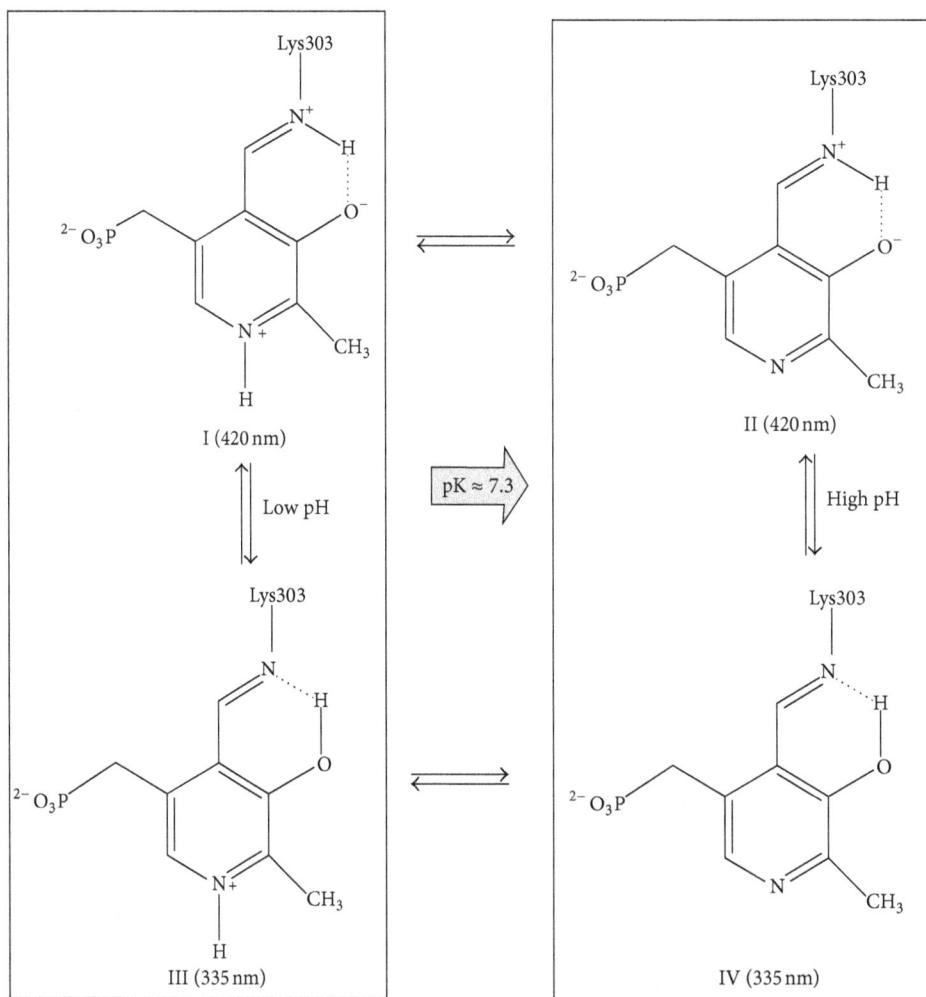

SCHEME 1: Putative model for the pH dependence of the internal aldimine.

explanation that accounts for the pH titration of coenzyme absorbance and fluorescence is that the ionization observed is not associated with any functional group on the Schiff base itself. Rather, it is an active site residue in close proximity to the coenzyme whose ionization alters the ratio between the two tautomers that absorb at 420 and 335 nm.

The k_{cat} profile of the decarboxylation reaction catalyzed by hDDC toward L-Dopa displays two pKa values at about 6.3 and 7.9, thus suggesting that one group is required to be unprotonated and a second group protonated to achieve maximum velocity. The pKa on the acidic side of the profile is similar to that seen in the k_{cat}/K_m profile, and, thus, it can be concluded that this group is involved in catalysis but not in binding. Since it has been proven that in pig kidney DDC CO_2 release is rate limiting, k_{cat} must report on ionization(s) of the external aldimine. Thus, the spectral changes taking place in the bound coenzyme upon addition of L-Dopa been analyzed as a function of pH by rapid scanning stopped-flow spectroscopy. The kinetic analyses of the reaction of hDDC with L-Dopa clearly demonstrate the presence of two intermediates absorbing at 420 and 390 nm. The 420 nm absorbing species is formed first, followed by formation of the second intermediate absorbing at 390 nm. The amplitude of the signal change at 390 nm increases with increasing pH above a single pK of ~6.4. The finding that the rate of appearance of the 420 nm band measured at pH 6.0 shows that a hyperbolic dependence on the concentration of L-Dopa is consistent with the assignment of this absorbance band to a 4′-N-protonated external aldimine. The time and pH dependence of the conversion of the 420 to the 390 nm absorbance band strongly suggest that the 390 nm band could be attributed to a 4′-N-unprotonated external aldimine. Even in the absence of the spectra of these intermediates, Minelli et al. [30] have predicted the presence of the step ESH$^+$ → ES, corresponding to the 420 → 390 nm conversion, along the reaction pathway of the decarboxylation of L-Dopa. Our results validate this proposal, thus ruling out that the 420 nm intermediate represents Michaelis complex, as previously suggested [25, 26]. The pK of this spectral transition (6.4) roughly coincident with those observed in both k_{cat} and k_{cat}/K_m profiles strengthens the argument that this ionization is associated with a catalytic event.

There is no evidence to support an assignment of the pK (7.9) observed on the basic side of the k_{cat} profile to a specific group. It is likely the same one seen in the pH profile for $K_{D(PLP)}$ (8.3). Its presence on the $K_{D(PLP)}$ profile could suggest that the phosphate ester of the coenzyme phosphate group is the likely origin of this pK, which has a value similar to that observed for the coenzyme phosphate group in *Treponema denticola* cystalysin [35]. In hDDC the effect of this ionization could be the loss of the hydrogen bond between the hydroxyl group of Ser149 and the coenzyme phosphate group oxygen. Considering the large conformational change accompanying the transition from the apo to the holo form of hDDC [27], it can be speculated that the loss of hydrogen bond not only decreases the PLP binding affinity but also could hamper a correct apo → holo conversion resulting in a less catalytically competent conformation as the pH increases above 8. Nonetheless, it cannot be ruled out that the pK observed in

$K_{D(PLP)}$ does not coincide with the pK of the k_{cat} profile. In this case, the effect on the k_{cat} could be exerted by the ionization of a residue that, although remote from the active site, could affect the active site. A good candidate could be Tyr332 for which a role in $C\alpha$ protonation of the quinonoid along the decarboxylation pathway has been identified [22]. Although this assignment should be taken with caution, it is not in contrast with the detection at pH higher than pK of a ~440 nm absorbing shoulder attributable to a quinonoid species. Taken together, these results indicate that the maximum k_{cat} value of hDDC is achieved when the deprotonation of the external aldimine and that of an unidentified residue take place.

In the light of these data, the absence of a band absorbing at 390 nm in the reaction with L-Dopa of four DDC variants responsible for AADC deficiency, an inherited rare neurometabolic disease, should be reconsidered. In these variants, mutations concerning amino acid residues that interact directly or indirectly with PLP and/or its microenvironment, cause a perturbation of the active site [11]. It can be postulated that in these variants the external aldimine absorbing at 420 nm is not in a proper position and/or orientation to transfer the proton at 4′N of the Schiff base. Taking into account that, according to Minelli et al. [30], the 390 nm form is about 5-fold more reactive than the 420 nm one, it can be speculated that their reduced catalytic activity could be ascribable, at least in part, to the lack of the 420 → 390 nm conversion.

5. Conclusions

A detailed investigation of the pH dependence of the steady-state kinetic parameters, of the spectroscopic titrations of the internal and external aldimine, as well as of the PLP binding affinity allows us to identify three observable ionizations in hDDC. The following tentative assignments for these have been made: pK_1 (6.3-6.4), the deprotonation of the 4′-N-protonated external aldimine occurring during the decarboxylation pathway, pK_2 (~7.2), a residue governing the equilibrium between the low- and the high-pH forms of the internal aldimine, and pK_3 (7.9, 8.3), two distinct groups (the coenzyme phosphate ester of the internal aldimine and a residue involved in the catalysis) or a unique residue affecting both PLP binding affinity and k_{cat} value: additional studies will be needed to sort out the various possibilities.

Abbreviations

hDDC: Human Dopa decarboxylase
PLP: Pyridoxal 5′-phosphate
AADC: L-Aromatic amino acid decarboxylase
$K_{D(PLP)}$: Equilibrium dissociation constant for PLP.

Acknowledgments

This work was supported by grants from M.I.U.R and the Consorzio Interuniversitario per le Biotecnologie CIB (IT) to C. B. Voltattorni and B. Cellini.

References

[1] A. Amadasi, M. Bertoldi, R. Contestabile et al., "Pyridoxal $5'$-phosphate enzymes as targets for therapeutic agents," *Current Medicinal Chemistry*, vol. 14, no. 12, pp. 1291–1324, 2007.

[2] B. Cellini, A. Lorenzetto, R. Montioli, E. Oppici, and C. B. Voltattorni, "Human liver peroxisomal alanine:glyoxylate aminotransferase: different stability under chemical stress of the major allele, the minor allele, and its pathogenic G170R variant," *Biochimie*, vol. 92, no. 12, pp. 1801–1811, 2010.

[3] B. Cellini, E. Oppici, A. Paiardini, and R. Montioli, "Molecular insights into primary hyperoxaluria type 1 pathogenesis," *Frontiers in Bioscience*, vol. 17, pp. 621–634, 2012.

[4] M. L. di Salvo, R. Contestabile, A. Paiardini, and B. Maras, "Glycine consumption and mitochondrial serine hydroxymethyltransferase in cancer cells: the heme connection," *Medical Hypotheses*, vol. 80, pp. 633–636, 2013.

[5] A. E. Pegg, L. M. Shantz, and C. S. Coleman, "Ornithine decarboxylase as a target for chemoprevention," *Journal of Cellular Biochemistry*, vol. 58, no. 22, pp. 132–138, 1995.

[6] N. A. Rao, R. Talwar, and H. S. Savithri, "Molecular organization, catalytic mechanism and function of serine hydroxymethyltransferase—a potential target for cancer chemotherapy," *International Journal of Biochemistry and Cell Biology*, vol. 32, no. 4, pp. 405–416, 2000.

[7] P. Storici, G. Capitani, D. De Biase et al., "Crystal structure of GABA-aminotransferase, a target for antiepileptic drug therapy," *Biochemistry*, vol. 38, no. 27, pp. 8628–8634, 1999.

[8] B. Cellini, R. Montioli, E. Oppici, and C. B. Voltattorni, "Biochemical and computational approaches to improve the clinical treatment of dopa decarboxylase-related diseases: an overview," *Open Biochemistry Journal*, vol. 6, pp. 131–138, 2012.

[9] F. Daidone, R. Montioli, A. Paiardini et al., "Identification by virtual screening and in vitro testing of human DOPA decarboxylase inhibitors," *PLoS ONE*, vol. 7, no. 2, Article ID e31610, 2012.

[10] R. Montioli, E. Oppici, B. Cellini, A. Roncador, M. Dindo, and C. B. Voltattorni, "S250F variant associated with aromatic amino acid decarboxylase deficiency: molecular defects and intracellular rescue by pyridoxine," *Human Molecular Genetics*, vol. 22, no. 8, pp. 1615–1624, 2013.

[11] R. Montioli, B. Cellini, and C. Borri Voltattorni, "Molecular insights into the pathogenicity of variants associated with the aromatic amino acid decarboxylase deficiency," *Journal of Inherited Metabolic Disease*, vol. 34, pp. 1213–1224, 2011.

[12] B. Maras, P. Dominici, D. Barra, F. Bossa, and C. Borri Voltattorni, "Pig kidney 3,4-dihydroxyphenylalanine (Dopa) decarboxylase. Primary structure and relationships to other amino acid decarboxylases," *European Journal of Biochemistry*, vol. 201, no. 2, pp. 385–391, 1991.

[13] C. B. Voltattorni, A. Minelli, and P. Dominici, "Interaction of aromatic amino acids in D and L forms with 3,4-dihydroxyphenylalanine decarboxylase from pig kidney," *Biochemistry*, vol. 22, no. 9, pp. 2249–2254, 1983.

[14] C. B. Voltattorni, A. Minelli, and C. Turano, "Spectral properties of the coenzyme bound to DOPA decarboxylase from pig kidney," *FEBS Letters*, vol. 17, no. 2, pp. 231–235, 1971.

[15] C. B. Voltattorni, A. Minelli, and P. Vecchini, "Purification and characterization of 3,4-dihydroxyphenylalanine decarboxylase from pig kidney," *European Journal of Biochemistry*, vol. 93, no. 1, pp. 181–187, 1979.

[16] P. Dominici, B. Tancini, D. Barra, and C. B. Voltattorni, "Purification and characterization of rat-liver 3,4-dihydroxyphenylalanine decarboxylase," *European Journal of Biochemistry*, vol. 169, no. 1, pp. 209–213, 1987.

[17] A. Fiori, C. Turano, and C. Borri Voltattorni, "Interaction of L DOPA decarboxylase with substrates. A spectrophotometric study," *FEBS Letters*, vol. 54, no. 2, pp. 122–125, 1975.

[18] M. Bertoldi and C. Borri Voltattorni, "Reaction of dopa decarboxylase with L-aromatic amino acids under aerobic and anaerobic conditions," *Biochemical Journal*, vol. 352, no. 2, pp. 533–538, 2000.

[19] M. Bertoldi, B. Cellini, B. Maras, and C. B. Voltattorni, "A quinonoid is an intermediate of oxidative deamination reaction catalyzed by Dopa decarboxylase," *FEBS Letters*, vol. 579, no. 23, pp. 5175–5180, 2005.

[20] M. Bertoldi, B. Cellini, R. Montioli, and C. B. Voltattorni, "Insights into the mechanism of oxidative deamination catalyzed by DOPA decarboxylase," *Biochemistry*, vol. 47, no. 27, pp. 7187–7195, 2008.

[21] M. Bertoldi, P. Frigeri, M. Paci, and C. B. Voltattorni, "Reaction specificity of native and nicked 3,4-dihydroxyphenylalanine decarboxylase," *Journal of Biological Chemistry*, vol. 274, no. 9, pp. 5514–5521, 1999.

[22] M. Bertoldi, M. Gonsalvi, R. Contestabile, and C. B. Voltattorni, "Mutation of tyrosine 332 to phenylalanine converts dopa decarboxylase into a decarboxylation-dependent oxidative deaminase," *Journal of Biological Chemistry*, vol. 277, no. 39, pp. 36357–36362, 2002.

[23] P. Dominici, P. S. Moore, S. Castellani, M. Bertoldi, and C. B. Voltattorni, "Mutation of cysteine 111 in Dopa decarboxylase leads to active site perturbation," *Protein Science*, vol. 6, no. 9, pp. 2007–2015, 1997.

[24] P. S. Moore, P. Dominici, and C. Borri Voltattorni, "Cloning and expression of pig kidney dopa decarboxylase: comparison of the naturally occurring and recombinant enzymes," *Biochemical Journal*, vol. 315, no. 1, pp. 249–256, 1996.

[25] H. Hayashi, H. Mizuguchi, and H. Kagamiyama, "Rat liver aromatic L-amino acid decarboxylase: spectroscopic and kinetic analysis of the coenzyme and reaction intermediates," *Biochemistry*, vol. 32, no. 3, pp. 812–818, 1993.

[26] H. Hayashi, F. Tsukiyama, S. Ishii, H. Mizuguchi, and H. Kagamiyama, "Acid base chemistry of the reaction of aromatic L-amino acid decarboxylase and dopa analyzed by transient and steady-state kinetics: preferential binding of the substrate with its amino group unprotonated," *Biochemistry*, vol. 38, no. 47, pp. 15615–15622, 1999.

[27] G. Giardina, R. Montioli, S. Gianni et al., "Open conformation of human DOPA decarboxylase reveals the mechanism of PLP addition to Group II decarboxylases," *Proceedings of the National Academy of Sciences of the United States of America*, vol. 108, no. 51, pp. 20514–20519, 2011.

[28] P. Burkhard, P. Dominici, C. Borri-Voltattorni, J. N. Jansonius, and V. N. Malashkevich, "Structural insight into Parkinson's disease treatment from drug-inhibited DOPA decarboxylase," *Nature Structural Biology*, vol. 8, no. 11, pp. 963–967, 2001.

[29] R. Singh, F. Spyrakis, P. Cozzini, A. Paiardini, S. Pascarella, and A. Mozzarelli, "Chemogenomics of pyridoxal $5'$-phosphate dependent enzymes," *Journal of Enzyme Inhibition and Medicinal Chemistry*, vol. 28, no. 1, pp. 183–194, 2013.

[30] A. Minelli, A. T. Charteris, C. B. Voltattorni, and R. A. John, "Reactions of DOPA (3,4-dihydroxyphenylalanine) decarboxylase with DOPA," *Biochemical Journal*, vol. 183, no. 2, pp. 361–368, 1979.

[31] A. F. Sherald, J. C. Sparrow, and T. R. F. Wright, "A spectrophotometric assay for Drosophila dopa decarboxylase," *Analytical Biochemistry*, vol. 56, no. 1, pp. 300–305, 1973.

[32] A. Charteris and R. John, "An investigation of the assay of dopamine using trinitrobenzensulphonic acid," *Analytical Biochemistry*, vol. 66, no. 2, pp. 365–371, 1975.

[33] C. M. Metzler, A. G. Harris, and D. E. Metzler, "Spectroscopic studies of quinonoid species from pyridoxal 5'-phosphate," *Biochemistry*, vol. 27, no. 13, pp. 4923–4933, 1988.

[34] M. Bertoldi and C. B. Voltattorni, "Multiple roles of the active site lysine of Dopa decarboxylase," *Archives of Biochemistry and Biophysics*, vol. 488, no. 2, pp. 130–139, 2009.

[35] B. Cellini, M. Bertoldi, R. Montioli, and C. B. Voltattorni, "Probing the role of Tyr 64 of Treponema denticola cystalysin by site-directed mutagenesis and kinetic studies," *Biochemistry*, vol. 44, no. 42, pp. 13970–13980, 2005.

[36] G. A. Hunter, J. Zhang, and G. C. Ferreira, "Transient kinetic studies support refinements to the chemical and kinetic mechanisms of aminolevulinate synthase," *Journal of Biological Chemistry*, vol. 282, no. 32, pp. 23025–23035, 2007.

[37] X. Zhou and M. D. Toney, "pH Studies on the mechanism of the pyridoxal phosphate-dependent dialkylglycine decarboxylase," *Biochemistry*, vol. 38, no. 1, pp. 311–320, 1999.

[38] W. E. Karsten, T. Ohshiro, Y. Izumi, and P. F. Cook, "Initial velocity, spectral, and pH studies of the serine-glyoxylate aminotransferase from Hyphomicrobiuim methylovorum," *Archives of Biochemistry and Biophysics*, vol. 388, no. 2, pp. 267–275, 2001.

[39] D. M. Kiick and P. F. Cook, "pH studies toward the elucidation of the auxiliary catalyst for pig heart aspartate aminotransferase," *Biochemistry*, vol. 22, no. 2, pp. 375–382, 1983.

[40] K. O. Honikel and N. B. Madsen, "Comparison of the absorbance spectra and fluorescence behavior of phosphorylase b with that of model pyridoxal phosphate derivatives in various solvents," *Journal of Biological Chemistry*, vol. 247, no. 4, pp. 1057–1064, 1972.

[41] M. T. Olmo, F. Sánchez-Jiménez, M. A. Medina, and H. Hayashi, "Spectroscopic analysis of recombinant rat histidine decarboxylase," *Journal of Biochemistry*, vol. 132, no. 3, pp. 433–439, 2002.

[42] W. C. Chu and D. E. Metzler, "Enzymatically active truncated cat brain glutamate decarboxylase: expression, purification, and absorption spectrum," *Archives of Biochemistry and Biophysics*, vol. 313, no. 2, pp. 287–295, 1994.

13

Structures and Properties of Naturally Occurring Polyether Antibiotics

Jacek Rutkowski and Bogumil Brzezinski

Department of Biochemistry, Faculty of Chemistry, Adam Mickiewicz University, Grunwaldzka 6, 60-780 Poznań, Poland

Correspondence should be addressed to Jacek Rutkowski; jacekr@amu.edu.pl

Academic Editor: Ivayla Pantcheva-Kadreva

Polyether ionophores represent a large group of natural, biologically active substances produced by *Streptomyces spp*. They are lipid soluble and able to transport metal cations across cell membranes. Several of polyether ionophores are widely used as growth promoters in veterinary. Polyether antibiotics show a broad spectrum of bioactivity ranging from antibacterial, antifungal, antiparasitic, antiviral, and tumour cell cytotoxicity. Recently, it has been shown that some of these compounds are able to selectively kill cancer stem cells and multidrug-resistant cancer cells. Thus, they are recognized as new potential anticancer drugs. The biological activity of polyether ionophores is strictly connected with their molecular structure; therefore, the purpose of this paper is to present an overview of their formula, molecular structure, and properties.

1. Introduction

Polyether ionophores are very large and important group of naturally occurring compounds. Increased interest in this type of compound has been observed in recent years. There are over 120 naturally occurring ionophores known [1]. Major commercial use of the ionophores is to control coccidiosis. They are also used as growth promoters in ruminants. These compounds specifically target the ruminal bacterial population increasing production efficiency. In 2003, the antimicrobials most commonly used in beef cattle production were ionophores. Lasalocid (Avatec, Bovatec), monensin (Coban, Rumensin, and Coxidin), salinomycin (Bio-cox, Sacox), narasin (Monteban, Maxiban), maduramycin (Cygro), and laidlomycin propionate (Cattlyst) were the ionophores of combined yearly sales of more than $150 million [2]. Ionophores can also be used in the production of ion-selective electrodes [3, 4].

All applications of ionophores mentioned above are closely related to their ability to form complexes with metal cations (host-guest complexes) and transport these complexes across lipid bilayers and cell membranes. According to the results of X-ray studies, the metal cation sites in a cage formed by oxygen atoms of the ionophore. The polyether antibiotics form neutral complexes with monovalent cations (i.e., monensin, salinomycin) or divalent metal cations (i.e., lasalocid, calcimycin) as well as with organic bases (i.e., lasalocid). The neutral complexes with cations are formed more preferably than charged complexes because the carboxylic group is deprotonated at physiological pH. However, recently it has been shown that polyether ionophores can also act as neutral ionophores and transport cations as charged complexes. Typically, the complexation of the cation is connected with formation of a pseudocyclic structure which is stabilized by intramolecular hydrogen bonds formed between the carboxylic group and the hydroxyl groups. The ring of ionophore molecule is wrapped around this hydrophilic cage rendering the whole complex lipophilic. The mechanism of transport of a cation by polyether ionophores is attributed to their ability to exchange protons and cations in an electroneutral process. In this type of transport of the cations (M^+), the polyether ionophore anion ($I{-}COO^-$) binds the metal cation or proton (H^+) to give a neutral salt ($I{-}COO^-M^+$) or a neutral polyether ionophore in acidic form

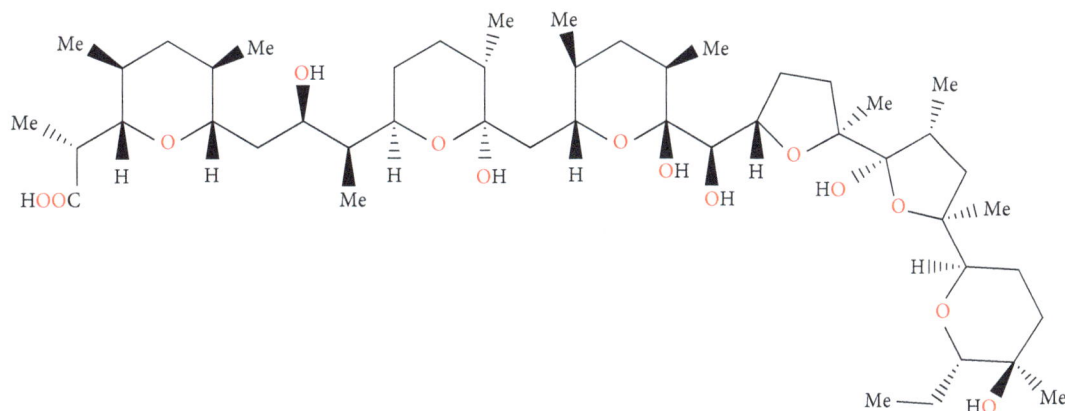

FIGURE 1: Structure of alborixine.

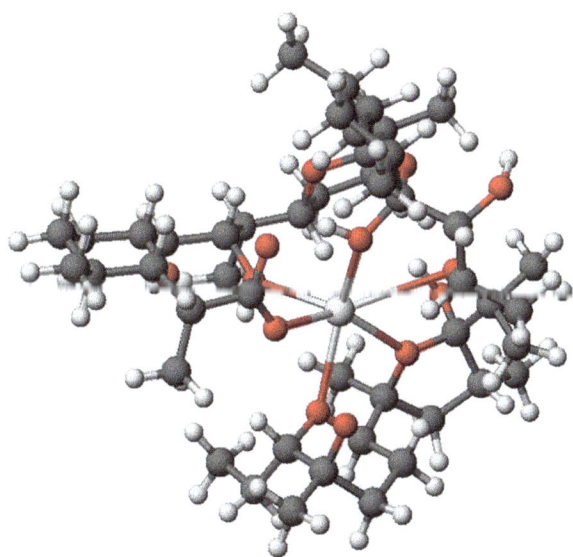

FIGURE 2: Structure of 6-demethyl-alborixin complex with sodium cation.

(I–COOH), respectively, and only uncharged molecules containing either the metal cation or proton can move through the cell membrane. The whole process leads to changes in Na^+/K^+ gradient and to an increase in the osmotic pressure inside the cell, causing swelling and vacuolization, and finally death of the bacteria cell. The remarkable selectivity of some ionophores is attributed to the size of the cage. Only cations with an appropriate radius fit the cavity perfectly, larger ones have to deform it, while smaller ones find a nonoptimal coordination geometry [5–10].

Ionophores are most effective against Gram-positive bacteria. The cell of these bacteria is surrounded by the peptidoglycan layer which is porous and allows small molecules to pass through, reaching the cytoplasmic membrane, where the lipophilic ionophore rapidly dissolves into the membrane. Conversely, Gram-negative bacteria are separated from the environment and antimicrobial agents by a lipopolsaccharide layer, outer membrane, and periplasmic space [2]. *Escherichia*

coli, Gram-negative bacteria of significant importance to food safety and human health, is insensitive to ionophore addition, unless the outer membrane is removed [11].

In this paper, we present structure and properties of several naturally occurring polyether antibiotics.

The molecular structures of polyether ionophores presented below are visualized using X-ray data deposited in Cambridge Structural Database.

2. Structures and Biological Properties of Several Ionophores

2.1. Alborixin. Alborixin (Figure 1) was first isolated from fermenting a fermentation of *Streptomyces hygroscopicus* [12]. Its structure was determined by X-ray analysis of its potassium salt [13]. A crystal structure of 6-demethyl-alborixin complex with sodium cation (Figure 2) has also been reported [14]. Alborixin is active against Gram-positive bacteria. The value of LD_{50} of alborixin determined in mice subcutaneously is 150 mg/kg [15].

2.2. Antibiotic 6016. Antibiotic 6016 (Figure 3) was first isolated from the culture of *Streptomyces albus* strain No. 6016 [16]. Molecular structure (Figure 4) was determined by X-ray analysis of its thalium salt [17]. Antibiotic 6016 is active against Gram-positive bacteria including mycobacteria. No activity was observed against Gram-negative bacteria and fungi. The toxicity of antibiotic 6016 in mice was also examined. The LD_{50} is 23 mg/kg intraperitoneally and 63 mg/kg orally. The anticoccidial evaluation of antibiotic 6016 was carried out on chickens infected with *Eimeria tenella* oocyst. It was effective in reducing mortality of chickens and increasing average body weight of treated infected chickens compared to untreated infected controls [16].

2.3. Calcimycin, Cezomycin, and X-14885A. Calcimycin (Figure 5) was first isolated from a fermentation of *Streptomyces chartreuses* [18]. This antibiotic is able to bind divalent cations with a preference for the complexation of calcium cation [19]. Complexes of calcimycin with Mg^{2+}, Ni^{2+}, and

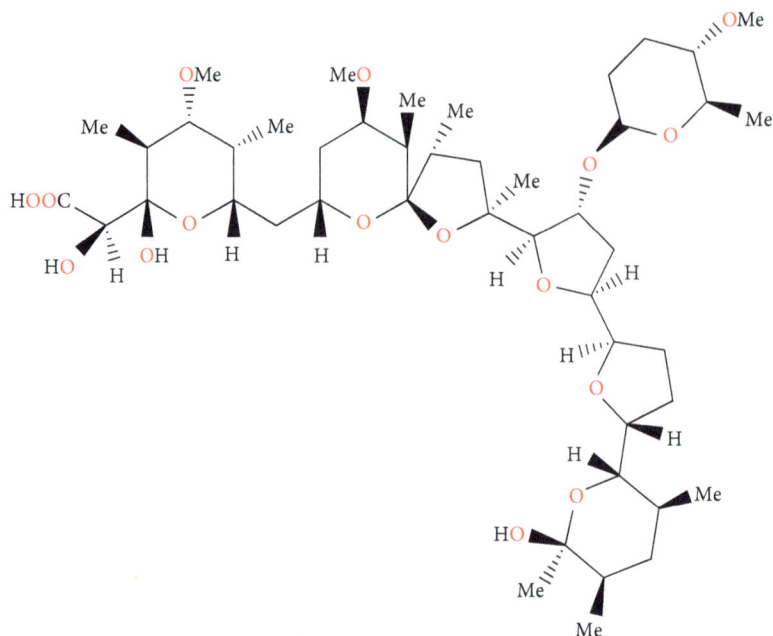

FIGURE 3: Structure of antibiotic 6016.

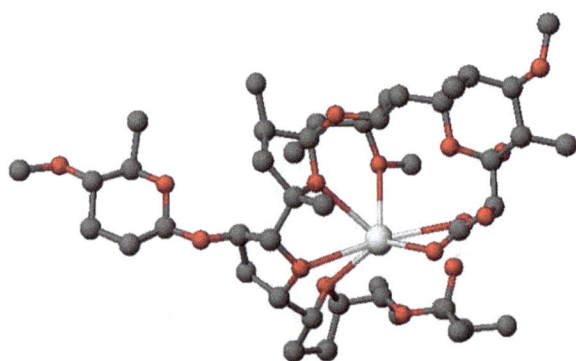

FIGURE 4: Crystal structure of antibiotic 6016 thallium salt.

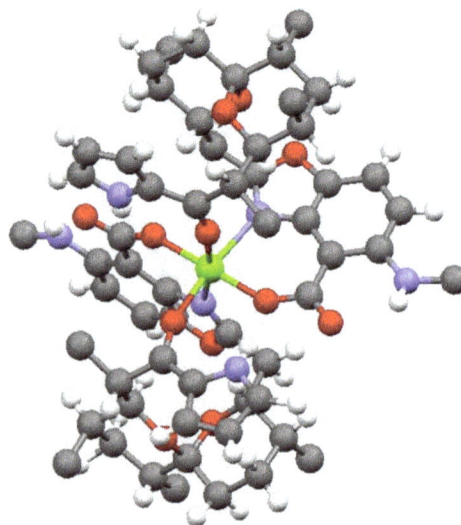

FIGURE 6: Crystal structure of 2:1 complex of calcimycin with the magnesium cation.

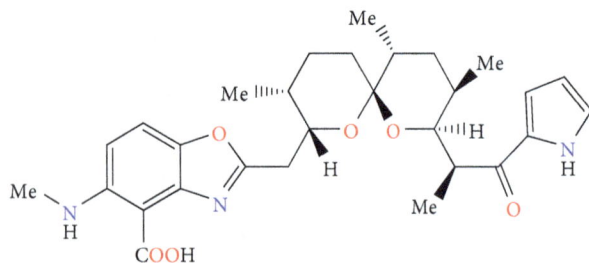

FIGURE 5: Structure of calcimycin.

Zn^{2+} cations on 2:1 stoichiometry are also described [20–22]. Structure of complex with Mg^{2+} is presented in Figure 6.

By equilibrating calcium ion across the inner membrane of mitochondria, calcimycin can uncouple oxidative phosphorylation. Various biological processes which require the

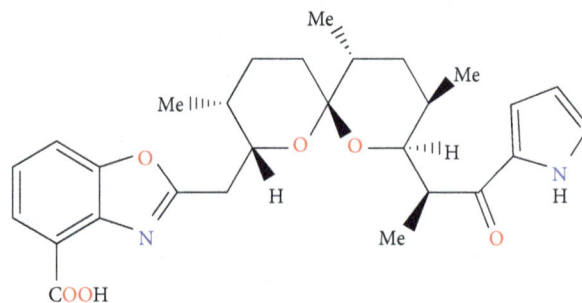

FIGURE 7: Structure of cezomycin.

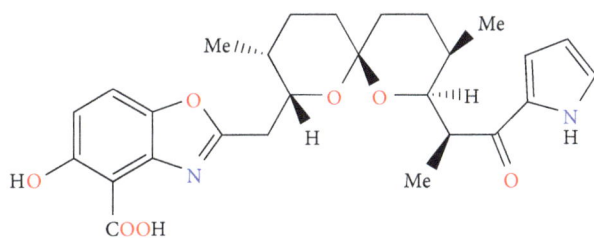

FIGURE 8: Structure of X-14885A.

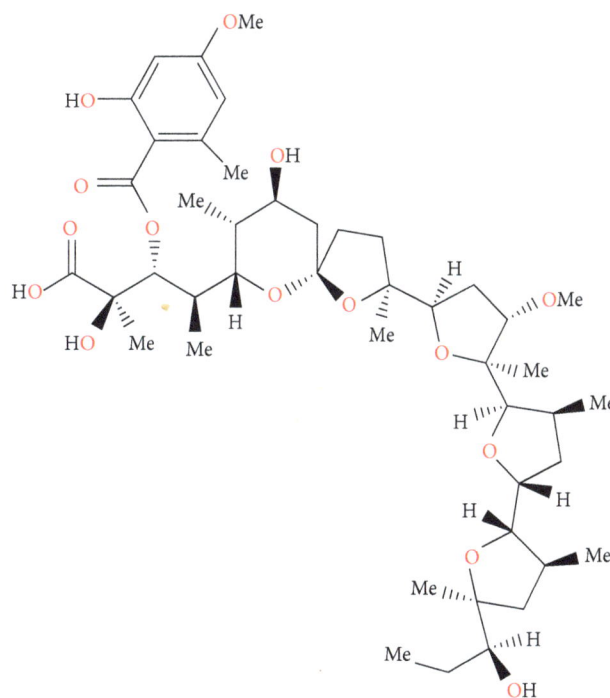

FIGURE 9: Crystal structure of 2 : 1 complex of 11-demethyl-cezomy-cin complex with the sodium cation.

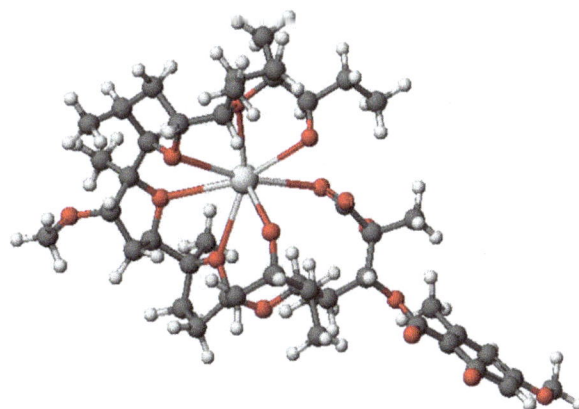

FIGURE 10: Structure of cationomycin.

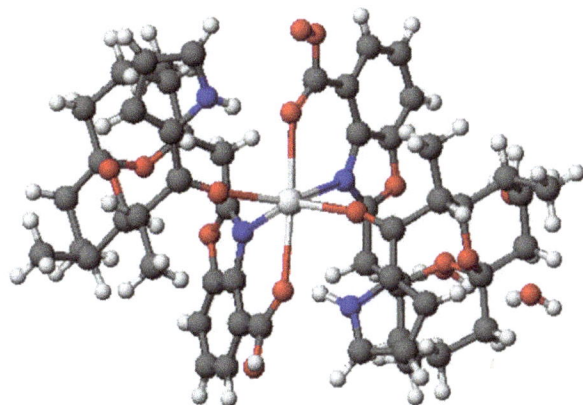

FIGURE 11: Crystal structure of cationomycin thallium salt.

presence of a calcium ion, such as thyroid secretion and insulin release, have been stimulated using this ionophore [23, 24]. Because of the fluorescent properties of calcimycin, it has been used also as a probe for divalent cations in artificial and biological membranes and to determine the mode of action of ionophore-mediated divalent cation transport [25].

Calcimycin is an efficient antibiotic against Gram-positive bacteria and inactive towards Gram-negative species. This difference in activity is attributed to the outer membrane of Gram-negative bacteria which is presumably impermeable to these highly hydrophobic compounds [26].

There are two compounds structurally related to cal-cimycin: cezomycin (Figure 7) also known as demethyloam-ino-calcimycin and X-14885A (Figure 8), which has one methyl group less on the spiroketal and a hydroxyl group instead of a methyloamino group present in calcimycin. Both compounds are isolated from the same strain as calcimycin. The crystal structure of cezomycin (Figure 9) was determined as its 11-demethyl complex with the sodium cation of the stoichiometry 2 : 1 [27].

Antibiotic X-14885A exhibits *in vitro* activity at con-centrations less than 1 μg/mL against such Gram-positive bacteria as *Staphylococcus aureus* and *Bacillus subtilis* and the spirochete responsible for swine dysentery *Treponema hyo-dysenteriae* [28].

2.4. Cationomycin. Cationomycin (Figure 10) is an ionophore isolated from *Actinomadura azurea*. Crystal structure of cationomycin (Figure 11) was determined by X-ray analysis

of its thallium salt [29]. Cationomycin contains an unusual aromatic side fragment, which is important for biological activity. When this chain is removed, the activity of cation-omycin is reduced. A large group of derivatives was prepared after deacylation, but only anisyl analogue was more active than cationomycin [30, 31].

Kinetic studies showed that cationomycin transported the potassium cation more rapidly than the sodium cation, and the more stable complex was formed with potassium at the water/membrane interface [32].

2.5. Endusamycin. Endusamycin (Figure 12) was isolated from *Streptomyces endus*. The crystal structure (Figure 13) was determined by X-ray analysis of its rubidium salt [33].

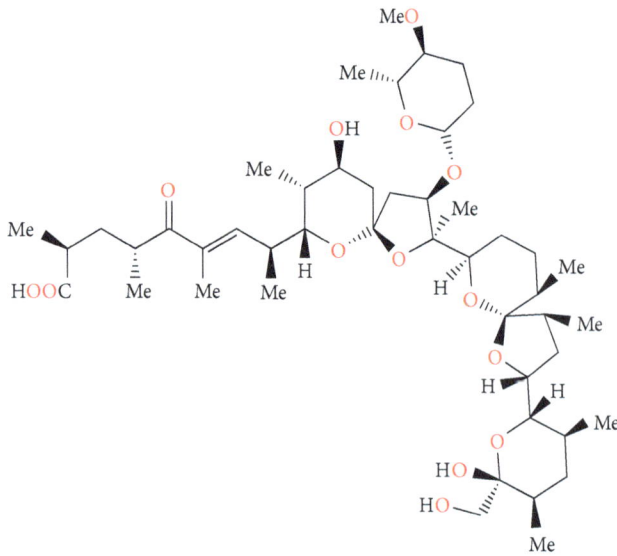

FIGURE 12: Structure of endusamycin.

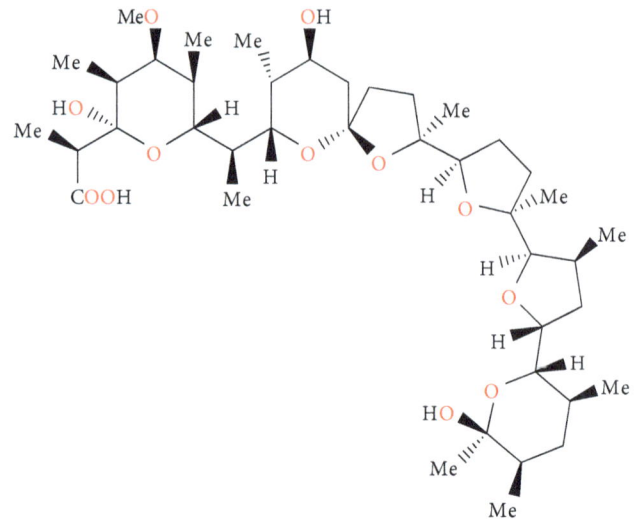

FIGURE 14: Structure of mutalomycin.

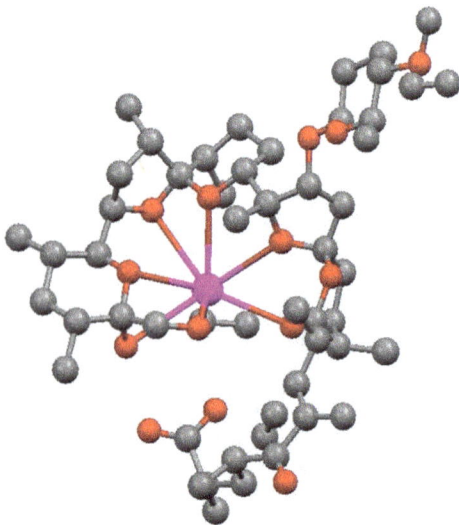

FIGURE 13: Crystal structure of endusamycin rubidium salt.

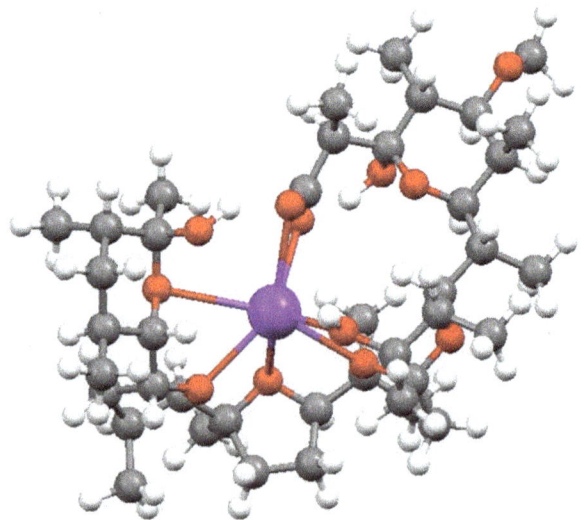

FIGURE 15: Crystal structure of 28-epimutalomycin potassium salt.

Endusamycin has a good spectrum of activity against Gram-positive bacteria and good activity against many anaerobes and organisms such as *Treponema hyodysenteriae*. Endusamycin was active against *Eimeria tenella* and *Eimeria acervulina* coccidia when administrated in feed from 10 to $40\,\mu g/g$. Chickens were protected from lesions but showed poor weight gains and feed intake. Endusamycin has LD_{50} of 7.5 mg/kg orally in male rats [33].

Endusamycin also induced a change in the proportion of volatile fatty acids (acetate, propionate, and butyrate) produced in the rumen by increasing the molar proportion of propionate in the rumen fluids [33].

2.6. Mutalomycin. Mutalomycin (Figure 14) was first isolated from a strain of *Streptomyces mutabilis* [34]. Crystal structure

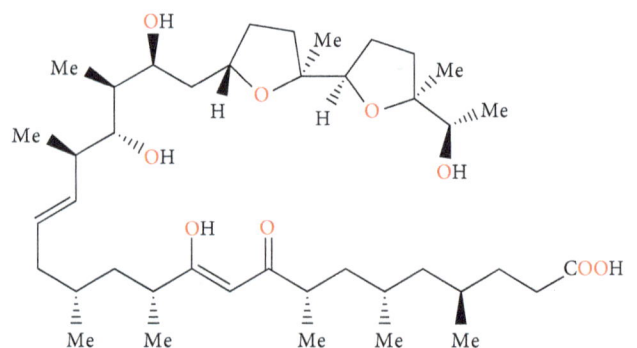

FIGURE 16: Structure of ionomycin.

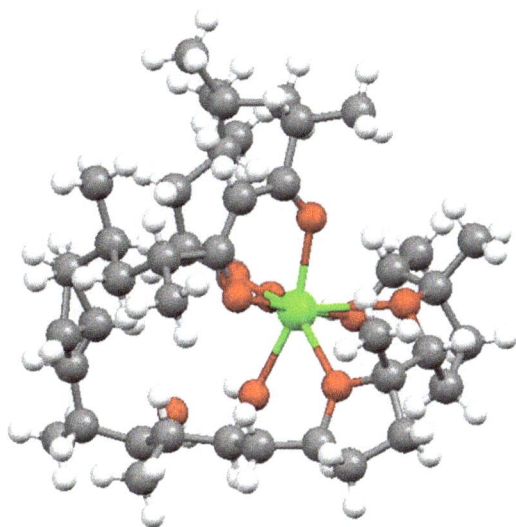

FIGURE 17: Crystal structure of ionomycin complex with the calcium cation.

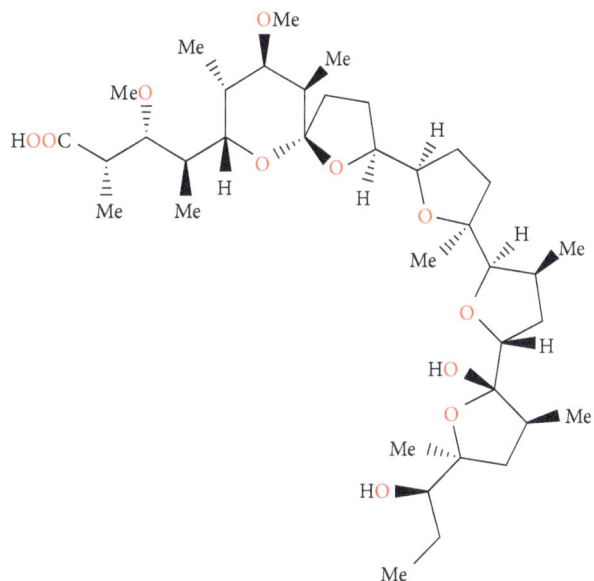

FIGURE 19: Structure of kijimicin.

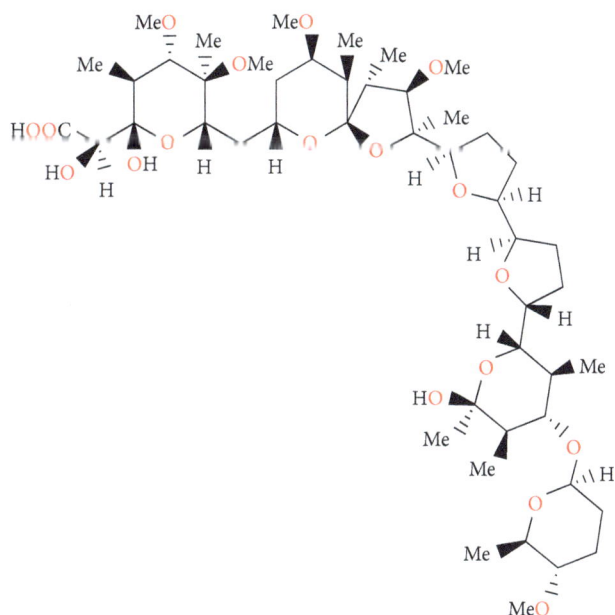

FIGURE 18: Structure of K-41.

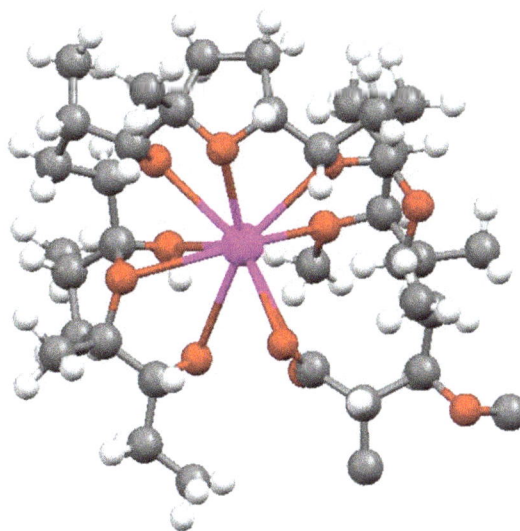

FIGURE 20: Crystal structure of kijimicin rubidium salt.

of its epimer, 28-epimutalomycin potassium salt (Figure 15), was reported [35].

Mutalomycin possesses antibiotic activity against Gram-positive bacteria and also exhibits an anticoccidial activity in chickens similar to other polyether antibiotics. It is effective in reducing mortality and in increasing the average body weight of chickens infected with *Eimeria tenella* and other *Eimeria* species [34].

2.7. Ionomycin. Ionomycin (Figure 16) was first isolated from *Streptomyces conglobatus* [36]. Crystal structure of its complex with the calcium cation (Figure 17) was reported [37].

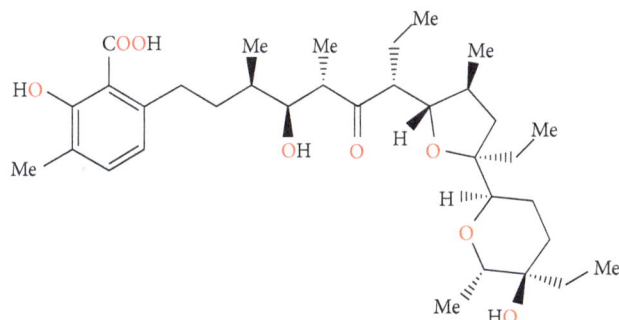

FIGURE 21: Structure of lasalocid.

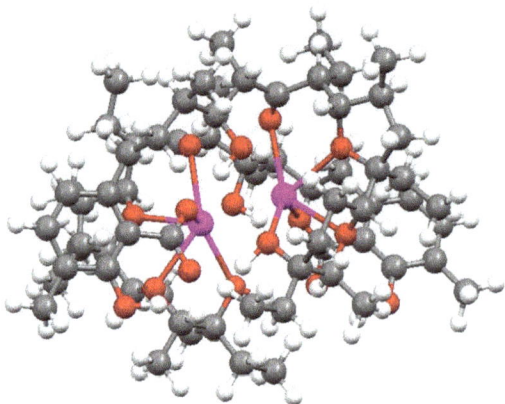

FIGURE 22: Crystal structure of lasalocid silver salt on 2:2 stoichiometry.

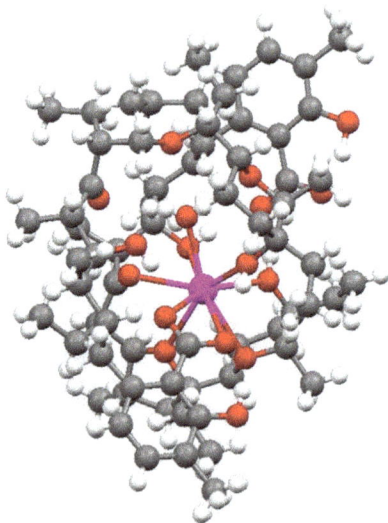

FIGURE 23: Crystal structure of 2:1 complex of lasalocid with the strontium cation.

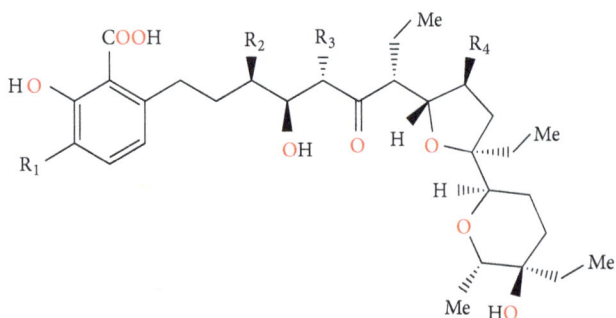

	R₁	R₂	R₃	R₄
Lasalocid A	Me	Me	Me	Me
Lasalocid B	Et	Me	Me	Me
Lasalocid C	Me	Et	Me	Me
Lasalocid D	Me	Me	Et	Me
Lasalocid E	Me	Me	Me	Et

FIGURE 24: Lasalocid acid analogues.

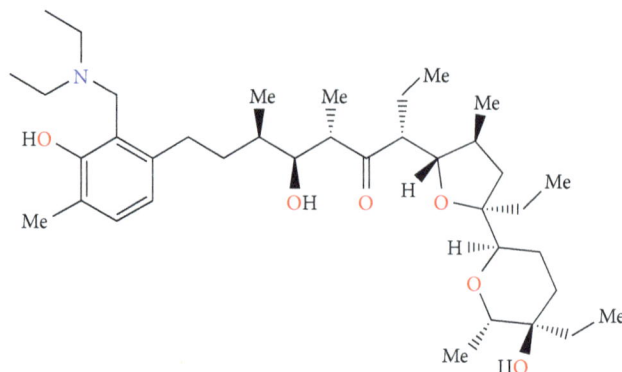

FIGURE 25: Structure of lasalocid acid Mannich base.

Ionomycin is capable of extracting calcium ion from the aqueous phase into an organic phase. This antibiotic also acts as a mobile ion carrier transporting the cation across a solvent barrier. The divalent cation selectivity order for ionomycin as determined by ion competition experiments was found to be $Ca^{2+} > Mg^{2+} \gg Sr^{2+} = Ba^{2+}$, where the binding of strontium and barium by the antibiotic is insignificant [36].

Ionomycin, like other polyether antibiotics, is active primarily against Gram-positive bacteria with no demonstrable activity against Gram-negative bacteria. The acute toxicity of ionomycin, LD_{50}, administered subcutaneously to mice is 28 mg/kg [38].

2.8. K-41. K-41 (Figure 18) was first isolated from *Streptomyces hygroscopicus* [39]. The molecular structure of this antibiotic was established by X-ray analysis of the sodium salt of its *p*-bromobenzoate [40].

K-41 is active against Gram-positive bacteria [40]. K-41 showed also antimalarial activity against the drug-resistant strain of *Plasmodium falciparum* and was more potent than clinically used antimalarial drugs: artemisinin, chloroquine, and pyrimethamine [41].

2.9. Kijimicin. Kijimicin (Figure 19) was found in the culture filtrate of an *Actinomadura* sp. MI215-NF3. Its crystal structure (Figure 20) was established by X-ray analysis of its rubidium salt [42].

Kijimicin inhibits mainly the growth of Gram-positive bacteria and shows activity against *Eimeria tenella*. The acute toxicity of the antibiotic in mice is 56 mg/kg [43]. Kijimicin was also examined for its HIV inhibitory activity, was proved to inhibit HIV replication in both T-cell and monocyte lineage cell lines, and was shown to be active in *in vitro* assays against both acute and chronic infections [43].

2.10. Lasalocid A and Its Analogues. Lasalocid (Figure 21) was first isolated from *Streptomyces lasaliensis* [44]. Crystal structures of lasalocid acid barium, silver (Figure 22), and strontium (Figure 23) salts were determined [45, 46]. Monomeric unit of Lasalocid thallium salt is stabilized by strong intramolecular aryl-Tl type-metal half sandwich bonding

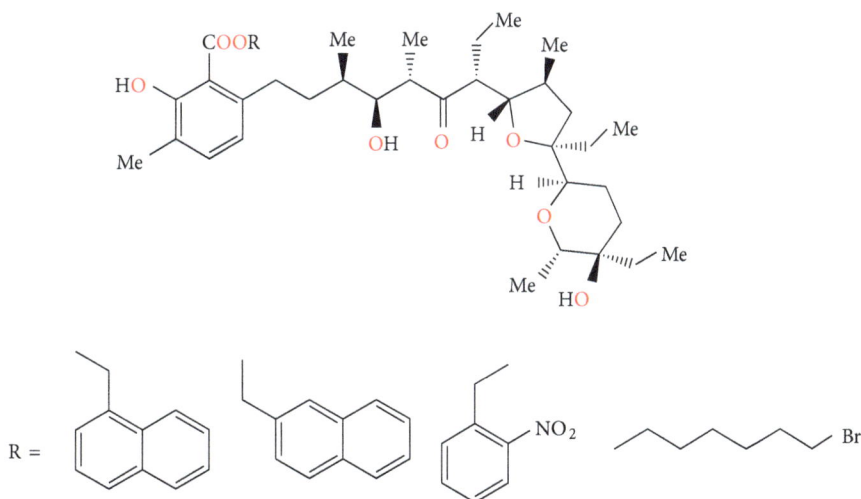

FIGURE 26: Structures of lasalocid acid esters.

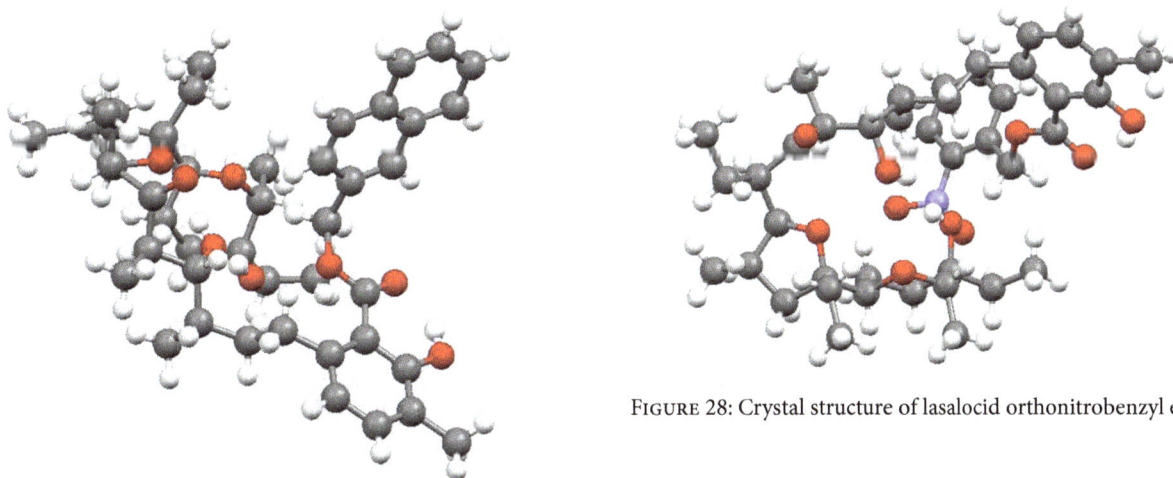

FIGURE 27: Structure of lasalocid 2-naphthylmethyl ester.

FIGURE 28: Crystal structure of lasalocid orthonitrobenzyl ester.

interactions [47]. Homologs of lasalocid (Figure 24) acid were also described [48].

Chemistry of lasalocid was extensively investigated. Treatment of the free acid with diethylamine and paraformaldehyde in toluene, employing Dean—Stark conditions, gave the Mannich base (Figure 25) [49]. Four ester derivatives of lasalocid were obtained (Figure 26), and their crystal structures (Figures 27 and 28) were also reported [50–53].

Lasalocid is able to form complexes with amines and its several complexes have been obtained (e.g., with allylamine, 1,1,3,3-tetramethylguanidine, TBD, phenylamine, and N-butylamine) (Figure 29), and their crystal structures (Figures 30 and 31) were reported. Lasalocid is active against Gram-positive bacteria. Complex of lasalocid acid with allylamine is even more active than pure lasalocid acid. Lasalocid sodium salt (Bovatec, Avate) is used in a veterinary to prevent

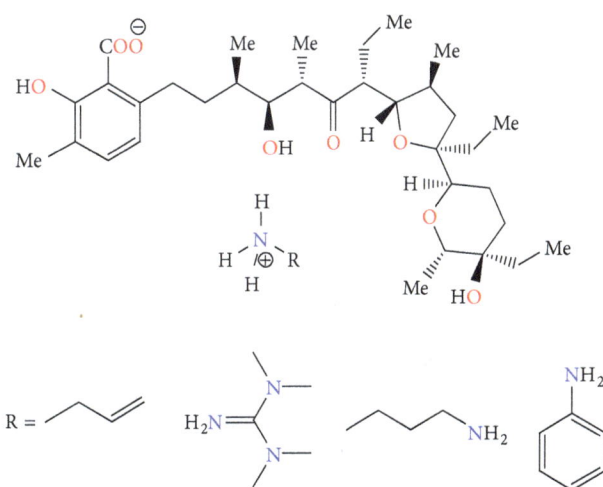

FIGURE 29: Lasalocid acid complexes with several amines: allylamine, 1,1,3,3-tetramethylguanidine, N-butylamine, and phenylamine.

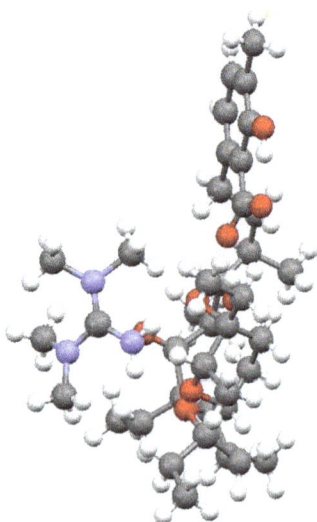

FIGURE 30: Crystal structure of lasalocid complex with tetramethylguanidine.

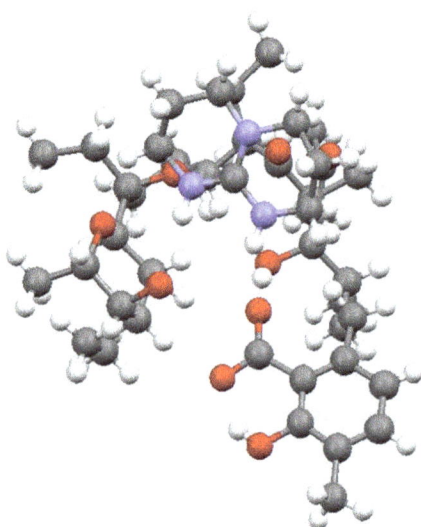

FIGURE 31: Crystal structure of lasalocid complex with TBD.

coccidiosis in poultry and to improve nutrient absorption and feed efficiency in ruminants [54–57].

Recently, two complexes of lasalocid (with phenylamine and butylamine) were tested *in vitro* for cytotoxic activity against human cancer cell lines: A-549 (lung), MCF-7 (breast), HT-29 (colon), and mouse cancer cell line P-388 (leukemia). It was found that lasalocid and its complexes are strong cytotoxic agents towards cell lines. The cytostatic activity of the compounds studied is greater than that of cisplatin, indicating that lasalocid and its complexes are promising candidates for new anticancer drugs [57].

2.11. Semduramicin and CP-120509. Semduramicin (Figure 32) was isolated from *Actinomadura roserufa* [58]. The anticoccidial activity tests of semduramicin against

laboratory isolates of five species of poultry Eimeria have shown that this antibiotic is active from 20 to 30 ppm concentrations. Semduramicin was well tolerated when coadministered with tiamulin [59].

Antibiotic CP-120509 (Figure 33) is also isolated from *Actinomadura roserufa* and is structurally related to semduramicin. Its crystal structure was reported [60]. CP-120509 exhibited *in vitro* activity against certain Grampositive bacteria and the spirochete *Serpulina hyodysenteriae*, the causative agent of swine dysentery, but was inactive against Gram-negative bacteria. It afforded excellent anticoccidial activity against *Eimeria acervulina* in chickens at levels between 30 and 60 mg/kg in fed [60].

2.12. Tetronasin. Tetronasin (Figure 34) was isolated from *Streptomyces longisporoflavus*. The structure of this antibiotic has been determined by X-ray analysis of the 4-bromo-3,5-dinitrobenzoyl derivative [61].

Gram-positive bacteria are sensitive to the tetronasin and were unable to adapt to grow in the presence of this antibiotic. Gram-negative bacteria were more resistant. An *in vivo* trial with cattle and *in vitro* growth experiments indicated that the effect of tetronasin on ciliate protozoa was minor. *In vitro* experiments measuring hydrogen production by *Neocallimastix frontalis* suggested that this fungus would be unable to survive in ruminants receiving tetronasin [62].

2.13. Zincophorin and CP-78545. Zincophorin (Figure 35) was first isolated from *Streptomyces griseus*. Its crystal structure was determined by X-ray analysis of its magnesium salt (Figure 36) [63].

Zincophorin is able to complex divalent cations, with the stability order of zinc ≈ cadmium > magnesium > strontium ≈ barium ≈ calcium.

Zincophorin showed good *in vitro* activity against Grampositive bacteria. It also inhibited methane production and favourably altered volatile fatty acid ratios in *in vitro* rumen fermentations. It showed some anticoccidial activity against *Eimeria tenella* in chicks [63].

Antibiotic CP-78545 (Figure 37) also isolated from *Streptomyces griseus* is structurally related to zincophorin. Additional unsaturated bond is present in CP-78545 compared to zincophorin. Crystal structure of CP-78545 was determined by X-ray analysis of its cadmium salt (Figure 38).

CP-78545 exhibited *in vitro* antibiotic activity against certain Gram-positive bacteria such as *Bacillus*, *Staphylococcus*, and *Streptococcus* as well as the anaerobe *Treponema hyodysenteriae* (the causative agent of swine dysentery), but no activity towards Gram-negative bacteria. It was active *in vitro* against a coccidium *Eimeria tenella* in a tissue culture assay, but it was inactive *in vivo* at levels between 100 and 200 mg/kg in feed versus *Eimeria tenella* coccidial infections in chickens [64].

2.14. Salinomycin SY-1, SY-2, SY-4, and SY-9. Salinomycin (Figure 39) was isolated from *Streptomyces albus*. Its crystal structure (Figure 40) has been established by X-ray analysis of its p-iodophenacyl ester [65]. Compounds structurally related to salinomycin have also been obtained and

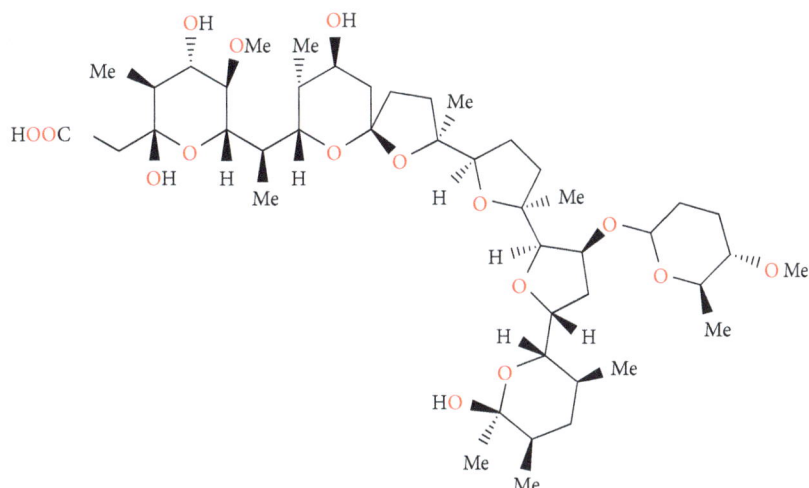

FIGURE 32: Structure of semduramicin.

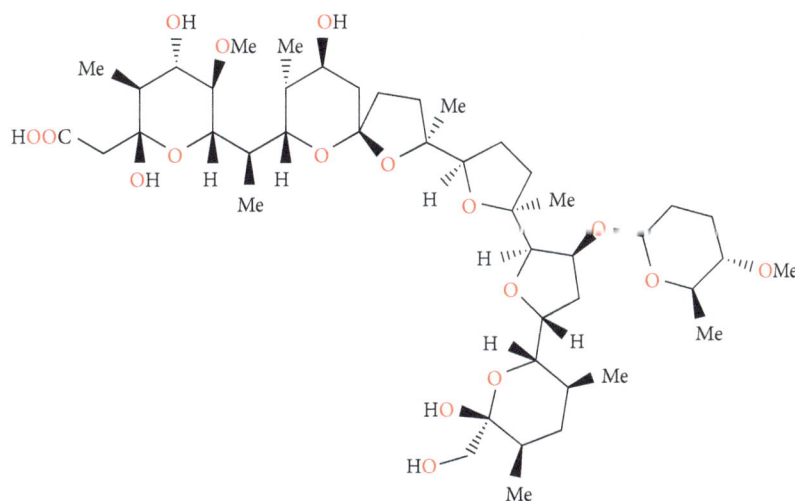

FIGURE 33: Structure of CP-120509.

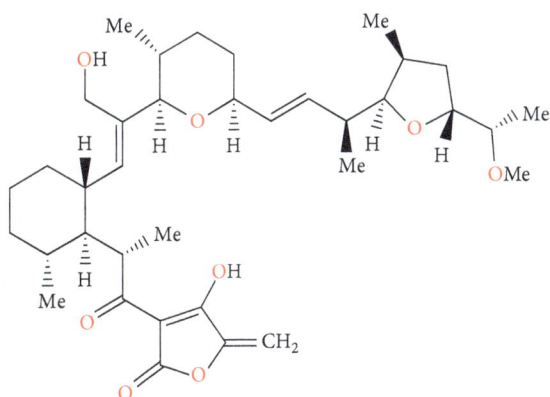

FIGURE 34: Structure of tetronasin.

described. SY-1 (Figure 41) is 20-deoxysalinomycin. SY-2 (Figure 42) is C-17 epimer of SY-1 [66]. SY-4 (Figure 43)

is 5-hydroxysalinomycin [67], and SY-9 (Figure 44) is 20-oxosalinomycin. Crystal structures of SY-1 (Figure 45) and SY-9 (Figure 46) were reported [68, 69]. Two other ester and amide derivatives of salinomycin (Figures 47 and 48) were obtained, and their crystal structures (Figures 49 and 50) were reported [70, 71].

The relative affinity of salinomycin for complex formation with various cations decreases in the order $K^+ > Na^+ > Cs^+ > Sr^{2+} > Ca^{2+}, Mg^{2+}$. Salinomycin has been shown to transport monovalent cations more effectively than divalent cations from an aqueous buffer into an organic solvent [72].

Salinomycin is active against Gram-positive bacteria including mycobacteria and some filamentous fungi. No activity was observed against Gram-negative bacteria and yeast. The acute toxicity of salinomycin in mice was examined, and its LD_{50} was 18 mg/kg intraperitoneally and 50 mg/kg orally. The anticoccidial estimation of salinomycin was carried out on chickens infected with *Eimeria tenella*. Salinomycin was effective in reducing mortality of chickens

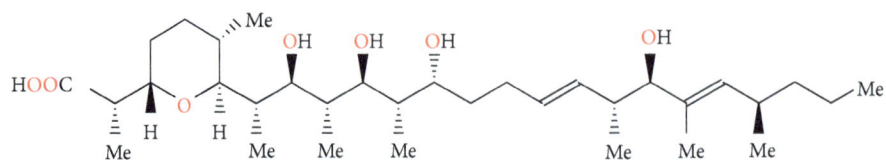

FIGURE 35: Structure of zincophorin.

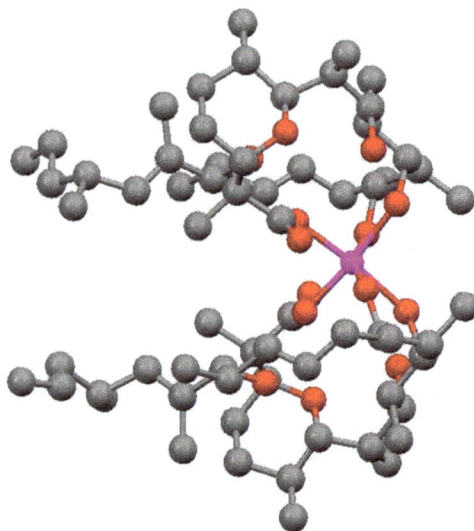

FIGURE 36: Crystal structure of zincophorin magnesium salt.

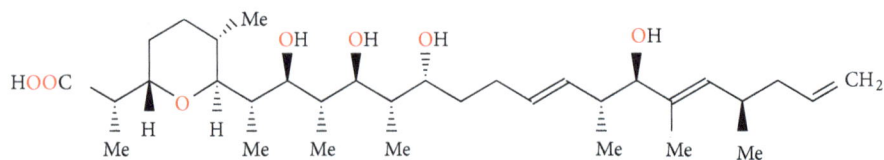

FIGURE 37: Structure of CP-78545.

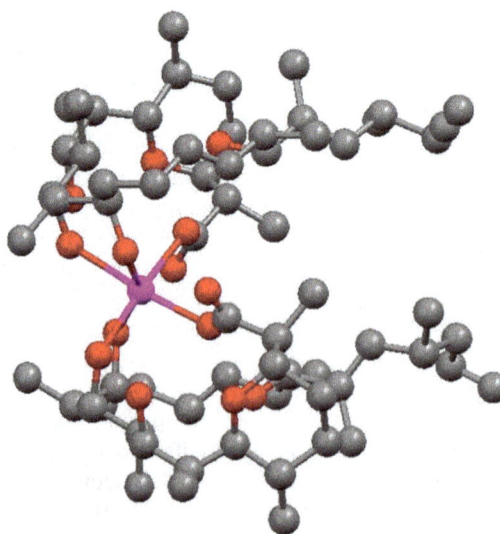

FIGURE 38: Crystal structure of CP-78545 cadmium salt.

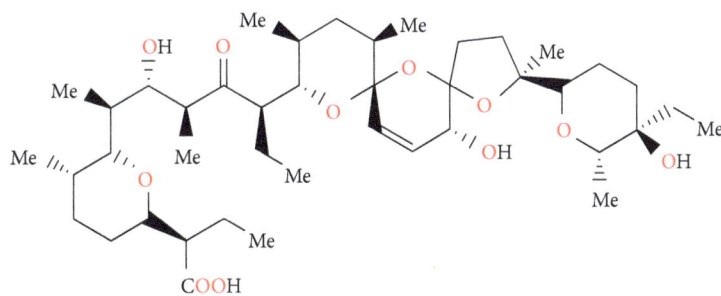

FIGURE 39: Structure of salinomycin.

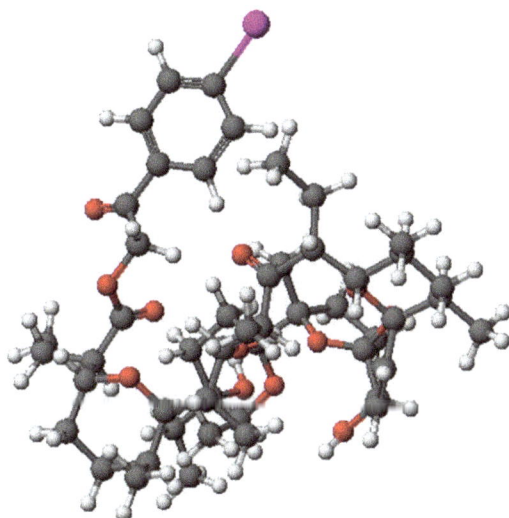

FIGURE 40: Crystal structure of salinomycin p-iodophenacyl ester.

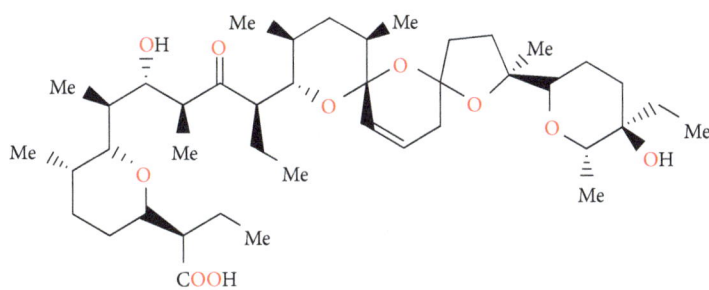

FIGURE 41: Structure of SY-1.

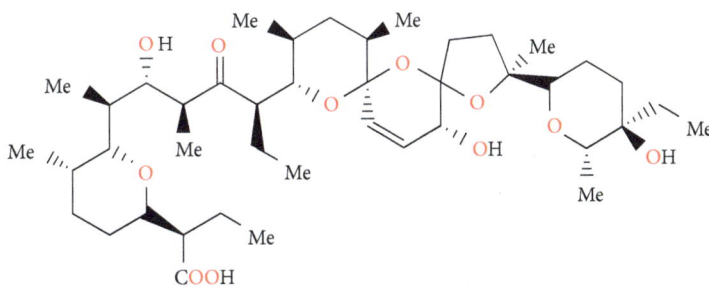

FIGURE 42: Structure of SY-2.

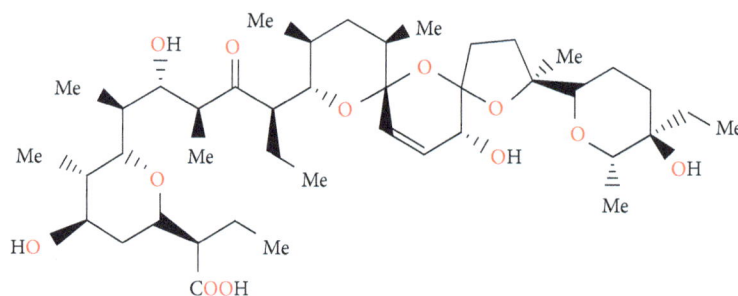

FIGURE 43: Structure of SY-4.

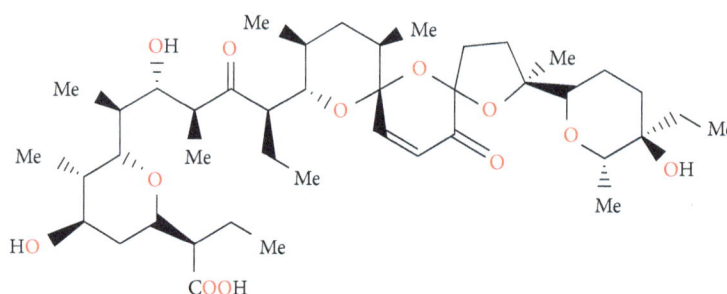

FIGURE 44: Structure of SY-9.

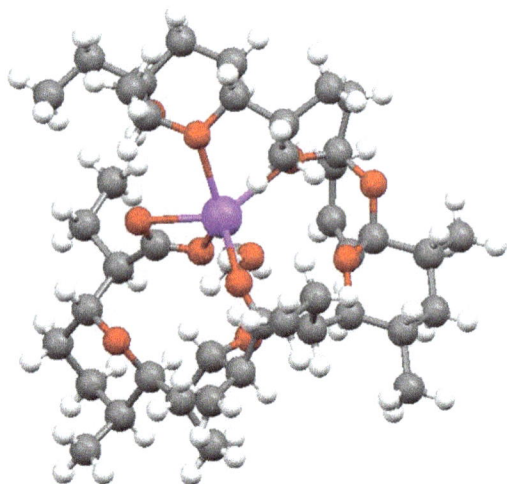

FIGURE 45: Crystal structure of SY-1 with the sodium cation.

rubidium (Figures 52, 53, and 54) were reported [76–79]. In the last years Pantcheva et al. have shown that monensin (MON) can form two types of salt complex species with divalent metal cations.

In the first type, monensin sodium salt forms complexes with metal dichloride of $[M(MON–Na)_2]Cl_2 \cdot H_2O$ stoichiometry, where $M = Co^{2+}$, Mn^{2+} and Cu^{2+}. In this type of structure, the divalent metal cation is tetrahedrally coordinated by oxygen atoms of two carboxylic groups of two monensin sodium salt molecules and by two chloride anions. The sodium cation remains in the hydrophilic cavity of the ligand and cannot be replaced by the transition metal cation. The second type of monensin complexes with the divalent metal cations is the neutral salt of the $[M(MON)_2 \cdot (H_2O)_2]$ formula ($M = Mg^{2+}$, Ca^{2+}, Zn^{2+}, Ni^{2+}, Cd^{2+}, and Hg^{2+}). These salts consist of two monoanionic ligands (monensinates) bound in a bidentate coordination mode to the metal cation. These types of monensin salt complexes with divalent metal cation are untypical, because the etheric oxygen atoms do not play any role in the complexation of the cations. In contrast, in the typical complexes of monensin with monovalent metal cations, the etheric oxygen atoms of the monensin ligand are always involved in the complexation process [80–86]. A large group of monensin ester, amide, and urethane derivatives (Figure 55) was obtained and reported. Crystal structures of several of them (Figures 56 and 57) were also reported [87, 88]. Monensin exhibited *in vitro* antibiotic activity against certain Gram-positive bacteria such as *Bacillus*, *Staphylococcus*, and *Streptococcus*. No activity was observed against Gram-negative bacteria [89]. It was shown that monensin phenylurethane sodium salt shows a higher antibacterial activity against human pathogenic

and in increasing the average body weight of treated infected chickens compared to those of untreated infected controls [73].

Very recently, it has been shown that it is possible to selectively kill breast cancer stem cells using salinomycin. Its ability to kill cancer stem cells and apoptosis-resistant cancer cells may define salinomycin as a novel anticancer drug [74].

2.15. Monensin. Monensin (Figure 51) was first isolated from *Streptomyces cinnamonensis* [75]. Crystal structures of monensin hydrate and its salts with sodium, lithium, and

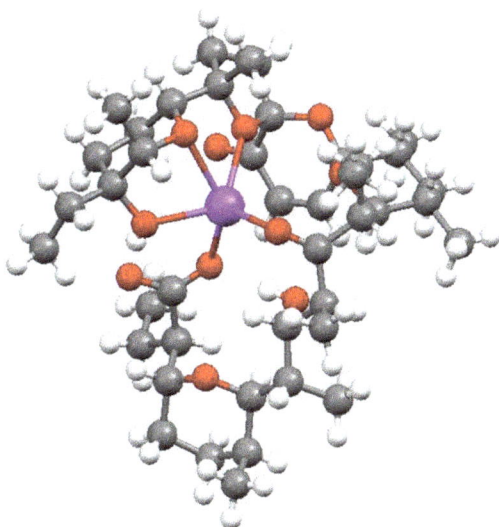

FIGURE 46: Crystal structure of SY-9 with the sodium cation.

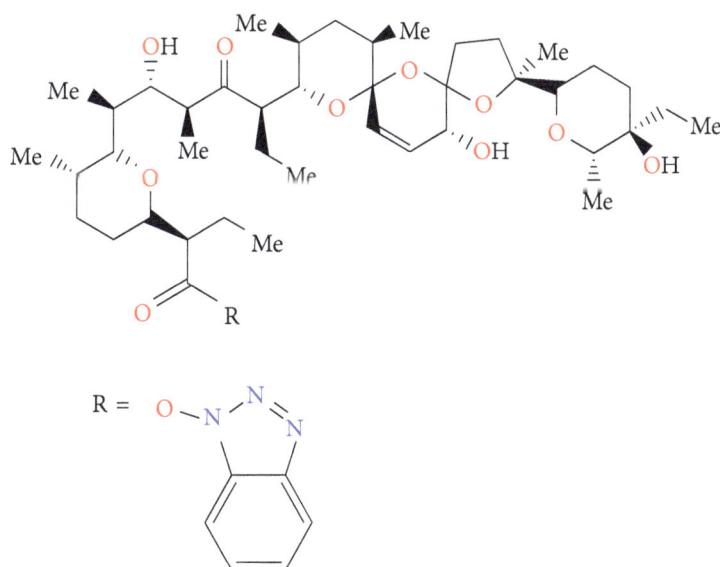

FIGURE 47: Structure of benzotriazole ester of salinomycin.

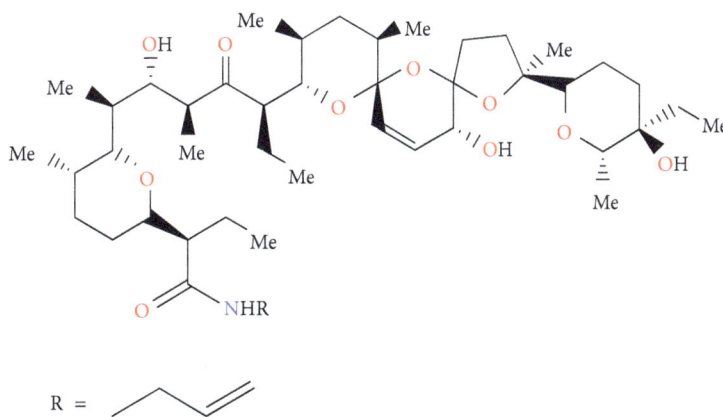

FIGURE 48: Structure of allyl amide of salinomycin.

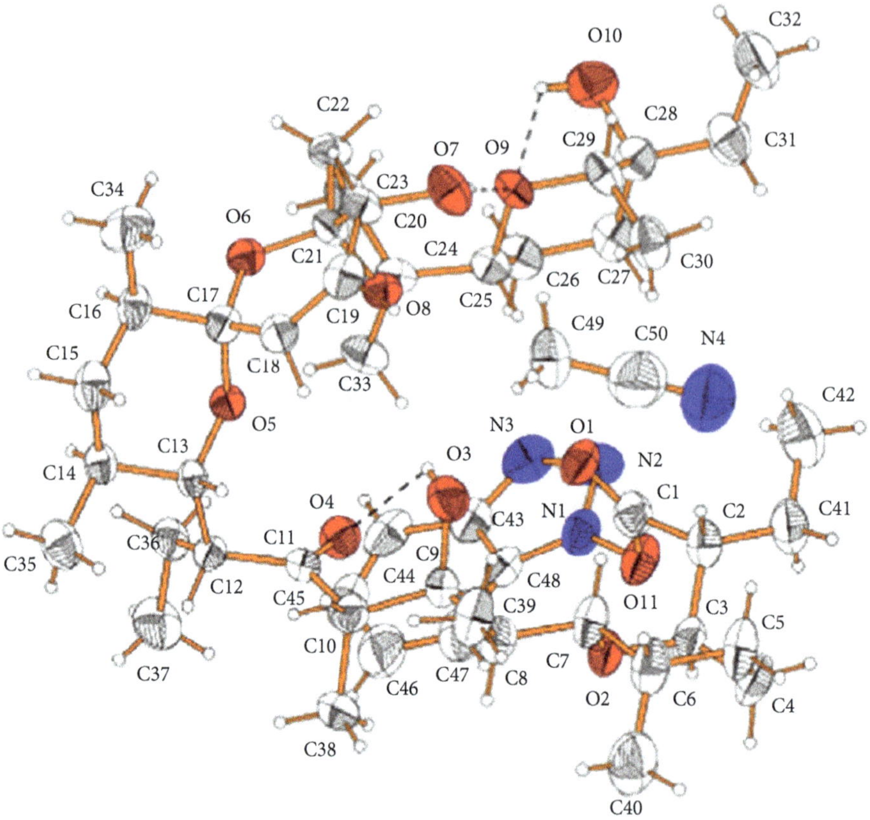

FIGURE 49: Crystal structure of salinomycin benzotriazole ester acetonitrile solvate.

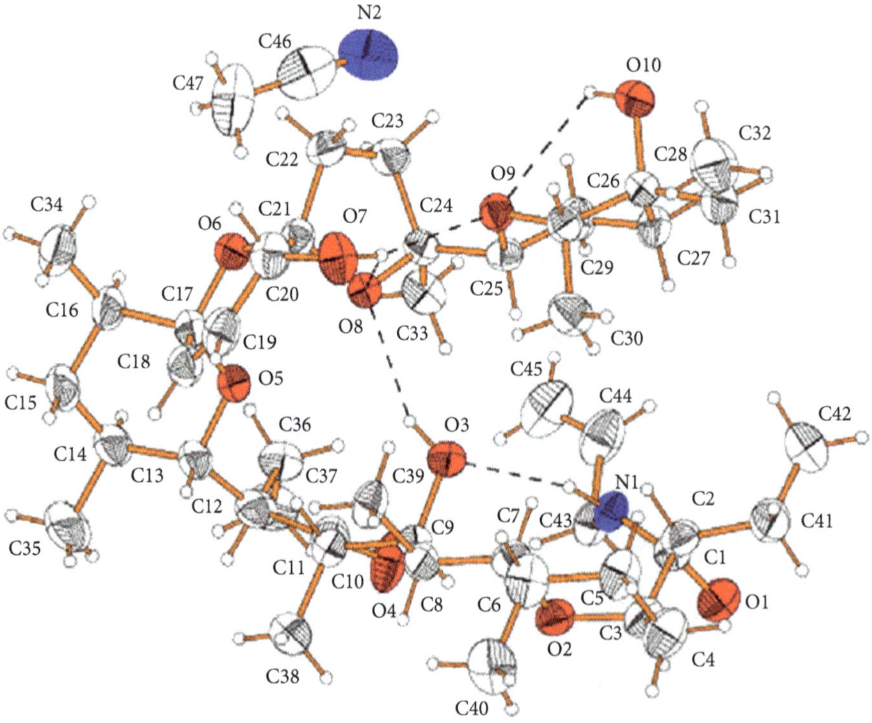

FIGURE 50: Crystal structure of salinomycin allyl amide acetonitrile solvate.

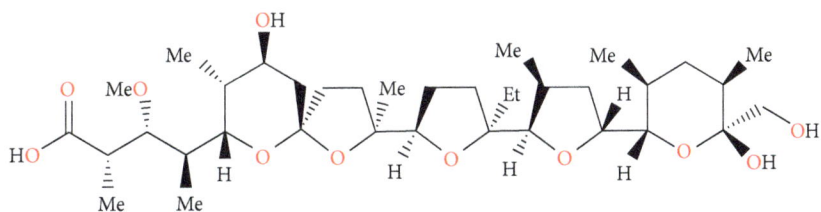

FIGURE 51: Structure of monensin.

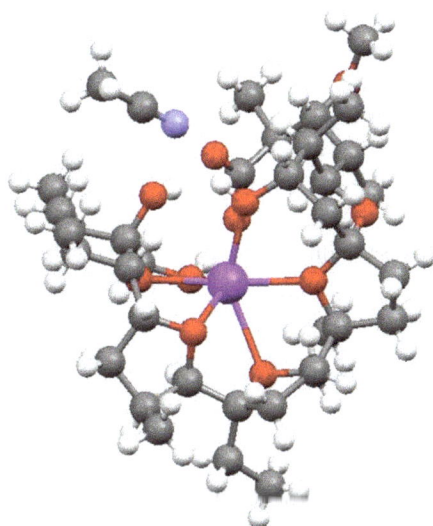

FIGURE 52: Crystal structure of monensin sodium salt acetonitrile solvate.

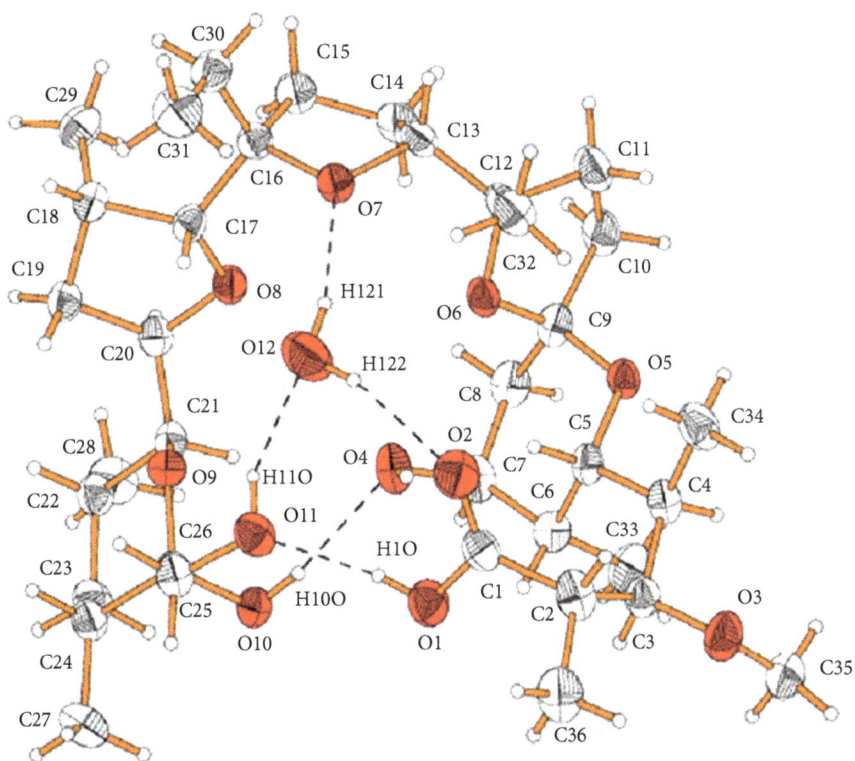

FIGURE 53: Structure of monensin hydrate.

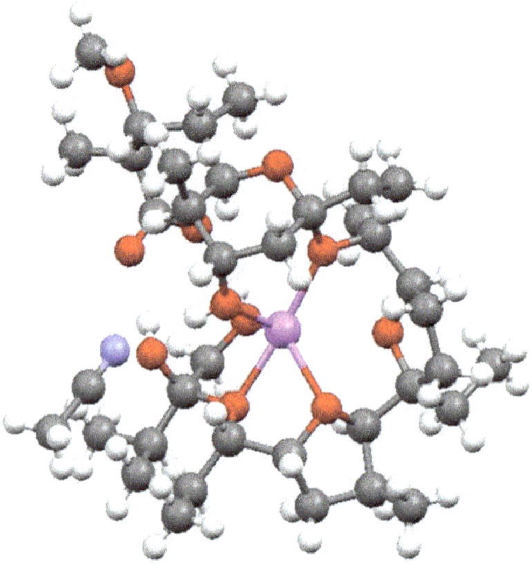

FIGURE 54: Crystal structure of monensin lithium salt acetonitrile solvate.

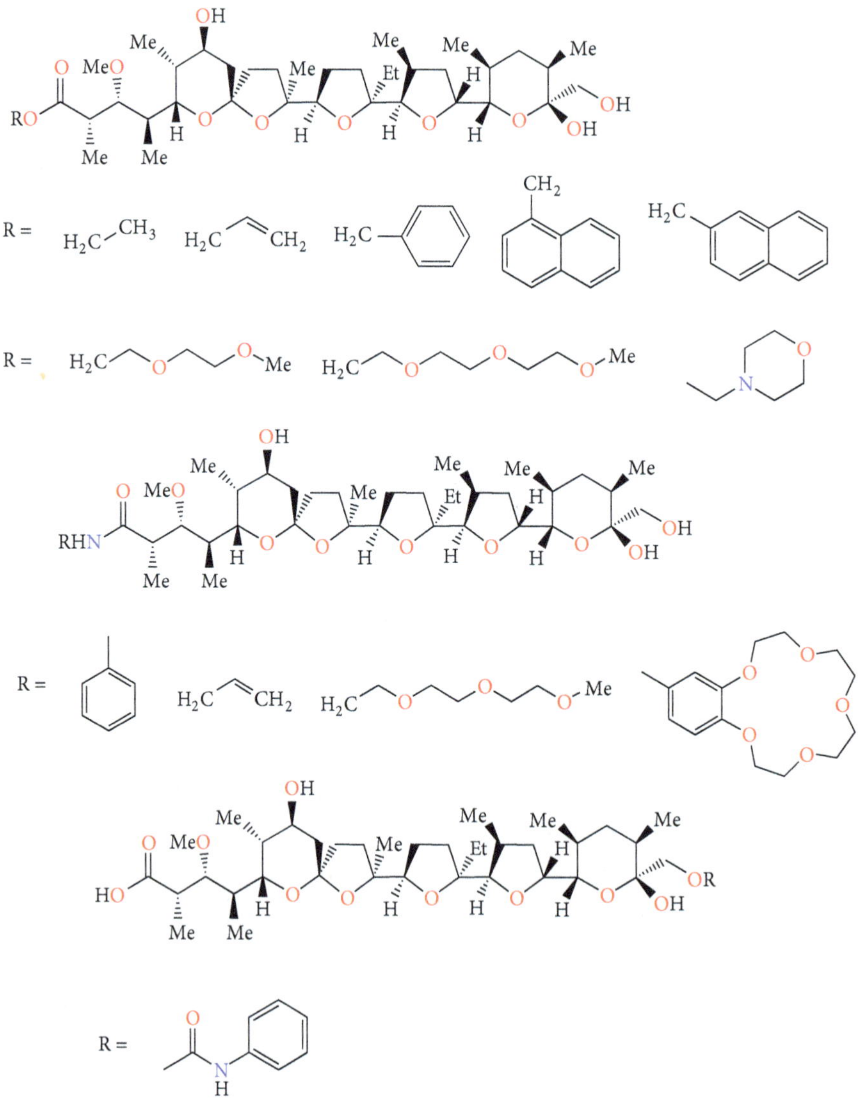

FIGURE 55: Structures of monensin ester amide and urethane derivatives.

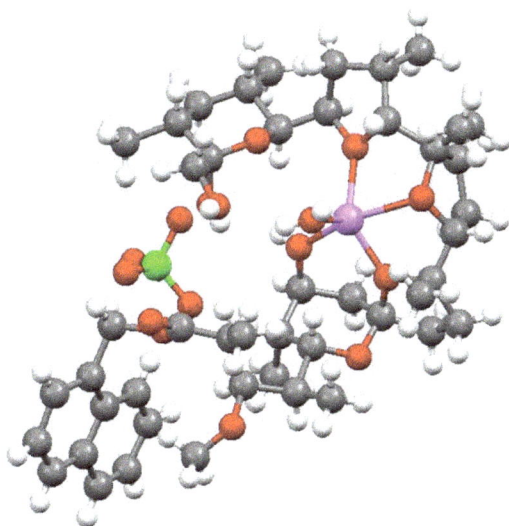

FIGURE 56: Crystal structure of monensin 1-naphtylmethyl ester with the lithium perchlorate.

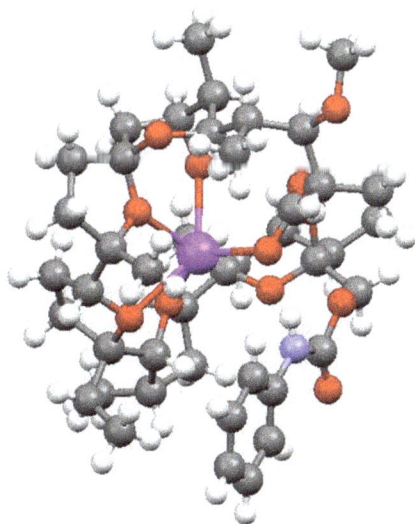

FIGURE 57: Crystal structure of monensin phenylurethane sodium salt.

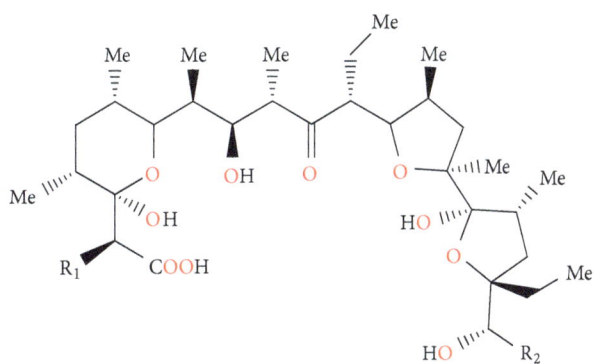

Ferensimycin A: $R_1 = CH_3$, $R_2 = CH_3$
Ferensimycin B: $R_1 = CH_3$, $R_2 = CH_2CH_3$

FIGURE 58: Structures of ferensimycins A and B.

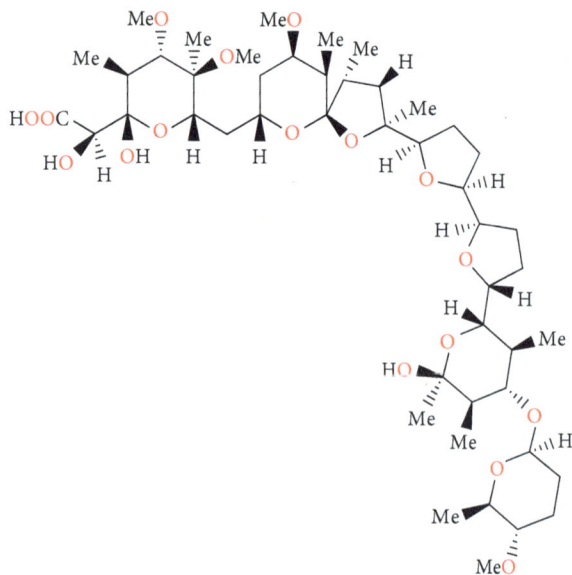

Figure 59: Structure of CP-96797.

bacteria, including antibiotic-resistant *Staphylococcus aureus* and *Staphylococcus epidermidis* than the parent unmodified antibiotic monensin A [88]. Monensin blocks endocytosis and, therefore, impedes entry of toxic molecules. The drug also inhibits viral proliferation of RNA and DNA viruses such as vesicular stomatitis, influenza, and human polyomaviruses. Monensin also effectively abolishes viral DNA replication of mouse polyomavirus [90].

2.16. Ferensimycins A and B. Ferensimycins A and B (Figure 58) were isolated as their sodium salts from the fermentation broth of *Streptomyces* sp. No. 5057. Both antibiotics are active against Gram-positive bacteria but inactive against Gram-negative bacteria and fungi. The acute toxicity of these compounds in mice was examined, and the LD_{50} values of ferensimycin A and ferensimycin B were 30–50 mg/kg and 50 mg/kg, respectively [91].

2.17. CP-96797. CP-96797 (Figure 59) was isolated from *Sterptomyces* sp. ATCC 55028. Crystal structure of the silver salt of this antibiotic was determined by X-ray analysis [92].

CP-96797 sodium salt showed good activity against a number of Gram-positive bacteria, as well as the spirochete *Serpulina hyodysenteriae* (the causative agent of swine dysentery), but was inactive against Gram-negative bacteria. It afforded anticoccidial activity against *Eimeria tenella* in chickens at levels between 60 and 90 mg/kg in feed [92].

2.18. Octacyclomycin. Octacyclomycin (Figure 60) was isolated from *Streptomyces* sp. No. 82. This antibiotic showed cytocidal activity against B16 melanoma cells, and its IC_{50} value was 0.23 μg/mL when the cells were exposed to the antibiotic for 3 days *in vitro*. On the other hand, octacyclomycin showed weak antimicrobial activity against Gram-positive bacteria such as *Staphylococcus aureus* and *Micrococcus luteus* at the concentration of 100 μg/mL, whereas *Bacillus subtilis* was not affected at this concentration [93].

2.19. CP-91243 and CP-91244. CP-91243 and CP-91244 (Figure 61) were isolated from *Actinomadura roseorufa*. Both compounds exhibited *in vitro* activity against certain Gram-positive bacteria and the spirochete *Treponema hyodysenteriae* (the causative agent of swine dysentery), but were not active against Gram-negative bacteria. CP-91243 afforded anticoccidial activity against *Eimeria tenella* in chickens at 60 mg/kg in feed, and the less polar CP-91244 was about twice as active, 25 mg/kg in feed [94].

2.20. W341C. W341C (Figure 62) was isolated from strains of *Streptomyces* W341. This antibiotic is K^+-selective ionophore that inhibits mitochondrial substrate oxidation. W341C transported K^+ ion at a greater rate than nigericin, but it transported Na^+ ion at a lower rate than monensin. W341C is able to induce potassium loss in *Bacillus subtilis* and *Streptococcus lactiae* and promote potassium uptake into *Escherichia coli* [95].

2.21. Laidlomycin. Laidlomycin (Figure 63) was isolated from *Streptomyces eurocidicus*. This antibiotic inhibited growth of some Gram-positive bacteria only at high concentrations such as 50–100 mg/mL, but was not active against Gram-negative bacteria, yeast, and fungi. In broth dilution laidlomycin was active against several *Mycoplasmas*. The acute toxicity of laidlomycin, expressed as LD_{50}, was 5 mg/kg (intraperitoneally) and 2.5 mg/kg (subcutaneously) in mice. Antitumor activity against Sarcoma 180 solid tumours in mice and antiviral activity against several viruses *in vitro* were examined but showed no significant effect [96].

2.22. CP-84657. CP-84657 (Figure 64) was isolated from *Actinomadura* sp. ATCC 53708. Its crystal structure was determined by X-ray analysis of its rubidium salt. CP-84657 was active against *Eimeria tenella* (major causative agent of chicken coccidiosis) at doses of 5 mg/kg or less in feed. It was also active *in vitro* against certain Gram-positive bacteria, as well as the spirochete *Treponema hyodysenteriae*. No activity was observed against *Escherichia coli* [97].

2.23. Grisorixin and Epigrisorixin. Grisorixin (Figure 65) was isolated from *Streptomyces griseus* [98]. Its crystal structure was determined by X-ray analysis of its thallium and silver salts [99, 100]. Grisorixin showed activity against Gram-positive bacteria, but was shown to be toxic. Determination of the toxicity of grisorixin in mice revealed LD_{50} of 15 mg/kg when the antibiotic was given subcutaneously [98].

Epigrisorixin (Figure 66) was isolated from *Streptomyces hygroscopicus*. Antimicrobial study showed that epigrisorixin is less toxic than grisorixin [101].

2.24. CP-54883. CP-54883 (Figure 67) was isolated from *Actinomadura routienii* [102]. Its crystal structure was determined by X-ray analysis of its benzoate derivative [103]. CP-54883 exhibited activity only against Gram-positive bacteria. It was not active against Gram-negative bacteria and yeasts. This antibiotic was active against *Eimeria tenella*, *Eimeria maxima*, and *Eimeria acervulina* coccidian when administrated in feed at 10 to 20 μg/g. Chickens were protected from lesions at the higher levels but suffered from poor weight

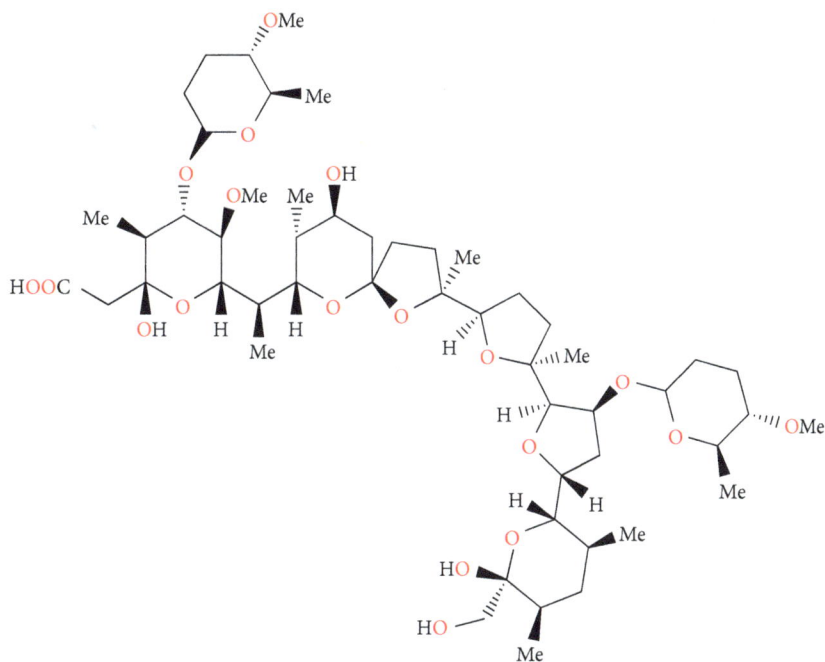

FIGURE 60: Structure of octacyclomycin.

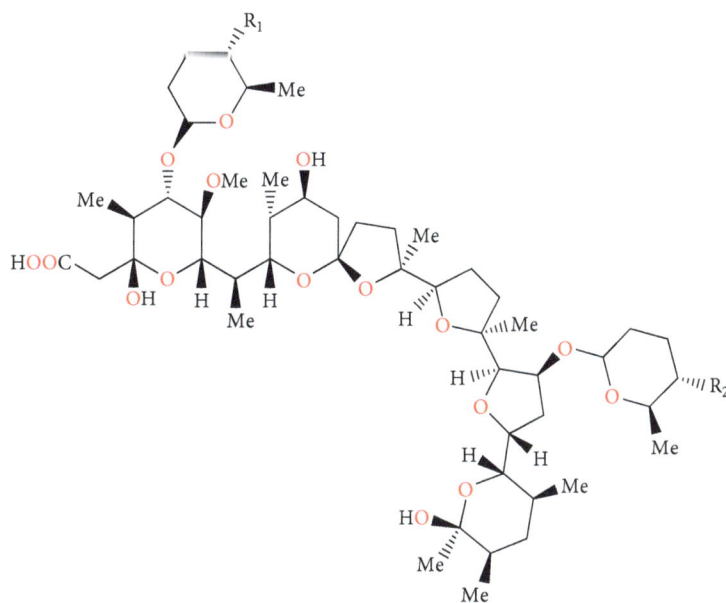

CP-91243 $R_1 = R_2 = OH$

CP-91244 $R_1 = OH$, $R_2 = OCH_3$

FIGURE 61: Structures of CP-91243 and CP-91244.

gains and feed intake [100]. CP-54883 also induced a change in the proportion of volatile fatty acids (acetate, propionate, and butyrate) produced in the rumen by increasing the molar proportion of propionate in the rumen fluids [102].

2.25. SF-2487. SF-2487 (Figure 68) was isolated from a culture broth of *Actinomadura* sp. SF2487. Its crystal structure was determined by X-ray analysis of its silver salt [104]. SF-2487 showed moderate activity against Gram-positive

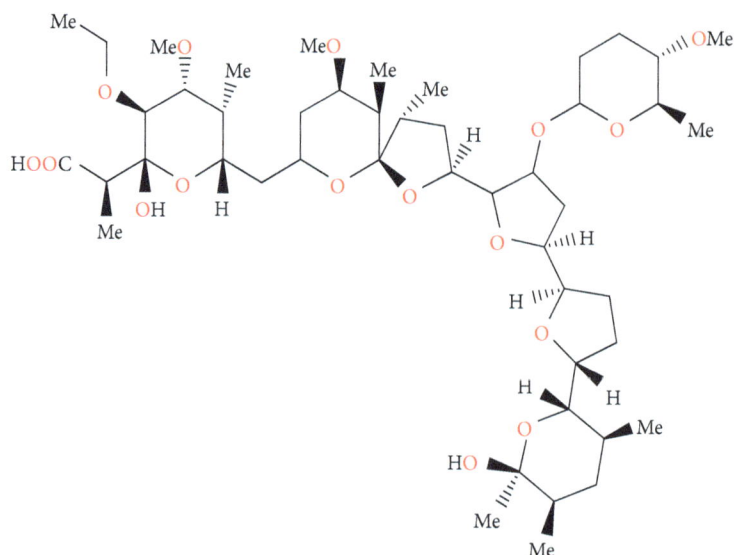

FIGURE 62: Structure of W341C.

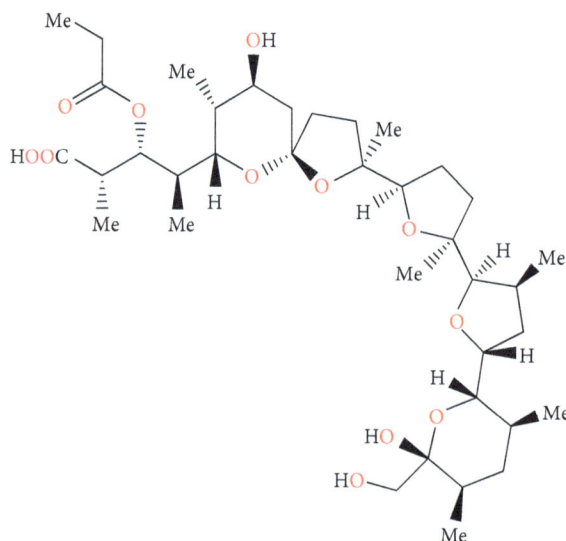

FIGURE 63: Structure of laidlomycin.

bacteria, but no activity against Gram-negative bacteria. SF-2487 exhibited *in vitro* antiviral activity against influenza virus. The LD_{50} value of SF-2487 was 25 mg/kg by injection in mice [104].

2.26. X-14868A, X-14868B, X-14868C, and X-14868D. Four polyether antibiotics (Figure 69) were isolated from a culture of *Nocardia* (strain X-14868A, X-14868B, X-14868C, and X-14868D). Crystal structures of each of these antibiotics were determined by X-ray analysis [105]. All four compounds were active against Gram-positive bacteria and exhibited no activity against Gram-negative bacteria and fungi. Antibiotic X-14868A was also active against *Treponema hyodysenteriae* [105].

2.27. CP-80219. Cp-80219 (Figure 70) was isolated from *Streptomyces hygroscopicus*. Its crystal structure was determined by X-ray analysis of its rubidium salt [106]. CP-80219 sodium salt showed good activity against Gram-positive bacteria as well as the spirochete *Treponema hyodysenteriae*. No activity was observed against Gram-negative bacteria including *Escherichia coli*. It afforded anticoccidial activity between 30 and 120 mg/kg in feed against *Eimeria tenella* in chickens [106].

2.28. Moyukamycin. Moyukamycin (Figure 71) was isolated from *Streptomyces hygroscopicus*. It showed activity against a wide range of Gram-positive bacteria, while no activity against Gram-negative bacteria [107].

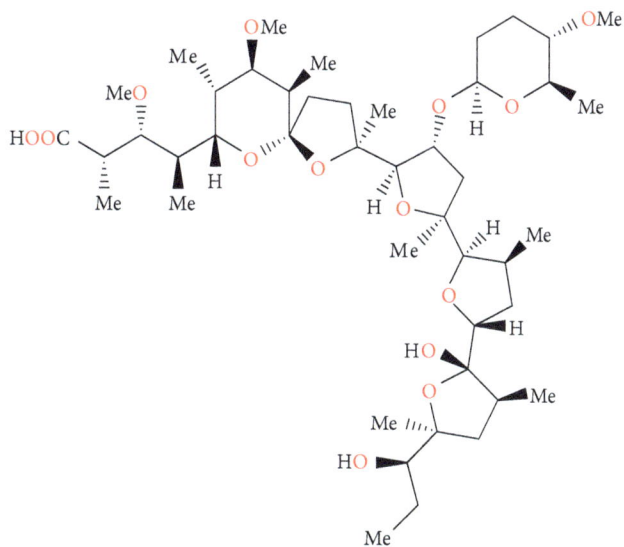

FIGURE 64: Structure of CP-84657.

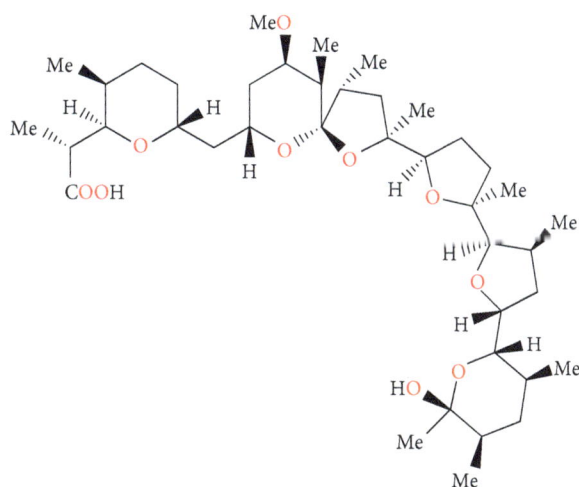

FIGURE 65: Structure of grisorixin.

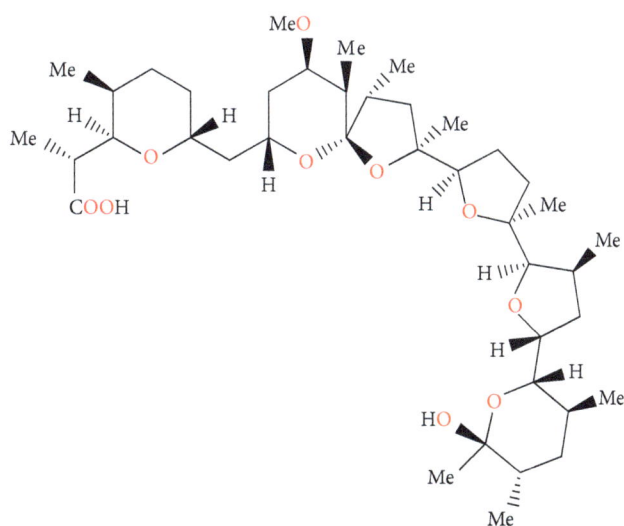

FIGURE 66: Structure of epigrisorixin.

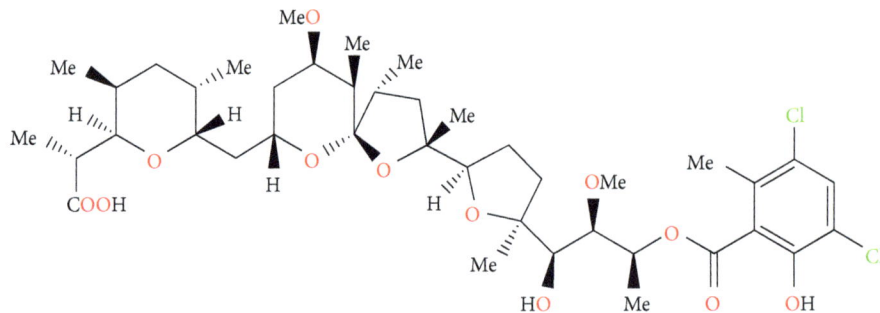

FIGURE 67: Structure of CP-54883.

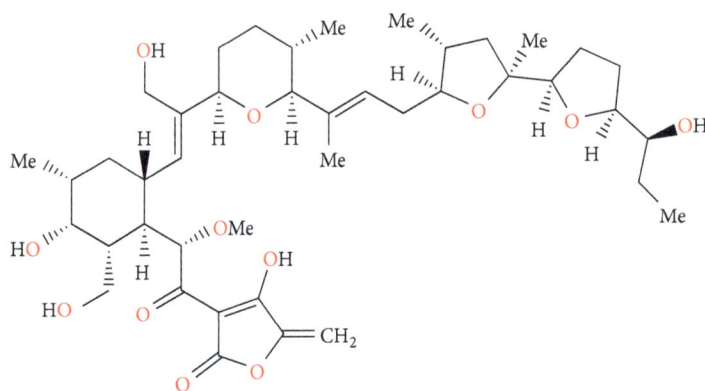

FIGURE 68: Structure of SF-2487.

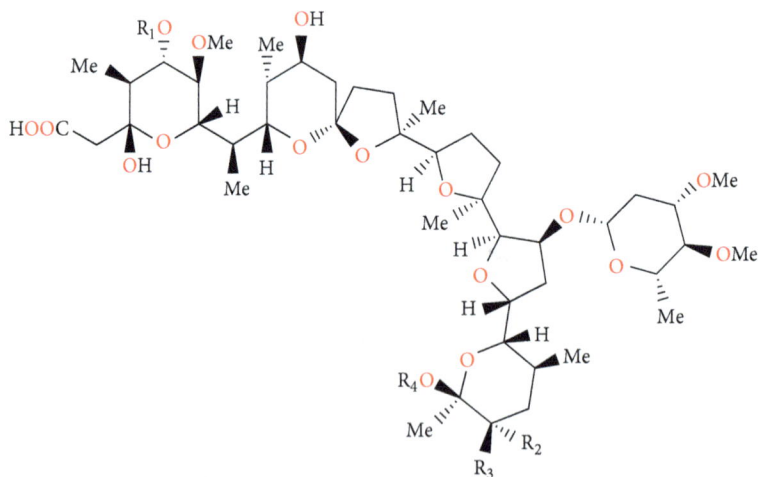

X-14868A: $R_1 = R_3 = CH_3$, $R_2 = R_4 = H$

X-14868B: $R_1 = R_3 = R_4 = CH_3$, $R_2 = H$

X-14868C: $R_1 = R_2 = R_4 = H$, $R_3 = CH_3$

X-14868D: $R_1 = R_2 = CH_3$, $R_3 = R_4 = H$

FIGURE 69: Structures of X-14868A, X-14868B, X-14868C, and X-14868D.

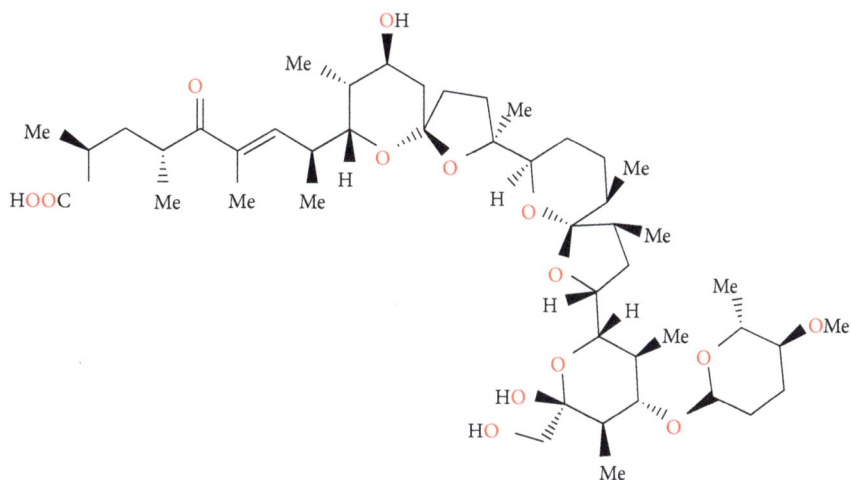

FIGURE 70: Structure of CP-80219.

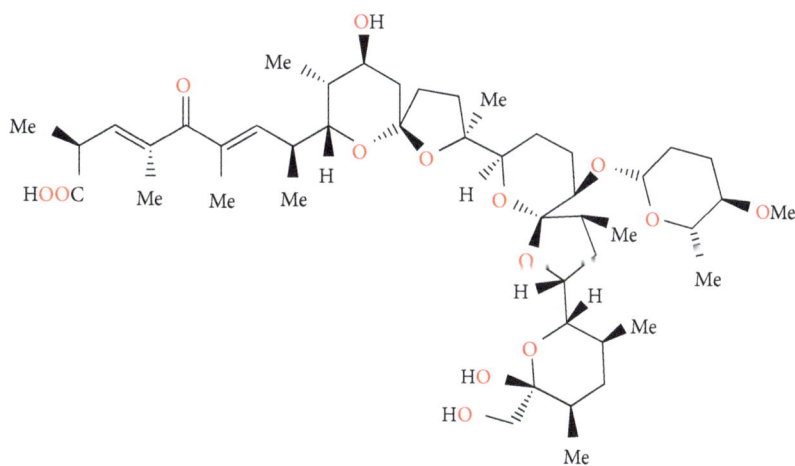

FIGURE 71: Structure of moyukamycin.

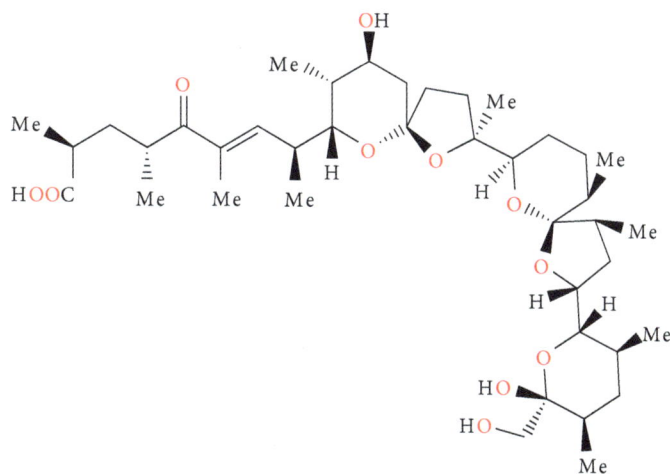

FIGURE 72: Structure of X-14931A.

X-14873A

X-1 4873 G

X 14873 H

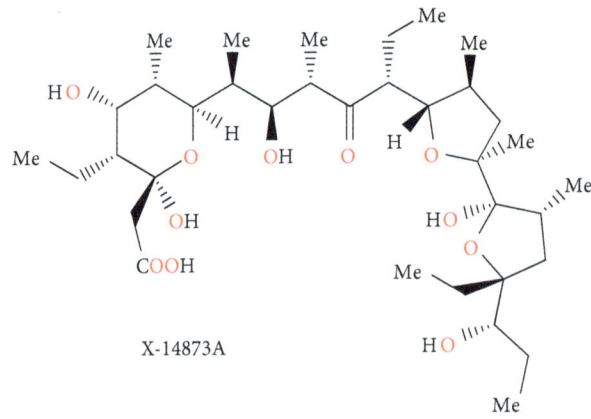

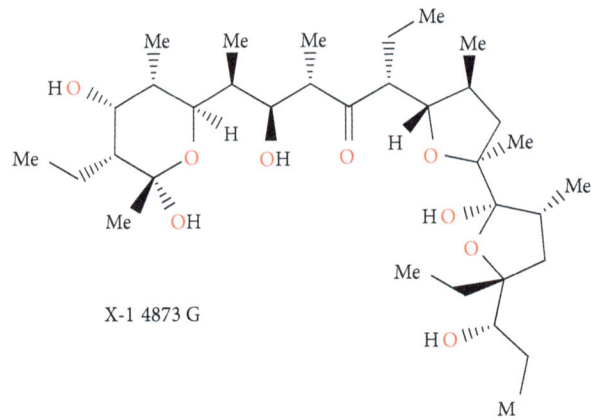

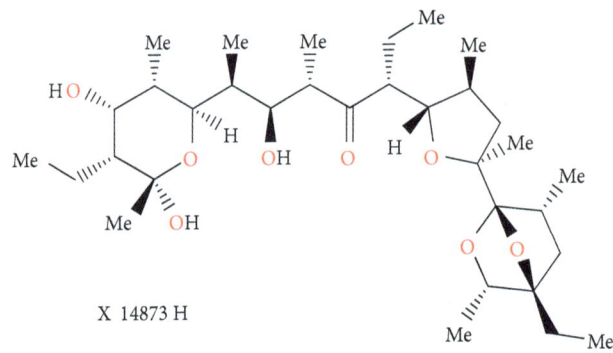

FIGURE 73: Structures of X-14873A, X-14873G, and X-14873H.

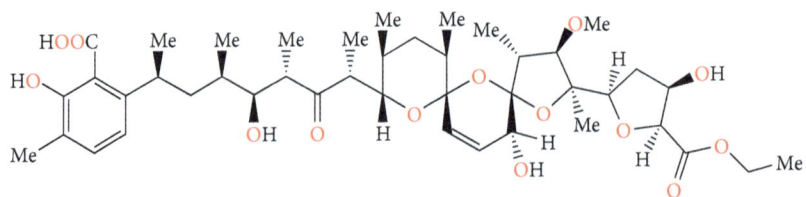

FIGURE 74: Structure of noboritomycin.

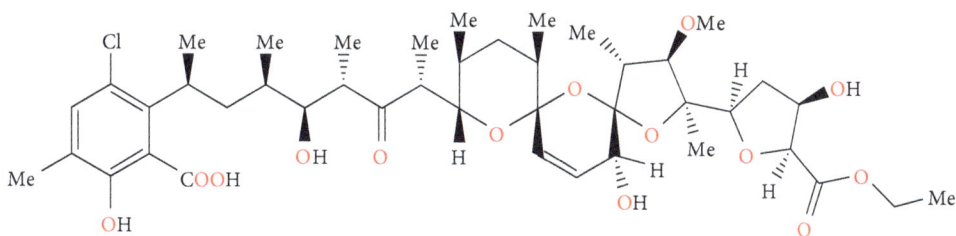

FIGURE 75: Structure of 6-chloronoboritomycin.

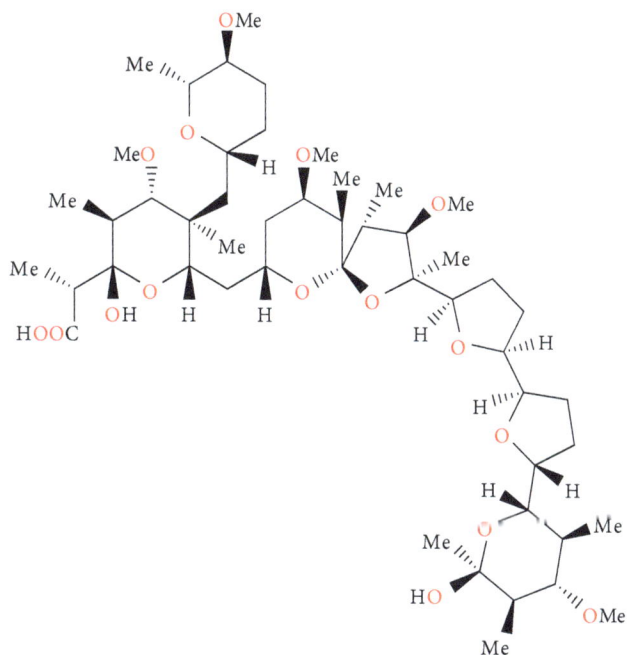

FIGURE 76: Structure of CP-82009.

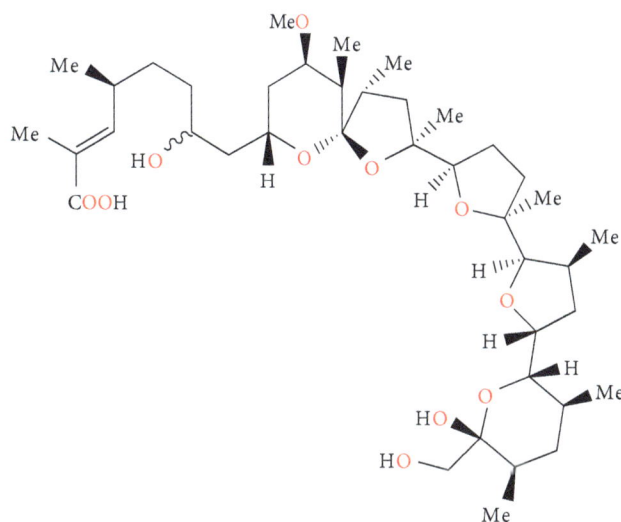

FIGURE 77: Structure of abierixin.

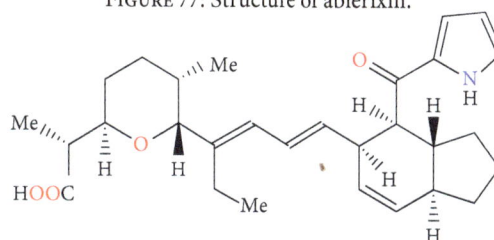

FIGURE 78: Structure of A-83094A.

2.29. X-14931A. X-14931 (Figure 72) was isolated from a culture of *Streptomyces* sp. X-14931. Its crystal structure was determined by X-ray analysis of its silver salt [108]. Antibiotic X-14931A showed *in vitro* activity against Gram-positive microorganisms and yeasts. It was also active against mixed *Eimeria* infection in chickens at 50 μg/g in feed. The antibiotic also exhibited activity in the rumen growth promotant test [108].

2.30. X-14873A, X-14873G, and X-14873H. Three polyether ionophores: X-14873A, X-14873G, and X-14873H (Figure 73) were isolated from the fermentation of *Streptomyces* sp. X-14873 (ATCC31679). Crystal structures of X-14873A and X-14873H were reported [109]. Antibiotic X-14873A was mainly active against Gram-positive bacteria and exhibited no activity against Gram-negative bacteria. It was interesting to note that X-14873H, the descarboxyl derivative of X-14873A, was also active against Gram-positive bacteria, while the other descarboxyl derivative, X-14873G, was practically inactive. These results indicated that the carboxylic function of the polyether antibiotic molecule was not required for the

antimicrobial activity, even though the ionized carboxylic group played an important role in complexation of cations by carboxylic acid polyether ionophores [110]. Antibiotic X-14873A also induced a change in the proportion of volatile fatty acids (acetate, propionate, and butyrate) produced in the rumen by increasing the molar proportion of propionate in the rumen fluid [110].

2.31. Noboritomycin. Noboritomycin (Figure 74) was isolated from *Streptomyces noboritoensis.* Crystal structure was determined by X-ray analysis of its silver salt. Noboritomycin was the first polyether ionophore possessing two carboxylic acid functions on the carbon backbone, namely, a free acid and an additional carboxylic acid ethyl ester group [111]. Noboritomycin was active against a wide range of Gram-positive bacteria, but was inactive against Gram-negative

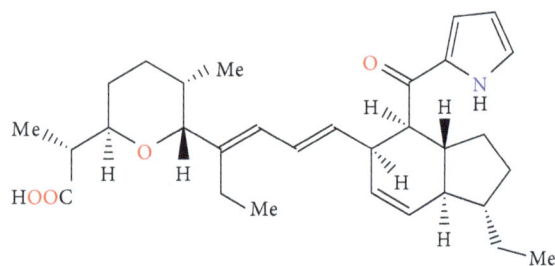

FIGURE 79: Structure of indanomycin.

bacteria and yeasts. It exhibited only weak anticoccidial activity (*Eimeria tenella*) in chicken [111].

2.32. 6-chloronoboritomycin. 6-chloronoboritomycin (Figure 75) was isolated from *Streptomyces malachitofuscus* [112]. Its crystal structure was determined by X-ray analysis of its thallium and rubidium salts [113]. This antibiotic is able to complex and transport monovalent as well as divalent metal cations. 6-chloronoboritomycin is active against Gram-positive bacteria and some anaerobes. In addition, it exhibits *in vitro* activity against several strains of *Treponema hyodysenteriae*, a causal agent of swine dysentery [112].

2.33. CP-82009. CP-82009 (Figure 76) was isolated by solvent extraction from the fermentation broth of *Actinomadura* sp. ATCC 53676. Its crystal structure was determined by X-ray analysis of its rubidium salt. CP-82009 exhibited activity against Gram-positive bacteria, as well as the spirochete *Treponema hyodysenteriae*. No activity was observed against *Escherichia coli* [114].

2.34. Abierixin. Abierixin (Figure 77) was isolated from *Streptomyces albus*. It exhibited weak activity against Gram-positive bacteria. The acute toxicity of abierixin in mice was examined. The LD_{50} value was 80–100 mg/kg. The anticoccidial evaluation of abrexin was carried out with chickens infected with *Eimeria tenella*. Abierixin, at 40 ppm, was effective in reducing the mortality of chickens and in increasing the average body weight of treated infected chickens compared to untreated infected controls [115].

2.35. A-83094A. A-83094A (Figure 78) was isolated from the strain of *Streptomyces setonii*. The antibiotic is active against Gram-positive bacteria. At a concentration of 0.31 μg/mL, it completely inhibits the development of *Eimeria tenella*. However, the compound showed no efficacy *in vivo* when chicks infected with *Eimeria tenella* or *Eimeria acervulina* were fed a diet containing 200 μg/g A-83094A [116].

2.36. Indanomycin. Indianomycin (Figure 79) was isolated from the strain of *Streptomyces antibioticus* [117]. Its crystal structure was determined by X-ray analysis of its bromophenethylamine salt [118]. Indianomycin was active *in vitro* against Gram-positive bacteria. It is also effective as a

growth promotant for ruminants, increasing feed utilization by these animals [117].

3. Conclusion

Polyether ionophores are generally active against Gram-positive bacteria. These compounds are able to form complexes with metal cations and transport them across lipid bilayer. In consequence, the whole process leads to changes in the osmotic pressure inside the cell, causing death of the bacteria cell. Some of them also show anticoccidial activity. Recently, it has been shown that several ionophores exhibit anticancer activities (monensin, salinomycin, and laidlomycin). Results of many studies have shown that both cancer stem cells (CSCs) and multidrug-resistant cancer cells (MDR) are effectively killed by polyether ionophores particularly by salinomycin. Polyether ionophores are currently well-recognized candidates to be clinically tested as anticancer drug candidates. The activity of these compounds should be further studied *in vitro* to specify their mechanisms of action and *in vivo* to assess their activities and tolerance in the different types of cancer. The studies performed so far have shown that these compounds affect cancer cells in a special way by increasing their sensitivity to chemotherapy (monensin, salinomycin, inostamycin) and reverse multidrug resistance (laidlomycin and monensin) in human carcinoma. Furthermore, these compounds have been found to be cytotoxic to the human carcinoma multidrug-resistant cells (monensin and salinomycin). Ionophore antibiotics also inhibit chemoresistant cancer cells by increasing apoptosis, but up to now, only salinomycin has been successfully able to kill human cancer stem cells (CSCs). Therefore, polyether antibiotics should be considered as new anticancer drugs for cancer prevention and cancer therapy, and the mechanism of their anticancer activity should be studied in detail.

References

[1] C. J. Dutton, B. J. Banks, and C. B. Cooper, "Polyether ionophores," *Natural Product Reports*, vol. 12, no. 2, pp. 165–181, 1995.

[2] T. R. Callaway, T. S. Edrington, J. L. Rychlik et al., "Ionophores: their use as ruminant growth promotants and impact on food safety," *Current Issues in Intestinal Microbiology*, vol. 4, no. 2, pp. 43–51, 2003.

[3] S. Rochefeuille, C. Jimenez, S. Tingry, P. Seta, and J. P. Desfours, "Mixed Langmuir-Blodgett monolayers containing carboxylic ionophores. Application to Na$^+$ and Ca^{2+} ISFET-based sensors," *Materials Science and Engineering C*, vol. 21, no. 1-2, pp. 43–46, 2002.

[4] C. Gabrielli, P. Hemery, P. Letellier et al., "Investigation of ion-selective electrodes with neutral ionophores and ionic sites by EIS. II. Application to K$^+$ detection," *Journal of Electroanalytical Chemistry*, vol. 570, no. 2, pp. 291–304, 2004.

[5] M. Dobler, "Natural cation-binding agents," in *Comprehensive Supramolecular Chemistry: Molecular Recognition: Receptors for Cationic Guests*, G. W. Gokel, Ed., vol. 1, pp. 267–313, Pergamon, New York, NY, USA, 2004.

[6] H. H. Mollenhauer, D. J. Morre, and L. D. Rowe, "Alteration of intracellular traffic by monensin; mechanism, specificity and

relationship to toxicity," *Biochimica et Biophysica Acta*, vol. 1031, no. 2, pp. 225–246, 1990.

[7] B. C. Pressman, *Antibiotics and Their Complexes*, Marcel Dekker, New York, 1985.

[8] L. F. Lindoy, "Outer-sphere and inner-sphere complexation of cations by the natural ionophore lasalocid A," *Coordination Chemistry Reviews*, vol. 148, pp. 349–368, 1996.

[9] M. Inabayashi, S. Miyauchi, N. Kamo, and T. Jin, "Conductance change in phospholipid bilayer membrane by an electroneutral ionophore, monensin," *Biochemistry*, vol. 34, no. 10, pp. 3455–3460, 1995.

[10] H. Tsukube, K. Takagi, T. Higashiyama, T. Iwachido, and N. Hayama, "Biomimetic membrane transport: interesting ionophore functions of naturally occurring polyether antibiotics toward unusual metal cations and amino acid ester salts," *Inorganic Chemistry*, vol. 33, no. 13, pp. 2984–2987, 1994.

[11] S. Ahmed and I. R. Booth, "Quantitative measurements of the proton-motive force and its relation to steady state lactose accumulation in *Escherichia coli*," *Biochemical Journal*, vol. 200, no. 3, pp. 573–581, 1981.

[12] M. Kuhn and H. D. King, Belgian Patent, 839355, 1976.

[13] M. Allèaume, B. Busetta, C. Farges, P. Gachon, A. Kergomard, and T. Staron, "X-Ray structure of alborixin, a new antibiotic ionophore," *Journal of the Chemical Society, Chemical Communications*, no. 11, pp. 411–412, 1975.

[14] P. Van Roey, W. L. Duax, P. D. Strong, and G. D. Smith, *Israel Journal of Chemistry*, vol. 24, p. 283, 1984.

[15] P. Gachon, C. Farges, and A. Kergomard, "Alborixin, a new antibiotic ionophore: isolation, structure, physical and chemical properties," *Journal of Antibiotics*, vol. 29, no. 6, pp. 603–610, 1976.

[16] Y. Kusakabe, S. Mitsuoka, Y. Omuro, and A. Seino, "Antibiotic No. 6016, a polyether antibiotic," *Journal of Antibiotics*, vol. 33, no. 12, pp. 1437–1442, 1980.

[17] N. Otake, T. Ogita, H. Nakayama, H. Miyamae, S. Sato, and Y. Saito, "X-Ray crystal structure of the thallium salt of antibiotic-6016, a new polyether ionophore," *Journal of the Chemical Society, Chemical Communications*, no. 20, pp. 875–876, 1978.

[18] R. M. Gale, C. E. Higgens, and M. M. Hoehn, US Patent, 3923823, 1975.

[19] G. D. Smith and W. L. Duax, "Crystal and molecular structure of the calcium ion complex of A23187," *Journal of the American Chemical Society*, vol. 98, no. 6, pp. 1578–1580, 1976.

[20] M. Alleaume and Y. Barrans, "Crystal structure of the magnesium complex of calcimycin," *Canadian Journal of Chemistry*, vol. 63, no. 12, pp. 3482–3485, 1985.

[21] M. Akkurt, A. Melhaoui, T. Ben Hadda, M. Mimouni, S. Öztürk Yíldírím, and V. McKeec, "Synthesis and crystal structure of the bis-calcimycin anion-Ni^{2+} complex," *Arkivoc*, vol. 2008, no. 11, pp. 154–164, 2008.

[22] S. Vila, I. Canet, J. Guyot, G. Jeminet, and L. Toupet, "Unusual structure of the dimeric 4-bromocalcimycin-Zn^{2+} complex," *Chemical Communications*, vol. 9, no. 4, pp. 516–517, 2003.

[23] G. Grenier, J. Van Sande, D. Glick, and J. E. Dumont, "Effect of ionophore A23187 on thyroid secretion," *FEBS Letters*, vol. 49, no. 1, pp. 96–99, 1974.

[24] R. C. Karl, W. S. Zawalich, J. A. Ferrendelli, and F. M. Matschinsky, "The role of Ca^{2+} and cyclic adenosine $3':5'$ monophosphate in insulin release induced *in vitro* by the divalent cation ionophore A23187," *The Journal of Biological Chemistry*, vol. 250, no. 12, pp. 4575–4579, 1975.

[25] G. D. Case, J. M. Vanderkooi, and A. Scarpa, "Physical properties of biological membranes determined by the fluorescence of the calcium ionophore A23187," *Archives of Biochemistry and Biophysics*, vol. 162, no. 1, pp. 174–185, 1974.

[26] J. Guyot, G. Jeminet, M. Prudhomme, M. Sancelme, and R. Meiniel, "Interaction of the calcium ionophore A.23187 (calcimycin) with Bacillus cereus and *Escherichia coli*," *Letters in Applied Microbiology*, vol. 16, no. 4, pp. 192–195, 1993.

[27] K. D. Klika, J. P. Haansuu, V. V. Ovcharenko et al., "Frankiamide: a structural revision to demethyl (C-11) cezomycin," *Zeitschrift für Naturforschung*, vol. 58, no. 12, pp. 1210–1215, 2003.

[28] J. W. Westley, C. M. Liu, and J. F. Blount, "Isolation and characterization of a novel polyether antibiotic of the pyrrolether class, antibiotic X-14885A," *Journal of Antibiotics*, vol. 36, no. 10, pp. 1275–1278, 1983.

[29] T. Sakurai, K. Kobayashi, G. Nakamura, and K. Isono, "Structure of the thallium salt of cationomycin," *Acta Crystallographica Section B*, vol. 38, no. 9, pp. 2471–2473, 1982.

[30] M. Ubukata, Y. Hamazaki, and K. Isono, "Chemical modification of cationomycin and its structure-activity relationship," *Agricultural and Biological Chemistry*, vol. 50, no. 5, pp. 1153–1160, 1986.

[31] M. Ubukata, T. Akama, and K. Isono, "Aromatic side chain analogs of cationomycin and their biological activities," *Agricultural and Biological Chemistry*, vol. 52, no. 7, pp. 1637–1641, 1988.

[32] A. M. Delort, G. Jeminet, S. Sareth, and F. G. Riddle, "Ionophore properties of cationomycin in large unilamellar vesicles studied by ^{23}Na- and ^{39}K-NMR," *Chemical and Pharmaceutical Bulletin*, vol. 46, no. 10, pp. 1618–1620, 1998.

[33] J. R. Oscarson, J. Bordner, W. D. Celmer et al., "Endusamycin, a novel polycyclic ether antibiotic produced by a strain of *Streptomyces endus subsp. aureus*," *Journal of Antibiotics*, vol. 42, no. 1, pp. 37–48, 1989.

[34] T. Fehr, H. D. King, and M. Kuhn, "Mutalomycin, a new polyether antibiotic. Taxonomy, fermentation, isolation and characterization," *Journal of Antibiotics*, vol. 30, no. 11, pp. 903–907, 1977.

[35] T. Fehr, M. Kuhn, H. R. Loosli, M. Ponelle, J. J. Boelsterli, and M. D. Walkinshaw, "2-Epimutalomycin and 28-epimutalomycin, two new polyether antibiotics from *Streptomyces mutabilis*. Derivatization of mutalomycin and the structure elucidation of two minor metabolites," *Journal of Antibiotics*, vol. 42, no. 6, pp. 897–902, 1989.

[36] C. M. Liu and T. E. Hermann, "Characterization of ionomycin as a calcium ionophore," *The Journal of Biological Chemistry*, vol. 253, no. 17, pp. 5892–5894, 1978.

[37] Z. Gao, Y. Li, J. P. Cooksey et al., "A synthesis of an ionomycin calcium complex," *Angewandte Chemie—International Edition*, vol. 48, no. 27, pp. 5022–5025, 2009.

[38] W. C. Liu, D. Smith-Slusarczyk, G. Astle, W. H. Trejo, W. E. Brown, and E. Meyers, "Ionomycin, a new polyether antibiotic," *Journal of Antibiotics*, vol. 31, no. 9, pp. 815–819, 1978.

[39] N. Tsuji, K. Nagashima, M. Kobayashi et al., "Two new antibiotics, A 218 and K 41 isolation and characterization," *Journal of Antibiotics*, vol. 29, no. 1, pp. 10–14, 1976.

[40] M. Shiro, H. Nakai, K. Nagashima, and N. Tsuji, "X-Ray determination of the structure of the polyether antibiotic K-41," *Journal of the Chemical Society, Chemical Communications*, no. 16, pp. 682–683, 1978.

[41] K. Otoguro, A. Ishiyama, H. Ui et al., "*In vitro* and *in vivo* antimalarial activities of the monoglycoside polyether antibiotic, K-41 against drug resistant strains of Plasmodia," *Journal of Antibiotics*, vol. 55, no. 9, pp. 832–834, 2002.

[42] Y. Takahashi, H. Nakamura, R. Ogata et al., "Kijimicin, a polyether antibiotic," *Journal of Antibiotics*, vol. 43, no. 4, pp. 441–443, 1990.

[43] T. Yamauchi, M. Nakamura, H. Honma, M. Ikeda, K. Kawashima, and T. Ohno, "Mechanistic effects of Kijimicin on inhibition of human immunodeficiency virus replication," *Molecular and Cellular Biochemistry*, vol. 119, no. 1-2, pp. 35–41, 1993.

[44] J. Berger, A. I. Rachlin, W. E. Scott, L. H. Sternbach, and M. W. Goldberg, "The isolation of three new crystalline antibiotics from *Streptomyces*," *Journal of the American Chemical Society*, vol. 73, no. 11, pp. 5295–5298, 1951.

[45] S. M. Johnson, J. Herrin, S. J. Liu, and I. C. Paul, "Crystal structure of a barium complex of antibiotic X-537A, Ba(C$_{34}$H$_{53}$O$_8$)2 · H$_2$O," *Journal of the Chemical Society D*, no. 2, pp. 72–73, 1970.

[46] I. H. Suh, K. Aoki, and H. Yamazaki, "Crystal structure of a silver salt of the antibiotic lasalocid A: a dimer having an exact 2-fold symmetry," *Inorganic Chemistry*, vol. 28, no. 2, pp. 358–362, 1989.

[47] M. Akkurt, S. Öztürk Yildirim, F. Z. Khardli, M. Mimouni, V. McKee, and T. Ben Haddab, "Crystal structure of a new polymeric thallium-lasalocid complex: iasalocide anion-thallium(I) containing aryl-Tl interactions," *Arkivoc*, vol. 2008, no. 15, pp. 121–132, 2008.

[48] J. W. Westley, W. Benz, and J. Donahue, "Biosynthesis of lasalocid. III Isolation and structure determination of four homologs of lasalocid A," *Journal of Antibiotics*, vol. 27, no. 10, pp. 744–753, 1974.

[49] D. A. Coffen and D. A. Katonak, "Chemical degradation of lasalocid: (A) The Mannich reaction (B) Bayer-Villiger oxidation of the retro-aldol ketone," *Helvetica Chimica Acta*, vol. 64, no. 5, pp. 1645–1652, 1981.

[50] A. Huczyński, I. Paluch, M. Ratajczak-Sitarz et al., "Spectroscopic studies, crystal structures and antimicrobial activities of a new lasalocid 1-naphthylmethyl ester," *Journal of Molecular Structure*, vol. 891, no. 1–3, pp. 481–490, 2008.

[51] A. Huczyński, I. Paluch, M. Ratajczak-Sitarz, A. Katrusiak, B. Brzezinski, and F. Bartl, "Structural and spectroscopic studies of a new 2-naphthylmethyl ester of lasalocid acid," *Journal of Molecular Structure*, vol. 918, no. 1–3, pp. 108–115, 2009.

[52] A. Huczyński, T. Pospieszny, R. Wawrzyn et al., "Structural and spectroscopic studies of new *o*-, *m*- and *p*-nitrobenzyl esters of lasalocid acid," *Journal of Molecular Structure*, vol. 877, no. 1–3, pp. 105–114, 2008.

[53] A. Huczyński, M. Ratajczak-Sitarz, A. Katrusiak, and B. Brzezinski, "X-ray, spectroscopic and semiempirical investigation of the structure of lasalocid 6-bromohexyl ester and its complexes with alkali metal cations," *Journal of Molecular Structure*, vol. 998, no. 1–3, pp. 206–215, 2011.

[54] A. Huczyński, J. Janczak, J. Rutkowski et al., "Lasalocid acid as a lipophilic carrier ionophore for allylamine: spectroscopic, crystallographic and microbiological investigation," *Journal of Molecular Structure*, vol. 936, no. 1–3, pp. 92–98, 2009.

[55] A. Huczyński, J. Janczak, J. Stefańska, J. Rutkowski, and B. Brzezinski, "X-ray, spectroscopic and antibacterial activity studies of the 1:1 complex of lasalocid acid with 1,1,3,3-tetramethylguanidine," *Journal of Molecular Structure*, vol. 977, no. 1–3, pp. 51–55, 2010.

[56] A. Huczyński, T. Pospieszny, M. Ratajczak-Sitarz, A. Katrusiak, and B. Brzezinski, "Structural and spectroscopic studies of the 1:1 complex of lasalocid acid with 1,5,7-triazabicyclo[4.4.0]dec-5-ene," *Journal of Molecular Structure*, vol. 875, no. 1–3, pp. 501–508, 2008.

[57] A. Huczynski, J. Rutkowski, J. Wietrzyk et al., "X-Ray crystallographic, FT-IR and NMR studies as well as anticancer and antibacterial activity of the salt formed between ionophor antibiotic Lasalocid acid and amines," *Journal of Molecular Structure*, vol. 1032, pp. 69–77, 2012.

[58] E. J. Tynan, T. H. Nelson, R. A. Davies, and W. C. Wernau, "The production of semduramicin by direct fermentation," *Journal of Antibiotics*, vol. 45, no. 5, pp. 813–815, 1992.

[59] A. P. Ricketts, E. A. Glazer, T. T. Migaki, and J. A. Olson, "Anticoccidial efficacy of semduramicin in battery studies with laboratory isolates of coccidia," *Poultry Science*, vol. 71, no. 1, pp. 98–103, 1992.

[60] J. P. Dirlam, J. Bordner, S. P. Chang et al., "The isolation and structure of CP-120,509, a new polyether antibiotic related to semduramicin and produced by mutants of *Actinomadura roseorufa*," *Journal of Antibiotics*, vol. 45, no. 9, pp. 1544–1548, 1992.

[61] D. H. Davies, E. W. Snape, P. J. Suter, T. J. King, and C. P. Falshaw, "Structure of antibiotic M139603; x-ray crystal structure of the 4-bromo-3,5-dinitrobenzoyl derivative," *Journal of the Chemical Society, Chemical Communications*, no. 20, pp. 1073–1074, 1981.

[62] C. J. Newbold, R. J. Wallace, N. D. Watt, and A. J. Richardson, "Effect of the novel ionophore tetronasin (ICI 139603) on ruminal microorganisms," *Applied and Environmental Microbiology*, vol. 54, no. 2, pp. 544–547, 1988.

[63] H. A. Brooks, D. Gardner, J. P. Pyser, and T. J. King, "The structure and absolute stereochemistry of zincophorin (antibiotic M144255): a monobasic carboxylic acid ionophore having a remarkable specificity for divalent cations," *Journal of Antibiotics*, vol. 37, no. 11, pp. 1501–1504, 1984.

[64] J. P. Dirlam, A. M. Belton, S. P. Shang et al., "CP-78,545, a new monocarboxylic acid ionophore antibiotic related to zincophorin and produced by a *Streptomyces*," *Journal of Antibiotics*, vol. 42, no. 8, pp. 1213–1220, 1989.

[65] H. Kinashi, N. Otake, and H. Yonehara, "The structure of salinomycin, a new member of the polyether antibiotics," *Tetrahedron Letters*, vol. 49, pp. 4955–4958, 1973.

[66] J. W. Westley, J. F. Blount, R. H. Evans, and C. M. Liu, "C-17 epimers of deoxy-(O-8)-salinomycin from *Streptomyces albus* (ATCC 21838)," *Journal of Antibiotics*, vol. 30, no. 7, pp. 610–612, 1977.

[67] Y. Miyazaki, A. Shibata, T. Yahagi et al., Japanese Patent, 61247398, 1986.

[68] E. F. Paulus and L. Vértesy, "Crystal structure of the antibiotic SY-1 (20-deoxy-salinomycin): sodium 2-(6-[2-(5-ethyl-5-hydroxy-6-methyl-tetrahydro-pyran-2-yl)-2,10,12-trimethyl-1, 6,8-trioxa-dispiro[4.1.5.3]pentadec-13-en-9-yl]-2-hydroxy-1,3-dimethyl-4-oxo- heptyl-5-methyl-tetrahydro-pyran-2-yl)-butyrate—methanol solvate (1:0.69), C$_{42}$H$_{69}$NaO$_{10}$ · 0.69CH$_3$OH," *Zeitschrift für Kristallographie*, vol. 218, no. 4, pp. 575–577, 2003.

[69] E. F. Paulus and L. Vértesy, "Crystal structure of 2-(6-[2-(5-ethyl-5-hydroxy-6-methyl-tetrahydro-pyran2- yl)-15-oxo-2,10, 12-trimethyl-1,6,8-trioxa-dispiro[4.1.5.3]pentadec-13-en-9-yl] -2-hydroxy-1,3-dimethyl-4-oxo-heptyl-5-methyl-tetrahydro-pyran-2-yl)-butyrate sodium, Na(C$_{42}$H$_{67}$O$_{11}$), SY-9—antibiotic 20-oxo-salinomycin," *Zeitschrift für Kristallographie*, vol. 219, no. 2, pp. 184–186, 2004.

[70] A. Huczyński, J. Janczak, M. Antoszczak, J. Stefanska, and B. Brzezinski, "X-ray, FT-IR, NMR and PM5 structural studies and antibacterial activity of unexpectedly stable salinomycin-benzotriazole intermediate ester," *Journal of Molecular Structure*, vol. 1022, pp. 197–203, 2012.

[71] A. Huczyński, J. Janczak, J. Stefanska, M. Antoszczak, and B. Brzezinski, "Synthesis and antimicrobial activity of amide derivatives of polyether antibiotic—salinomycin," *Bioorganic and Medicinal Chemistry Letters*, vol. 22, no. 14, pp. 4697–4702, 2012.

[72] M. Mitani, T. Yamanishi, and Y. Miyazaki, "Salinomycin: a new monovalent cation ionophore," *Biochemical and Biophysical Research Communications*, vol. 66, no. 4, pp. 1231–1236, 1975.

[73] Y. Myiazaki, M. Shibuya, H. Sugawara et al., "Salinomycin, a new polyether antibiotic," *Journal of Antibiotics*, vol. 27, no. 11, pp. 814–821, 1974.

[74] A. Huczyński, "Salinomycin—a new cancer drug candidate," *Chemical Biology and Drug Design*, vol. 79, no. 3, pp. 235–238, 2012.

[75] A. Agtarap, J. W. Chamberlin, M. Pinkerton, and L. Steinrauf, "The structure of monensic acid, A new biologically active compound," *Journal of the American Chemical Society*, vol. 89, no. 22, pp. 5737–5739, 1967.

[76] A. Huczyński, M. Ratajczak-Sitarz, A. Katrusiak, and B. Brzezinski, "Molecular structure of the 1 : 1 inclusion complex of monensin A sodium salt with acetonitrile," *Journal of Molecular Structure*, vol. 832, no. 1–3, pp. 84–89, 2007.

[77] A. Huczyński, J. Janczak, D. Łowicki, and B. Brzezinski, "Monensin a acid complexes as a model of electrogenic transport of sodium cation," *Biochimica et Biophysica Acta*, no. 9, pp. 2108–2119, 1818.

[78] A. Huczyński, M. Ratajczak-Sitarz, A. Katrusiak, and B. Brzezinski, "Molecular structure of the 1 : 1 inclusion complex of monensin A lithium salt with acetonitrile," *Journal of Molecular Structure*, vol. 871, no. 1–3, pp. 92–97, 2007.

[79] A. Huczyński, M. Ratajczak-Sitarz, A. Katrusiak, and B. Brzezinski, "Molecular structure of rubidium six-coordinated dihydrate complex with monensin A," *Journal of Molecular Structure*, vol. 888, no. 1–3, pp. 224–229, 2008.

[80] I. N. Pantcheva, R. Zhorova, M. Mitewa, S. Simova, H. Mayer-Figge, and W. S. Sheldrick, "First solid state alkaline-earth complexes of monensic acid A (MonH): crystal structure of $[M(Mon)_2(H_2O)_2]$ (M = Mg, Ca), spectral properties and cytotoxicity against aerobic Gram-positive bacteria," *BioMetals*, vol. 23, no. 1, pp. 59–70, 2010.

[81] I. N. Pantcheva, J. Ivanova, R. Zhorova et al., "Nickel(II) and zinc(II) dimonensinates: single crystal X-ray structure, spectral properties and bactericidal activity," *Inorganica Chimica Acta*, vol. 363, no. 8, pp. 1879–1886, 2010.

[82] P. Dorkov, I. N. Pantcheva, W. S. Sheldrick, H. Mayer-Figge, R. Petrova, and M. Mitewa, "Synthesis, structure and antimicrobial activity of manganese(II) and cobalt(II) complexes of the polyether ionophore antibiotic Sodium Monensin A," *Journal of Inorganic Biochemistry*, vol. 102, no. 1, pp. 26–32, 2008.

[83] I. N. Pantcheva, P. Dorkov, V. N. Atanasov et al., "Crystal structure and properties of the copper(II) complex of sodium monensin A," *Journal of Inorganic Biochemistry*, vol. 103, no. 10, pp. 1419–1424, 2009.

[84] J. Ivanova, I. N. Pantcheva, M. Mitewa, S. Simova, H. Mayer-Figge, and W. S. Sheldrick, "Crystal structures and spectral properties of new Cd(II) and Hg(II) complexes of monensic

acid with different coordination modes of the ligand," *Central European Journal of Chemistry*, vol. 8, no. 4, pp. 852–860, 2010.

[85] I. N. Pantcheva, M. I. Mitewa, W. S. Sheldrick, I. M. Oppel, R. Zhorova, and P. Dorkov, "First divalent metal complexes of the polyether ionophore monensin A: x-ray structures of $[Co(Mon)_2(H_2O)_2]$ and $[Mn(Mon)_2(H2O)_2]$ and their bactericidal properties," *Current Drug Discovery Technologies*, vol. 5, no. 2, pp. 154–161, 2008.

[86] A. Huczyński, D. Łowicki, B. Ratajczak-Sitarz, A. Katrusiak, and B. Brzezinski, "Structural investigation of a new complex of N-allylamide of Monensin A with strontium perchlorate using X-ray, FT-IR, ESI MS and semiempirical methods," *Journal of Molecular Structure*, vol. 995, no. 1–3, pp. 20–28, 2011.

[87] A. Huczyński, J. Janczak, and B. Brzezinski, "Crystal structure and FT-IR study of aqualithium 1-naphthylmethyl ester of monensin A perchlorate," *Journal of Molecular Structure*, vol. 985, no. 1, pp. 70–74, 2011.

[88] A. Huczyński, M. Ratajczak-Sitarz, J. Stefańska, A. Katrusiak, B. Brzezinski, and F. Bartl, "Reinvestigation of the structure of monensin A phenylurethane sodium salt based on X-ray crystallographic and spectroscopic studies, and its activity against hospital strains of methicillin-resistant S. epidermidis and S. aureus," *Journal of Antibiotics*, vol. 64, no. 3, pp. 249–256, 2011.

[89] A. Huczyński, J. Stefańska, P. Przybylski, B. Brzezinski, and F. Bartl, "Synthesis and antimicrobial properties of Monensin A esters," *Bioorganic and Medicinal Chemistry Letters*, vol. 18, no. 8, pp. 2585–2589, 2008.

[90] A. Iacoangeli, G. Melucci-Vigo, and G. Risuleo, "The ionophore monensin inhibits mouse polyomavirus DNA replication and destabilizes viral early mRNAs," *Biochimie*, vol. 82, no. 1, pp. 35–39, 2000.

[91] Y. Kusakabe, T. Mizuno, and S. Kawabata, "Ferensimycins A and B, two polyether antibiotics," *Journal of Antibiotics*, vol. 35, no. 9, pp. 1119–1129, 1982.

[92] J. P. Dirlam, J. Bordner, W. P. Cullen, M. T. Jefferson, and L. Presseau-Linabury, "The structure of CP-96,797, polyether antibiotic related to K-41A and produced by *Streptomyces sp*," *Journal of Antibiotics*, vol. 45, no. 7, pp. 1187–1189, 1992.

[93] S. Funayama and S. Nozoe, "Isolation and structure of a new polyether antibiotic, octacyclomycin," *Journal of Antibiotics*, vol. 45, no. 10, pp. 1686–1691, 1992.

[94] J. P. Dirlam, W. P. Cullen, L. H. Huang et al., "CP-91,243 and CP-91,244, novel diglycosine polyether antibiotics related to UK-58,852 and produced by mutants of *Actinomadura roseorufa*," *Journal of Antibiotics*, vol. 44, no. 11, pp. 1262–1266, 1991.

[95] R. S. Wehbie, C. Runsheng, and H. A. Lardy, "The antibiotic W341C, its ion transport properties and inhibitory effects on mitochondrial substrate oxidation," *Journal of Antibiotics*, vol. 40, no. 6, pp. 887–893, 1987.

[96] F. Kitame, K. Utsushikawa, and T. Kohama, "Laidlomycin, a new antimycoplasmal polyether antibiotic," *Journal of Antibiotics*, vol. 27, no. 11, pp. 884–888, 1974.

[97] J. P. Dirlam, A. M. Belton, J. Bordner et al., "CP-84,657, a potent polyether anticoccidial related to portmicin and produced by *Actinomadura sp*," *Journal of Antibiotics*, vol. 43, no. 6, pp. 668–679, 1990.

[98] P. Gachon, A. Kergomard, T. Staron, and C. Esteve, "Grisorixin, an ionophorous antibiotic of the nigericin group. I. Fermentation, isolation, biological properties and structure," *Journal of Antibiotics*, vol. 28, no. 5, pp. 345–350, 1975.

[99] M. Alleaume and D. Hickel, "The crystal structure of grisorixin silver salt," *Journal of the Chemical Society D*, no. 21, pp. 1422–1423, 1970.

[100] M. Alleaume and D. Hickel, "Crystal structure of the thallium salt of the antibiotic grisorixin," *Journal of the Chemical Society, Chemical Communications*, pp. 175–176, 1972.

[101] J. Mouslim, A. Cuer, L. David, and J. C. Tabet, "Epigrisorixin, a new polyether carboxylic antibiotic," *Journal of Antibiotics*, vol. 46, no. 1, pp. 201–203, 1993.

[102] W. P. Cullen, W. D. Celmer, L. R. Chappel et al., "CP-54,883 a novel chlorine-containing polyether antibiotic produced by a new species of *Actinomadura*: taxonomy of the producing culture, fermentation, physico-chemical and biological properties of the antibiotic," *Journal of Antibiotics*, vol. 40, no. 11, pp. 1490–1495, 1987.

[103] J. Bordner, P. C. Watts, and E. B. Whipple, "Structure of the natural antibiotic ionophore CP-54,883," *Journal of Antibiotics*, vol. 40, no. 11, pp. 1496–1505, 1987.

[104] M. Hatsu, T. Sasaki, S. Miyadoh et al., "SF2487, a new polyether antibiotic produced by *Actinomadura*," *Journal of Antibiotics*, vol. 43, no. 3, pp. 259–266, 1990.

[105] C. M. Liu, T. E. Hermann, and A. Downey, "Novel polyether antibiotics X-14868A, B, C, and D produced by a Nocardia. Discovery, fermentation, biological as well as ionophore properties and taxonomy of the producing culture," *Journal of Antibiotics*, vol. 36, no. 4, pp. 343–350, 1983.

[106] J. P. Dirlam, L. Presseau-Linabury, and D. A. Koss, "The structure of CP-80,219, a new polyether antibiotic related to dianemycin," *Journal of Antibiotics*, vol. 43, no. 6, pp. 727–730, 1990.

[107] H. Nakayama, H. Seto, and N. Otake, "Studies on the ionophorous antibiotics. XXVIII. Moyukamycin, a new glycosylated polyether antibiotic," *Journal of Antibiotics*, vol. 38, no. 10, pp. 1433–1436, 1985.

[108] J. W. Westley, Chao-Min Liu, and L. H. Sello, "Isolation and characterization of antibiotic X-14931A, the naturally occurring 19-deoxyaglycone of dianemycin," *Journal of Antibiotics*, vol. 37, no. 7, pp. 813–815, 1984.

[109] J. W. Westley, C. M. Liu, and J. F. Blount, "Isolation and characterization of three novel polyether antibiotics and three novel actinomycins as cometabolites of the same *Streptomyces sp.* X-14873, ATCC 31679," *Journal of Antibiotics*, vol. 39, no. 12, pp. 1704–1711, 1986.

[110] C. M. Liu, J. W. Westley, and T. E. Hermann, "Novel polyether antibiotics X-14873A, G and H produced by a *Streptomyces*: taxonomy of the producing culture, fermentation, biological and ionophorous properties of the antibiotics," *Journal of Antibiotics*, vol. 39, no. 12, pp. 1712–1718, 1986.

[111] C. Keller-Juslen, H. D. King, and M. Kuhn, "Noboritomycins A and B, new polyether antibiotics," *Journal of Antibiotics*, vol. 31, no. 9, pp. 820–828, 1978.

[112] C. Liu, T. E. Hermann, and T. B. La Prosser, "X-14766A, a halogen containing polyether antibiotic produced by *Streptomyces malachitofuscus subsp.* downeyi ATCC 31547. Discovery, fermentation, biological properties and taxonomy of the producing culture," *Journal of Antibiotics*, vol. 34, no. 2, pp. 133–138, 1981.

[113] J. W. Westley, R. H. Evans, and L. H. Sello, "Isolation and characterization of the first halogen containing polyether antibiotic X-14766A, a product of *Streptomyces malachitofuscus subsp.* downeyi," *Journal of Antibiotics*, vol. 34, no. 2, pp. 139–147, 1981.

[114] J. P. Dirlam, A. M. Belton, J. Bordner et al., "CP-82,009, a potent polyether anticoccidial related to septamycin and produced by *Actinonladura sp*," *Journal of Antibiotics*, vol. 45, no. 3, pp. 331–340, 1992.

[115] L. David, H. L. Ayala, and J. C. Tabet, "Abierixin, a new polyether antibiotic. Production, structural determination and biological activities," *Journal of Antibiotics*, vol. 38, no. 12, pp. 1655–1663, 1985.

[116] S. H. Larsen, L. V. D. Boeck, F. P. Mertz, J. W. Paschal, and J. L. Occolowitz, "16-Deethylindanomycin (A83094A), a novel pyrrole-ether antibiotic produced by a strain of *Streptomyces setonii*. Taxonomy, fermentation, isolation and characterization," *Journal of Antibiotics*, vol. 41, no. 9, pp. 1170–1177, 1988.

[117] C. M. Liu, T. E. Hermann, and M. Liu, "X-14547A, a new ionophorous antibiotic produced by *Streptomyces antibioticus* NRRL 8167. Discovery, fermentation, biological properties and taxonomy of the producing culture," *Journal of Antibiotics*, vol. 32, no. 2, pp. 95–99, 1979.

[118] J. W. Westley, R. H. Evans, and L. H. Sello, "Isolation and characterization of antibiotic X-14547A, a novel monocarboxylic acid ionophore produced by *Streptomyces antibioticus* NRRL 8167," *Journal of Antibiotics*, vol. 32, no. 2, pp. 100–107, 1979.

14

Multimolecular Salivary Mucin Complex Is Altered in Saliva of Cigarette Smokers: Detection of Disulfide Bridges by Raman Spectroscopy

Motoe Taniguchi,[1,2] **Junko Iizuka,**[3] **Yukari Murata,**[4] **Yumi Ito,**[5,6] **Mariko Iwamiya,**[7] **Hiroshi Mori,**[2] **Yukio Hirata,**[4] **Yoshiharu Mukai,**[3] **and Yuko Mikuni-Takagaki**[1]

[1] *Department of Functional Biology, Kanagawa Dental College, 82 Inaokacho, Yokosuka 238-8580, Japan*

[2] *Department of Maxillofacial Diagnostic Science, Kanagawa Dental College, 82 Inaokacho, Yokosuka 238-8580, Japan*

[3] *Department of Oral Medicine, Kanagawa Dental College, 82 Inaokacho, Yokosuka 238-8580, Japan*

[4] *Department of Dental Sociology, Kanagawa Dental College, 82 Inaokacho, Yokosuka 238-8580, Japan*

[5] *Yokohama Training Center, Kanagawa Dental College, 3-31-6 Turuyacho, Kanagawa-ku, Yokohama 221-0835, Japan*

[6] *Department of Pathology, Tsurumi University School of Dental Medicine, Yokohama 230-8501, Japan*

[7] *Clinical Laboratory, Kanagawa Dental College, 82 Inaokacho, Yokosuka 238-8580, Japan*

Correspondence should be addressed to Motoe Taniguchi; motoe@kdcnet.ac.jp

Academic Editor: Y. James Kang

Saliva contains mucins, which protect epithelial cells. We showed a smaller amount of salivary mucin, both MG1 and MG2, in the premenopausal female smokers than in their nonsmoking counterparts. Smokers' MG1, which contains almost 2% cysteine/half cystine in its amino acid residues, turned out to be chemically altered in the nonsmoker's saliva. The smaller acidic glycoprotein bands were detectable only in smoker's saliva in the range of 20–25 kDa and at 45 kDa, suggesting that degradation, at least in part, caused the reduction of MG1 mucin. This is in agreement with the previous finding that free radicals in cigarette smoke modify mucins in both sugar and protein moieties. Moreover, proteins such as amylase and albumin are bound to other proteins through disulfide bonds and are identifiable only after reduction with DTT. Confocal laser Raman microspectroscopy identified a disulfide stretch band of significantly stronger intensity per protein in the stimulated saliva of smokers alone. We conclude that the saliva of smokers, especially stimulated saliva, contains significantly more oxidized form of proteins with increased disulfide bridges, that reduces protection for oral epithelium. Raman microspectroscopy can be used for an easy detection of the damaged salivary proteins.

1. Introduction

Cigarette smoke contains free radicals, which can damage tissues [1, 2]. Saliva plays a role in the general defense system of the oral environment, and in addition to antioxidants, it contains immunoglobulins, antibacterial enzymes, and growth factors. Saliva also contains a mucous secretion to protect epithelial cells from mechanical as well as chemical challenges [3]. The secreted mucins MG1 and MG2 [4], which make large complexes with amylase, proline-rich proteins, statherin, histatin, and other proteins, form the first line of epithelial protection [5, 6]. Previous reports showed that free radicals degrade proteins [7, 8] and that mucins are modified in both sugar and protein moieties [9]. In addition, surface-exposed cysteine residues of proteins are particularly sensitive to oxidation by almost all forms of reactive oxygen species (ROS), and the oxidation of these sulfur-containing amino acid residues is reversible [10]. These proteins therefore serve as antioxidants [8]. In the airway of smokers, mucin expression/secretion is upregulated [11–14]. However, there is no test or assay by which to easily detect oxidized proteins in the saliva of smokers, and there is no good way to determine to what extent they are altered. We, therefore, collected protein components of saliva from both nonsmokers and smokers by immediately precipitating them with ethanol to separate them from low-molecular-weight sulfhydryl donors. We then examined actual disulfide bonds in the protein components in the saliva of smokers.

2. Material and Methods

2.1. Subjects and Populations, Collection and Storage of Saliva. Premenopausal females between 35 and 49 years of age were recruited after gaining the approval of the ethics committee of Kanagawa Dental College (number 10–04, 2010). We selected healthy volunteers with no significant medical history who were either nonsmokers, who had never smoked, or current smokers. The average ages of the 48 nonsmokers and the 10 smokers were 41.8 ± 3.9 and 40.0 ± 4.8 years, respectively. Subjects did not smoke for 3 hours after they ate lunch. Then whole saliva was collected by draining in a single session until 7.5 min had elapsed or until the volume reached 20 mL, whichever came first. Saliva was collected either under an unstimulated (resting) condition (R) or a stimulated condition by having subjects chew a 5-g piece of paraffin wax for 5 min immediately before collection (S). Saliva was maintained on ice and centrifuged within 1 hr of collection at 12,000 g for 30 min to remove cellular and other debris. The samples were immediately either subjected to 70% ethanol precipitation of proteins or to measurements for sulfhydryl residues.

2.2. Measurements of Sulfhydryl Residues. To estimate the concentration of sulfhydryl groups in saliva, the dithioni-trobenzoic acid (DTNB) assay method was used as reported [15, 16] with L-cysteine as a standard. Fifty μL of stock DTNB solution (10 mM in ethanol) was added to 1 mL of a solution containing 250 μL of fresh saliva in a 0.25 mM Tris-HCl buffer at pH 8.3 in the presence or absence of L-ascorbic acid 2-phosphate (Asc2P) at 125 μM. The sample was left at room temperature ($22°C$) to allow maximum color development, which was stable for at least 24 h. Absorption of the assay at 412 nm was determined before and after incubation, and the baseline values were subtracted.

2.3. Protein Preparation, SDS-PAGE, and Staining. Seven volumes of ice-cold ethanol were mixed with 3 volumes of saliva, kept on ice for up to 1 hr, and centrifuged to collect the pellet, which was dried under vacuum, immediately weighed and dissolved in SDS sample buffer (0.3 M Tris-HCl buffer, pH 6.8, 6.25% SDS, 25% glycerol, and 0.1% bromophenol blue) containing 2 M urea at 10 mg/mL. Aliquots were heated at $94°C$ for 10 min in the presence or absence of 50 mM DTT. We purchased chemicals from Wako Pure Chemical Industries, Ltd., and SDS from BDH Chemicals Ltd. (London, UK). Five-μL samples were electrophoresed on polyacry-lamide gel with a gradient of either 5–20% or 10–20% and with Precision Plus Protein Standard Kaleidoscope (Bio-Rad Laboratories, Hercules, CA) as a standard. Gels were fixed

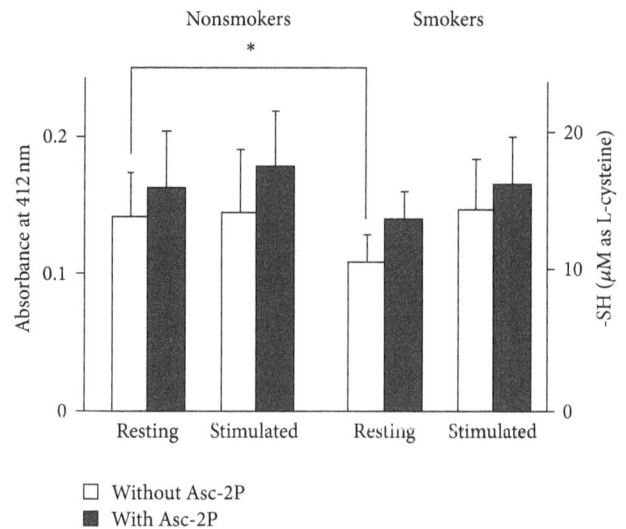

FIGURE 1: Sulfhydryl residues in the saliva of smokers and non-smokers collected under stimulated and unstimulated (resting) conditions were compared separately with or without Asc-2P. Asterisks represent significant differences ($P < 0.05$).

(50% methanol and 10% acetic acid), stained with Stains-All (Bio-Rad Laboratories, Ltd.) [17], washed overnight with two changes of 25% 2-propanol, and rinsed with two changes of H_2O. After recording scanned images, gels were treated with fixative again to remove the dye and restained for protein with Coomassie Brilliant Blue-R250 (Sigma Japan, Ltd., Tokyo, Japan). We selected four samples at random from each group (Figures 3 and 4).

2.4. Western Blot Analysis. After the SDS-PAGE, proteins were electrotransferred and incubated for 1 h with pri-mary antibodies [18]. Goat polyclonal antibody for human serum albumin at 1 : 10,000 (ab19183, Abcam PLC, Cam-bridge, UK), rabbit monoclonal antibody for human α-amylase at 1 : 24,000 (3796S, Cell Signaling Technology, Inc., Danvers, MA), and mouse polyclonal antibody for human MG1 mucin (MUC5B gene product) at 1 : 500 (H00727897-A01, Abnova Corporation, Taipei, Taiwan) were used. Images recorded with LAS-3000 (Fuji Photo Film Co. Ltd., Kaisei-Machi, Japan) were digitized with ImageJ software (http://rsb.info.nih.gov/ij/).

2.5. Detection of Disulfide Bonds in Saliva Protein by Confocal Laser Raman Microspectroscopy. A portion of the wet pellet described in Section 2.3 above was placed between two glass slides before drying as a thin film in a container filled with N_2 gas. A Nicolet Almega XR Dispersive Raman microscope system (not a transmission type as FT-IR) equipped with the OMNIC Atlμs imaging software program (Thermo Fisher Scientific, Inc., MA) and a high-brightness, low-intensity laser operating at 780 nm was used. Intensities of the amide I peak at 1740–1550 cm^{-1} [19], the S-S stretch at 600–470 cm^{-1} [20], and SH at 2,480–2,620 cm^{-1} [21] were measured.

Multimolecular Salivary Mucin Complex Is Altered in Saliva of Cigarette Smokers: Detection of Disulfide Bridges by
Raman Spectroscopy

147

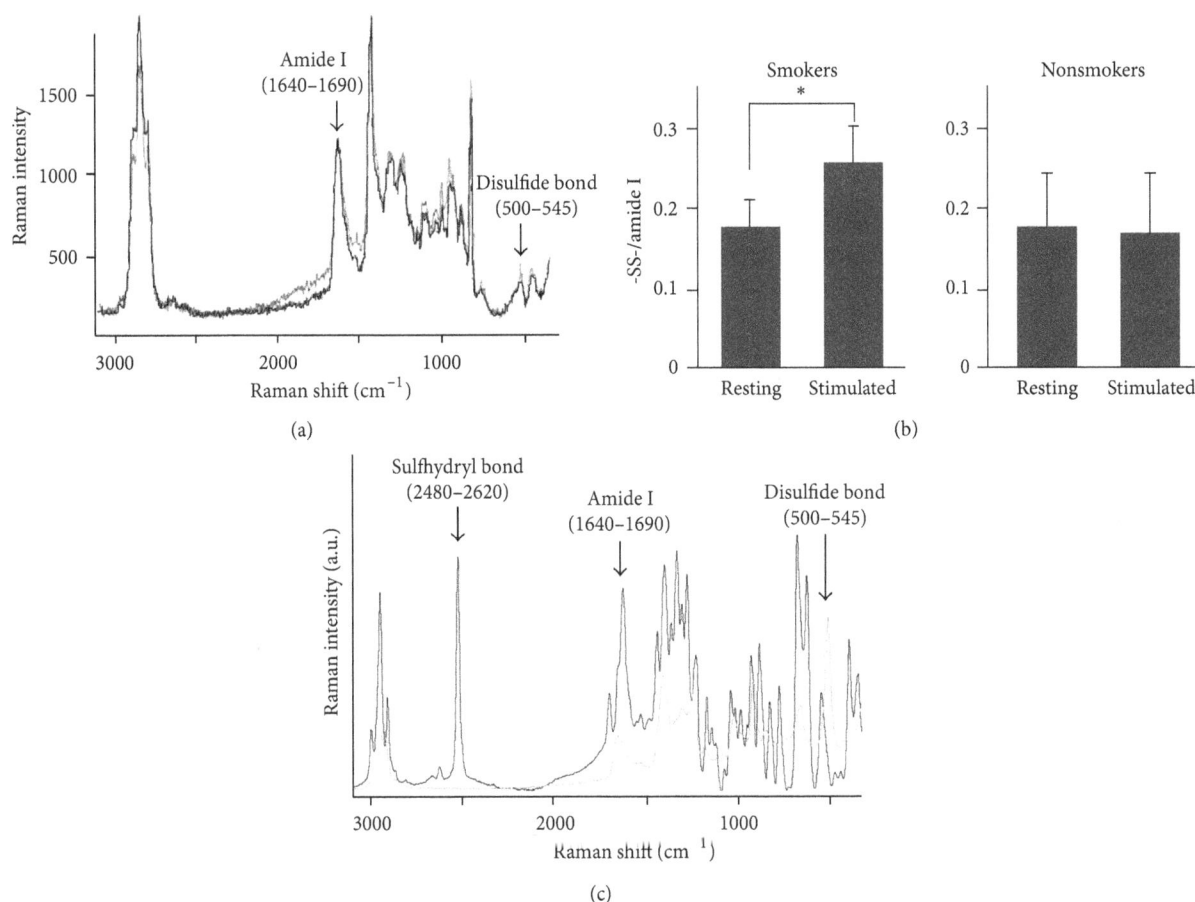

FIGURE 2: Raman spectra of S-S stretch and amide I derived from disulfide bonds of saliva proteins (a) and the areal ratio of disulfide residues per amide I (b). The black line and grey line correspond to stimulated saliva and unstimulated (resting) saliva, respectively. (c) Spectra similar to that of (a) of crystalline glutathione, GSH and GSSG forms, are presented with black and grey lines, respectively.

2.6. Statistical Analysis. To test the statistical significance of all measurements, we used Fisher's exact probability test for the smokers and for the nonsmokers. We judged a difference to be statistically significant when $P < 0.05$.

3. Results

3.1. Sulfhydryl Content of Salivary Proteins. The DTNB assay showed that the content of sulfhydryl residues in the saliva of nonsmokers and stimulated saliva (S) was greater than that in the saliva of smokers and unstimulated saliva (R), respectively (Figure 1). Among untreated saliva, smokers' unstimulated (resting) saliva gave significantly lower values than that of nonsmokers. The increments by reduction with Asc2P were not significantly different from each other (data not shown).

From the Raman profiles (Figure 2(a)), provided the intensity of the S-S stretch bands divided by the corresponding amide I, -SS-/protein ratio (Figure 2(b)), which was significantly higher in the smokers' stimulated saliva smokers' saliva than in all others (at $P < 0.005$ by t-test). The spectrum of glutathione crystals showed in Figure 2(c) a conspicuous peak at around $2,500 \, \text{cm}^{-1}$ for GSH and at $510 \, \text{cm}^{-1}$ for GSSG. If such GSH is weighed and dissolved

in water in the absence of Asc2P, about 90% of the expected absorbance at 412 nm was attained by the DTNB method. By leaving the GSH crystals in the air, both the $2,500\text{-cm}^{-1}$ peak in the Raman spectrum and the 412-nm absorbance were significantly reduced. Peaks at $2,500 \, \text{cm}^{-1}$ and $510 \, \text{cm}^{-1}$ can be used as a measure of oxidation.

3.2. Identification of Oxidized Salivary Proteins . All the samples were analyzed by SDS-PAGE either with or without reducing agent. Figure 3 shows the results of 4 smokers and 4 nonsmokers. Among the blue bands of acidic glycoproteins stained with Stains-All (Figure 3(a)), the MG2 band appeared around 150 kDa regardless of reduction or smoking background. On the other hand, the MG1 band of nonsmokers was apparent above 250 kDa after reduction with DTT. The MG1 band of smokers both in (S) and (R) saliva was much less significant. Instead, an alternative smaller blue band of about 20 to 25 kDa was distinct only in saliva of smokers. In samples of some smokers, staining of another distinct blue band was intense around 45 kDa. A common feature of the two major salivary mucins, MG1 and MG2, is that the intensity of mucin bands derived from the unstimulated saliva is almost always much higher than that from stimulated

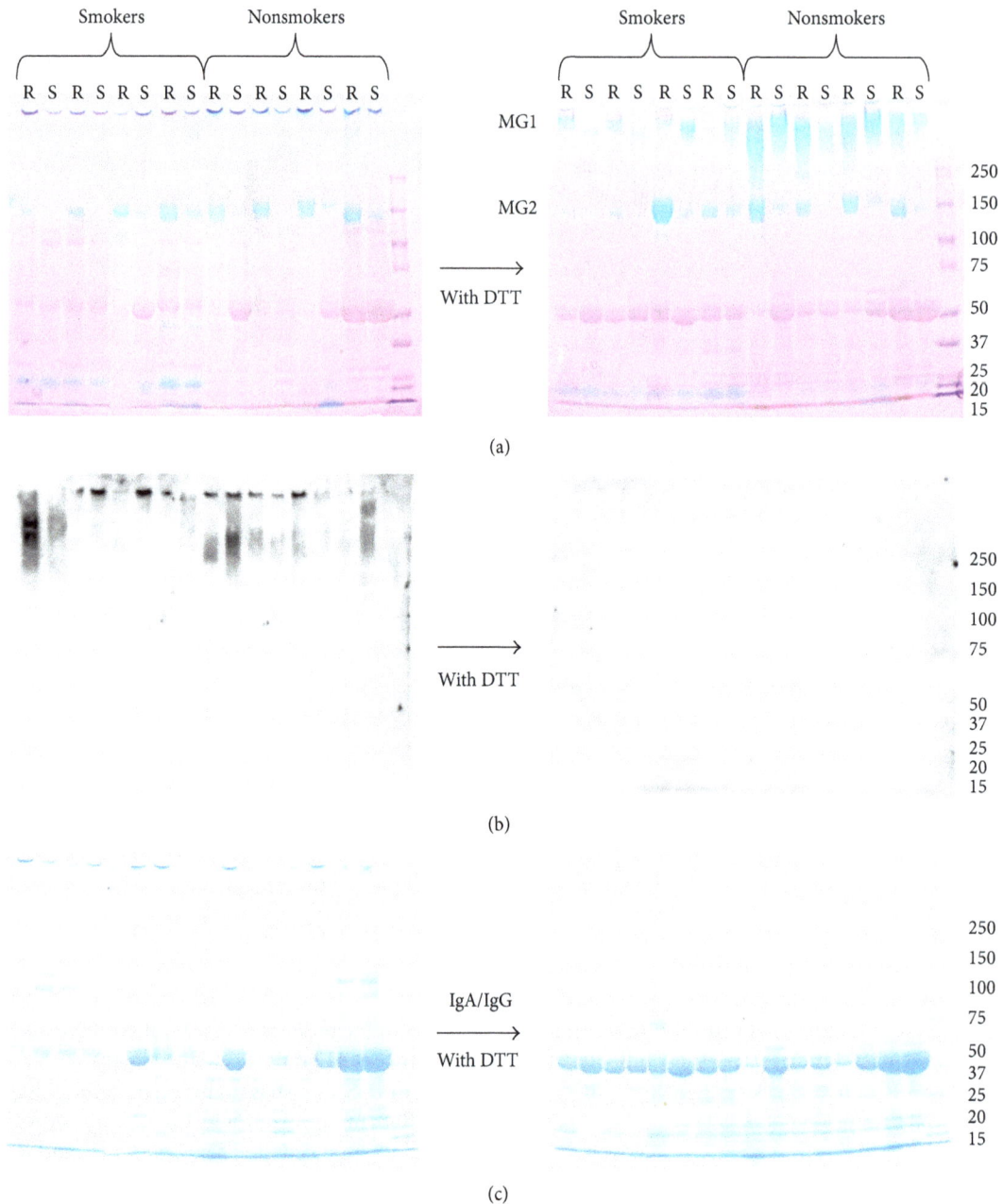

FIGURE 3: SDS-PAGE profiles of saliva proteins from smokers and nonsmokers on a 4–20% gradient gel visualized by Stains-All staining (a), Western blot with MUC5B antibody (b), and CBB protein staining (c). Portions from the same samples were treated with (right panels) and without (left panels) reducing agent DTT. Sizes of standard molecules run in the last lane are shown to the right of the gel. Saliva samples collected under unstimulated (resting) and stimulated conditions are shown in lanes R and S, respectively.

saliva. This also holds true with regard to the blue staining of the nonreduced samples, too large to be included in the gel. Commercially available MUC5B antibody detected MG1 of smaller size only without DTT (Figure 3(b)). Blue-staining MG1 with sulfate residues was not detectable. CBB staining, on the other hand, showed another distinct protein band of ~50 kDa (Figure 3(c)).

While the intensity of the protein band of nonsmokers was comparable with and without DTT, it increased by reduction of the saliva samples of smokers. In general, more proteins, that is, stronger bands, were found in stimulated saliva, a result opposite to that of mucins. To further confirm the aforementioned results of potential oxidation and binding to other proteins in saliva, two major salivary proteins, albumin and amylase, were characterized by Western blotting with and without DTT reduction (Figure 4). In Figure 4(a), amylase antibody is bound to the 50 kDa band, which is seen in Figures 3(a) and 3(c) together with additional bands. Without reduction, antigenic reaction was seen in the area where immunoglobulins appeared. While the staining

Multimolecular Salivary Mucin Complex Is Altered in Saliva of Cigarette Smokers: Detection of Disulfide Bridges by Raman Spectroscopy

149

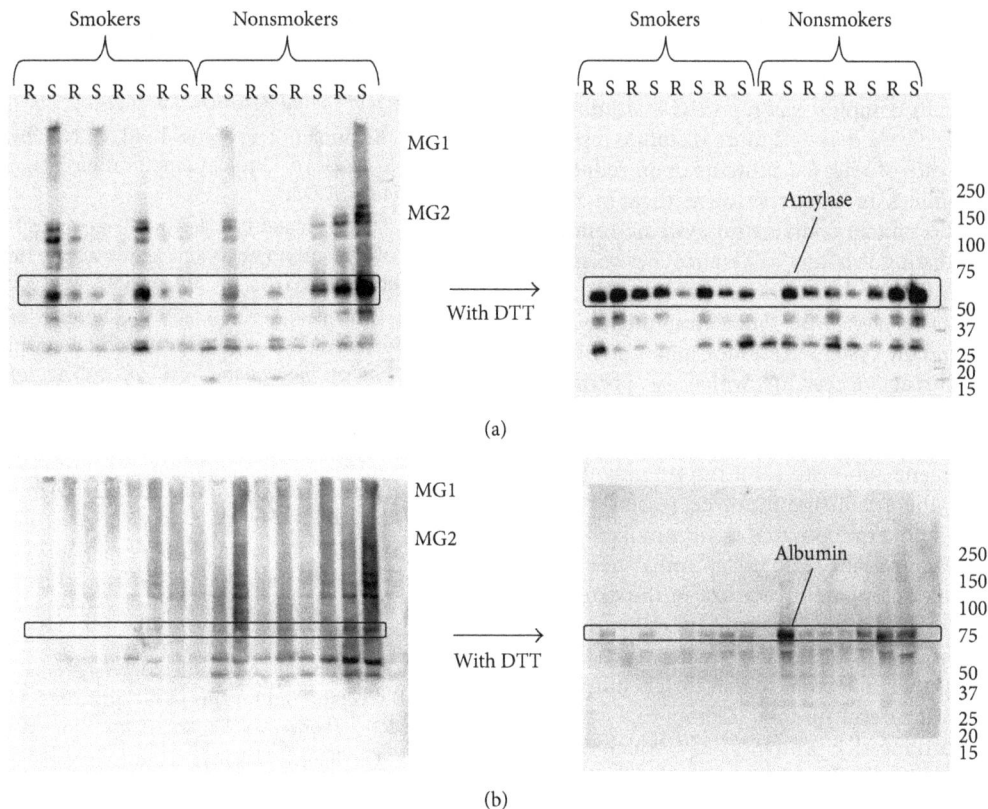

FIGURE 4: Western blot profiles detected with antibodies against (a) salivary amylase (top panels) and (b) serum albumin (bottom panels). Samples were treated either with (right panels) or without (left panels) reducing agent DTT.

intensity of amylase bands in a few samples showed no increase by reduction with DTT, increased staining was seen in most samples. The increase in smokers was more significant than that of nonsmokers. There was no major difference, however, between the results of smokers and nonsmokers. On the other hand, specific albumin staining after reduction was distinct from that without reduction (Figure 4(b)). Without reduction, the albumin band around 67 kDa was not apparent. Instead, intense smears including the areas of MG1 were found. Also, the overall staining was stronger in the samples from nonsmokers regardless of reduction. Therefore, we compared the increments of the specific albumin band intensity by the inclusion of DTT to that of the band after reduction (see Figure 4(b), right panel). The intensities are 0.73 ± 0.21 for smokers and 0.48 ± 0.35 for nonsmokers. The P value of 0.038 showed that the difference was significant.

4. Discussion and Conclusions

Although saliva contains antioxidant defense systems to counteract the toxic effect of radical species formed by superoxide dismutase (SOD), glutathione peroxidase (GSH-Px), and other enzymes, cigarette smoke contains oxidants of other types as well, including oxygen-free radicals and volatile aldehydes, which damage biomolecules [22–25]. Mucins are susceptible to attack by reactive oxygen species

during which terminal sugars are lost and both protein and sugar moieties are fragmented [9]. We studied components of saliva including mucins and showed that there is less multimolecular mucin complex in the saliva of smokers. Although blue staining of MG1 band decreases, the levels of MG2 and MUC5B antibody-reacting materials under nonreducing conditions are not that different between smokers and nonsmokers (Figures 3(a) and 3(b)). Therefore, acidic residues such as sulfated sugars may be specifically decreased in smokers' MG1 by degradation resulting in much less MG1 in saliva of smokers, especially in stimulated saliva. Lower MG1/MG2 levels in the stimulated saliva samples can be explained by increased contributions of parotid saliva, which is prevalent under that condition. Although we lack information regarding saliva flow rate, no significant correlations were previously found between mucin levels of stimulated saliva and age or flow rate [4]. Of note, a significant amount of smaller acidic glycoprotein bands appeared only in saliva of smokers. Also mainly in smokers' saliva, CBB staining of a distinct protein band of 50 kDa, which turned out to be amylase, was intensified together with several other protein bands after reduction with DTT. In addition, the albumin band detected by antibody binding showed not only that the albumin content decreased significantly in the saliva of smokers but also that the ratio of albumin bound to other proteins including MG1 by the disulfide bridges is significantly higher in the smokers' saliva. By using confocal laser Raman microspectroscopy, we confirmed the increased S-S stretch band derived from protein disulfide bonds in the stimulated saliva (S) of smokers. While significantly altered multimolecular mucin complex was reported in stimulated saliva previously [5, 6], we detected more disulfide residues as well as a higher rate of released proteins upon reduction by DTT in the samples of smokers' saliva than in those of nonsmokers'. MG2 mucin contains no cysteine residues, and we did not distinguish any difference between the characteristics of MG2 in the saliva of smokers and that of nonsmokers. It was also true with immunoglobulins. MG1 with the abundant cysteine residues may function as the surface-exposed reactive cysteine, which was previously reported [8, 10]. Our result agrees with that of Levine et al. who reported that DTT increases binding of MG1 alone to 1-anilino-8-naphthalene sulfonate [26]. Importantly, Brock et al. reported that there is a significant decrease of GSH-Px activity in male smokers compared to nonsmokers ($P < 0.05$) resulting in an oxidant/antioxidant imbalance [24]. The reason why such an apparent increase in the extent of protein oxidation was detectable in smokers' saliva in our study, we believe, is that we generally treated each saliva sample immediately after collection so that the chemical and enzymatic modifications are minimal. Complex formation with statherin and proline-rich proteins, PRPs, which do not contain cysteine, reported by Iontcheva et al. [6] is yet to be determined. In our future studies, we will address the actual degradation and oxidation mechanisms of salivary mucin MG1 by the radical species in the saliva of female smokers.

Acknowledgments

The authors greatly appreciate the technical assistance by Y. Funayama. This study was supported by JSPS Grants-in-Aid for Scientific Research no. 21592436 to Y. Mvrata None of the authors have any conflict of interests to state.

References

[1] I. Kondakova, E. A. Lissi, and M. Pizarro, "Total reactive antioxidant potential in human saliva of smokers and non-smokers," *Biochemistry and Molecular Biology International*, vol. 47, no. 6, pp. 911–920, 1999.

[2] B. Zappacosta, S. Persichilli, P. De Sole, A. Mordente, and B. Giardina, "Effect of smoking one cigarette on antioxidant metabolites in the saliva of healthy smokers," *Archives of Oral Biology*, vol. 44, no. 6, pp. 485–488, 1999.

[3] B. L. Slomiany, V. L. N. Murty, J. Piotrowski, and A. Slomiany, "Salivary mucins in oral mucosal defense," *General Pharmacology*, vol. 27, no. 5, pp. 761–771, 1996.

[4] S. A. Rayment, B. Liu, G. D. Offner, F. G. Oppenheim, and R. F. Troxler, "Immunoquantification of human salivary mucins MG1 and MG2 in stimulated whole saliva: factors influencing mucin levels," *Journal of Dental Research*, vol. 79, no. 10, pp. 1765–1772, 2000.

[5] P. C. Denny, P. A. Denny, D. K. Klauser, S. H. Hong, M. Navazesh, and L. A. Tabak, "Age-related changes in mucins from human whole saliva," *Journal of Dental Research*, vol. 70, no. 10, pp. 1320–1327, 1991.

[6] I. Iontcheva, F. G. Oppenheim, and R. F. Troxler, "Human salivary mucin MG1 selectively forms heterotypic complexes with amylase, proline-rich proteins, statherin, and histatins," *Journal of Dental Research*, vol. 76, no. 3, pp. 734–743, 1997.

[7] E. R. Stadtman, P. E. Starke-Reed, C. N. Oliver, J. M. Carney, and R. A. Floyd, "Protein modification in aging," *EXS*, vol. 62, pp. 64–72, 1992.

[8] E. R. Stadtman and R. L. Levine, "Free radical-mediated oxidation of free amino acids and amino acid residues in proteins," *Amino Acids*, vol. 25, no. 3-4, pp. 207–218, 2003.

[9] I. A. Brownlee, J. Knight, P. W. Dettmar, and J. P. Pearson, "Action of reactive oxygen species on colonic mucus secretions," *Free Radical Biology and Medicine*, vol. 43, no. 5, pp. 800–808, 2007.

[10] S. M. Marino, Y. Li, D. E. Fomenko, N. Agisheva, R. L. Cerny, and V. N. Gladyshev, "Characterization of surface-exposed reactive cysteine residues in saccharomyces cerevisiae," *Biochemistry*, vol. 49, no. 35, pp. 7709–7721, 2010.

[11] E. Gensch, M. Gallup, A. Sucher et al., "Tobacco smoke control of mucin production in lung cells requires oxygen radicals AP-1 and JNK," *Journal of Biological Chemistry*, vol. 279, no. 37, pp. 39085–39093, 2004.

[12] R. A. O'Donnell, A. Richter, J. Ward et al., "Expression of ErbB receptor and mucins in the airways of long term current smokers," *Thorax*, vol. 59, no. 12, pp. 1032–1040, 2004.

[13] M. X. G. Shao, T. Nakanaga, and J. A. Nadel, "Cigarette smoke induces MUC5AC mucin overproduction via tumor necrosis factor-α-converting enzyme in human airway epithelial (NCI-H292) cells," *American Journal of Physiology*, vol. 287, no. 2, pp. L420–L427, 2004.

[14] S. M. Casalino-Matsuda, M. E. Monzon, A. J. Day, and R. M. Forteza, "Hyaluronan fragments/CD44 mediate oxidative stress-induced MUC5B up-regulation in airway epithelium," *American Journal of Respiratory Cell and Molecular Biology*, vol. 40, no. 3, pp. 277–285, 2009.

[15] R. M. Nagler, I. Klein, N. Zarzhevsky, N. Drigues, and A. Z. Reznick, "Characterization of the differentiated antioxidant profile of human saliva," *Free Radical Biology and Medicine*, vol. 32, no. 3, pp. 268–277, 2002.

[16] Y. Suzuki, V. Lyall, T. U. L. Biber, and G. D. Ford, "A modified technique for the measurement of sulfhydryl groups oxidized by reactive oxygen intermediates," *Free Radical Biology and Medicine*, vol. 9, no. 6, pp. 479–484, 1990.

[17] P. A. Denny, P. C. Denny, and K. Jenkins, "Purification and biochemical characterization of a mouse submandibular sialomucin," *Carbohydrate Research*, vol. 87, no. 2, pp. 265–274, 1980.

[18] H. Watabe, T. Furuhama, N. Tani-Ishii, and Y. Mikuni-Takagaki, "Mechanotransduction activates alphabeta integrin and PI3K/Akt signaling pathways in mandibular osteoblasts," *Experimental Cell Research*, vol. 317, no. 18, pp. 2642–2649, 2011.

[19] Y. Matsumoto, Y. Mikuni-Takagaki, Y. Kozai et al., "Prior treatment with vitamin K2 significantly improves the efficacy of risedronate," *Osteoporosis International*, vol. 20, no. 11, pp. 1863–1872, 2009.

[20] M. C. Chen, R. C. Lord, and R. Mendelsohn, "Laser-excited Raman spectroscopy of biomolecules—V. Conformational changes associated with the chemical denaturation of lysozyme," *Journal of the American Chemical Society*, vol. 96, no. 10, pp. 3038–3042, 1974.

Multimolecular Salivary Mucin Complex Is Altered in Saliva of Cigarette Smokers: Detection of Disulfide Bridges by
Raman Spectroscopy

151

[21] S. W. Raso, P. L. Clark, C. Haase-Pettingell, J. King, and G. J. Thomas Jr., "Distinct cysteine sulfhydryl environments detected by analysis of raman S-H markers of Cys → Ser mutant proteins," *Journal of Molecular Biology*, vol. 307, no. 3, pp. 899–911, 2001.

[22] R. Agnihotri, P. Pandurang, S. U. Kamath et al., "Association of cigarette smoking with superoxide dismutase enzyme levels in subjected with chonic periodontitis," *Journal of Periodontology*, vol. 80, no. 4, pp. 657–662, 2009.

[23] M. T. Ashby, "Inorganic chemistry of defensive peroxidases in the human oral cavity," *Journal of Dental Research*, vol. 87, no. 10, pp. 900–914, 2008.

[24] G. R. Brock, C. J. Butterworth, J. B. Matthews, and I. L. C. Chapple, "Local and systemic total antioxidant capacity in periodontitis and health," *Journal of Clinical Periodontology*, vol. 31, no. 7, pp. 515–521, 2004.

[25] M. R. Giuca, E. Giuggioli, M. R. Metelli et al., "Effects of cigarette smoke on salivary superoxide dismutase and glutathione peroxidase activity," *Journal of Biological Regulators and Homeostatic Agents*, vol. 24, no. 3, pp. 359–366, 2010.

[26] M. J. Levine, M. S. Reddy, L. A. Tabak et al., "Structural aspects of salivary glycoproteins," *Journal of Dental Research*, vol. 66, no. 2, pp. 436–441, 1987.

The Serine Protease Plasmin Triggers Expression of the CC-Chemokine Ligand 20 in Dendritic Cells *via* Akt/NF-κB-Dependent Pathways

Xuehua Li, Tatiana Syrovets, and Thomas Simmet

Institute of Pharmacology of Natural Products and Clinical Pharmacology, Universitat Ulm, Helmholtzstraße 20, 89081 Ulm, Germany

Correspondence should be addressed to Thomas Simmet, thomas.simmet@uni-ulm.de

Academic Editor: Lindsey A. Miles

The number of dendritic cells is increased in advanced atherosclerotic lesions. In addition, plasmin, which might stimulate dendritic cells, is generated in atherosclerotic lesions. Here, we investigated cytokine and chemokine induction by plasmin in human dendritic cells. In human atherosclerotic vessel sections, plasmin colocalized with dendritic cells and the CC-chemokine ligand 20 (CCL20, MIP-3α), which is important for homing of lymphocytes and dendritic cells to sites of inflammation. Stimulation of human dendritic cells with plasmin, but not with catalytically inactivated plasmin, induced transcriptional regulation of CCL20. By contrast, proinflammatory cytokines such as TNF-α, IL-1α, and IL-1β were not induced. The plasmin-mediated CCL20 expression was preceded by activation of Akt and MAP kinases followed by activation of the transcription factor NF-κB as shown by phosphorylation of its inhibitor IκBα, by nuclear localization of p65, its phosphorylation, and binding to NF-κB consensus sequences. The plasmin-induced CCL20 expression was dependent on Akt- and ERK1/2-mediated phosphorylation of IκBα on Ser32/36 and of p65 on Ser276, whereas p38 MAPK appeared to be dispensable. Thus, plasmin triggers release of the chemokine CCL20 from dendritic cells, which might facilitate accumulation of CCR6+ immune cells in areas of plasmin generation such as inflamed tissues including atherosclerotic lesions.

1. Introduction

The serine protease plasmin is mainly recognized for its central role in fibrinolysis. In addition, however, plasmin may also be generated at inflammatory sites from ubiquitously distributed plasminogen [1]. Indeed, generation of plasmin has been shown in a number of chronic inflammatory conditions including arthritis and atherosclerosis [1]. Specifically in unstable atherosclerotic lesions, plasminogen and plasmin appear to be associated with clinical complications [2–5]. Local plasmin generation at sites of inflammation might aggravate inflammatory processes by triggering proinflammatory effects. *In vitro*, plasmin is capable of stimulating lipid mediator release and of eliciting chemotaxis of human monocytes [6, 7]. In addition, plasmin is a potent inducer of proinflammatory cytokines in human macrophages [8] and monocytes, where it also causes expression of procoagulant tissue factor [9].

Dendritic cells play a crucial role in innate and adaptive immune responses [10]. Dendritic cells are crucial for immune diseases including rheumatoid arthritis, where they accumulate in synovial tissue and activate T cells [11]. Likewise, the number of dendritic cells is strongly increased in advanced atherosclerotic lesions, where they colocalize with T cells [12–14]. Dendritic cells induce differentiation of T cells into different T-cell subsets through direct interaction with T-cell receptors and the release of cytokines. Dendritic cells are heterogeneous in their origin and their ability to activate either tolerogenic or immunogenic T-cell responses [15]. A distinct dendritic cell type is monocyte-derived dendritic cells, which arise in the course of inflammation [15, 16]. We have recently shown that plasmin is a potent chemoattractant for immature dendritic cells, and that it activates dendritic cells to produce interleukin-12 (IL-12) and to promote polarization of CD4+ T cells towards the interferon-γ (IFN-γ-) producing, proinflammatory Th1 phenotype [17].

Chemokines orchestrate the homing of lymphocytes and dendritic cells to lymphoid tissues as well as the recruitment

The Serine Protease Plasmin Triggers Expression of the CC-Chemokine Ligand 20 in Dendritic
Cells via Akt/NF-κB-Dependent Pathways

153

of leukocytes to sites of infection or tissue damage [18]. As a result, chemokines play crucial roles in the pathogenesis of diseases that are characterized by inflammatory cell accumulation, such as atherosclerosis [12, 14, 18, 19].

CCL20 (also known as liver- and activation-regulated chemokine, LARC, or macrophage inflammatory protein-3α, MIP-3α) is a CC-type chemokine, which activates chemokine receptor CCR6 and therefore plays an important role in homing CCR6+ lymphocytes and dendritic cells into secondary lymphoid organs and to sites of inflammation [20]. Memory T lymphocytes, naïve and memory B cells, Langerhans cells, and subsets of immature dendritic cells all express CCR6 [20] and migrate to sites of CCL20 expression, for example, in atherosclerosis [20], inflammatory bowel disease [21], arthritis [19], chronic obstructive pulmonary disease [22], psoriasis [23], and tumor tissues [24]. Accordingly, an important role for CCL20 has been postulated in atherosclerosis, skin, and mucosal immunity, in rheumatoid arthritis, and in cancer [20].

Immunohistochemical colocalization of immature dendritic cells and CCL20 indicates a link between the accumulation of immature dendritic cells and the local production of CCL20 by epithelial and tumor cells [20, 22]. Indeed, subsets of immature dendritic cells express CCR6 and are able to migrate towards CCL20 [25]. However, dendritic cells might also be a source of CCL20 on their own. Thus, CCL20 secretion can be induced in dendritic cells either by stimulation with LPS [26] or extracellular nucleotides [27]

Here, we investigated whether plasmin might affect the expression of cytokines by human monocyte-derived dendritic cells and whether this might occur in human atherosclerotic lesions.

2. Materials and Methods

2.1. Materials.
Antibodies used are phospho-IκBα, phospho-ERK1/2, phospho-p38, phospho-Akt (Ser473), phospho-p65 Ser536, and phospho-p65 Ser276-Cell Signaling Technology (Danvers, MA); CCL20-R&D Systems (Minneapolis, MN); p65-Santa Cruz Biotechnology (Santa Cruz, CA); actin-Chemicon International (Chemicon, Temecula, CA); HLA-DR, CD80, CD86, and CD1a-BD Biosciences (Heidelberg, Germany); S100 [28]-AbD SeroTec (Oxford, UK). Antibodies against Phycoerythrin- (PE-) conjugated donkey anti-mouse, anti-rabbit, and anti-goat F(ab')₂ were from Dianova (Hamburg, Germany). The catalytic inhibitor of plasmin, D-Val-Phe-Lys chloromethyl ketone (VPLCK), the kinase inhibitors SB203580, U0126, Akt inhibitor VIII, and controls to the inhibitors SB202474 and U0124 were from Calbiochem (San Diego, CA). GM-CSF was from Berlex (Bayer HealthCare). Human recombinant IL-4, proteome profiler array, and CCL20 ELISA were from R&D Systems. Endotoxic lipopolysaccharide (LPS; *Escherichia coli* serotype 055 : B5) and Histopaque 1077 were from Sigma (St. Louis, MO). Purified human plasmin (lot no. 2008-01L) was from Athens Research & Technology (Athens, GA). The plasmin lot used in this study contained no detectable LPS contamination as measured by the Limulus amoebocyte lysate

assay (Sigma, sensitivity 0.05–0.1 EU/mL). The plasmin substrate S-2251 (H-D-valyl-leucyl-L-lysine-*P*-nitroanilide dihydrochloride) was supplied by Diapharma Group Inc. (Columbus, OH). Catalytically inactivated plasmin (VPLCK plasmin) was prepared by incubation of 4 mg/mL human plasmin with 200 μM VPLCK for 30 min at 37°C. Aliquots of the mixture were used to assure the complete loss of any residual proteolytic activity of plasmin using the chromogenic substrate S-2251. VPLCK was separated from VPLCK plasmin by NAP-5 Sephadex G25 chromatography (GE Healthcare), and the concentration of VPLCK plasmin was determined by the BCA protein assay (Pierce). The NF-κB/p65 transcription factor ELISA was from Active Motif (Carlsbad, CA).

2.2. Methods

2.2.1. Immunohistochemical Staining.
Sections of surgical specimens from human abdominal aorta from 3 patients were stained with antibodies recognizing plasmin, CCL20, or the DC marker S100, which is exclusively expressed by dendritic cells in the arterial wall [28, 29]. For immunohistochemical double staining, HRP- and AP-conjugated secondary antibodies were visualized by DAB, Fast Red, or AEC substrates (PicTure kit, Invitrogen). The images were digitally recorded with an Axiophot microscope and a Sony MC-3249 CCD camera using Visupac 22.1 software (Carl Zeiss, Göttingen, Germany) [17, 30]. The study was approved by the University Ethics Review Board (approval reference number 114/10) and complied with the principles of the Declaration of Helsinki.

2.2.2. Cell Preparation and Differentiation.
Immature dendritic cells were differentiated from human monocytes obtained from buffy coats with 1000 U/mL GM-CSF and 25 ng/mL IL-4 for 6 days in RPMI 1640 containing 10% FCS [30]. The differentiation was confirmed by flow cytometric analysis of HLA-DR, CD80, CD86, and CD1a using a FACScan (BD Biosciences, Franklin Lakes, NJ). Dendritic cells were used on day 6 of differentiation. The cells (1 × 10⁶ cells/mL) were kept for 12 h in AIM-V medium (Invitrogen, Carlsbad, CA) without cytokines and FCS before treatment with plasmin. In some experiments, the dendritic cells were treated with catalytically inactivated plasmin, equivalent to 0.143 CTA U/mL of native plasmin. Catalytically inactivated plasmin (VPLCK plasmin) was prepared as described [31].

2.2.3. Analysis of mRNA Expression.
mRNA was isolated from dendritic cells stimulated with plasmin or equivalent amounts of active site-blocked plasmin (VPLCK plasmin) [9, 30] and analyzed by RT-PCR and quantitative real-time PCR. Primer pairs for CCL20 were sense 5′-GACATAGCCCAAGAACAGAAA-3′, antisense 5′-TAATTGGACAAGTCCAGTGAGG-3′ [32]; GAPDH served as control [33]. The identity of the PCR products was confirmed by direct sequencing (Abi Prism 310, Applied

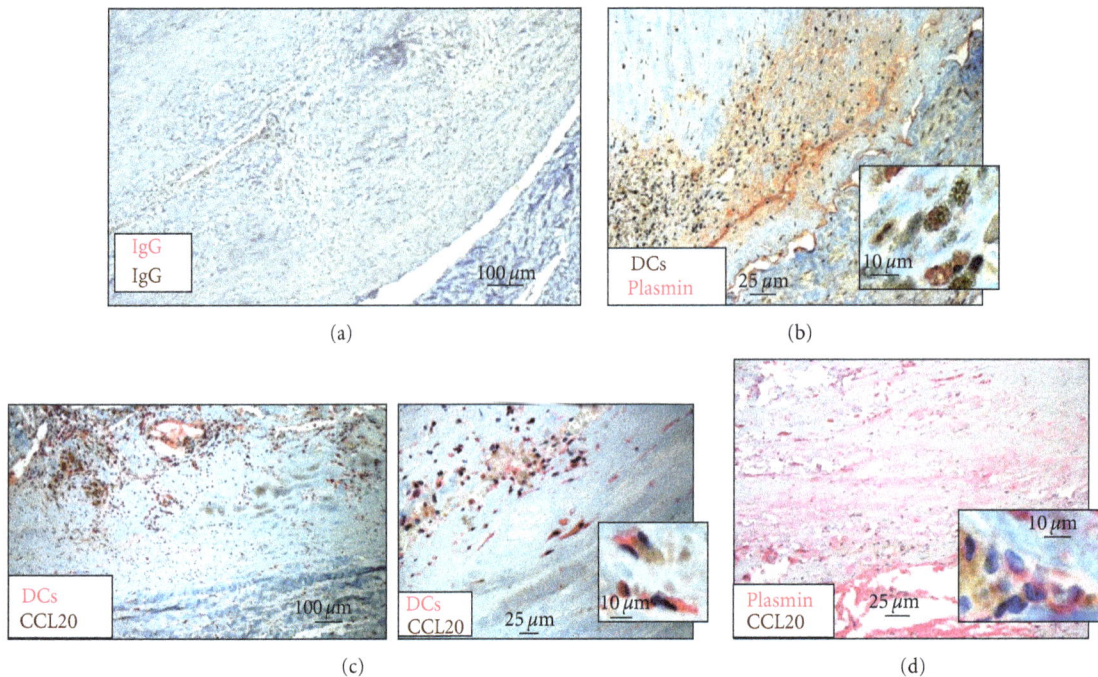

FIGURE 1: CCL20 is present in human atherosclerotic lesions, where it colocalizes with plasmin and dendritic cells. (a) Negative control. Sections of human atherosclerotic abdominal aorta specimens were stained with control antibodies and visualized with double immunostaining using AP- and HRP-conjugated secondary antibodies and FastRed (pink) and DAB (brown) substrates. Original magnification is ×100. (b) Plasmin colocalizes with dendritic cells. Sections of human atherosclerotic abdominal aorta specimens were stained with antibodies against plasmin (AEC, red) and dendritic cell marker S100 (DAB, brown). Original magnifications ×200 and ×400. (c) CCL20 colocalizes with dendritic cells. Sections of human atherosclerotic abdominal aorta specimens were stained with antibodies against CCL20 (DAB, brown) and dendritic cell marker S100 (FastRed, pink). Original magnifications are ×100, ×200, and ×400. (d) Plasmin colocalizes with CCL20. Sections of human atherosclerotic abdominal aorta specimens were stained with antibodies against plasmin (FastRed, pink) and CCL20 (DAB, brown). Original magnifications are ×200 and ×400.

Biosystems, Foster City, CA). Quantitative PCR was performed using real-time PCR system (7300 Real-Time PCR, Applied Biosystems), and the relative gene expression was determined by normalizing to GAPDH using the $\Delta\Delta C_T$ method.

2.2.4. Analysis of Protein Expression. Protein expression was analyzed by western immunoblotting, proteome profiler array, ELISA, and flow cytometry [30, 34]. Dendritic cells were kept in AIM-V medium for 12 h prior to stimulation. For the analysis of phosphorylated IκBα and p65, whole cell lysates were analyzed by western immunoblotting [9]. CCL20 secretion was measured by ELISA (R&D Systems) in supernatants of dendritic cells stimulated for 24 h with plasmin or the positive control LPS (0.5 μg/mL). For flow cytometric analysis, dendritic cells were pretreated with 1 μg/mL brefeldin A (Sigma) for 4 h prior to analysis to prevent release of CCL20 from the cells. Dendritic cells were fixed with paraformaldehyde, permeabilized with 0.5% saponin, stained with antibodies against CCL20 or control IgG and analyzed by FACScan (BD Biosciences). TNF-α, IL-1α, and IL-1β were analyzed by proteome profiler array (R&D Systems) in the supernatants of dendritic cells stimulated with plasmin (0.143 CTA U/mL) for 24 h.

2.2.5. NF-κB ELISA. Activation of transcription factor NF-κB p65/RelA was quantified in nuclear extracts (5 μg) using TransAM ELISA (Active Motif, Carlsbad, CA) [35]. Nuclear extracts were prepared from dendritic cells treated with plasmin (0.143 CTA U/mL) or LPS (0.5 μg/mL) for 60 min [9]. Results are expressed as fold activation compared to the control samples.

2.2.6. Statistical Analysis. Data shown represent mean ± SEM where applicable. Statistical significances were calculated with the Newman-Keuls test. Differences were considered significant for $P < 0.05$.

3. Results

3.1. Plasmin and Dendritic Cells Colocalize with CCL20 in the Human Atherosclerotic Vessel Wall. Immunohistochemical analysis of sections from atherosclerotic tissue specimens obtained from human abdominal aorta confirmed that plasmin is abundant in the atherosclerotic vessel wall, where it colocalizes with clusters of dendritic cells (Figures 1(a) and 1(b)). In addition, these immunohistochemical studies revealed that plasmin and dendritic cells are in close proximity to CCL20 (Figures 1(a), 1(c), and 1(d)) suggesting that dendritic cells might be activated by locally generated

The Serine Protease Plasmin Triggers Expression of the CC-Chemokine Ligand 20 in Dendritic Cells via Akt/NF-κB-Dependent Pathways

155

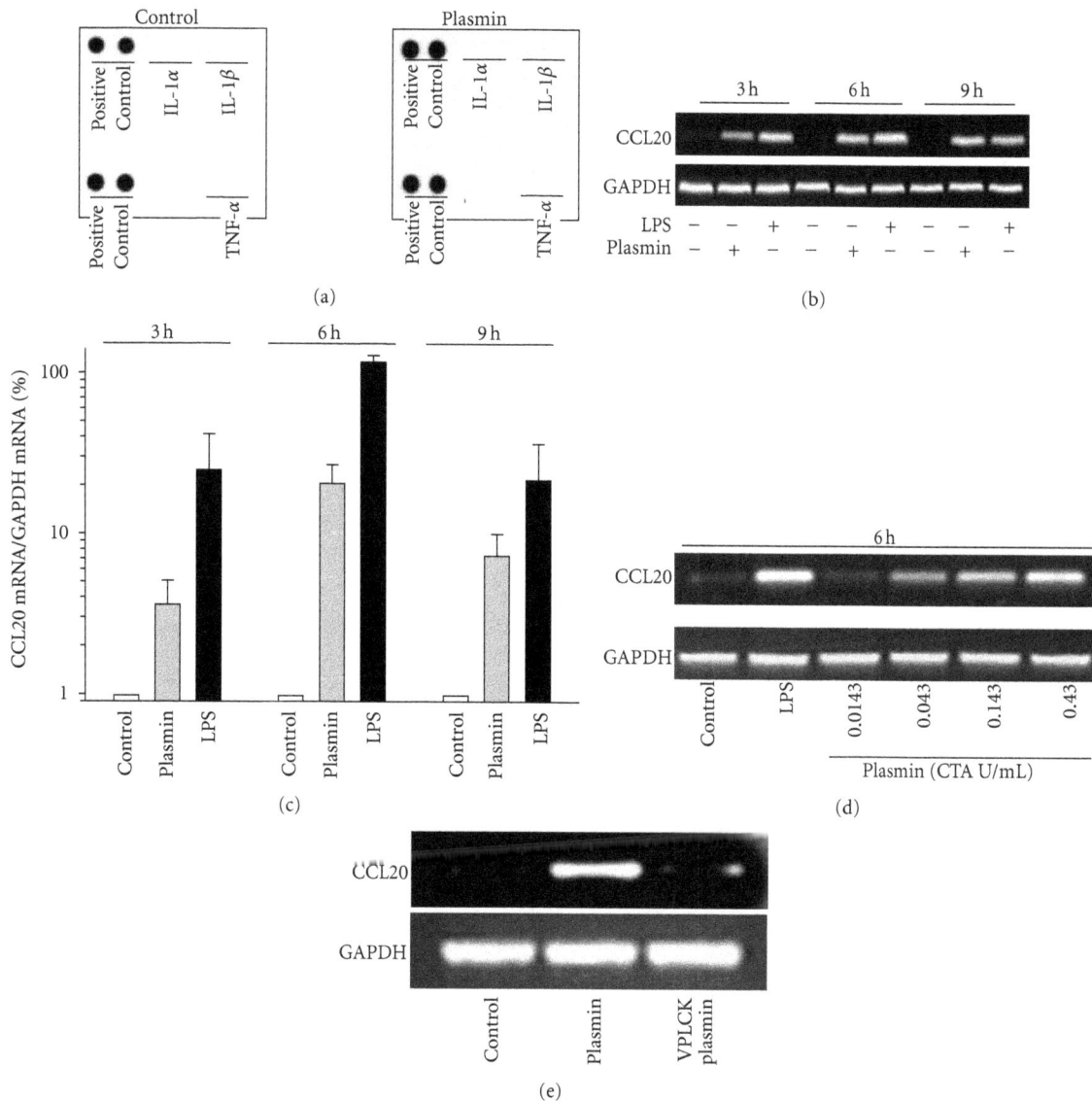

FIGURE 2: Plasmin induces time- and concentration-dependent expression of CCL20 mRNA in dendritic cells. (a) Plasmin does not induce release of the proinflammatory cytokines TNF-α, IL-1α, and IL-1β. At day 6, dendritic cells were left untreated or stimulated with plasmin (0.143 CTA U/mL) for 24 h. Release of cytokines in the culture media was analyzed with the proteome profiler array. (b) Plasmin induces time-dependent expression of CCL20 mRNA in dendritic cells. Dendritic cells were treated with plasmin (0.143 CTA U/mL) or the positive control LPS (0.5 μg/mL) for the indicated time. mRNA was isolated and subjected to RT-PCR (b) and real time qPCR (c) analysis using CCL20-specific primers; GAPDH served as control. Results are mean ± SEM of 3 experiments. (d) Dendritic cells were treated for 6 h with the indicated concentrations of plasmin, and CCL20 mRNA expression was analyzed by RT-PCR. (e) Proteolytic activity of plasmin is required for the induction of CCL20 mRNA expression. Dendritic cells were stimulated either with plasmin (0.143 CTA U/mL) or the equivalent amount of catalytically inactivated plasmin (VPLCK plasmin) for 6 h. CCL20 mRNA expression was analyzed by RT-PCR. All data are representative of at least 3 independent experiments.

plasmin, and that dendritic cells could serve as a source of CCL20.

3.2. Plasmin Induces CCL20 mRNA Expression in Dendritic Cells.
To address the possible generation of cytokines and chemokines by plasmin-activated dendritic cells, we stimulated monocyte-derived dendritic cells with plasmin *in vitro*. Analysis of the supernatants of such cells revealed that in contrast to human monocytes [9] and macrophages [8], dendritic cells do not release proinflammatory cytokines,

such as TNF-α, IL-1α and β (Figure 2(a)), or IL-16 (LCF), nor did they release chemokines such as CXCL10 (IP-10), CXCL11 (I-TAC), CXCL12 (SDF-1), CCL1 (I-309), CCL2 (MCP-1), or CCL5 (RANTES). Control dendritic cells produced CXCL8 (IL-8) and small amounts of CXCL1 (GRO), but the release of these chemotactic cytokines remained unaffected by plasmin treatment (data not shown). However, stimulation of dendritic cells with human plasmin (0.143 CTA U/mL) elicited a time-dependent increase of CCL20 mRNA expression as analyzed by RT-PCR

(Figure 2(b)) and real-time qPCR (Figure 2(c)). The maximum of the CCL20 mRNA expression was observed 6 h after stimulation with either plasmin (0.143 CTA U/mL) or the positive control LPS (0.5 μg/mL). The stimulatory effect of plasmin was concentration dependent with a maximum at 0.143–0.43 CTA U/mL (Figure 2(d)).

Previous studies had suggested that the proteolytic activity of plasmin might be required for cell activation [7, 8, 31, 36]. To test whether this is also true for the plasmin-induced activation of human dendritic cells, we generated catalytically inactivated plasmin (VPLCK plasmin) [9]. In contrast to active plasmin, catalytically inactivated plasmin did not trigger any CCL20 induction in dendritic cells (Figure 2(e)) indicating that the plasmin-mediated dendritic cell activation depends on a proteolytic signaling mechanism.

3.3. Plasmin Induces Release of CCL20 in Dendritic Cells. The transcription of CCL20 mRNA by plasmin was followed by a concentration-dependent release of CCL20 with a maximum at 0.143 CTA U/mL (Figure 3(a)). Similar to the mRNA expression levels, higher plasmin concentrations did not further increase the amount of secreted CCL20. The positive control LPS (0.5 μg/mL) induced release of higher amounts of CCL20 (792.9 ± 129.9 pg/mL, $n = 5$) compared to 0.143 CTA U/mL plasmin (132.6 ± 26.1 pg/mL, $P < 0.01$, $n = 8$); control cells released 35.8 ± 10.4 pg/mL CCL20.

Consistently, flow cytometric analysis of the CCL20-expressing cells revealed that about 41% of the dendritic cells treated with plasmin expressed CCL20 within 24 h after treatment (Figure 3(b)). Thus, plasmin triggers production of chemotactic CCL20 by human dendritic cells.

3.4. Plasmin Elicits Activation of Akt, ERK1/2 and p38 MAP Kinases, and NF-κB Signaling. Expression of cytokines and chemokines is regulated primarily at the level of transcription. The promoter region of CCL20 is known to contain an NF-κB consensus sequence indicating that the expression of CCL20 might be regulated by NF-κB [37]. In addition, it has been previously shown that NF-κB can be activated by Akt-dependent IκBα kinase phosphorylation [38, 39], and Akt mediates an IL-17A-induced expression of CCL20 in human airway epithelial cells [40]. Moreover, ERK1/2 and p38 MAP kinases have been implicated in the regulation of the NF-κB activation via MSK1/2 activation and the phosphorylation of p65 [39, 41].

Taking into account that NF-κB is involved in the expression of various proinflammatory genes including chemokines [42], and that plasmin in turn activates NF-κB in monocytes and macrophages [8, 9, 43], we investigated whether plasmin might activate Akt, MAP kinases, and NF-κB in dendritic cells.

Western immunoblot analysis of plasmin-stimulated dendritic cells indicated that plasmin triggers a rapid phosphorylation of Akt, ERK1/2, and p38 MAP kinases (Figure 4(a)). In addition, the phosphorylation of IκBα was increased with a maximum response at 15–30 min after stimulation (Figure 4(b)) indicating activation of NF-κB.

Phosphorylation of IκBα by IκB kinases is a prerequisite for IκB ubiquitination and degradation required for the release of p65 and other NF-κB subunits, and their subsequent nuclear translocation and NF-κB-dependent gene induction [42]. Among different NF-κB subunits, the p50/p65 heterodimer is the most abundant. Only the p65 subunit of the p50/p65 heterodimer contains a domain initiating transcriptional activation essential for the expression of the NF-κB-dependent genes [41]. In addition, p65 overexpression significantly increased the CCL20 mRNA expression in HeLa cells stimulated with TNF-α [44]. Therefore, we analyzed activation of p65 in the nuclear extracts of dendritic cells that had been stimulated for 1 h with either plasmin or the positive control LPS (0.5 μg/mL). Plasmin induced a significant increase in the p65 NF-κB activity (2.00 ± 0.27-fold compared to control, $P < 0.05$) (Figure 4(c)); LPS induced a higher NF-κB activation (8.20 ± 0.59-fold, $P < 0.01$), which is consistent with the higher amounts of CCL20 released by the LPS-stimulated dendritic cells (Figure 3(a)).

3.5. Plasmin Induces CCL20 Expression in Dendritic Cells through Akt- and ERK1/2 MAPK-Dependent NF-κB Activation. To analyze the role of Akt and MAPK in the plasmin-induced CCL20 expression, dendritic cells were pretreated with pharmacological inhibitors of Akt (Akt inhibitor VIII) [45], MEK/ERK1/2 (U0126), p38 (SB203580) [46], and NF-κB (AKβBA) [35, 47–49] before addition of plasmin. AKβBA is an NF-κB inhibitor targeting IκB kinases (IKK) thereby inhibiting NF-κB-dependent signaling in monocytes [48] and tumor cells [35]. In preliminary tests, we ensured that the used concentrations induced specific inhibition of the respective pathways, yet did not impair cell viability. The Akt inhibitor VIII, the MEK/ERK1/2 inhibitor U0126, and the IκB kinase inhibitor AKβBA, but not the p38 MAPK inhibitor SB203580, abolished the plasmin-induced expression of CCL20 mRNA and CCL20 protein release (Figures 5(a) and 5(b)) indicating that plasmin-induced activation of Akt, ERK1/2, and NF-κB is indispensable for the CCL20 expression.

To address whether the plasmin-induced activation of Akt and ERK1/2 would be located upstream of the NF-κB activation, we analyzed protein phosphorylation in the presence of the inhibitors. Inhibition of either Akt or ERK1/2 impaired the plasmin-induced IκBα phosphorylation and the phosphorylation of p65 at Ser276, whereas the phosphorylation of p65 at Ser536 remained unaffected (Figure 6). These data indicate that Akt and ERK1/2 activation is indispensable for the plasmin-induced NF-κB activation and the subsequent expression of CCL20.

4. Discussion

The serine protease plasmin is activated under physiological and pathological conditions. Plasmin is locally generated during tissue damage or thrombus formation, but also in the context of contact activation during inflammatory processes [1, 50–53]. It has been shown that the plasminogen activator uPA and its receptor are present on the surface of

The Serine Protease Plasmin Triggers Expression of the CC-Chemokine Ligand 20 in Dendritic
Cells via Akt/NF-κB-Dependent Pathways

157

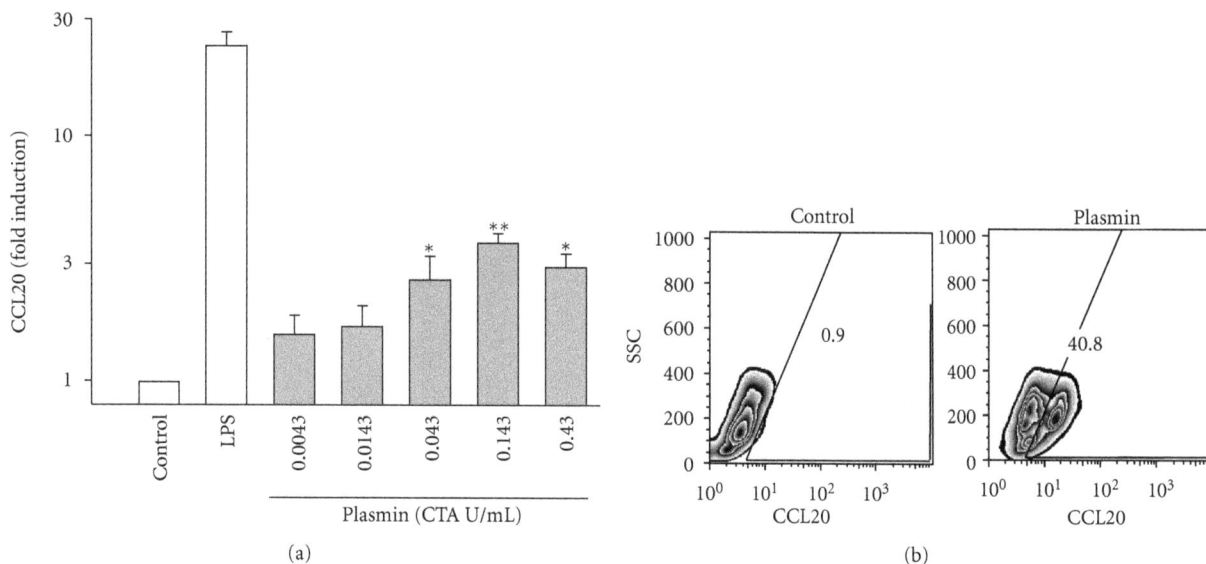

(a)

(b)

FIGURE 3: Plasmin elicits CCL20 protein expression in dendritic cells. (a) Plasmin induces a concentration-dependent release of CCL20. Dendritic cells were stimulated with various concentrations of plasmin or LPS (0.5 μg/mL) for 24 h before being analyzed by ELISA. The results are mean ± SEM of 8 experiments, *$P < 0.05$, **$P < 0.01$ versus control. (b) Flow cytometric analysis of CCL20 expression by dendritic cells. Dendritic cells were either unstimulated or stimulated with plasmin (0.143 CTA U/mL) for 24 h. Brefeldin A was added to the cell culture 4 h before the end of the incubation, and the cells were fixed, permeabilized, stained, and analyzed by flow cytometry. Representative data of 3 independent experiments are shown.

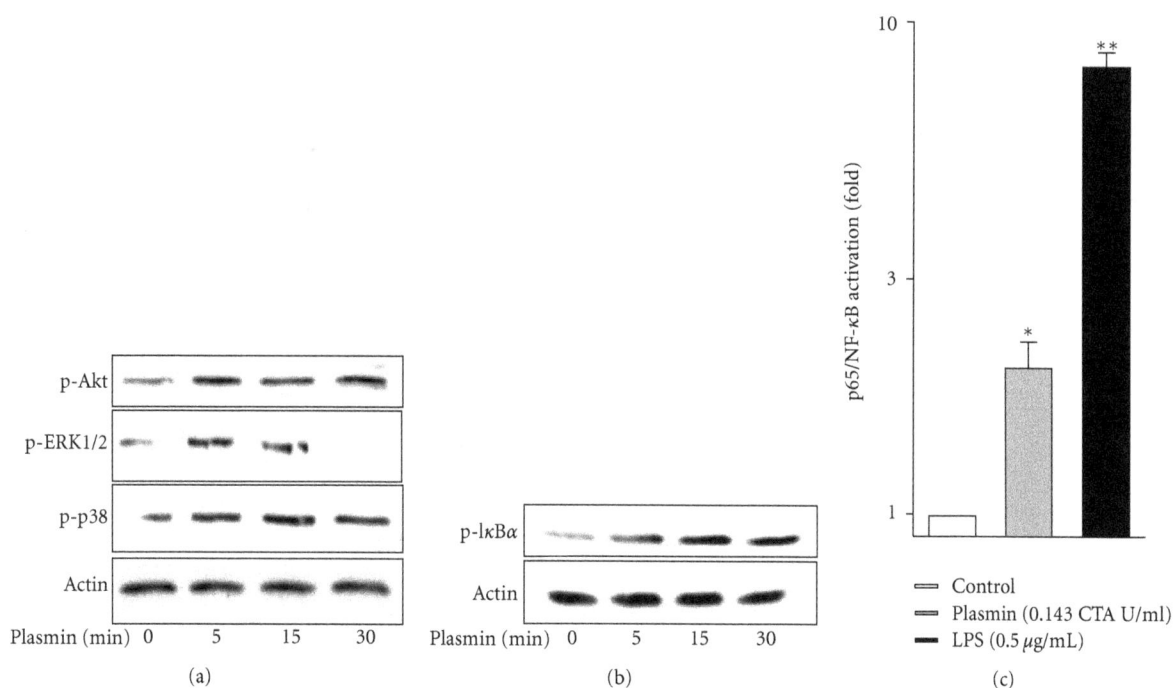

(a)

(b)

(c)

FIGURE 4: Plasmin activates Akt, MAPK, and NF-κB signaling in dendritic cells. (a) Time-dependent phosphorylation of Akt (Ser473) and MAP kinases in dendritic cells after stimulation with plasmin (0.143 CTA U/mL). Dendritic cells were stimulated with plasmin, and whole cell lysates were analyzed by western immunoblotting; actin served as loading control. Representative data of three experiments are shown. (b) Time-dependent phosphorylation of IκBα (Ser32/Ser36) in dendritic cells after stimulation with plasmin (0.143 CTA U/mL). Representative data of three experiments are shown. (c) Activation of p65 NF-κB as analyzed with the NF-κB TransAM ELISA. Nuclear extracts were obtained from dendritic cells stimulated with plasmin (0.143 CTA U/mL) or LPS (0.5 μg/mL) for 1 h. The results are mean ± SEM of 3 independent experiments, *$P < 0.05$, **$P < 0.01$ versus control.

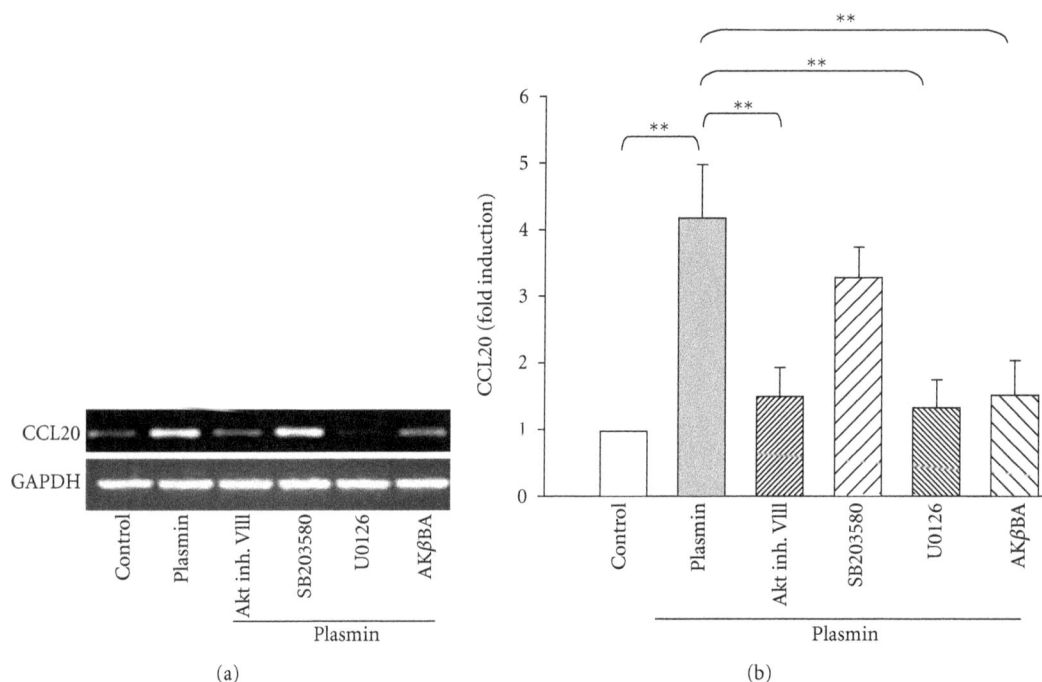

(a)   (b)

FIGURE 5: Activation of Akt and ERK1/2 is indispensable for the plasmin-induced CCL20 expression. (a) RT-PCR analysis of CCL20 mRNA expression. Dendritic cells were pretreated with the Akt inhibitor VIII, the MEK inhibitor U0126, the p38 inhibitor SB203580 (each at $1\,\mu$M), or the IκB kinase inhibitor AKβBA ($10\,\mu$M) for 15 min and then stimulated with plasmin (0.143 CTA U/mL) for 6 h. mRNA was isolated and subjected to RT-PCR using CCL20-specific primers; GAPDH served as control. Representative data of 3 independent experiments are shown. (b) Release of CCL20 by plasmin-stimulated dendritic cells. Dendritic cells were treated as in A, but for 24 h, and CCL20 release into the supernatants was analyzed by ELISA. Results are mean $\pm$ SEM of 5 experiments, $**P < 0.01$.

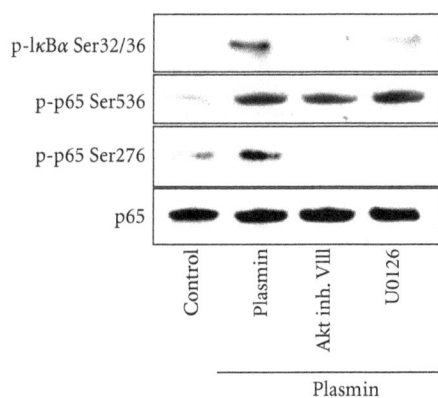

FIGURE 6: Akt and ERK1/2 activation is indispensable for the plasmin-induced activation of NF-κB. Analysis of IκBα and p65 phosphorylation. Dendritic cells were pretreated with the Akt inhibitor VIII or the MEK/ERK1/2 inhibitor U0126 (each at $1\,\mu$M) for 15 min and then stimulated with plasmin (0.143 CTA U/mL) for 40 min. Dendritic cells were collected, lysed, and subjected to western blot analysis with antibodies against the phosphorylated forms of IκBα (Ser32/Ser36) and p65 (Ser276 and Ser536). Staining with p65 antibody-loading control. Results are representative of 3 independent experiments.

immature dendritic cells derived from myeloid progenitors [54]. Plasmin generated at the cell surface is protected from inactivation by its physiological inhibitor α_2-antiplasmin and can, therefore, trigger cell activation [1].

CCL20 is constitutively expressed by lymphoid and nonlymphoid tissue, where it contributes to homeostatic functions and immunity [20]. Thus, mucosa-associated lymphoid tissues and different tumors constitutively express CCL20 [20]. Under inflammatory conditions, CCL20 can be rapidly induced by proinflammatory cytokines, bacterial and viral infections of epithelial cell, keratinocytes, fibroblasts, or endothelial cells [20, 27, 40, 55]. Recent studies have shown that neutrophils produce CCL20 in response to treatment with LPS or TNF-α [56]. Human monocytes express CCL20 when activated with LPS, extracellular nucleotides [20, 27], or under hypoxic conditions [37]. Similarly, dendritic cells can produce CCL20 when stimulated with LPS, CD40L [26], or extracellular nucleotides [27], but not TNF-α [26].

Here, we show for the first time that plasmin elicits CCL20 expression in dendritic cells. The plasmin-induced expression of CCL20 is very rapid and is not dependent on the release of proinflammatory TNF-α. Moreover, we show that plasmin does not induce expression of TNF-α by dendritic cells, and TNF-α does not induce expression of CCL20 in dendritic cells [26]. Similar to the plasmin-induced activation of monocytes and macrophages [8, 9, 31], the proteolytic activity of plasmin is essential for the induction of the CCL20 expression in dendritic cells.

Chemokines are regulated primarily at the level of gene transcription. The CCL20 promoter region contains binding sites for different transcription factors such as activator protein-1 (AP-1) and AP-2, CAAT/enhancer-binding protein

The Serine Protease Plasmin Triggers Expression of the CC-Chemokine Ligand 20 in Dendritic
Cells via Akt/NF-κB-Dependent Pathways

159

(C-EBP), stimulating protein 1 (SP1), and the epithelium-specific Ets nuclear factor ESE-1 [20]. However, activation of the NF-κB transcription factor family is indispensable for the CCL20 gene expression in several tissues and in response to various agonists [20, 37, 44, 55]. Plasmin induces phosphorylation of IκBα, nuclear translocation, and phosphorylation of p65 at Ser276 and Ser536, as well as binding of activated p65 to the NF-κB consensus sequence. All those events concur with NF-κB activation induced in dendritic cells by plasmin. Consistently, using an NF-κB inhibitor, we demonstrated that the NF-κB pathway is indispensable for plasmin-induced CCL20 expression in dendritic cells.

Akt and MAPK pathways have been shown to be involved in the plasmin-induced gene expression in monocytes and macrophages [8, 9]. In this study, we found that inhibitors of Akt and ERK1/2, but not of p38/MAPK, inhibited the plasmin-induced CCL20 mRNA and protein expression. Others also reported that the CCL20 expression might depend on the activation of Akt, ERK1/2, and p38 MAPK. However, the involvement of different pathways in the CCL20 gene expression strongly depends on the cell type and stimulus. Thus, stimulation of intestinal epithelial cells with IL-21 resulted in enhanced phosphorylation of ERK1/2 and p38 and increased synthesis of CCL20, but only inhibition of ERK1/2, but not of p38 MAPK, suppressed the IL-21-induced CCL20 production [57]. On the other hand, when human monocyte-derived dendritic cells were stimulated with nucleotides, the CCL20 expression was NF-κB, ERK1/2, and p38 MAPK dependent. By contrast, the release of CCL20 by LPS-stimulated dendritic cells was NF-κB and p38 dependent, yet ERK1/2 was independent [27]. These data indicate that the expression of CCL20 is differentially regulated in distinct cell types and in response to different activators.

Similar to human airway epithelial cells stimulated with IL-17A [40, 55], in plasmin-stimulated dendritic cells, the CCL20 expression was dependent on NF-κB, Akt, and ERK1/2, but not on p38 MAPK activation. Plasmin-induced ERK1/2 signaling might contribute to NF-κB activation via several independent mechanisms. In melanoma cells, constitutive ERK1/2 activation has been shown to increase the IκBα phosphorylation and the NF-κB activity [58]. On the other hand, ERK1/2 could facilitate the engagement of transcriptional cofactors CBP/p300, which may increase the transcriptional activity of NF-κB. Thus, ERK1/2 has been shown to activate nuclear kinases MSK1/2 [39, 41], which are potent activators of CREB, whose activity, in turn, is essential for the recruitment of CBP/p300. Interestingly, the CREB site phosphorylated by MSK1/2 is very similar to the site surrounding Ser276 in the sequence of p65. This led to the finding that MSK1/2 can effectively increase the transcriptional activity of p65 via phosphorylation at Ser276 [59]. ERK1/2-mediated MSK activation might also contribute to enhanced gene expression via histone 3 phosphorylation creating a more accessible chromatin structure [59]. We have observed that the inhibition of ERK1/2 activity inhibited the plasmin-induced phosphorylation of IκBα and the phosphorylation of p65 at Ser276 indicating that plasmin-induced ERK1/2 activation might contribute to the CCL20 induction through increased phosphorylation of both IκBα and p65/Ser276, which would result in increased activation of NF-κB and enhanced recruitment of transcriptional cofactors.

The role of Akt in the plasmin-induced NF-κB activation is more complex. The ability of Akt to regulate NF-κB activity might occur through the phosphorylation of IκB kinase, which in turn phosphorylates IκB and allows the release of NF-κB [38], and/or by stimulating transactivation of the p65 subunit by IκB kinase-dependent phosphorylation of p65 on Ser536 [60, 61]. However, the later process is p38 dependent. Consistent with the fact that plasmin-activated dendritic cells did not utilize the p38 MAPK pathway to induce CCL20, we did not observe any effects of p38 inhibition on p65 phosphorylation. However, the Akt inhibition impaired the plasmin-induced IκBα and p65 Ser276 phosphorylation, indicating the Akt-dependent activation of IKK. We have previously shown that plasmin-induced ERK1/2 activation in dendritic cells is Akt dependent [17]. Therefore, Akt might induce the p65 Ser276 phosphorylation via ERK1/2. The activation pathway triggered in dendritic cells by plasmin is different to the IL-17A-induced CCL20 expression in human airway epithelial cells, which is Akt and NF-κB dependent, although both pathways act independently [40]. The plasmin-induced expression of CCL20 in dendritic cells also differed from that in a transformed T-cell line, where Akt inhibition resulted in reduced phosphorylation of p65 on Ser536, whereas the IκBα phosphorylation remained unaffected [62]. Akt might also positively regulate the NF-κB activity through GSK3β inhibition. GSK3β regulates the phosphorylation and function of certain transcriptional coactivators, such as C/EBP and β-catenin, and some transcriptional repressors [40]. Therefore, it is possible that plasmin-induced PI3K/Akt/GSK3β pathway is involved in the modulation of transcriptional activators and/or repressors, which might contribute to the plasmin-induced expression of CCL20.

In summary, the present study demonstrates that plasmin and dendritic cells colocalize with CCL20 in human atherosclerotic vessels. We also show that plasmin is a potent activator of dendritic cells triggering CCL20 expression by the coordinated activation of Akt, ERK1/2, and NF-κB signaling pathways. Hence, by activating dendritic cells to produce CCL20, locally generated plasmin might control the composition of the cellular infiltrate and modulate inflammatory and immune reactions in atherosclerotic lesions. By contrast, such effects might be rather unlikely during conditions of fibrinolysis, where plasmin in the plasma phase would be spatially separated from inflammatory dendritic cells and rapidly bound to fibrin or quickly inactivated by plasmin inhibitors such as α_2-antiplasmin and α_2-macroglobulin [1].

Acknowledgments

The authors thank Dr. K. H. Orend, Department of Thoracic and Vascular Surgery, Ulm University, for atherosclerotic blood vessel specimens and Felicitas Genze for expert technical assistance. This work was supported by a grant from the Deutsche Forschungsgemeinschaft (to T. Syrovets and T. Simmet).

References

[1] T. Syrovets and T. Simmet, "Novel aspects and new roles for the serine protease plasmin," *Cellular and Molecular Life Sciences*, vol. 61, no. 7-8, pp. 873–885, 2004.

[2] J. L. Martin-Ventura, V. Nicolas, X. Houard et al., "Biological significance of decreased HSP27 in human atherosclerosis," *Arteriosclerosis, Thrombosis, and Vascular Biology*, vol. 26, no. 6, pp. 1337–1343, 2006.

[3] A. Leclercq, X. Houard, S. Loyau et al., "Topology of protease activities reflects atherothrombotic plaque complexity," *Atherosclerosis*, vol. 191, no. 1, pp. 1–10, 2007.

[4] J. Le Dall, B. Ho-Tin-Noé, L. Louedec et al., "Immaturity of microvessels in haemorrhagic plaques is associated with proteolytic degradation of angiogenic factors," *Cardiovascular Research*, vol. 85, no. 1, pp. 184–193, 2010.

[5] X. Houard, F. Rouzet, Z. Touat et al., "Topology of the fibrinolytic system within the mural thrombus of human abdominal aortic aneurysms," *Journal of Pathology*, vol. 212, no. 1, pp. 20–28, 2007.

[6] I. Weide, B. Tippler, T. Syrovets, and T. Simmet, "Plasmin is a specific stimulus of the 5-lipoxygenase pathway of human peripheral monocytes," *Thrombosis and Haemostasis*, vol. 76, no. 4, pp. 561–568, 1996.

[7] T. Syrovets, B. Tippler, M. Rieks, and T. Simmet, "Plasmin is a potent and specific chemoattractant for human peripheral monocytes acting via a cyclic guanosine monophosphate-dependent pathway," *Blood*, vol. 89, no. 12, pp. 4574–4583, 1997.

[8] Q. Li, Y. Laumonnier, T. Syrovets, and T. Simmet, "Plasmin triggers cytokine induction in human monocyte-derived macrophages," *Arteriosclerosis, Thrombosis, and Vascular Biology*, vol. 27, no. 6, pp. 1383–1389, 2007.

[9] T. Syrovets, M. Jendrach, A. Rohwedder, A. Schüle, and T. Simmet, "Plasmin-induced expression of cytokines and tissue factor in human monocytes involves AP-1 and IKKβ-mediated NF-κB activation," *Blood*, vol. 97, no. 12, pp. 3941–3950, 2001.

[10] R. M. Steinman and H. Hemmi, "Dendritic cells: translating innate to adaptive immunity," *Current Topics in Microbiology and Immunology*, vol. 311, pp. 17–58, 2006.

[11] S. Sarkar and D. A. Fox, "Dendritic cells in rheumatoid arthritis," *Frontiers in Bioscience*, vol. 10, pp. 656–665, 2005.

[12] E. Galkina and K. Ley, "Immune and inflammatory mechanisms of atherosclerosis," *Annual Review of Immunology*, vol. 27, pp. 165–197, 2009.

[13] O. Soehnlein, M. Drechsler, M. Hristov, and C. Weber, "Functional alterations of myeloid cell subsets in hyperlipidaemia: relevance for atherosclerosis," *Journal of Cellular and Molecular Medicine*, vol. 13, no. 11-12, pp. 4293–4303, 2009.

[14] C. Weber and H. Noels, "Atherosclerosis: current pathogenesis and therapeutic options," *Nature Medicine*, vol. 17, no. 11, pp. 1410–1422, 2011.

[15] K. Shortman and S. H. Naik, "Steady-state and inflammatory dendritic-cell development," *Nature Reviews Immunology*, vol. 7, no. 1, pp. 19–30, 2007.

[16] M. Merad and M. G. Manz, "Dendritic cell homeostasis," *Blood*, vol. 113, no. 15, pp. 3418–3427, 2009.

[17] X. Li, T. Syrovets, F. Genze et al., "Plasmin triggers chemotaxis of monocyte-derived dendritic cells through an Akt2-dependent pathway and promotes a T-helper type-1 response,"

[18] R. Bonecchi, E. Galliera, E. M. Borroni, M. M. Corsi, M. Locati, and A. Mantovani, "Chemokines and chemokine receptors: an overview," *Frontiers in Bioscience*, vol. 14, no. 2, pp. 540–551, 2009.

[19] P. Miossec, "Dynamic interactions between T cells and dendritic cells and their derived cytokines/chemokines in the rheumatoid synovium," *Arthritis Research and Therapy*, vol. 10, supplement 1, article S2, 2008.

[20] E. Schutyser, S. Struyf, and J. Van Damme, "The CC chemokine CCL20 and its receptor CCR6," *Cytokine and Growth Factor Reviews*, vol. 14, no. 5, pp. 409–426, 2003.

[21] A. Kaser, O. Ludwiczek, S. Holzmann et al., "Increased expression of CCL20 in human inflammatory bowel disease," *Journal of Clinical Immunology*, vol. 24, no. 1, pp. 74–85, 2004.

[22] I. K. Demedts, K. R. Bracke, G. Van Pottelberge et al., "Accumulation of dendritic cells and increased CCL20 levels in the airways of patients with chronic obstructive pulmonary disease," *American Journal of Respiratory and Critical Care Medicine*, vol. 175, no. 10, pp. 998–1005, 2007.

[23] E. G. Harper, C. Guo, H. Rizzo et al., "Th17 cytokines stimulate CCL20 expression in keratinocytes in vitro and in vivo: implications for psoriasis pathogenesis," *Journal of Investigative Dermatology*, vol. 129, no. 9, pp. 2175–2183, 2009.

[24] K. Beider, M. Abraham, M. Begin et al., "Interaction between CXCR4 and CCL20 pathways regulates tumor growth," *PLoS ONE*, vol. 4, no. 4, Article ID e5125, 2009.

[25] M. Le Borgne, N. Etchart, A. Goubier et al., "Dendritic cells rapidly recruited into epithelial tissues via CCR6/CCL20 are responsible for CD8+ T cell crosspriming in vivo," *Immunity*, vol. 24, no. 2, pp. 191–201, 2006.

[26] M. Vulcano, S. Struyf, P. Scapini et al., "Unique regulation of CCL18 production by maturing dendritic cells," *Journal of Immunology*, vol. 170, no. 7, pp. 3843–3849, 2003.

[27] B. Marcet, M. Horckmans, F. Libert, S. Hassid, J. M. Boeynaems, and D. Communi, "Extracellular nucleotides regulate CCL20 release from human primary airway epithelial cells, monocytes and monocyte-derived dendritic cells," *Journal of Cellular Physiology*, vol. 211, no. 3, pp. 716–727, 2007.

[28] D. N. J. Hart, "Dendritic cells: unique leukocyte populations which control the primary immune response," *Blood*, vol. 90, no. 9, pp. 3245–3287, 1997.

[29] Y. V. Bobryshev and R. S. A. Lord, "55-kD actin-bundling protein (p55) is a specific marker for identifying vascular dendritic cells," *Journal of Histochemistry and Cytochemistry*, vol. 47, no. 11, pp. 1481–1486, 1999.

[30] X. Li, T. Syrovets, S. Paskas, Y. Laumonnier, and T. Simmet, "Mature dendritic cells express functional thrombin receptors triggering chemotaxis and CCL18/pulmonary and activation-regulated chemokine induction," *Journal of Immunology*, vol. 181, no. 2, pp. 1215–1223, 2008.

[31] Y. Laumonnier, T. Syrovets, L. Burysek, and T. Simmet, "Identification of the annexin A2 heterotetramer as a receptor for the plasmin-induced signaling in human peripheral monocytes," *Blood*, vol. 107, no. 8, pp. 3342–3349, 2006.

[32] B. Sperandio, B. Regnault, J. Guo et al., "Virulent Shigella flexneri subverts the host innate immune response through manipulation of antimicrobial peptide gene expression," *Journal of Experimental Medicine*, vol. 205, no. 5, pp. 1121–1132, 2008.

[33] N. Katoh, S. Kraft, J. H. M. Weßendorf, and T. Bieber, "The high-affinity IgE receptor (FcεRI) blocks apoptosis in normal

The Serine Protease Plasmin Triggers Expression of the CC-Chemokine Ligand 20 in Dendritic
Cells via Akt/NF-κB-Dependent Pathways

161

human monocytes," *Journal of Clinical Investigation*, vol. 105, no. 2, pp. 183–190, 2000.

[34] T. Syrovets, A. Schüle, M. Jendrach, B. Büchele, and T. Simmet, "Ciglitazone inhibits plasmin-induced proinflammatory monocyte activation via modulation of p38 MAP kinase activity," *Thrombosis and Haemostasis*, vol. 88, no. 2, pp. 274–281, 2002.

[35] T. Syrovets, J. E. Gschwend, B. Büchele et al., "Inhibition of IκB kinase activity by acetyl-boswellic acids promotes apoptosis in androgen-independent PC-3 prostate cancer cells in vitro and in vivo," *Journal of Biological Chemistry*, vol. 280, no. 7, pp. 6170–6180, 2005.

[36] I. Weide, J. Romisch, and T. Simmet, "Contact activation triggers stimulation of the monocyte 5-lipoxygenase pathway via plasmin," *Blood*, vol. 83, no. 7, pp. 1941–1951, 1994.

[37] F. Battaglia, S. Delfino, E. Merello et al., "Hypoxia transcriptionally induces macrophage-inflammatory protein-3α/CCL-20 in primary human mononuclear phagocytes through nuclear factor (NF)-κB," *Journal of Leukocyte Biology*, vol. 83, no. 3, pp. 648–662, 2008.

[38] O. N. Ozes, L. D. Mayo, J. A. Gustin, S. R. Pfeffer, L. M. Pfeffer, and D. B. Donner, "NF-κB activation by tumour necrosis factor requires tie Akt serine-threonine kinase," *Nature*, vol. 401, no. 6748, pp. 82–85, 1999.

[39] N. D. Perkins, "Post-translational modifications regulating the activity and function of the nuclear factor kappa B pathway," *Oncogene*, vol. 25, no. 51, pp. 6717–6730, 2006.

[40] F. Huang, C. Y. Kao, S. Wachi, P. Thai, J. Ryu, and R. Wu, "Requirement for both JAK-mediated PI3K signaling and ACT1/TRAF6/TAK1- dependent NF-κB activation by IL-17A in enhancing cytokine expression in human airway epithelial cells," *Journal of Immunology*, vol. 179, no. 10, pp. 6504–6513, 2007.

[41] M. S. Hayden and S. Ghosh, "Shared principles in NF-κB signaling," *Cell*, vol. 132, no. 3, pp. 344–362, 2008.

[42] M. S. Hayden, A. P. West, and S. Ghosh, "NF-κB and the immune response," *Oncogene*, vol. 25, no. 51, pp. 6758–6780, 2006.

[43] L. Burysek, T. Syrovets, and T. Simmet, "The serine protease plasmin triggers expression of MCP-1 and CD40 in human primary monocytes via activation of p38 MAPK and Janus kinase (JAK)/STAT signaling pathways," *Journal of Biological Chemistry*, vol. 277, no. 36, pp. 33509–33517, 2002.

[44] S. Sugita, T. Kohno, K. Yamamoto et al., "Induction of macrophage-inflammatory protein-3α gene expression by TNF-dependent NF-κB activation," *Journal of Immunology*, vol. 168, no. 11, pp. 5621–5628, 2002.

[45] C. W. Lindsley, Z. Zhao, W. H. Leister et al., "Allosteric Akt (PKB) inhibitors: discovery and SAR of isozyme selective inhibitors," *Bioorganic and Medicinal Chemistry Letters*, vol. 15, no. 3, pp. 761–764, 2005.

[46] S. P. Davies, H. Reddy, M. Caivano, and P. Cohen, "Specificity and mechanism of action of some commonly used protein kinase inhibitors," *Biochemical Journal*, vol. 351, no. 1, pp. 95–105, 2000.

[47] C. Cuaz-Pérolin, L. Billiet, E. Baugé et al., "Antiinflammatory and antiatherogenic effects of the NF-κB inhibitor acetyl-11-Keto-β-boswellic acid in LPS-challenged ApoE-/- mice," *Arteriosclerosis, Thrombosis, and Vascular Biology*, vol. 28, no. 2, pp. 272–277, 2008.

[48] T. Syrovets, B. Büchele, C. Krauss, Y. Laumonnier, and T. Simmet, "Acetyl-boswellic acids inhibit lipopolysaccharide-mediated TNF-α induction in monocytes by direct interaction with IκB kinases," *Journal of Immunology*, vol. 174, no. 1, pp. 498–506, 2005.

[49] H. Wang, T. Syrovets, D. Kess et al., "Targeting NF-κB with a natural triterpenoid alleviates skin inflammation in a mouse model of psoriasis," *Journal of Immunology*, vol. 183, no. 7, pp. 4755–4763, 2009.

[50] A. P. Kaplan, K. Joseph, Y. Shibayama, S. Reddigari, B. Ghebrehiwet, and M. Silverberg, "The intrinsic coagulation/kinin-forming cascade: assembly in plasma and cell surfaces in inflammation," *Advances in Immunology*, vol. 66, pp. 225–272, 1997.

[51] A. H. Schmaier, "Contact activation: a revision," *Thrombosis and Haemostasis*, vol. 78, no. 1, pp. 101–107, 1997.

[52] R. W. Colman and A. H. Schmaier, "Contact system: a vascular biology modulator with anticoagulant, profibrinolytic, antiadhesive, and proinflammatory attributes," *Blood*, vol. 90, no. 10, pp. 3819–3843, 1997.

[53] A. H. Schmaier and K. R. McCrae, "The plasma kallikrein-kinin system: its evolution from contact activation," *Journal of Thrombosis and Haemostasis*, vol. 5, no. 12, pp. 2323–2329, 2007.

[54] E. Ferrero, K. Vettoretto, A. Bondanza et al., "uPA/uPAR system is active in immature dendritic cells derived from CD14+CD34+ precursors and is down-regulated upon maturation," *Journal of Immunology*, vol. 164, no. 2, pp. 712–718, 2000.

[55] C. Y. Kao, F. Huang, Y. Chen et al., "Up-regulation of CC chemokine ligand 20 expression in human airway epithelium by IL-17 through a JAK-independent but MEK/NF-κB-dependent signalling pathway," *Journal of Immunology*, vol. 175, no. 10, pp. 6676–6685, 2005.

[56] P. Scapini, C. Laudanna, C. Pinardi et al., "Neutrophils produce biologically active macrophage inflammatory protein-3a (MIP-3a)/CCL20 and MIP-3β/CCL19," *European Journal of Immunology*, vol. 31, no. 7, pp. 1981–1988, 2001.

[57] R. Caruso, D. Fina, I. Peluso et al., "A functional role for interleukin-21 in promoting the synthesis of the T-Cell chemoattractant, MIP-3α, by gut epithelial cells," *Gastroenterology*, vol. 132, no. 1, pp. 166–175, 2007.

[58] P. Dhawan and A. Richmond, "A novel NF-κB-inducing kinase-MAPK signaling pathway up-regulates NF-κB activity in melanoma cells," *Journal of Biological Chemistry*, vol. 277, no. 10, pp. 7920–7928, 2002.

[59] L. Vermeulen, G. De Wilde, S. Notebaert, W. Vanden Berghe, and G. Haegeman, "Regulation of the transcriptional activity of the nuclear factor-κB p65 subunit," *Biochemical Pharmacology*, vol. 64, no. 5-6, pp. 963–970, 2002.

[60] L. V. Madrid, M. W. Mayo, J. Y. Reuther, and A. S. Baldwin, "Akt stimulates the transactivation potential of the RelA/p65 subunit of NF-κB through utilization of the IκB kinase and activation of the mitogen-activated protein kinase p38," *Journal of Biological Chemistry*, vol. 276, no. 22, pp. 18934–18940, 2001.

[61] N. Sizemore, N. Lerner, N. Dombrowski, H. Sakurai, and G. R. Stark, "Distinct roles of the IκB kinase α and β subunits in liberating nuclear factor κB (NF-κB) from IκB and in phosphorylating the p65 subunit of NF-κB," *Journal of Biological Chemistry*, vol. 277, no. 6, pp. 3863–3869, 2002.

[62] S. J. Jeong, C. A. Pise-Masison, M. F. Radonovich, H. U. Park, and J. N. Brady, "Activated AKT regulates NF-κB activation, p53 inhibition and cell survival in HTLV-1-transformed cells," *Oncogene*, vol. 24, no. 44, pp. 6719–6728, 2005.

16

Production of Ethanol from Sugars and Lignocellulosic Biomass by *Thermoanaerobacter* J1 Isolated from a Hot Spring in Iceland

Jan Eric Jessen and Johann Orlygsson

Faculty of Natural Resource Sciences, University of Akureyri, Borgir, Nordurslod 2, 600 Akureyri, Iceland

Correspondence should be addressed to Johann Orlygsson, jorlygs@unak.is

Academic Editor: Anuj K. Chandel

Thermophilic bacteria have gained increased attention as candidates for bioethanol production from lignocellulosic biomass. This study investigated ethanol production by *Thermoanaerobacter* strain J1 from hydrolysates made from lignocellulosic biomass in batch cultures. The effect of increased initial glucose concentration and the partial pressure of hydrogen on end product formation were examined. The strain showed a broad substrate spectrum, and high ethanol yields were observed on glucose (1.70 mol/mol) and xylose (1.25 mol/mol). Ethanol yields were, however, dramatically lowered by adding thiosulfate or by cocultivating strain J1 with a hydrogenotrophic methanogen with acetate becoming the major end product. Ethanol production from 4.5 g/L of lignocellulosic biomass hydrolysates (grass, hemp stem, wheat straw, newspaper, and cellulose) pretreated with acid or alkali and the enzymes Celluclast and Novozymes 188 was investigated. The highest ethanol yields were obtained on cellulose (7.5 mM·g^{-1}) but the lowest on straw (0.8 mM·g^{-1}). Chemical pretreatment increased ethanol yields substantially from lignocellulosic biomass but not from cellulose. The largest increase was on straw hydrolysates where ethanol production increased from 0.8 mM·g^{-1} to 3.3 mM·g^{-1} using alkali-pretreated biomass. The highest ethanol yields on lignocellulosic hydrolysates were observed with hemp hydrolysates pretreated with acid, 4.2 mM·g^{-1}.

1. Background

More than 95% of the ethanol produced today is from simple biomass like mono- and disaccharides and starch [1]. The use of this type of biomass has been increasingly debated due to its impact on food and feed prices as well as for environmental reasons [2]. Therefore, complex (lignocellulosic) biomass has been put forward as a feasible alternative due to its abundance in nature and the large quantities generated as waste from agricultural activities [2, 3]. Lignocellulosic biomass is primarily composed of cellulose, hemicellulose, and lignin. Cellulose and hemicellulose are the main substrates used for ethanol production, but lignin is composed of aromatic lignols that need to be separated and removed before enzymatic hydrolysis. Today, expensive pretreatments are the main reason for unsuccessful implementation of complex lignocellulosic biomasses as a starting material for ethanol production [2].

The best-known microorganisms used for ethanol production today are the yeast *Saccharomyces cerevisiae* and the bacterium *Zymomonas mobilis*. Both organisms have very high yields of ethanol (>1.9 mol ethanol/mol hexose) but very narrow substrate spectra and thus are not suitable for ethanol production from complex substrates. Therefore, the use of thermophilic bacteria with broad substrate range and high yields may be a better option for ethanol production from complex biomasses. It has been known for some time now that many thermophilic bacteria are highly efficient ethanol producers [4]. After the oil crisis in the 1980s, there was a peak in investigations on thermophilic ethanol-producing bacteria; bacteria within the genera of *Thermoanaerobacterium*, *Thermoanaerobacter*, and *Clostridium* have demonstrated good ethanol yields and fast growth rates [5–8]. There are several advantages in using these thermophilic bacteria: the increased temperature deters contamination from mesophilic bacteria and fungi,

Production of Ethanol from Sugars and Lignocellulosic Biomass by Thermoanaerobacter J1 Isolated from a Hot Spring in Iceland

163

possible self-distillation of ethanol avoiding the generally low ethanol tolerance problem with those bacteria, and broad substrate spectrum [9, 10]. Some of these strains produce more than 1.5 mol ethanol/mol hexose [11–16], whereas the theoretical maximum yield is 2.0 mol ethanol/mol hexose degraded. The main reasons for low yields are the formation of other end products such as acetate, butyrate, and CO_2 [11–16].

The present study focuses on a recently isolated thermophilic bacterium, strain J1, which is most closely related to species within the genus *Thermoanaerobacter*. Bacteria within this genus seem to be among the most efficient ethanol producers known and show very high yields from simple sugar fermentations [12–14, 16] as well as from complex lignocellulosic biomass [10, 13, 17–19]. These bacteria are Gram-variable rods with broad substrate spectrum (mostly sugars) and produces ethanol, acetate, lactate, hydrogen, and carbon dioxide during anaerobic fermentation [20, 21]. The physiological characteristics of *Thermoanaerobacter* strain J1, isolated from Icelandic hot spring, were investigated in detail with the main aim of exploring the ethanol production capacity both from simple sugars as well as from various lignocellulosic biomass.

2. Methods

2.1. Medium. The composition and preparation of the medium used has been described earlier [12]. This medium, referred to as basal medium (BM) hereafter, contains yeast extract (? g/L) in addition to glucose or other carbon sources. All experiments were performed at 65°C at pH 7.0 without agitation with the exception of the temperature and pH optimum experiments. The inoculum volume was 2% (v/v) in all experiments which were always performed in duplicates.

2.2. Isolation of Strain J1. The strain was isolated in BM with glucose (20 mM) from a hot spring (69°C, pH 7.5) in Grensdalur in Southwest of Iceland. Samples were enriched on glucose, and positive samples (increase in growth and production of hydrogen) were reinoculated five times. From the final enrichment series, end point dilutions were performed by using BM containing agar ($30\,g\cdot L^{-1}$). Colonies were picked from final positive dilution and reinoculated to liquid BM with glucose. Isolation of the hydrogenotrophic methane producing strain has been described elsewhere [22].

2.3. Optimum pH and Temperature Growth Experiments. To determine the strain's growth characteristics at various pHs and temperatures, the strain was cultivated on glucose (20 mM), and cell concentration was measured by increase in absorbance at 600 nm by a Perkin-Elmer Lambda 25 UV-Vis spectrophotometer. Maximum (specific) growth rate (μ_{max}) for each experiment was derived from absorbance data. For pH optimum experiments, the initial pH was set to various levels in the range from 3.0 to 9.0 with increments of 1.0 pH unit. The experimental bottles were supplemented with acid (HCl) and alkali (NaOH) to set the pH accordingly.

To determine the optimum temperature for growth, the incubation temperature varied from 35°C to 80°C. For the pH optimum experiments, the strain was cultivated at 65°C, and for the temperature optimum experiments, the pH was 7.0. Optimal pH and temperature were used in all experiments performed. Experiments were done in 117.5 mL serum bottles with 50 mL liquid medium.

2.4. Phylogenetic Analysis. Full 16S rRNA analysis of 1479-nucleotide long sequence was done according to Orlygsson and Baldursson [23] and references therein. Sequences from 16S rRNA analysis were compared to sequences in the NCBI database using the nucleotide-nucleotide BLAST (BLAST-N) tool. The most similar sequences were aligned with the sequencing results in the programs BioEdit [24] and CLUSTAL_X [25]. Finally, the trees were displayed with the program TreeView. *Caloramator viterbiensis* was used as an outgroup.

2.5. Effect of Initial Glucose Concentration on End Product Formation. The effect of initial glucose concentration on strain J1, by varying the concentration from 5 to 200 mM, was tested. Control samples contained only yeast extract. Glucose, hydrogen, acetate, and ethanol concentrations were measured at the beginning and at the end of incubation time (7 days). Experiments were done in 117.5 mL serum bottles with 60 mL liquid medium, and the pH was measured at the end of incubation time.

2.6. Substrate Utilization Spectrum. The ability of strain J1 to utilize different substrates was tested using the BM medium supplemented with various carbon substrates (xylose, arabinose, glucose, mannose, galactose, fructose, rhamnose, maltose, cellobiose, sucrose, lactose, trehalose, raffinose, starch, cellulose, CMC, avicel, xylan (from oat spelt), glycerol, pyruvate, serine, and threonine). All substrates were added from filter-sterilized (0.45 μm) substrates except for xylan, starch, CMC, cellulose, and avicel which were autoclaved with the medium. In all cases, the concentration of substrates was 20 mM except for xylan, starch, CMC, cellulose, and avicel when $2\,g\cdot L^{-1}$ was used. Hydrogen, acetate, and ethanol concentrations were analysed after one week of incubation. Experiments were performed in 24.5 mL serum bottles with 10 mL liquid medium.

2.7. Pretreatment of Biomass and Hydrolysates Preparation. Hydrolysates (HLs) were made from different biomasses: Whatman no. 1 filter paper, newspaper, hemp stem (*Cannabis sativa*), barley straw (*Hordeum vulgare*), and grass (*Phleum pratense*). Hydrolysates were prepared according to Sveinsdottir et al. [12], and the final concentration of each biomass type was $22.5\,g\cdot L^{-1}$. Biomass was pretreated chemically by using 0.50% (v/v) of acid (H_2SO_4) or alkali (NaOH) (control was without chemical pretreatment) before heating (121°C, 60 min). Two commercial enzyme solutions, Celluclast (Novozyme, $750\,U\cdot g^{-1}$) and Novozyme 188 (Sigma C6105, $200\,U\cdot g^{-1}$), were added to each bottle after chemical pretreatment; the bottles were cooled down to

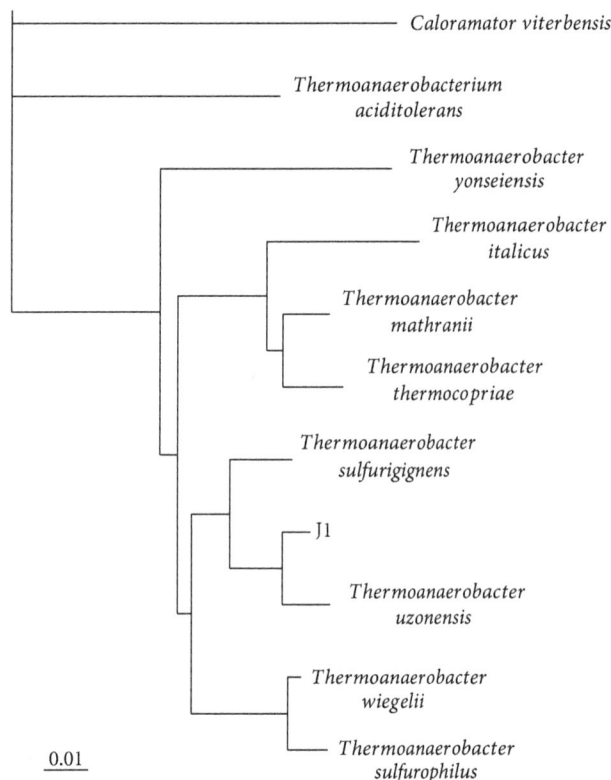

FIGURE 1: Phylogeny of strain J1 based on partial 16S rRNA sequence analysis. The phylogenetic tree was generated by using distance matrix and neighbor-joining algorithms. *Caloramator viterbensis* was selected as outgroup. The bar indicates 0.01 substitutions per nucleotide position.

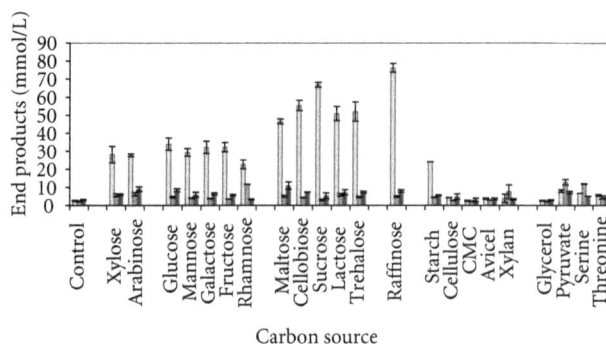

FIGURE 2: End product formation from various substrates by strain J1. Data represents average of two replicate experiments. Standard deviation are shown as error bars. From left to right; ethanol, acetate and hydrogen.

room temperature and the pH adjusted to 5.0 before enzymes were added. The hydrolysates were incubated in water bath at 45°C for 68 h. After the enzyme treatment, the pH was adjusted with NaOH or HCl to pH 7.0 which is the pH optimum of the strain. The hydrolysates were then filtered (Whatman-WeiBrand; 0.45 μm) into sterile bottles.

2.8. Fermentation during External Electron-Scavenging Systems. In one set of experiments, strain J1 was incubated on glucose (20 mM) in the presence of sodium thiosulfate (40 mM) and in coculture with a hydrogenotrophic methanogen. The methanogen was precultivated in BM medium with a gas phase consisting of 80% of H_2 and 20% of CO_2 for one week. Then the experimental culture bottles were flushed with nitrogen prior to the addition of glucose (20 mM) and strain J1. The coculture was incubated at 65°C for one week.

2.9. Fermentation of Hydrolysates. Fermentation of carbohydrates present in the hydrolysates after chemical and enzymatic pretreatment was performed in 24.5 mL serum bottles. The BM medium and inoculum (8.0 mL) were supplemented with different hydrolysates (2.0 mL, total liquid volume of 10 mL) giving a final hydrolysate concentration of 4.5 g·L^{-1}. Control samples did not contain hydrolysate; the only carbon source was yeast extract.

2.10. Analytical Methods. Hydrogen, ethanol, and volatile fatty acids were measured by gas chromatography as previously described [23]. Glucose was determined by slight modification of the method from Laurentin and Edwards [26]; supernatant broth (400 μL) was mixed with 2 mL of anthrone solution (0.2% (w/v) of anthrone in 72% (v/v) of sulphuric acid). The sample was boiled for 11 minutes and then cooled down on ice. Absorbance was then measured at 600 nm by using Perkin-Elmer Lambda 25 UV-Vis spectrophotometer.

3. Results and Discussion

3.1. Phylogeny. Figure 1 shows that strain J1 belongs to the genus *Thermoanaerobacter* with its closest neighbours being *T. uzonensis* (97.7% homology) and *T. sulfurigenes* (95.5%). The genus *Thermoanaerobacter* falls into clusters V in the phylogenetic interrelationship of *Clostridium* according to Collins and coworkers [27]. All species within the genus are obligate anaerobes and ferment various carbohydrates to ethanol, acetate, lactate, hydrogen, and carbon dioxide [20], while some species can degrade amino acids [28]. Most strains can reduce thiosulfate to hydrogen sulphide [20, 28]. Today, the genus consists of 18 species according to the Euzeby list of prokaryotes.

3.2. Optimum Growth Conditions. The strain was able to grow between 55.0°C and 75.0°C with optimal temperature being 65.0°C (μ_{max}; 0.23 h). The pH optimum was 7.0 (μ_{max}; 0.19 h). No growth was observed below pH 4.0 and above pH 9.0.

3.3. End Product Production from Sugars and Other Substrates. One of the main reasons for increased interest in using thermophilic bacteria for second-generation ethanol production is because of their broad substrate spectrum. Therefore, it was decided to cultivate the strain on the most common sugars present in lignocellulosic biomass as well as pyruvate, glycerol, serine, and threonine (Figure 2). Clearly, the strain is a very powerful ethanol producer; it produces 1.70 mol ethanol/mol glucose and 1.25 mol ethanol/mol

Production of Ethanol from Sugars and Lignocellulosic Biomass by Thermoanaerobacter J1 Isolated from a Hot Spring in Iceland

165

xylose (control values subtracted) or 85.0 and 75.0% of theoretical yields, respectively. The following stoichiometry from glucose and xylose was observed:

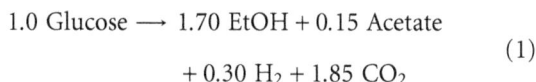

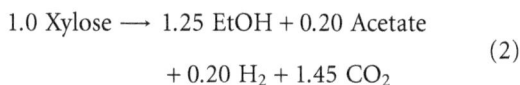

$$1.0 \text{ Glucose} \longrightarrow 1.70 \text{ EtOH} + 0.15 \text{ Acetate}$$
$$+ 0.30 \text{ H}_2 + 1.85 \text{ CO}_2 \quad (1)$$

$$1.0 \text{ Xylose} \longrightarrow 1.25 \text{ EtOH} + 0.20 \text{ Acetate}$$
$$+ 0.20 \text{ H}_2 + 1.45 \text{ CO}_2 \quad (2)$$

Lactate was not analysed in the present paper, but high carbon recoveries from analysed end products from glucose and xylose (92.5 and 87.4%, resp.) indicate that if it was produced, its significance is very little. The substrate spectrum of the strain shows a broad capacity in degrading pentoses (xylose, arabinose), hexoses (glucose, mannose, galactose, fructose, and rhamnose), disaccharides (maltose, cellobiose, lactose, trehalose, and sucrose) the trisaccharide raffinose, and starch, pyruvate, and serine. In all the cases, the major end product is ethanol except for serine and pyruvate in which acetate is the primary end product. The highest ethanol concentrations were produced from the trisaccharide raffinose (75.2 mM). As earlier mentioned, the strain is most closely related to *T. uzonensis* (strain JW/IW010) which also produces ethanol and acetate as the only volatile end products, but the ratio between ethanol and acetate is 1.35 in that strain [28]. However, *T. uzonensis* has a more narrow sugar degradation spectrum as compared to strain J1; it cannot degrade arabinose and rhamnose. Other well-known ethanol producers within the genus are *T. ethanolicus*, *T. thermohydrosulfuricus*, and *T. finnii* with yields between 1.5 and 1.9 mol ethanol/mol glucose [11, 13, 14, 29].

During growth on serine and pyruvate, the carbon flow was shifted away from ethanol to acetate and hydrogen. This can be explained by the oxidation state of these substrates as compared to sugars; the oxidation state of the carbon in glucose is zero, and during its oxidation to pyruvate, the electrons are transferred to NAD$^+$ leading to the formation of NADH. Reoxidation of NADH to NAD$^+$ by the strain occurs most likely through acetaldehyde dehydrogenase and alcohol dehydrogenase rendering ethanol as the main product. However, both pyruvate and serine are more oxidized substrates as compared to sugars (glucose), and there is no need to reoxidize NADH. Instead, the strain deaminates serine directly to pyruvate which is decarboxylated to acetyl phosphate (by phosphotransacetylase) and further to acetate (by acetate kinase) resulting in ATP formation. However, since hydrogen production is less as compared to acetate, it is likely that the strain is also producing formate (not analyzed) instead of hydrogen from these substrates.

3.4. Effect of Initial Glucose Loadings on Ethanol Production. High initial substrate concentrations may inhibit substrate utilization and/or decrease end product yields [5, 10, 30]. In closed systems, such as batch cultures, the limited buffer capacity of the medium may be overloaded by the accumulation of organic acids resulting in a pH drop and the inhibition of substrate fermentation utilization

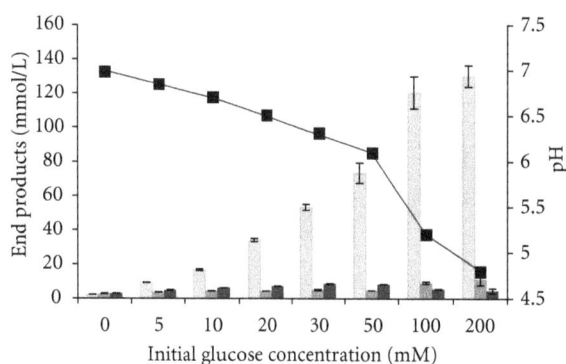

FIGURE 3: End product formation from different initial glucose concentrations. Also shown are percent of glucose degraded. Values represent means of two replicates and standard deviation are shown as error bars. Columns from left to right; ethanol, acetate, hydrogen. pH measered after fermentation (■).

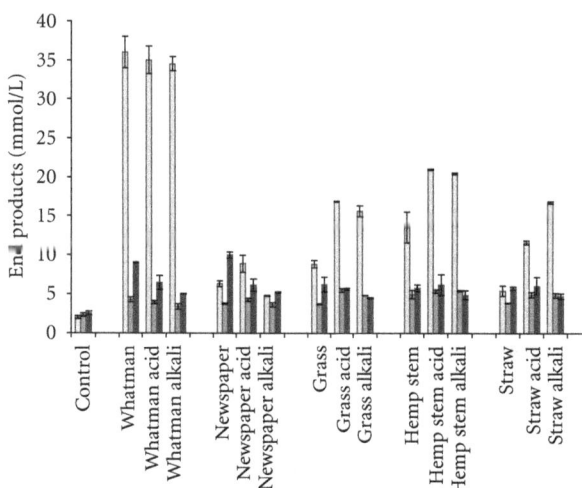

FIGURE 4: Production of end products from hydrolysates (4.5 g·L^{-1}) from different biomasses. Values represent mean of two replicates (±standard deviation). From left to right: ethanol, acetate, and hydrogen.

[30]. To investigate the influence of initial substrate concentration on end product formation, changes in pH, and substrate degradation, strain J1 was cultivated with different concentration of glucose (0 to 200 mM). The strain completely degraded glucose in all experiments, except for the highest (200 mM) initial glucose loadings, and ethanol yields were between 1.2 and 1.7 mol ethanol/mol glucose (Figure 3). Acetate formation increased from 2.7 mM in control bottles (without glucose) to 9.5 mM at 100 mM glucose concentrations which was directly linked to a decrease from pH 7.0 (control) to 5.2 (100 mM glucose). At 200 mM glucose concentrations, acetate was only slightly higher as compared to 100 mM glucose concentrations, the pH dropped from 5.2 to 4.8, and only 110 mM of glucose was degraded. Thus, the limit of glucose seems to be pH related, because of the formation of acetate, rather than

TABLE 1: Utilization of glucose by strain J1 in the presence of thiosulfate or a hydrogenotrophic methanogen. Data represents average of two replicate experiments ± standard deviation.

	Concentration (mmol·L^{-1})			
	Ethanol	Acetate	Hydrogen	Methane
Control	3.0 ± 0.1	2.9 ± 0.1	2.0 ± 0.1	0.0 ± 0.0
Control + S$_2$O$_3$	1.1 ± 0.1	5.2 ± 0.2	0.3 ± 0.0	0.0 ± 0.0
Control + methanogen	0.9 ± 0.5	4.9 ± 0.4	0.0 ± 0.0	2.4 ± 0.0
Glucose	29.0 ± 1.5	4.2 ± 0.3	7.2 ± 0.5	0.0 ± 0.0
Glucose + S$_2$O$_3$	20.0 ± 0.3	15.5 ± 2.1	0.3 ± 0.1	0.0 ± 0.0
Glucose + methanogen	4.1 ± 0.2	29.5 ± 1.2	0.5 ± 0.0	7.4 ± 1.2

substrate inhibition. The strain seems to be more tolerant for initial substrate concentrations as compared to many other thermophilic bacteria where often a concentration between 20 and 30 mM is too high for a complete degradation [7, 8]. In those cases, however, more acetate was produced as compared to ethanol and may be crucial for lowering the pH at lower substrate concentrations.

3.5. Effect of Hydrogen-Scavenging Systems on End Product Formation. It is well known that *Thermoanaerobacter* species are highly flexible concerning end product formation depending on the culture conditions. Fardeau et al. [31] showed a dramatic shift in end product formation by *Thermoanaerobacter finnii* when grown on glucose in the presence and absence of thiosulfate. In that case, both ethanol and lactate decreased during thiosulfate reduction to hydrogen sulphide, whereas the acetate concentration increased. The influence of using biological hydrogen-scavenging systems has also been investigated throughout *Thermoanaerobacter brockii* during amino acid degradation [27]. Both thiosulfate and the presence of a hydrogen-scavenging methanogen were crucial for the oxidative deamination of the branched chain amino acids by this strain. However, degradation of a substrate that is thermodynamically easier to degrade, for example, the amino acid serine, was completely degraded in the presence and absence of thiosulfate and *Methanobacterium* sp. although a shift occurred between ethanol and acetate formation [27]. To investigate the influence of low partial pressure (pH_2) on end product formation, strain J1 was cultivated in the presence of thiosulfate and in coculture with a hydrogenotrophic methanogen. As observed earlier, strain J1 produced ethanol as the main end product during glucose fermentation only (Table 1). The addition of thiosulfate to glucose fermentations resulted in a shift towards acetate from ethanol where the ratio between ethanol and acetate changed from 6.90 to 1.29. Cocultivating strain J1 with a hydrogenotrophic methanogen led even to more dramatic shift towards acetate (and methane), and the ratio of ethanol and acetate was 0.14. This difference in end product formation by using thiosulfate or a hydrogenotrophic methanogen is surprisingly big considering that the concentration of hydrogen is very low at the end of experimental time (0.3 to 0.5 mmol·L^{-1}) in both cases. This difference could be caused by more rapid uptake of hydrogen in the coculture experiment, but end products were only analysed at the end of the experimental time.

3.6. Fermentation of Hydrolysates from Lignocellulosic Biomass. The strain is producing maximally 33.9 mM (1.56 g/L) of ethanol from 4.5 g/L of hydrolysates made from cellulose (Figure 4). The yields on cellulose pretreated only with enzymes and heat are 7.5 mM·g^{-1} dry weight (dw) which is considered lower as compared to glucose degradation alone (1.70 mol ethanol/mol glucose; 9.4 mM·g^{-1} glucose). No glucose was analysed in the cellulose hydrolysate after fermentation. Thus, the lower ethanol yields on cellulose as compared to glucose indicate that the cellulose was not completely degraded during enzymatic hydrolysis. Chemical pretreatment of cellulose by the addition of acid or alkali did not increase the end product formation yields on cellulose. The highest ethanol yields on the more complex biomass types (without chemical pretreatment) were observed on hemp (11.6 mM; 2.6 mM·g^{-1} dw) but lowest on straw (3.5 mM; 0.8 mM·g^{-1} dw). Chemical pretreatment by adding either acid or alkali increased yields substantially on most of the lignocellulosic biomasses tested. The increase was most profound on hydrolysates from straw pretreated with alkali where ethanol production was increased from 3.5 to 14.8 mM (controls subtracted). The highest ethanol yields were however observed on hemp, 4.3 mM·g^{-1} dw (19.0 mM). The highest ethanol yields by *Thermoanaerobacter* species have been reported by continuous cultures of *Thermoanaerobacter* strain BG1L1 on wheat straw [17] and corn stover [18], or 8.5–9.2 mM·g^{-1} sugar consumed. *Thermoanaerobacter ethanolicus* has been reported to produce 4.5 and 4.8 mM ethanol·g^{-1} hexose equivalent degraded from wood hydrolysate and beet molasses, respectively [13, 32]. *Thermoanaerobacter mathranii,* isolated from the same geographical area in Iceland [33] as strain J1 produced 5.3 mM·g^{-1} sugar from wheat straw hydrolysate [34]. Recently, a new *Thermoanaerobacter* strain, AK$_5$ closely related to *T. thermohydrosulfuricus* and *T. ethanolicus,* was isolated from a hot spring in Iceland and has similar yields on cellulose (7.7 mM·g^{-1}), hemp (3.1 mM·g^{-1}), and grass (4.1 mM·g^{-1}) hydrolysates [22].

Production of Ethanol from Sugars and Lignocellulosic Biomass by Thermoanaerobacter J1 Isolated from a
Hot Spring in Iceland

167

4. Conclusion

Ethanol production was studied by *Thermoanaerobacter* J1 isolated from hot spring in Iceland. The main aim of the study was to investigate the importance of various factors on ethanol production from both sugars and complex lignocellulosic biomass. The strain produces 1.70 mol ethanol/mol glucose and 1.25 mol ethanol/mol xylose and shows a broad substrate spectrum, degrading various sugars and starch but not cellulosic substrates. High ethanol yields were observed at initial glucose concentrations up to 100 mM. During growth under hydrogen removal, a shift from ethanol to acetate formation occurs. The strain produces up to 7.5 mM ethanol·g^{-1} cellulose and 4.2 mM·g^{-1} hemp hydrolysate.

Authors' Contribution

J. E. Jessen carried out all experimental procedures. J. Orlygsson planned the experimental procedure and drafted the paper. Both authors read and approved the final paper.

Conflict of Interests

The authors declare that there is no conflict of interests.

Acknowledgments

This work was sponsored by RANNÍS, Technology Development Fund, projects 081303408 (BioEthanol) and RAN091016-2376 (BioFuel), and the Research Fund of the University of Akureyri. Special thanks are due to Margret Audur Sigurbjornsdottir for aligning the 16S rRNA sequences and building the phylogenetic tree and to Sean M. Scully for proofreading the paper.

References

[1] Renewable Fuels Association, "Choose ethanol," 2011, http://chooseethanol.com/what-is-ethanol/entry/ethanol-at-a-glance/.

[2] Ó. J. Sánchez and C. A. Cardona, "Trends in biotechnological production of fuel ethanol from different feedstocks," *Bioresource Technology*, vol. 99, no. 13, pp. 5270–5295, 2008.

[3] J. Zaldivar, J. Nielsen, and L. Olsson, "Fuel ethanol production from lignocellulose: a challenge for metabolic engineering and process integration," *Applied Microbiology and Biotechnology*, vol. 56, no. 1-2, pp. 17–34, 2001.

[4] J. Wiegel, "Formation of ethanol by bacteria. A pledge for the use of extreme thermophilic anaerobic bacteria in industrial ethanol fermentation processes," *Experientia*, vol. 36, no. 12, pp. 1434–1446, 1980.

[5] L. S. Lacis and H. G. Lawford, "Ethanol production from xylose by *Thermoanaerobacter ethanolicus* in batch and continuous culture," *Archives of Microbiology*, vol. 150, no. 1, pp. 48–55, 1988.

[6] R. Lamed and J. G. Zeikus, "Ethanol production by thermophilic bacteria: relationship between fermentation product yields of and catabolic enzyme activities in *Clostridium thermocellum* and *Thermoanaerobium brockii*," *Journal of Bacteriology*, vol. 144, no. 2, pp. 569–578, 1980.

[7] A. R. Almarsdottir, M. A. Sigurbjornsdottir, and J. Orlygsson, "Effect of various factors on ethanol yields from ligncellulosic biomass by *Thermoanaerobacterium* AK17," *Biotechnology and Bioengineering*, vol. 109, no. 3, pp. 686–694, 2012.

[8] M. A. Sigurbjornsdottir and J. Orlygsson, "Combined hydrogen and ethanol production from sugars and lignocellulosic biomass by *Thermoanaerobacterium* AK54," *Applied Energy*, vol. 97, pp. 785–791, 2012.

[9] M. P. Taylor, K. L. Eley, S. Martin, M. I. Tuffin, S. G. Burton, and D. A. Cowan, "Thermophilic ethanologenesis: future prospects for second-generation bioethanol production," *Trends in Biotechnology*, vol. 27, no. 7, pp. 398–405, 2009.

[10] P. Sommer, T. Georgieva, and B. K. Ahring, "Potential for using thermophilic anaerobic bacteria for bioethanol production from hemicellulose," *Biochemical Society Transactions*, vol. 32, no. 2, pp. 283–289, 2004.

[11] J. Wiegel and L. G. Ljungdahl, "*Thermoanaerobacter ethanolicus* gen. nov., spec. nov., a new, extreme thermophilic, anaerobic bacterium," *Archives of Microbiology*, vol. 128, no. 4, pp. 343–348, 1981.

[12] M. Sveinsdottir, S. R. B. Baldursson, and J. Orlygsson, "Ethanol production from monosugars and lignocellulosic biomass by thermophilic bacteria isolated from Icelandic hot springs," *Icelandic Agricultural Sciences*, vol. 22, pp. 45–58, 2009.

[13] A. Avci and S. Dönmez, "Effect of zinc on ethanol production by two *Thermoanaerobacter* strains," *Process Biochemistry*, vol. 41, no. 4, pp. 984–989, 2006.

[14] L. H. Carreira, J. Wiegel, and L. G. Ljungdahl, "Production of ethanol from biopolymers by anaerobic, thermophilic, and extreme thermophilic bacteria: I. Regulation of carbohydrate utilization in mutants of *Thermoanaerobacter ethanolicus*," *Biotechnology Bioengineering Symposium*, vol. 13, no. 13, pp. 183–191, 1983.

[15] P. E. P. Koskinen, S. R. Beck, J. Örlygsson, and J. A. Puhakka, "Ethanol and hydrogen production by two thermophilic, anaerobic bacteria isolated from Icelandic geothermal areas," *Biotechnology and Bioengineering*, vol. 101, no. 4, pp. 679–690, 2008.

[16] R. W. Lovitt, G. J. Shen, and J. G. Zeikus, "Ethanol production by thermophilic bacteria: biochemical basis for ethanol and hydrogen tolerance in *Clostridium Thermohydrosulfuricum*," *Journal of Bacteriology*, vol. 170, no. 6, pp. 2809–2815, 1988.

[17] T. I. Georgieva, M. J. Mikkelsen, and B. K. Ahring, "Ethanol production from wet-exploded wheat straw hydrolysate by thermophilic anaerobic bacterium *Thermoanaerobacter* BG1L1 in a continuous immobilized reactor," *Applied Biochemistry and Biotechnology*, vol. 145, no. 1–3, pp. 99–110, 2008.

[18] T. I. Georgieva and B. K. Ahring, "Evaluation of continuous ethanol fermentation of dilute-acid corn stover hydrolysate using thermophilic anaerobic bacterium *Thermoanaerobacter* BG1L1," *Applied Microbiology and Biotechnology*, vol. 77, no. 1, pp. 61–68, 2007.

[19] K. S. Rani, M. V. Swamy, and G. Seenayya, "Production of ethanol from various pure and natural cellulosic biomass by *Clostridium Thermocellum* strains SS21 and SS22," *Process Biochemistry*, vol. 33, no. 4, pp. 435–440, 1998.

[20] Y. E. Lee, M. K. Jain, C. Lee et al., "Taxonomic distinction of saccharolytic thermophilic anaerobes: description of *Thermoanaerobacterium xylanolyticum* gen. nov., sp. nov., and *Thermoanaerobacterium saccharolyticum* gen. nov., sp. nov.,

reclassification of *Thermoanaerobium brockii*, *Clostridium thermosulfurogenes* , and *Clostridium thermohydrosulfuricum* E100-69 as *Thermoanaerobacter brockii* comb. nov., *Thermoanaerobacter thermosulfurigenes* comb. nov., and *Thermoanaerobacter thermohydrosulfuricus* comb. nov., respectively, and transfer of *Clostridium Thermohydrosulfuricum* 39E to *Thermoanaerobacter ethanolicus*," *International Journal of Systematic Bacteriology*, vol. 43, no. 1, pp. 41–51, 1993.

[21] M. L. Fardeau, M. B. Salinas, S. L'Haridon et al., "Isolation from oil resorvoirs of novel thermophilic anaerobes phylogenetically related to *Thermoanaerobacter subterraneus*: reassignment of *T. subterraneus*, *Thermoanaerobacter yonseiensis*, *Thermoanaerobacter tengcongensis* and *Carboxydibrachium pacificum* to *Caldanaerobacter subterraneus* gen. nov., sp. nov., comb. nov. as four novel subspecies," *International Journal of Systematic and Evolutionary Microbiology*, vol. 54, no. 2, pp. 467–474, 2004.

[22] H. Brynjarsdottir, B. Wawiernia, and J. Orlygsson, "Ethanol production from sugars and complex biomass by *Thermoanaerobacter* AK5: the effect of electron scavenging systems on end product formation," *Energy & Fuels*, vol. 26, no. 7, pp. 4568–4574, 2012.

[23] J. Orlygsson and S. R. B. Baldursson, "Phylogenetic and physiological studies of four hydrogen-producing thermoanaerobes from Icelandic geothermal areas," *Icelandic Agricultural Sciences*, vol. 93, pp. 93–106, 2007.

[24] T. A. Hall, "BioEdit: a user-friendly biological sequence alignment editor and analysis program for Windows 95/98/NT," *Nucleic Acids Symposium Series*, vol. 41, pp. 95–98, 1999.

[25] J. D. Thompson, T. J. Gibson, F. Plewniak, F. Jeanmougin, and D. G. Higgins, "The CLUSTAL X windows interface: flexible strategies for multiple sequence alignment aided by quality analysis tools," *Nucleic Acids Research*, vol. 25, no. 24, pp. 4876–4882, 1997.

[26] A. Laurentin and C. A. Edwards, "A microtiter modification of the anthrone-sulfuric acid colorimetric assay for glucose-based carbohydrates," *Analytical Biochemistry*, vol. 315, no. 1, pp. 143–145, 2003.

[27] M. D. Collins, P. A. Lawson, A. Willems et al., "The phylogeny of the genus Clostridium: proposal of five new genera and eleven new species combinations," *International Journal of Systematic Bacteriology*, vol. 44, no. 4, pp. 812–826, 1994.

[28] M. L. Fardeau, B. K. C. Patel, M. Magot, and B. Ollivier, "Utilization of serine, leucine, isoleucine, and valine by *Thermoanaerobacter brockii* in the presence of thiosulfate or *Methanobacterium* sp. as electron acceptors," *Anaerobe*, vol. 3, no. 6, pp. 405–410, 1997.

[29] I. D. Wagner, W. Zhao, C. L. Zhang, C. S. Romanek, M. Rohde, and J. Wiegel, "*Thermoanaerobacter uzonensis* sp. nov., an anaerobic thermophilic bacterium isolated from a hot spring within the Uzon Caldera, Kamchatka, Far East Russia," *International Journal of Systematic and Evolutionary Microbiology*, vol. 58, no. 11, pp. 2565–2573, 2008.

[30] S. Van Ginkel, S. Sung, and J. J. Lay, "Biohydrogen production as a function of pH and substrate concentration," *Environmental Science and Technology*, vol. 35, no. 24, pp. 4726–4730, 2001.

[31] M. L. Fardeau, C. Faudon, J. L. Cayol, M. Magot, B. K. C. Patel, and B. Ollivier, "Effect of thiosulphate as electron acceptor on glucose and xylose oxidation by *Thermoanaerobacter finnii* and a *Thermoanaerobacter* sp. isolated from oil field water," *Research in Microbiology*, vol. 147, no. 3, pp. 159–165, 1996.

[32] J. Wiegel, L. H. Carreira, C. P. Mothershed, and J. Puls, "Production of ethanol from biopolymers by anaerobic, thermophilic, and extreme thermophilic bacteria. II. *Thermoanaerobacter ethanolicus* JW200 and its mutants in batch cultures and resting cell experiments," *Biotechnology Bioengineering Symposium*, vol. 13, no. 13, pp. 193–205, 1983.

[33] L. Larsen, P. Nielsen, and B. K. Ahring, "*Thermoanaerobacter mathranii* sp. nov., an ethanol-producing, extremely thermophilic anaerobic bacterium from a hot spring in Iceland," *Archives of Microbiology*, vol. 168, no. 2, pp. 114–119, 1997.

[34] B. K. Ahring, D. Licht, A. S. Schmidt, P. Sommer, and A. B. Thomsen, "Production of ethanol from wet oxidised wheat straw by *Thermoanaerobacter mathranii*," *Bioresource Technology*, vol. 68, no. 1, pp. 3–9, 1999.

17

Gene Silencing of 4-1BB by RNA Interference Inhibits Acute Rejection in Rats with Liver Transplantation

Yang Shi,[1] Shuqun Hu,[1] Qingwei Song,[1] Shengcai Yu,[1] Xiaojun Zhou,[2] Jun Yin,[2] Lei Qin,[2] and Haixin Qian[2]

[1] *Department of General Surgery, The Affiliated Hospital of Xuzhou Medical College, Xuzhou, Jiangsu 221002, China*
[2] *Department of General Surgery, The First Affiliated Hospital of Soochow University, Suzhou, Jiangsu 215006, China*

Correspondence should be addressed to Haixin Qian; qianhaixin228@126.com

Academic Editor: Andrew St. John

The 4-1BB signal pathway plays a key role in organ transplantation tolerance. In this study, we have investigated the effect of gene silencing of 4-1BB by RNA interference (RNAi) on the acute rejection in rats with liver transplantation. The recombination vector of lentivirus that contains shRNA targeting the 4-1BB gene (LV-sh4-1BB) was constructed. The liver transplantation was performed using the two-cuff technique. Brown-Norway (BN) recipient rats were infected by the recombinant LVs. The results showed that gene silencing of 4-1BB by RNAi downregulated the 4-1BB gene expression of the splenic lymphocytes *in vitro*, and the splenic lymphocytes isolated from the rats with liver transplantation. LV-sh4-1BB decreased the plasma levels of liver injury markers including AST, ALT, and BIL and also decreased the level of plasma IL-2 and IFN-γ in recipient rats with liver transplantation. Lentivirus-mediated delivery of shRNA targeting 4-1BB gene prolonged the survival time of recipient and alleviated the injury of liver morphology in recipient rats with liver transplantation. In conclusion, our results demonstrate that gene silencing of 4-1BB by RNA interference inhibits the acute rejection in rats with liver transplantation.

1. Introduction

Costimulatory pathways between antigen-presenting cells (APCs) and Tcells play a crucial role in T-cells activation and alloimmune responses [1, 2]. 4-1BB is a member of the tumor necrosis factor receptor superfamily. The functions of receptor mainly as a costimulatory molecule in T lymphocytes. 4-1BB is expressed in activated T cells and antigen presenting cells such as dendritic cells [3]. 4-1BB/4-1BB ligand (4-1BBL) signaling provides T-cells activation with costimulation, that is, dependent or independent of CD28 [4]. 4-1BB can supply sufficient costimulatory signals for T-cells activation [5]. Under the condition of repeated antigen stimulation, the downregulationg expresion of CD28 protein in activated T cells results in activation-induced celll death (AICD), whereas very few 4-1BB molecules may supply sufficient costimulatory signals to sustain T-cells activation and inhibit AICD [6].

As we all know, the liver transplantation is an effective therapy for end-stage liver disease. But the transplantation rejective response is a formidable problem after liver transplantation, especially acute rejection which is the main cause of early dysfunction and transplantation failure [7]. The incidence rate of acute rejection after liver transplantation is up to 50% to 70% [8]. The acute rejection after the organ transplantation is mainly caused by the activation of the immune system in the host. During the acute rejection, T cells are the most important effectors involved in antigen recognition, lymphcyte proliferation and differentiation, and immune regulation, cytolysis. 4-1BB plays a critical role in allograft rejection [9, 10]. Our previous study illuminated that the blockade of the 4-1BB/4-1BBL costimulatory pathway with 4-1BBL monoclonal antibody decreased the acute rejection in rat with liver transplantation [11].

RNA interference- (RNAi-) mediated gene silencing can be generated by the expression of a vector-mediated RNAi anywhere in the genome and is already being tested as a potential therapy in clinical trials for a number of diseases [12, 13]. The viral-based vectors have been developed as an

TABLE 1: Sense and antisense oligomers of shRNA targeting the 4-1BB gene of rats.

shRNA symbol	Target sequence	Start position	shRNA strand
shRNA1	GCTGTTACAAC ATGGTGGTCA	11	sense 5′-CACCGCTGTTACAACATGGTGGTCA TTCAAGAGA TGA CCACCATGTTGTAACAGCTTTTTTG-3′ antisense 5′-ACCTTGATTCCAACGGTAACCATCA TCTCTTGAA GGA TACCTTCAAGCATTCCGGAGGATTA-3′
shRNA2	GCAGTAAATAC CCTCCGGTCT	110	sense 5′-CACCGCAGTAAATACCCTCCGGTCT TTCAAGAGA AG ACCGGAGGGTATTTACTGCTTTTTTG -3′ antisense 5′-AACCCCTAGGGCTTAAGCAATCCAGT CTCTTGAA CCA TTTGGCACTTAAGCCATTGACTGAC 3′
shRNA3	GCAGAGTGTGT CAAGGCTATT	191	sense 5′-CACCGCAGAGTGTGTCAAGGCTAT TTCAAGAGA AT AGCCTTCGAGCACACTCTGCTTTTTTG -3′ antisense 5′-AGGTTACCGGGATCCTTGAACAATA CTCTTGAA CA ACCCGGTTTACCAAATACCGGAATTGG-3′
shRNANC	ATTGGACCAAG TGGTTCATAGC		sense 5′-CCTGACGATTAAGGACTAGGTCAGC TTCAAGAGA CCC CTTAAAGCTTTCGGACGTCGCGAAC-3′ antisense 5′-AAGTACCAACTACATCTGGGAACTT CTCTTGAA ACCC TGAATGGGTCATCGTAACTACAAC-3′

alternative strategy of gene therapy. Therefore, the aim of present study was to investigate the effect of gene silencing of 4-1BB by RNA interference on the acute rejection in rats with liver transplantation.

2. Materials and Methods

2.1. Lentiviral Vectors. The design of siRNA was performed according to the previously published guidelines by using the Ambion siRNA-finding software [14]. To minimize off-target effects, a BLAST homology search was systematically performed to ensure that a single-mRNA sequence was targeted. Replication-deficient, self-inactivating lentiviral vectors pcDNA-CMV-sh4-1BB-Lentivector (LV-sh4-1BB) and the empty vector (LV-NC) were generated as follows. As shown in Table 1, the target sequence of siRNA sequences targeting 4-1BB of rat (GenBank: NM-001025773) and control were as follows: shRNA1 (11–31), GCTGTTACA-ACATGGTGGTCA; shRNA2 (110–130), GCAGTAAAT-ACCCTCCGGTCT; shRNA3 (191–211), GCAGAGTGTGTC-AAGGCTATT; nonsilencing siRNA, ATTGGACCAAGT-GGTTCATAGC. The oligonucleotides were designed according to the structure of the siRNA sense strand-loop-siRNA antisense strand. The siRNA sequences targeting 4-1BB of rat were shown in Table 1. The oligonucleotides were synthesized and used for the construction of pcDNA-CMV-sh4-1BB-Lentivector. A negative control (LV-NC) with DNA oligos targeting 4-1BB of rat was also designed.

The shRNAs were cloned into lentiviral work vector pcDNA-CMV-Lentivector (Shanghai GeneChem Co. Ltd., Shanghai, China), which was linearized using restriction

endonucleases BamH I and Pst I. All constructs were verified by sequence analysis. The recombinant lentiviral vectors were designated as LV-sh4-1BB1, LV-shP4-1BB2, LV-shP4-1BB3, and LV-NC. The recombinant work vector and package plasmids were cotransduced into 293T cells using Lipofectamine 2000 (Invitrogen, Carlsbad, CA, USA) for generating the lentivirus. After the cells were cultivated for 48 h, the culture medium was collected, concentrated by ultracentrifugation, aliquoted, and stored at −80°C until used. Virus titer is the number of cells expressing green fluorescent protein (GFP) multiplied by the corresponding dilution. The titer of lentivirus was determined by hole-by-dilution titer assay. The final titer of recombinant virus was 2×10^9 transducing units (TU)/mL.

2.2. 293T-Cell Culture and Transduction. The 293T packaging cell line (Academy of Life Science, Shanghai, China) was cultivated in Dulbecco's modified Eagle's medium (DMEM) (Invitrogen, Carlsbad, CA, USA) supplemented with 10% fetal bovine serum (FBS) (Invitrogen, Carlsbad, CA, USA), 100 mg/mL streptomycin and 100 U/mL penicillin at 37°C in a humidified incubator containing 5% CO_2. Cells were seeded in 24-well plates at 50–70% confluence 24 h prior to transduction. To analyze the transduction efficiency, 293T cells were gated to determine the percentage of GFP-positive cells. Cells with >85% viability were cultivated for additional experiments.

2.3. T-Lymphocytes Isolation, Culture, and Transduction. Spleens were harvested from Brown-Norway (BN) rats

(Animal Center of Soochow University, Suzhou, China) euthanazed by cervical dislocation. The procedures were performed according to the local guidelines for animal research approved by the Administrative Committee of Experimental Animal Care and Use of Soochow University. Single-cell suspensions were prepared by forcing tissue through a fine wire mesh using a syringe plunger followed by repeated pipetting in culture medium. RBC depletion involved cell lysis in 5 mL lysing buffer [0.14 mol/L NH_4Cl, 0.017 mol/L Tris-base (pH 7.5)] for 5 min at 20°C followed by three washes in ice-cold medium. To activated T cell, phytohemagglutinin (50 μg/mL) was added in the culture medium [15].

The T lymphocytes from BN rats were cultivated in RPMI 1640 culture medium (Invitrogen, Carlsbad, CA, USA) supplemented with 10% fetal calf serum (FCS) (Invitrogen, Carlsbad, CA, USA). Transient transduction was performed using Lipofectamine 2000 (Invitrogen, Carlsbad, CA, USA) according to the manufacturer's instructions in six- or 12-well plates with cells at 70–90% confluence.

2.4. Real-Time Quantitative PCR Analysis. Total RNA was extracted by using Trizol reagent (Invitrogen, Carlsbad, CA, USA) according to the manufacturer's instructions. Briefly, the splenic lymphocyte was treated with 1.5 mL Trizol reagent and kept at room temperature for 5 min. After the cells were incubated with 600 μL chloroform for 3 min, the total RNA was centrifuged for 20 min at 12,000 g and 4°C. The total RNA was precipitated with 1000 μL isopropanol. After incubation of 10 min at room temperature, the RNA was precipitated for 10 min at 16.000 g and 4°C. The resulting RNA-pellet was washed with 2000 μL isopropanol. The concentration of RNA was assessed photometrically at 260 nm. The reactions were run on a Roche light Cycler Run 5.32 Real-Time PCR System (Bio-Rad, Hercules, CA, USA) with the following cycle conditions: 95°C for 15 sec, 45 cycles at 95°C for 10 sec, and at 60°C for 30 sec. Melt curve analyses of all real-time PCR products were performed and shown to produce a single-DNA duplex. Quantitative measurements were determined using the Δ Δ Ct method and expression of β-actin was used as the internal control. The mRNA expression of the control group was expressed as 100%. Fold-induction of mRNA expression was calculated [16]. The sequences of the 4-1BB gene primers were as follows: forward primer, 5′-GTGTCAAGGCTATTTCAG-3′; reverse primer, 5′-AGACCACGTCTTTCTCC-3′. the product was 275 bp. β-Actin served as a control for normalization. The primers were as follows: forward primer, 5′-CCTCAT-GAAGATCCTGACCG-3′; reverse primer, 5′-AGCCAG-GGCAGTAATCTCCT-3′. The product was 488 bp.

2.5. Western Blot Analysis. The splenic lymphocytes were collected. Cells were washed with cold phosphate-buffered saline (PBS) containing 2 mmoL/L EDTA and lysed with denaturing SDS-PAGE sample buffer using standard methods. Protein extract was prepared. The protein concentration in supernatant was determined by the BCA assay. The final concentration of protein in each sample was adjusted to 2 mg/mL. Protein lysates were separated by 12% SDS-PAGE

and transferred onto polyvinylidene fluoride membrane (Millipore, Bedford, MA). The membranes were blocked with 5% skim milk for 1 h at 37°C. Then the membranes were incubated with rabbit anti-4-1BB monoclonal antibody (dilution at 1 : 200) and anti-β-actin antibody (dilution at 1 : 300) (Santa Cruz Biotechnology, Santa Cruz, CA, USA) at 48°C overnight. After the membranes were washed three times with Tris-buffered saline, they were incubated with horseradish peroxidase (HRP)—conjugated goat anti-rabbit immunoglobulin G (IgG) antibody (dilution at 1 : 1,000) (Santa Cruz) at room temperature for 4 h. Immunoreactivity was detected by enhanced chemiluminescence (ECL). The band density was measured with Quantity One analysis software (Bio-Rad, Hercules, CA, USA), and the quantification of 4-1BB to β-actin levels was done by densitometry analysis [17].

2.6. Animals and Rat Liver Transplantation Models. Inbred male Lewis (LEW) and BN rats weighing 200–250 g were purchased from the Vital River Laboratories (Beijing, China). All animals were housed under conditions of constant temperature (22°C) and humidity in a specific-pathogen-free facility. The rats were fed with the commercial rat chow pellets. Male LEW rats were used as the liver donors and BN rats as the recipients.

Orthotopic liver transplantation was performed using the "two-cuff technique" as previously described [11, 18]. Briefly, the donors and recipients were anesthetized with the ether. The suprahepatic vena cava was reconstructed using continuous 8-0 polypropylene sutures. The hepatic artery was not reconstructed. The portal vein was reanastomosed using a polyethylene cuff (8F). When the anastomosis of the portal vein and suprahepatic vena cava was completed, the liver was reperfused. The anastomosis of the infrahepatic vena cava was then completed by the same cuff technique (6F). The bile duct was anastomosed with an intraluminal polyethylene stent (22G). The transplantation procedure required less than 60 min, during which the portal vein was clamped for 13 to 15 min. The procedures were performed according to the local guidelines for animal research approved by the Administrative Committee of Experimental Animal Care and Use of Soochow University.

2.7. Experimental Animal Grouping. Sixty four BN rats were randomly divided into four groups: sham group (S), liver transplantation model group (M), LV-sh4-1BB1 group (LV-sh4-1BB1), and LV-NC group (LV-NC). The rats of the sham and model group were injected 1 mL saline via the dorsal penis vein, respectively, before operation. The rats of the LV-sh4-1BB1 group and LV-NC group were injected 1 mL recombinant LVs and empty LVs via the dorsal penis vein, respectively, before liver transplantation.

On the 7th day after transplantation, 8 rats in each group were killed by cervical dislocation. The lymphocytes were obtained from spleen and used for measurement of 4-1BB expression by real-time RT-PCR and Western blot analysis. Blood samples were gathered by the inferior vena cava and used for biochemistry tests and cytokine assay. The liver lobes were excised to study the pathological changes. The remanent rats in each group were used to recored the survival time of posttransplantation.

2.8. Analysis of Plasma Liver Function Markers and Cytokines. On the 7th day after transplantation, serum aspartate transaminase (AST), alanine aminotransferase (ALT), lactate dehydrogenase (LDH), and bilirubin (BIL) were measured by using the automatic biochemical meter (TMS1024, Tokyo Bokei, Japan). The concentrations of interleukin-2 (IL-2), IL-10, and interferon-γ (IFN-γ) in plasma were tested using enzyme-linked immunosorbent assay (ELISA) kits (Biosource International, Inc. Camarillo, CA, USA). All the procedures were performed according to the instruction of the manufacturers.

2.9. Histological and Morphometric Analysis of Liver Grafts. On the 7th day posttransplantation, recipient rats were sacrificed. The grafted liver samples were fixed in 10% formalin and embedded in paraffin. Five micrometer thick sections were affixed to slides, deparaffinized, and stained with hematoxylin and eosin. The severity of acute rejection was assessed in a blinded fashion with a rejection activity index (RAI) according to Banff criterion [11, 19, 20].

Grafted livers were fixed in 4% glutaraldehyde, dehydrated with ethanol, and then embedded in Epon812. Liver sections were sliced and stained with uranium lead. Microstructure was read using Hitachi H-600 electron microscopy (Hitachi, Japan).

2.10. Statistical Analysis. Data are presented as mean ± standard deviation (SD). The statistical evaluation of differences in recipient survival was performed using the log rank test applied to Kaplan-Meier plots. All other statistical comparisons among groups were conducted using analysis of variance (ANOVA) with subsequent Dunnett's t-test. Significance was defined as $P < 0.05$.

3. Results

3.1. PCR Identification of Constructed shRNA Expression Plasmids and Titer Determination and Packaging of the Lentiviral Vector. The pcDNA-CMV-Lentivector 4-1BB shRNA expression plasmids were identified using a restriction endonuclease digestion. The lengths of pcDNA-CMV-Lentivector is 7.8 kb and pcDNA-CMV-Lentivector contains BamH I and Pst I two restriction sites. The product of pcDNA-CMV-Lentivector is linear large fragment after the double enzyme. The product of the recombination vector of lentivirus containing shRNA targeting the 4-1BB gene were two fragments, one was 7.8 kb and the other was 326 bp. As shown in Figure 1, results of DNA sequencing were as expected. The recombination vector of lentivirus encoding the specially designed shRNA against the 4-1BB gene was named pcDNA-CMV-sh4-1BB-Lentivector (LV-sh4-1BB). The empty vector was used as blank-control and named pcDNA-CMV-Lentivector (LV-BC). The negative-control shRNA was named LV-NC.

Forty-eight hours after cotransfection of the three-plasmid lentiviral vector into 293T cells, strong green fluorescence was observed using an inverted fluorescence

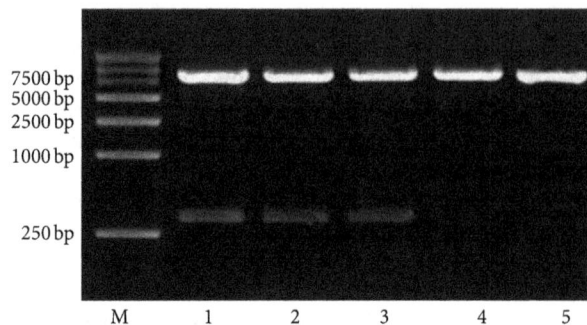

FIGURE 1: The identification of constructed shRNA expression plasmids by restriction enzyme. Lane 1: LV-sh4-1BB1; lane 2: LV-sh4-1BB2; lanes 3: LV-sh4-1BB3; lanes 4: empty pcDNA-CMV-Lentivector (blank-control); 5: negative-control shRNA (LV-NC).

microscope. After a single exposure of 293T cells to the lentivirus, a high percentage (>90%) of transfectants expressed GFP at 48 h after the transduction, indicating a high and stable transduction of the lentiviral vector system (data not shown).

3.2. Efficiency of Lentiviral Transfection into the Splenic Lymphocytes. To demonstrate the efficiency of siRNA delivery into the splenic lymphocytes, the splenic lymphocytes were infected by LV-sh4-1BB and LV-NC labeled with GFP. Successful lentiviral transfection was evidenced by green fluorescence under fluorescence microscopy 72 h after transduction (Figure 2).

3.3. Lentivirus-Mediated Delivery of shRNA Inhibits 4-1BB Expression in the Splenic Lymphocytes In Vitro. The mRNA expression of 4-1BB in the splenic lymphocytes was measured by real-time PCR 72 h after transduction. As shown in Figure 3(a), the mRNA expression of 4-1BB in the splenic lymphocytes was decreased approximately 96.8%, 94.1, and 95.2%, respectively, by LV-sh4-1BB1, LV-sh4-1BB2, and LV-sh4-1BB3 transduction compared to the empty LV-NC (all $P < 0.05$). These results indicated that the mRNA sequences corresponding to the 4-1BB gene shRNA were specific RNAi targets.

Western blot analysis was performed 72 h after transduction. The results showed that LV-sh4-1BB1, LV-sh4-1BB2, and LV-sh4-1BB3 induced an 85.3%, 84.6, and 84.5% downregulation of the 4-1BB protein level, respectively, compared with the empty LV-NC (all $P < 0.05$) (Figures 3(b) and 3(c)). So among three 4-1BB siRNAs, LV-sh4-1BB1 was selected as the best performing siRNA for use in the next study.

3.4. Lentivirus-Mediated Delivery of shRNA Targeting 4-1BB Gene Downregulated the 4-1BB Expression of the Splenic Lymphocytes Isolated from the Rats with Liver Transplantation. To evaluate the inhibition of 4-1BB mRNA expression in the splenic lymphocytes isolated from the recipient rats, real-time PCR was performed 7 days after transduction. As shown in Figure 4(a), the mRNA expression of 4-1BB in the model of liver transplantation group was significantly

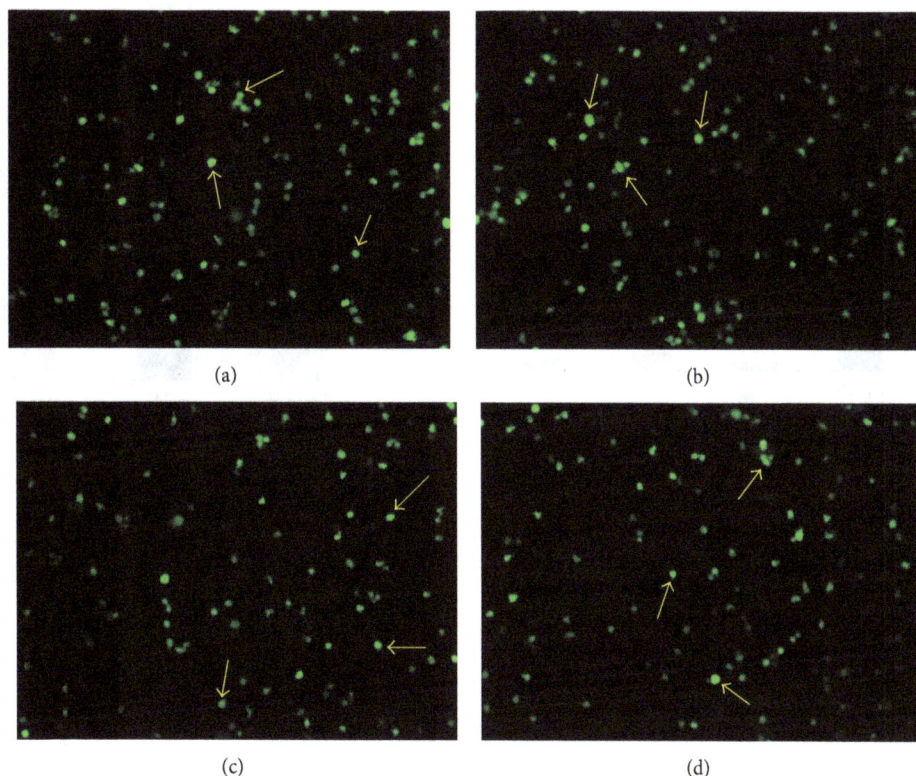

FIGURE 2: Representative fluorescence microscopy images of splenic lymphocytes 72 h after transduction with pcDNA-CMV-sh4-1BB-Lentivector1-3 (LV-sh4-1BB1-3) (a–c) and negative-control shRNA group (LV-NC) (d) (×100). Arrows indicate GFP fluorescence in the splenic lymphocytes.

TABLE 2: The levels of AST, LDH, ALT, and T-BIL in plasma of BN recipients rats 7 days after liver transplantation.

Groups	ALT (U/L)	AST (U/L)	LDH (U/L)	T-BIL (mg/L)
Sham (Control)	292.43 ± 32.41	428.39 ± 34.66	859.34 ± 94.36	75.39 ± 8.12
Model	580.52 ± 51.96#	738.11 ± 75.63#	1404.65 ± 168.74#	83.94 ± 9.53
LV-NC	577.21 ± 61.14	784.7 ± 47.85	1553.19 ± 125.83	84.38 ± 10.46
LV-sh4-1BB1	333.26 ± 39.92*	457.7 ± 43.61*	942.81 ± 83.17*	76.84 ± 8.43

The recombination vector of lentivirus containing shRNA targeting the 4-1BB gene was constructed. The liver transplantation was performed using the two-cuff technique. BN recipient rats were infected by the recombinant LVs. The blood samples were gathered 7 days after liver transplantation. The liver function markers were tested by using the automatic biochemical meter. Model: liver transplantation model group; LV-NC: negative-control shRNA group; LV-sh4-1BB1: pcDNA-CMV-sh4-1BB-Lentivector1 group. Data were expressed as means ± SD, $n = 8$. #$P < 0.05$ compared with the sham (control) group; *$P < 0.05$ compared with the LV-NC group.

upregulated compared with the sham group ($P < 0.05$). LV-sh4-1BB1 transduction resulted in a 91.8% reduction of 4-1BB mRNA expression compared to the LV-NC groups ($P < 0.05$). Western blot analysis was performed 7 days after transduction. The results showed that the protein expression of 4-1BB was significantly upregulated in the model of liver transplantation group compared with the sham group ($P < 0.05$). LV-sh4-1BB1 transduction resulted in a 90.3% reduction of 4-1BB protein expression compared to the LV-NC groups ($P < 0.05$) (Figures 4(b) and 4(c)).

3.5. Lentivirus-Mediated Delivery of shRNA Targeting 4-1BB Gene Prolonged Recipient Survival in Rats with Liver Transplantation. As shown in Figure 5, the survival time of BN recipients rats after liver transplantation was significantly

increased in LV-sh4-1BB1 group compared with the LV-NC group ($P < 0.05$). The mean survival time (MST) of rats in LV-NC group was 12 days (range 8–14 days) and the MST of rats in LV-sh4-1BB1 group was 34.5 days (range 15–48 days). 12-day survival rate was 62.5% and 100.0% in LV-NC and LV-sh4-1BB1 group, respectively.

3.6. Lentivirus-Mediated Delivery of shRNA Targeting 4-1BB Gene Decreased the Level of Liver Function Damage in Recipient Rats with Liver Transplantation. As shown in Table 2, compared with the sham group, the plasma concentrations of AST, LDH, and ALT in liver transplantation model group were significantly increased 7 days after liver transplantation ($P < 0.05$). But the plasma concentrations of AST, LDH, and ALT in LV-sh4-1BB1 group were significantly

(a)

(b)

(c)

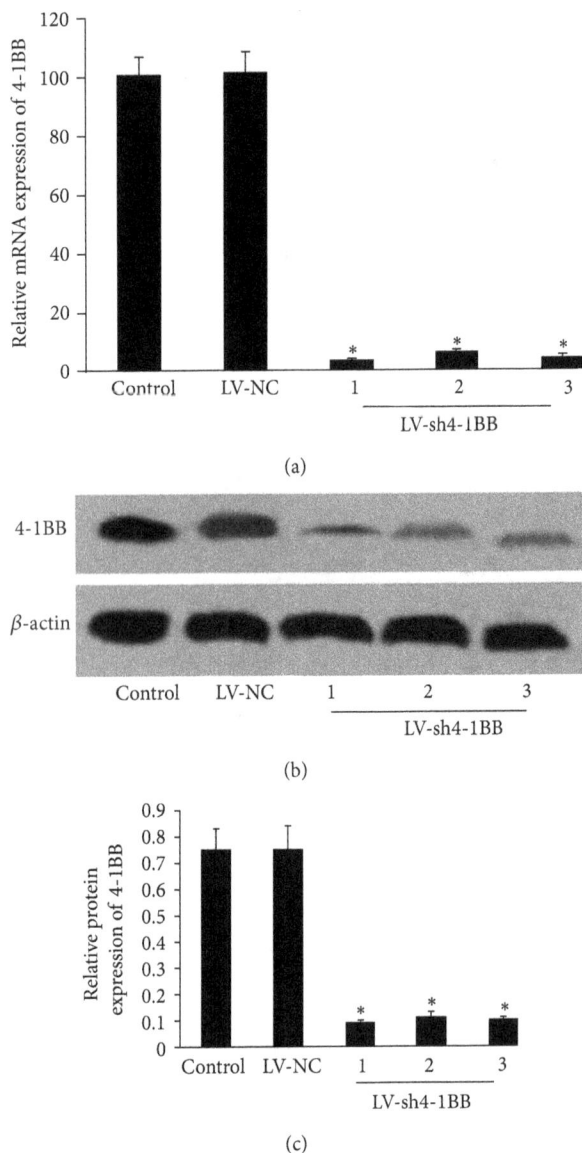

FIGURE 3: Effect of 4-1BB siRNA transduction on the expression of 4-1BB in the splenic lymphocytes. The recombination vector of lentivirus containing shRNA targeting the 4-1BB gene was constructed. Lymphocytes from BN rats were infected by recombinant LVs. ddH$_2$O was used as the control. (a) Real-time PCR analysis of the mRNA level of 4-1BB gene in the splenic lymphocytes 72 h after transduction. (b and c) The protein level of 4-1BB gene in the splenic lymphocytes was showed by Western blot analysis 72 h after transduction. β-actin was used as an internal standard. LV-NC: negative-control shRNA group; LV-sh4-1BB1-3: pcDNA-CMV-sh4-1BB1-3-Lentivector group. Data were expressed as means ± SD. *$P < 0.05$ compared with the LV-NC group.

decreased compared to the LV-NC group 7 days after liver transplantation ($P < 0.05$).

3.7. Lentivirus-Mediated Delivery of shRNA Targeting 4-1BB Gene Decreased the Level of Plasma IL-2, IL-10, and IFN-γ in Rats with Liver Transplantation. The concentrations of plasma IL-2 and IFN-γ in liver transplantation model group

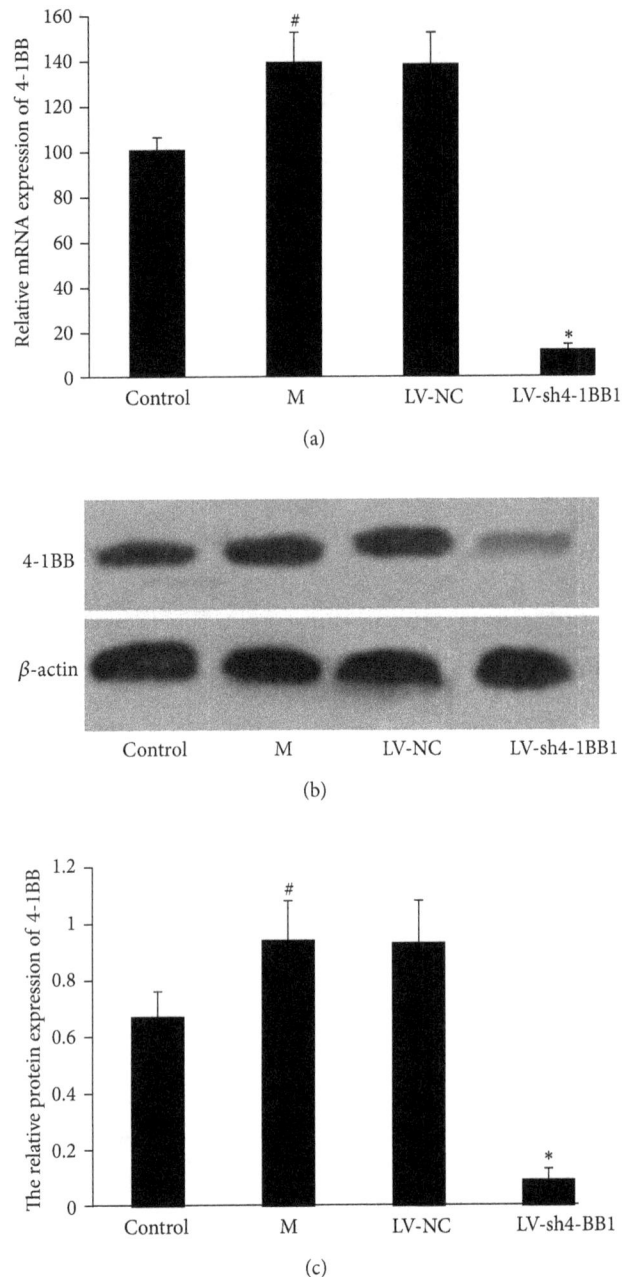

(a)

(b)

(c)

FIGURE 4: Effect of lentivirus-mediated delivery of shRNA targeting 4-1BB gene on 4-1BB expression of the splenic lymphocytes isolated from the recipient rats with liver transplantation. The recombination vector of lentivirus contains shRNA targeting the 4-1BB gene was constructed. The liver transplantation was performed using the two-cuff technique. BN recipient rats were infected by the recombinant LVs. The splenic lymphocytes from the recipient rats were gathered 7 days after transduction. Sham was used as the control. (a) Real-time PCR analysis of the mRNA level of 4-1BB gene in the splenic lymphocytes 7 days after transduction. (b and c) The protein level of 4-1BB gene in the splenic lymphocytes was showed by Western blot analysis 7 days after transduction. β-actin was used as an internal standard. M: the liver transplantation group; LV-NC: negative-control shRNA group; LV-sh4-1BB1: pcDNA-CMV-sh4-1BB1-Lentivector group. Data were expressed as means ± SD. #$P < 0.05$ compared with the sham (control) group; *$P < 0.05$ compared with the LV-NC group.

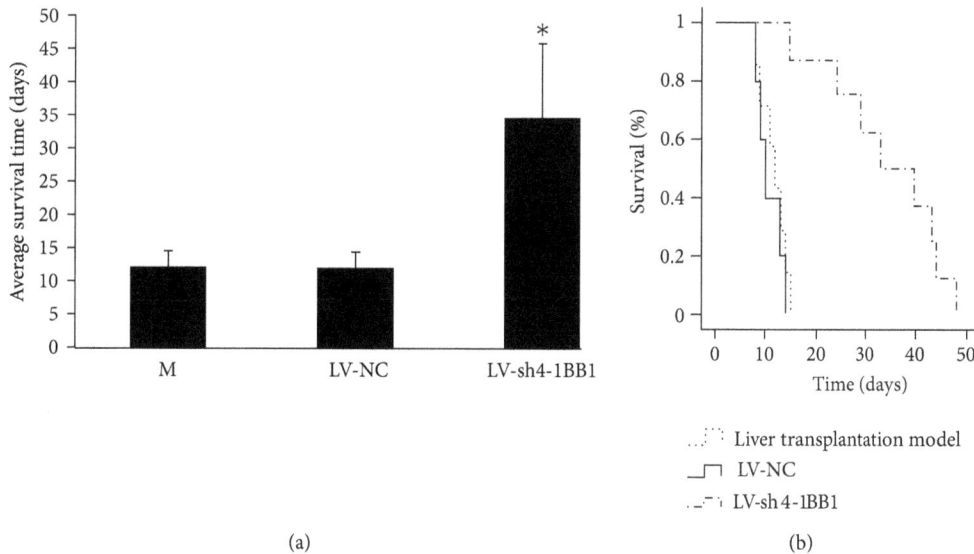

(a)   (b)

FIGURE 5: Effect of lentivirus-mediated delivery of shRNA targeting 4-1BB gene on the survival time in BN recipients rats after liver transplantation. The recombination vector of lentivirus contains shRNA targeting the 4-1BB gene (LV-sh4-1BB1) was constructed. The liver transplantation was performed using the two-cuff technique. BN recipient rats were infected by the recombinant LVs. The survival time of recipients rats was recorded and the percentage of survival was calculated. M: the liver transplantation model group; LV-NC: negative-control shRNA group; LV-sh4-1BB1: pcDNA-CMV-sh4-1BB1-Lentivector group. Data were expressed as means ± SD, $n = 8$. $^*P < 0.05$ compared with the LV-NC group.

TABLE 3: The levels of IL-2, IL-10, and IFN-γ in plasma of BN recipients rats 7 days after liver transplantation.

Groups	IL-2 (pg/mL)	IL-10 (pg/mL)	IFN-γ (pg/mL)
Sham (Control)	71.32 ± 8.67	16.85 ± 2.03	52.49 ± 6.75
Model	139.15 ± 13.73#	15.47 ± 1.27	86.2 ± 8.27#
LV-NC	142.86 ± 16.58	18.35 ± 2.06	80.3 ± 11.64
LV-sh4-1BB1	78.05 ± 8.64*	16.59 ± 1.85	57.86 ± 8.33*

The recombination vector of lentivirus contains shRNA targeting the 4-1BB gene was constructed. The liver transplantation was performed using the two-cuff technique. BN recipient rats were infected by the recombinant LVs. The blood samples were gathered 7 days after liver transplantation. The concentrations of interleukin-2 (IL-2), IL-10, and interferon-γ (IFN-γ) in plasma were tested using enzyme-linked immunosorbent assay (ELISA). Model: liver transplantation model group; LV-NC: negative-control shRNA group; LV-sh4-1BB1: pcDNA-CMV-sh4-1BB-Lentivector1 group. Data were expressed as means ± SD, $n = 8$. #$P < 0.05$ compared with the sham (control) group; *$P < 0.05$ compared with the LV-NC group.

were significantly increased 7 days after liver transplantation compared with the sham group. However, the plasma concentrations of plasma IL-2 and IFN-γ in LV-sh4-1BB1 group were significantly decreased compared with the LV-NC group 7 days after liver transplantation ($P < 0.05$) (Table 3).

3.8. Lentivirus-Mediated Delivery of shRNA Targeting 4-1BB Gene Alleviated the Injury of Liver Morphology in Recipient Rats with Liver Transplantation. As shown in Figure 6, the severe portal lymphocyte infiltration, the injuries of the portal area and interlobular bile duct, cholangitis, and bridged necrosis in liver parenchyma were observed detected 7 days after liver transplantation in the liver transplantation model group. The rejection was significantly inhibited in the LV-sh4-1BB1 group compared with the LV-NC group. Mild-to-moderate portal inflammatory infiltration, lower grade of endothelialitis, mild bile duct injuries, and no evident hepatocyte necrosis were detected 7 days after liver transplantation in the LV-sh4-1BB1 group.

The histological grade in Banff score in liver transplantation model group was significantly increased 7 days after liver transplantation compared with the sham control group (7.11 ± 0.78 versus 5.17 ± 0.68, $P < 0.05$) (Figure 7). Compared with the LV-NC group, the histological grade in Banff score in LV-sh4-1BB1 group was significantly decreased 7 days after liver transplantation (7.15 ± 0.84 versus 5.89 ± 0.79, $P < 0.05$).

Electron micrographs showed that a typical early stage apoptosis of hepatocytes, swollen mitochondria, and dilatation of endoplasmic reticulum were observed in a majority of hepatocytes 7 days after liver transplantation in liver transplantation model group. But these phenomenons were not observed in LV-sh4-1BB1 group (Figure 8).

4. Discussion

The orthotopic liver transplantation has become the most effective therapy for the patients with end-stage liver disease. However, organ rejection is a thorny problem after transplantation, especially acute rejection which is the main cause

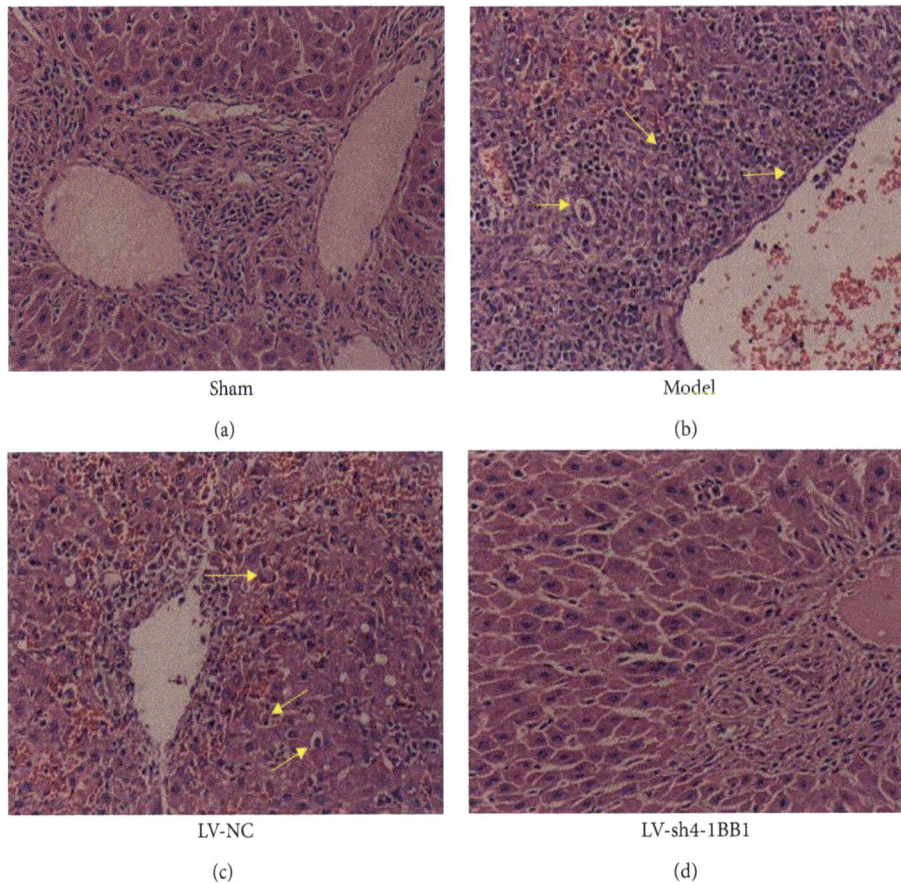

FIGURE 6: The pathology of liver in recipient rats with liver transplantation (HE, original magnification ×400). The recombination vector of lentivirus contains shRNA targeting the 4-1BB gene was constructed. The liver transplantation was performed using the two-cuff technique. BN recipient rats were infected by the recombinant LVs. The pathology of liver in recipient rats was observed by H-E staining 7 days after liver transplantation. Model: liver transplantation model group; LV-NC: negative-control shRNA group; LV-sh4-1BB1: pcDNA-CMV-sh4-1BB1-Lentivector group.

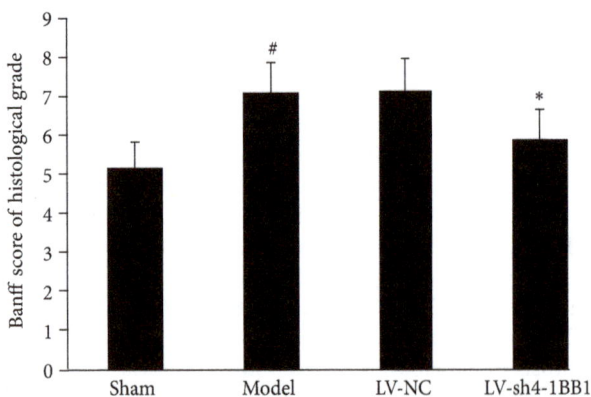

FIGURE 7: The Banff score of histological grade. Model: liver transplantation model group; LV-NC: negative-control shRNA group; LV-sh4-1BB1: pcDNA-CMV-sh4-1BB1-Lentivector group. Data were expressed as means ± SD, $n = 8$. $^{\#}P < 0.05$ compared with the sham group; $^{*}P < 0.05$ compared with the LV-NC group.

of early dysfunction and retransplantation [21]. Therefore, it is very important to search new and effective methods to prevent and inhibit the organ rejection response in organ transplantation.

Transplantation rejective response is principally mediated by T cells in the peripheral circulation and a variety of inflammatory stimuli that induce T cells to infiltrate the transplanted tissue [22]. CD4$^+$ T cells secrete a number of cytokines which may induce cell infiltration in the graft after allogeneic recognition, and CD8$^+$ T cells are involved in the direct cytotoxicity towards the liver graft [23]. Deletion, anergy, regulation/suppression, ignorance, or induction of activation induced T-cell death will be ways to suppress acute rejection and induce immune tolerance [24].

4-1BB is a family member of tumor necrosis factor receptor. 4-1BB is an important T-cell costimulatory molecule and expressed in activated cytolytic and helper T cells, as well as natural killers (NK) cells. The ligand of 4-1BB (4-1BBL) is expressed in B cells, macrophages, and dendritic cells. 4-1BB signaling preferentially promotes the proliferation and survival of CD8$^+$ T cells and promotes the production of IL-2 in CD4$^+$ T cells and prevents activation-induced cell death. 4-1BB is involved in the activation and survival of CD4$^+$, CD8$^+$, and NK cells [25]. The activation of T cells in the absence of costimulation is futile because T cells deprived of costimulatory signals enter a state of unresponsiveness or anergy.

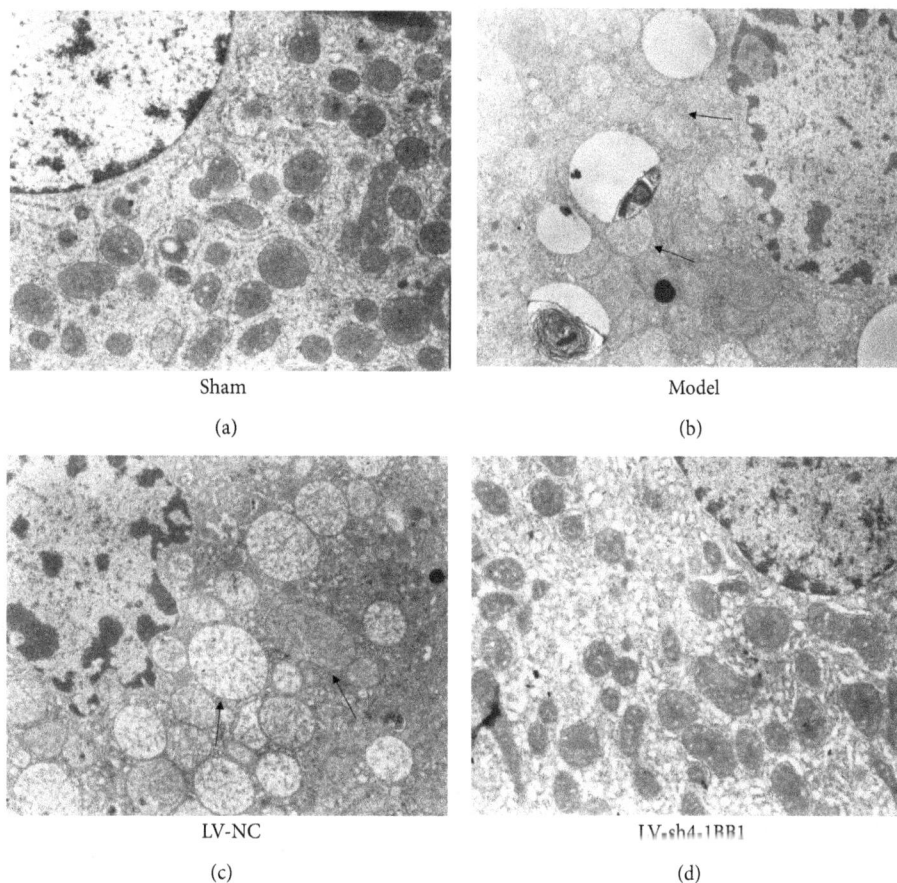

FIGURE 8: The electron micrographs of liver in recipient rats with liver transplantation (original magnification ×12000). The recombination vector of lentivirus contains shRNA targeting the 4-1BB gene was constructed. The liver transplantation was performed using the two-cuff technique. BN recipient rats were infected by the recombinant LVs. The ultrastructure of liver in recipient rats was observed by electron microscope 7 days after liver transplantation. Model: liver transplantation model group; LV-NC: negative-control shRNA group; LV-sh4-1BB1: pcDNA-CMV-sh4-1BB1-Lentivector group.

The interaction of 4-1BB and 4-1BBL can activate an important costimulatory pathway which plays the diverse and important roles in immune response and organ transplantation tolerance. The previous studies showed that blocking the 4-1BB/4-1BBL signal pathway might modulate the secretion of Th1/Th2 cytokines and prolong the survival of the grafts [26]. It was reported that administration of agonistic anti-4-1BB monoclonal antibody (mAB) prevented the development of various autoimmune and nonautoimmune conditions *in vivo* [27, 28]. Our previous results also suggested the blockade of the 4-1BB/4-1BBL costimulatory pathway with 4-1BBL monoclonal antibody attenuated the acute rejection in recipient rats with the liver transplantation [11].

RNAi-based gene silencing is more rapid and cost-effective compared with the gene knockout techniques. RNA interference (RNAi) is a powerful tool to induce loss-of-function phenotypes by posttranscriptional silencing of gene expression. Viral delivery of short hairpin RNA (shRNA) expression cassettes allows efficient transduction in tissues such as immunological cell *in vivo* [29]. The knockdown of gene expression has been achieved using lentiviral vector constructs that express shRNAs within vector-infected cells.

The lentiviral vector system provided useful tools for elucidating gene function by analysis of loss-of-function phenotypes and for exploring the application of RNAi in gene therapy [30]. In our study, The recombination vector of lentivirus containing shRNA targeting the 4-1BB gene was successfully constructed. The liver transplantation was performed using the two-cuff technique. BN recipient rats were infected by the recombinant LVs. The results showed that gene silencing of 4-1BB by RNA interference downregulated the 4-1BB gene expression of the splenic lymphocytes isolated from the recipient rats with liver transplantation. Our results also showed that lentivirus-mediated delivery of shRNA targeting 4-1BB gene prolonged the survival time of recipient rats. These results suggested that the silencing of 4-1BB gene by RNA interference was successful in the recipient rats and useful for the liver transplantation.

The acute transplantation rejective response is the important cause of retransplantation and is principally mediated by T cells [31]. Cytokines including IL-2, IL-10, and INF-γ can accelerate the T-cell mediated immune response, while IL-4, 5, and 6 can be helpful for B-cell mediated humoral immunity. Increase of cytokines is the marker of the acute transplantation rejective response [32, 33]. Our results showed that gene silencing of 4-1BB by RNA interference decreased the levels of plasma IL-2 and INF-γ seven days after liver transplantation. These illuminated that gene silencing of 4-1BB inhibited the T cell-mediated acute rejection in recipient rats with liver transplantation.

Histopathology is the gold standard for diagnosing graft rejection after transplantation [34]. There are the infiltration of inflammatory cells to portal area including activated lymphocytes, neutrophils and acidophils, inflammation of endotheliocytes under the portal vein and central vein, and the inflammation and injury of the bile duct in the acute rejection of liver transplantation [35, 36]. The dysfunction of the liver should be observed in the acute rejection of liver transplantation. In our study, the results showed that gene silencing of 4-1BB by RNA interference decreased the plasma levels of liver injury markers including AST, ALT, and BIL and alleviated the injury of liver morphology in recipient rats with liver transplantation. Our experiment strongly suggested gene silencing of 4-1BB by RNA interference prevented the liver injury induced by the acute rejection in rats with liver transplantation.

In conclusion, we have demonstrated that gene silencing of 4-1BB by RNA interference inhibits the acute rejection in rats with liver transplantation and it is a promising strategy to prevent progression of graft rejection.

References

[1] J. J. Goronzy and C. M. Weyand, "T-cell co-stimulatory pathways in autoimmunity," *Arthritis Research and Therapy*, vol. 10, no. 1, article S3, 2008.

[2] M. Kornete and C. A. Piccirillo, "Critical co-stimulatory pathways in the stability of Foxp3$^+$ Treg cell homeostasis in type I diabetes," *Autoimmunity Reviews*, vol. 11, no. 2, pp. 104–111, 2011.

[3] A. Palazón, I. Martínez-Forero, A. Teijeira et al., "The HIF-1α hypoxia response in tumor-infiltrating T lymphocytes induces functional CD137 (4-1BB) for immunotherapy," *Cancer Discovery*, vol. 2, no. 7, pp. 608–623, 2012.

[4] D. Teschner, G. Wenzel, E. Distler et al., "In vitro stimulation and expansion of human tumour-reactive CD8$^+$ cytotoxic t lymphocytes by anti-CD3/CD28/CD137 magnetic beads," *Scandinavian Journal of Immunology*, vol. 74, no. 2, pp. 155–164, 2011.

[5] J. A. Hernandez-Chacon, Y. Li, R. C. Wu et al., "Costimulation through the CD137/4-1BB pathway protects human melanoma tumor-infiltrating lymphocytes from activation-induced cell death and enhances antitumor effector function," *Journal of Immunotherapy*, vol. 34, no. 3, pp. 236–250, 2011.

[6] H. W. Chen, C. H. Liao, C. Ying, C. J. Chang, and C. M. Lin, "Ex vivo expansion of dendritic-cell-activated antigen-specific CD4$^+$ T cells with anti-CD3/CD28, interleukin-7, and interleukin-15: potential for adoptive T cell immunotherapy," *Clinical Immunology*, vol. 119, no. 1, pp. 21–31, 2006.

[7] C. Stefanidis, G. Callebaut, W. Ngatchou et al., "The role of biventricular assistance in primary graft failure after heart transplantation," *Hellenic Journal of Cardiology*, vol. 53, no. 2, pp. 160–162, 2012.

[8] M. Glanemann, G. Gaebelein, N. Nussler et al., "Transplantation of monocyte-derived hepatocyte-like cells (NeoHeps)

[9] improves survival in a model of acute liver failure," *Annals of Surgery*, vol. 249, no. 1, pp. 149–154, 2009.

[9] T. Asai, B. K. Choi, P. M. Kwon et al., "Blockade of the 4-1BB (CD137)/4-1BBL and/or CD28/CD80/CD86 costimulatory pathways promotes corneal allograft survival in mice," *Immunology*, vol. 121, no. 3, pp. 349–358, 2007.

[10] B. J. Huang, H. Yin, Y. F. Huang et al., "Gene therapy using adenoviral vector encoding 4-1BBIg gene significantly prolonged murine cardiac allograft survival," *Transplant Immunology*, vol. 16, no. 2, pp. 88–94, 2006.

[11] L. Qin, H. G. Guan, X. J. Zhou, J. Yin, J. Lan, and H. X. Qian, "Blockade of 4-1BB/4-1BB ligand interactions prevents acute rejection in rat liver transplantation," *Chinese Medical Journal*, vol. 123, no. 2, pp. 212–215, 2010.

[12] N. Wu, A. B. Yu, H. B. Zhu, and X. K. Lin, "Effective silencing of Sry gene with RNA interference in developing mouse embryos resulted in feminization of XY gonad," *Journal of Biomedicine and Biotechnology*, vol. 2012, Article ID 343891, 11 pages, 2012.

[13] T. Yang, B. Zhang, B. K. Pat, M. Q. Wei, and G. C. Gobe, "Lentiviral-mediated RNA interference against TGF-beta receptor type II in renal epithelial and fibroblast cell populations in vitro demonstrates regulated renal fibrogenesis that is more efficient than a nonlentiviral vector," *Journal of Biomedicine and Biotechnology*, vol. 2010, Article ID 859240, 12 pages, 2010.

[14] K. Ui-Tei, Y. Naito, F. Takahashi et al., "Guidelines for the selection of highly effective siRNA sequences for mammalian and chick RNA interference," *Nucleic Acids Research*, vol. 32, no. 3, pp. 936–948, 2004.

[15] Y. C. Kuo, S. C. Weng, C. J. Chou, T. T. Chang, and W. J. Tsai, "Activation and proliferation signals in primary human T lymphocytes inhibited by ergosterol peroxide isolated from Cordyceps cicadae," *British Journal of Pharmacology*, vol. 140, no. 5, pp. 895–906, 2003.

[16] L. A. Muscarella, V. Guarnieri, M. Coco et al., "Small deletion at the 7q21.2 locus in a CCM family detected by real-time quantitative PCR," *Journal of Biomedicine and Biotechnology*, vol. 2010, Article ID 854737, 7 pages, 2010.

[17] C. S. Kim, J. G. Kim, B. J. Lee et al., "Deficiency for costimulatory receptor 4-1BB protects against obesity-induced inflammation and metabolic disorders," *Diabetes*, vol. 60, no. 12, pp. 3159–93168, 2011.

[18] T. Hori, L. B. Gardner, F. Chen et al., "Impact of hepatic arterial reconstruction on orthotopic liver transplantation in the rat," *Journal of Investigative Surgery*, vol. 25, no. 4, pp. 242–252, 2012.

[19] A. J. Demetris, K. P. Batts, A. P. Dhillon et al., "Banff schema for grading liver allograft rejection: an international consensus document," *Hepatology*, vol. 25, no. 3, pp. 658–663, 1997.

[20] M. H. Sanei, T. D. Schiano, C. Sempoux, C. Fan, and M. I. Fiel, "Acute cellular rejection resulting in sinusoidal obstruction syndrome and ascites postliver transplantation," *Transplantation*, vol. 92, no. 10, pp. 1152–1158, 2011.

[21] B. Fosby, T. H. Karlsen, and E. Melum, "Recurrence and rejection in liver transplantation for primary sclerosing cholangitis," *World Journal of Gastroenterology*, vol. 18, no. 1, pp. 1–5, 2012.

[22] S. Gras, L. Kjer-Nielsen, Z. Chen, J. Rossjohn, and J. McCluskey, "The structural bases of direct T-cell allorecognition: implications for T-cell-mediated transplant rejection," *Immunology and Cell Biology*, vol. 89, no. 3, pp. 388–395, 2011.

[23] I. Pérez-Flores, A. Sánchez-Fructuoso, J. L. Santiago et al., "Intracellular ATP levels in CD4$^+$ lymphocytes are a risk marker of rejection and infection in renal graft recipients," *Transplantation Proceedings*, vol. 41, no. 6, pp. 2106–2108, 2009.

[24] R. J. McKallip, Y. Do, M. T. Fisher, J. L. Robertson, P. S. Nagarkatti, and M. Nagarkatti, "Role of CD44 in activation-induced cell death: CD44-deficient mice exhibit enhanced T cell response to conventional and superantigens," *International Immunology*, vol. 14, no. 9, pp. 1015–1026, 2002.

[25] B. K. Choi, J. S. Bae, E. M. Choi et al., "4-1BB-dependent inhibition of immunosuppression by activated CD4$^+$, CD25$^+$ T cells," *Journal of Leukocyte Biology*, vol. 75, no. 5, pp. 785–791, 2004.

[26] K. Xu, C. Li, X. Pan, and B. Du, "Study of relieving graft-versus-host disease by blocking CD137-CD137 ligand costimulatory pathway in vitro," *International Journal of Hematology*, vol. 86, no. 1, pp. 84–90, 2007.

[27] J. Kim, W. S. Choi, S. La et al., "Stimulation with 4-1BB (CD137) inhibits chronic graft-versus-host disease by inducing activation-induced cell death of donor CD4$^+$ T cells," *Blood*, vol. 105, no. 5, pp. 2206–2213, 2005.

[28] S. Ganguly, J. Liu, V. B. Pillai, R. S. Mittler, and R. R. Amara, "Adjuvantive effects of anti-4-1BB agonist Ab and 4-1BBL DNA for a HIV-1 Gag DNA vaccine: different effects on cellular and humoral immunity," *Vaccine*, vol. 28, no. 5, pp. 1300–1309, 2010.

[29] V. A. Meliopoulos, L. E. Andersen, K. F. Birrer et al., "Host gene targets for novel influenza therapies elucidated by high-throughput RNA interference screens," *The FASEB Journal*, vol. 26, no. 4, pp. 1372–1386, 2012.

[30] T. H. Hutson, E. Foster, J. M. Dawes, R. Hindges, R. J. Yáñez-Muñoz, and L. D. Moon, "Lentiviral vectors encoding short hairpin RNAs efficiently transduce and knockdown LINGO-1 but induce an interferon response and cytotoxicity in central nervous system neurones," *Journal of Gene Medicine*, vol. 11, no. 5, pp. 299–313, 2012.

[31] M. L. del Rio, J. Kurtz, C. Perez-Martinez, A. Ghosh, J. A. Perez-Simon, and J. I. Rodriguez-Barbosa, "B- and T-lymphocyte attenuator targeting protects against the acute phase of graft versus host reaction by inhibiting donor anti-host cytotoxicity," *Transplantation*, vol. 92, no. 10, pp. 1085–1093, 2011.

[32] G. Chen, J. Mi, M. Z. Xiao, and Y. R. Fu, "PDIA3 mRNA expression and IL-2, IL-4, IL-6, and CRP levels of acute kidney allograft rejection in rat," *Molecular Biology Reports*, vol. 39, no. 5, pp. 5233–35238, 2012.

[33] M. H. Karimi, S. Daneshmandi, A. A. Pourfathollah et al., "Association of IL-6 promoter and IFN-γ gene polymorphisms with acute rejection of liver transplantation," *Molecular Biology Reports*, vol. 38, no. 7, pp. 4437–4443, 2011.

[34] P. Kulkarni, M. S. Uppin, A. K. Prayaga, U. Das, and K. V. D. Murthy, "Renal allograft pathology with C4d immunostaining in patients with graft dysfunction," *Indian Journal of Nephrology*, vol. 21, no. 4, pp. 239–244, 2011.

[35] Y. Sanada, K. Mizuta, T. Urahashi et al., "Co-occurrence of nonanastomotic biliary stricture and acute cellular rejection in liver transplant," *Experimental and Clinical Transplantation*, vol. 10, no. 2, pp. 176–179, 2012.

[36] J. I. Wyatt, "Liver transplant pathology—messages for the non-specialist," *Histopathology*, vol. 57, no. 3, pp. 333–341, 2010.

18

Effect of Linseed Oil Dietary Supplementation on Fatty Acid Composition and Gene Expression in Adipose Tissue of Growing Goats

M. Ebrahimi,[1] M. A. Rajion,[1] Y. M. Goh,[1,2] A. Q. Sazili,[3] and J. T. Schonewille[4]

[1] Department of Veterinary Preclinical Sciences, Faculty of Veterinary Medicine, Universiti Putra Malaysia,
43400 Serdang, Selangor, Malaysia
[2] Institute of Tropical Agriculture, Universiti Putra Malaysia, 43400 Serdang, Selangor, Malaysia
[3] Department of Animal Science, Faculty of Agriculture, Universiti Putra Malaysia, 43400 Serdang, Selangor, Malaysia
[4] Division of Nutrition, Department of Farm Animal Health, Faculty of Veterinary Medicine, Utrecht University,
P.O. Box 80151, 3508 TD Utrecht, The Netherlands

Correspondence should be addressed to M. A. Rajion; mohdali@vet.upm.edu.my

Academic Editor: Andre Van Wijnen

This study was conducted to determine the effects of feeding oil palm frond silage based diets with added linseed oil (LO) containing high α-linolenic acid (C18:3n-3), namely, high LO (HLO), low LO (LLO), and without LO as the control group (CON) on the fatty acid (FA) composition of subcutaneous adipose tissue and the gene expression of peroxisome proliferator-activated receptor (PPAR)α, PPAR-γ, and stearoyl-CoA desaturase (SCD) in Boer goats. The proportion of C18:3n-3 in subcutaneous adipose tissue was increased ($P < 0.01$) by increasing the LO in the diet, suggesting that the FA from HLO might have escaped ruminal biohydrogenation. Animals fed HLO diets had lower proportions of C18:1 trans-11, C18:2n-6, CLA cis-9 trans-11, and C20:4n-6 and higher proportions of C18:3n-3, C22:5n-3, and C22:6n-3 in the subcutaneous adipose tissue than animals fed the CON diets, resulting in a decreased n-6:n-3 fatty acid ratio (FAR) in the tissue. In addition, feeding the HLO diet upregulated the expression of PPAR-γ ($P < 0.05$) but downregulated the expression of SCD ($P < 0.05$) in the adipose tissue. The results of the present study show that LO can be safely incorporated in the diets of goats to enrich goat meat with potential health beneficial FA (i.e., n-3 FA).

1. Introduction

The usually high content of saturated fatty acids (SFA) in ruminant meat can increase the risk of cardiovascular diseases [1]. However, ruminant meat may also be a good dietary source of some nutrients with health benefits including some FA such as long chain (C20) polyunsaturated fatty acids (LC-PUFA) and conjugated linoleic acid (CLA) isomers [2]. The decrease of SFA and the increase of health-beneficial FA have been an important topic in ruminant meat research. The inclusion of sources of C18:3n-3 in lamb diets, such as forages [3–5], pastures [6], linseed [7, 8], or linseed oil [9, 10], had increased the concentration of n-3 PUFA in the

meat. Ruminant fats are among the richest natural sources of CLA isomers, particularly of rumenic acid (CLA cis-9 trans-11), and are the main sources of these isomers in the human diet [11]. The CLA cis-9 trans-11 is produced during ruminal biohydrogenation of C18:2n-6 to stearic acid [12] and by endogenous conversion of C18:1 trans-11 by Δ9-desaturase in tissues [13]. Feeding animals with diets rich in linoleic acid (C18:2n-6) and α-linolenic acid (C18:3n-3) acids increased the CLA cis-9 trans-11 content of ruminant meat [3, 6, 9, 14, 15]. However, feeding linseed oil (rich in C18:3n-3) seems to be less effective in the increase of CLA cis-9 trans-11 in intramuscular fat than sunflower oil (rich in C18:2n-6) [9, 16]. Bessa et al. [9] observed that a blend of

Effect of Linseed Oil Dietary Supplementation on Fatty Acid Composition and Gene Expression in Adipose
Tissue of Growing Goats

181

sunflower and linseed oil might be a good approach to obtain simultaneously an enrichment in n-3 PUFA and CLA in lamb intramuscular fat.

Both n-3 and n-6 PUFA appear to suppress the genes that encode for several enzymes, which are involved in carbohydrate and lipid metabolism, whereas saturated, trans-, and monounsaturated fatty acids (MUFA) fail to suppress [17, 18]. The PPAR are activated by FA and regulate FA uptake and oxidation in the liver [19]. The PUFA activates PPAR and stimulates peroxisomal β-oxidation and peroxisome proliferation [20]. The expression of the SCD gene in the ruminants is regulated by the transcription factors SREBP1 [21, 22], PPAR-α [22], and PPAR-γ [21, 22]. Furthermore, unsaturated FA are important because they play a role in the cellular activities, metabolism, and nuclear events that govern gene transcription [23]. In this sense, the dietary n-6 and n-3 PUFA, especially arachidonic acid (C20:4n-6), have been shown to repress SCD gene expression [23]. In most cases these studies have been carried out in the rodents, where lipid metabolism is different from that in ruminants.

This study was conducted to determine the effects of feeding oil palm frond (OPF) silage-based diets containing different levels of linseed oil on the FA composition of subcutaneous adipose tissue and the PPAR-α, PPAR-γ, and SCD gene expression in Boer goats.

2. Materials and Methods

2.1. Animals, Diets, and Management. Twenty-one five-month-old male Boer goats weighing 13.66 ± 1.07 Kg (mean initial body weight ± standard error) were initially drenched against parasites and randomly assigned to different dietary treatment groups. Goats were housed individually in wooden pens measuring 1.2 m × 1 m each, built inside a shed with slatted flooring 0.5 meter above the ground. The experimental diets were formulated to have a high LO (HLO), low LO (LLO), and no LO as the control group (CON). The feed ingredients and composition of experimental diets are shown in Table 2. Sunflower oil (SFO) (Lam Soon Edible Oils Sdn. Bhd.) which contained 6.40% of C16:0, 3.66% of C18:0, 28.32% of C18:1n-9, 61.23% of C18:2n-6, and 0.39% of C18:3n-3 expressed as a percentage of total identified FA, palm kernel oil (PKO) (Malaysian Palm Oil Board) which contained 52.55% of C12:0, 16.75% of C14:0, 9.00% of C16:0, 2.36% of C18:0, 16.62% of C18:1n-9, and 18.47% of C18:2n-6 expressed as a percentage of total identified FA, and linseed oil (LO) (Brenntag Canada, Inc., Montreal, QC, Canada) which contained 5.15% of C16:0, 3.17% of C18:0, 16.62% of C18:1n-9, 16.12% of C18:2n-6, and 57.09% of C18:3n-3 expressed as a percentage of total identified FA were used as oil sources to incorporate different levels of C18:3n-3. The linseed oil was used as the main source of α-linolenic acid (C18:3n-3) while sunflower oil was used as the main source of linoleic acid (C18:2n-6). The animals were fed twice daily at 3.7% of BW (DM basis), with adjustments made weekly according to the changing body weight. Fresh OPF were chopped to 2-3 cm length and mixed with cellulase enzyme (Onozuka R-10; Yakult Ltd, Tokyo, Japan; 2 g/Kg of fresh

matter) and lactic acid bacteria (*Lactobacillus plantarum* MTD1, Ecosyl, Stokesley, Yorkshire, UK) calculated to contain at least 1×10^6 colonies forming units (CFU) per gram as per manufacturer's instructions and ensiled in 200-liter plastic drums for 12 weeks. The concentrations (70% DM basis) and OPF silage (30% DM basis) were mixed and offered in 2 equal meals at 0800 and 1700. The diets were adjusted to be isonitrogenous and isocaloric and to meet the energy and protein requirements of growing goats (Table 2) [24]. All the goats had free access to drinking water and a mineral block. The feeding trial lasted for 100 days with a three-week adaptation period.

2.2. Chemical Analyses. Samples (500 g) of concentrate and OPF silage were collected every 7 d and stored at 4°C. Individual goat refusals of feed were weighed daily and stored at 4°C until analyzed for dry matter (DM). Concentrates and OPF silage samples were dried at 60°C for 48 h to determine the DM content, ground to pass a 1 mm screen and analyzed for crude protein (CP), ether extract (EE), ash, organic matter (OM), neutral detergent fiber (NDF), and acid detergent fiber (ADF) according to standard methods. Crude protein (CP, total nitrogen × 6.25) was determined by the method (number 990.03) of the [25]. Neutral detergent fiber (NDF) and acid detergent fiber (ADF) were determined according to van Soest et al. [26].

2.3. Measurement of FA. The total FA were extracted from oils, experimental feeds, and subcutaneous adipose tissue based on the method of [27] modified by [28] and described by [29], using chloroform : methanol 2 : 1 (v/v) containing butylated hydroxytoluene to prevent oxidation during sample preparation. The extracted fatty acids were transmethylated to their fatty acid methyl esters (FAME) using 0.66 N KOH in methanol and 14% methanolic boron trifluoride (BF_3) (Sigma Chemical Co., St. Louis, MO, USA) according to the methods by AOAC (1990). The FAME was separated by gas liquid chromatography on an Agilent 7890A GC system (Agilent, Palo Alto, CA, USA) using a 100 m × 0.25 mm ID (0.20 μm film thickness) Supelco SP-2560 capillary column (Supelco, Inc., Bellefonte, PA, USA). One microliter of FAME was injected by an autosampler into the chromatograph, equipped with a flame ionization detector (FID). The carrier gas was He, and the split ratio was 10 : 1 after injection of the FAME. The injector temperature was programmed at 250°C, and the detector temperature was 300°C. The column temperature program initiated runs at 120°C held for 5 min, increased by 2°C/min up to 170°C, held at 170°C for 15 min, increased again by 5°C/min up to 200°C, held at 200°C for 5 min, then increased again by 2°C/min to a final temperature of 235°C, and held for 10 min. The FA concentrations are expressed as percent of total identified FA. A reference standard (mix C4–C24 methyl esters; Sigma-Aldrich, Inc., St. Louis, MO, USA) and CLA standard mix (CLA cis-9 trans-11 and CLA trans-10, cis-12, Sigma-Aldrich, Inc., St. Louis, MO, USA) were used to determine recoveries and correction factors for the determination of individual FA composition.

TABLE 1: Names and sequences of the primers used in this study.

Target group		Sequence 5'-3'	Length, nt	Reference
β-actin	F	CGC CAT GGA TGA TGA TAT TGC3	123	[22]
	R	AAG CGG CCT TGC ACA T3		
PPAR-α	F	TGC CAA GAT CTG AAA AAG CA	101	[30]
	R	CCT CTT GGC CAG AGA CTT GA		
PPAR-γ	F	CTT GCT GTG GGG ATG TCT C	121	[30]
	R	GGT CAG CAG ACT CTG GGT TC		
SCD	F	CCC AGC TGT CAG AGA AAA GG		
	R	GAT GAA GCA CAA CAG CAG GA	115	[30]

[1]F: forward; [2]R: reverse.

2.4. Tissue Collection and RNA Extraction and Purification. Immediately after slaughtering the animals the subcutaneous adipose tissue was quickly excised and snap-frozen in liquid nitrogen and stored at −80°C until RNA extraction.

Total RNA was extracted from 100 mg of frozen tissue using the RNeasy lipid tissue mini kit (Cat. no. 74804, Qiagen, Hilden, Germany), and DNase digestion was completed during RNA purification using the RNase-Free DNase set (Qiagen, Hilden, Germany) according to the manufacturer's instructions. Total RNA purity was determined by the 260/280 nm ratio of absorbance readings using NanoDrop ND-1000 UV-Vis Spectrophotometer (NanoDrop Technologies, Wilmington, DE, USA).

2.5. Complementary DNA Synthesis. Purified total RNA (1 μg) was reverse transcribed using a QuantiTect reverse transcription kit (Qiagen, Hilden, Germany) in accordance with the manufacturer's recommended procedure.

2.6. Real-Time Polymerase Chain Reaction (PCR). Real-time PCR was performed with the Bio-Rad CFX96 Touch (Bio-Rad Laboratories, Hercules, CA, USA) using optical grade plates using QuantiFast SYBR green PCR kit (Cat. no. 204054, Qiagen, Hilden, Germany). The sequences of primers are shown in Table 1.

The β-actin was used as the reference gene to normalize the tested genes. All primers were purchased through 1st BASE oligonucleotide synthesis (1st Base, Singapore). Each reaction (20 μL) contained 8.5 μL SYBR green PCR mix, 1 μL cDNA, 1 μL each of forward and reverse primers, and 8.5 μL RNase free water. Target genes were amplified through the following thermocycling program: 95°C for 10', 40 PCR cycles at 95°C for 30", 60°C for 20", and 72°C for 20". Fluorescence was measured at every 15" to construct the melting curve. A real-time PCR was conducted for each primer pair in which cDNA samples were substituted with dH$_2$O to verify that exogenous DNA was not present. Additionally, 1 μg of RNA isolated by the procedure described above was substituted for cDNA in a real-time PCR reaction to confirm that there were no genomic DNA contaminants in the RNA samples. Both negative controls showed no amplification after 40 cycles. Efficiency of amplification was determined for each primer pair using serial dilutions. The cycle numbers at which amplified DNA samples exceeded a computer generated fluorescence threshold level were normalized and compared to determine relative gene expression. Higher cycle number values indicated lower initial concentrations of cDNA and thus lower levels of mRNA expression. Each sample was run in triplicate, and averaged triplicates were used to assign cycle threshold (CT) values. The ΔCT values were generated by subtracting experimental CT values from the CT values for β-actin targets amplified with each sample. The group with the highest mean ΔCT value (lowest gene expression) per amplified gene target was set to zero, and the mean ΔCT values of the other groups were set relative to this calibrator (ΔΔCT). The ΔΔCT values were calculated as powers of 2 (−2 ΔΔCT), to account for the exponential doubling of the PCR.

2.7. Statistical Analysis. Results were analyzed using analysis of variance with the different LO content as the main effects. FA data and all the gene expression data were analyzed by one-way ANOVA, using the MIXED procedure of the SAS software package, version 9.1 (SAS Inst. Inc., Cary, NC). The statistical models used the following equation:

$$Y_{ijk} = \mu + T_i + F_k + e_{ijk}, \tag{1}$$

where μ was the overall mean, T was the different dietary LO, F was the animal effect, and e was the residual error. The random effect was the animals. Means were separated using the "PDIFF" option of the "least-squares means (LSMEANS)" statement of the MIXED procedure. Differences of $P <$ 0.05 were considered to be significant. Linear and quadratic contrasts were used to determine the effect of increasing amounts of LO on the response variables. The data were checked for normality using the UNIVARIATE procedure of SAS software, and the results in the tables are presented as means ± standard error of the mean.

3. Results

3.1. Composition of Experimental Diets. The three experimental diets which were isocaloric are shown in Table 2. The average metabolizable energy content ranged from 2.41 to 2.51 Mcal/Kg of the dry matter (DM) content, whilst the protein (13% of DM) and crude fat content (7% of DM) of the treatment diets were also similar.

Effect of Linseed Oil Dietary Supplementation on Fatty Acid Composition and Gene Expression in Adipose
Tissue of Growing Goats

183

TABLE 2: Ingredients and chemical composition of the experimental diets.

| | Treatment diets | | |
	HLO	LLO	CON
Ingredients (% DM)			
OPF silage	30.00	30.00	30.00
Corn, grain	17.00	17.00	17.00
Soybean meal	13.30	13.30	13.30
Palm kernel cake	25.11	25.11	25.11
Rice bran	8.18	8.18	8.18
Linseed oil	1.30	0.40	0.00
Palm kernel oil	0.10	1.00	1.10
Sunflower oil	2.00	2.00	2.30
Mineral premix	0.50	0.50	0.50
Vitamin premix	0.50	0.50	0.50
Ammonium chloride	1.00	1.00	1.00
Limestone	1.00	1.00	1.00
Chemical composition			
ME (Mcal/Kg DM)[1]	2.51	2.51	2.51
CP (% DM)	13.00	13.00	13.00
EE (% DM)	7.00	7.00	7.00
NDF (% DM)	48.90	48.90	48.90
ADF (% DM)	33.00	33.00	33.00
CA (% DM)	0.68	0.68	0.68
P (% DM)	0.36	0.36	0.36

HLO: high LO; LLO: low LO; CON: without LO.
[1]Calculated values.

As expected the FA composition of linseed oil contained the highest amount of α-linolenic acid and sunflower oil contained the highest amount of linoleic acid.

3.2. FA Composition of Experimental Diets.

Oleic and linoleic acids were the predominant FA in all treatment diets, whereas the HLO diet showed a more balanced supply of the three main FA of plant origin, namely, oleic, linoleic, and α-linolenic, although the latter was quantitatively the most abundant (Table 3).

3.3. Adipose Tissue FA Composition.

The FA composition of the goat subcutaneous adipose tissue fed different dietary LO is presented in Table 4. The proportion of subcutaneous adipose tissue FA having 18 carbons was quite consistent across the three treatment groups, averaging between 56.77 and 58.23%. Mean concentrations of C18:0, C18:1n-9, C18:2n-6, and C18:3n-3 were 7.58, 39.72, 5.52, and 0.98%, respectively. On the other hand, C18:0 increased in a linear manner with increasing LO. The different levels of dietary LO had significant ($P < 0.05$) effects on some of the polyunsaturated fatty acids (PUFA) in the subcutaneous adipose tissue of especially the HLO treatment group. However, oleic acid and other monounsaturated fatty acid (MUFA) were not affected by the different LO. The C18:1 trans-11 (vaccenic acid) (VA) showed a linear increase in the CON diet compared to the LLO and HLO diets. The CLA cis-9 trans-11 (rumenic acid) also showed a positive linear effect with decreased LO.

Decreasing the LO increased the C20:4n-6 ($P < 0.01$), while the HLO diet increased ($P < 0.01$) the C18:3n-3 concentration in the subcutaneous adipose tissue when compared to the CON group.

The total n-3 PUFA content in the HLO diet significantly increased ($P < 0.01$) compared to the CON diet and the n-6 : n-3 fatty acid ratios (FAR) in the subcutaneous adipose tissue of the HLO treatment group were significantly ($P < 0.01$) decreased compared to the CON treatment group (Table 4). Incremental amounts of linseed oil in the diet resulted in dose-dependent increases in most 22-carbon PUFA in the adipose tissue. The concentrations of C22:5n-3 and C22:6n-3 were significantly ($P < 0.01$) increased by increasing the LO in the diet.

The total SFA, MUFA, and PUFA were not affected by different dietary LO. However total n-3 PUFA increased with increasing dietary LO. In contrast, the sum of CLA isomers showed a positive linear decrease to increasing LO in the diet. The major increase in this ratio was observed in the CON diet with the highest level of n-6 : n-3 FAR (Table 4). Increasing the dietary LO by 3.96-fold (from 13.63 to 3.44; Table 2) via oil supplementation changed by almost the same ratio by 3.31-fold in the subcutaneous adipose tissue from 1.59 to 0.48 (Table 4).

3.4. Adipose Tissue Gene Expression.

The relative expression of the genes in the subcutaneous adipose tissue of the HLO group compared to CON group is shown in Figures 1, 2, and 3. The PPAR-α gene showed a similar level of expression in

TABLE 3: Fatty acid composition (percentage of total identified fatty acids) of the experimental diets[1].

Fatty acids	Treatment diets		
	HLO	LLO	CON
C10:0, capric	0.53	0.90	0.90
C12:0, lauric	3.53	7.02	7.24
C14:0, myristic	1.79	3.83	3.00
C16:0, palmitic	15.65	15.36	16.00
C16:1n-7, palmitoleic	0.21	0.24	0.22
C17:0, margaric	0.29	0.28	0.29
C18:0, stearic	5.68	5.88	5.85
C18:1n-9, oleic	27.78	27.62	27.40
C18:2n-6, linoleic	30.92	32.40	35.68
C18:3n-3, linolenic	13.63	6.47	3.44
SFA[2]	27.18	32.99	32.97
UFA[3]	72.53	66.73	66.74
MUFA[4]	20.45	27.99	27.86
n-3 PUFA[5]	13.63	6.47	3.44
n-6 PUFA[6]	30.92	32.40	35.68
n-6 : n-3 FAR[7]	2.27	5.01	10.38

HLO: high LO; LLO: low LO; CON: without LO.
[1] The data are expressed as the percentage of total identified fatty acids.
[2] SFA = sum of C10:0 + C12:0 + C14:0 + C16:0 + C17:0 + C18:0.
[3] UFA = sum of C16:1 + C18:1n-9 + C18:2n-6 + C18:3n-3.
[4] MUFA = sum of C16:1 + C18:1n-9.
[5] PUFAn-3 = sum of C18:3n-3.
[6] PUFAn-6 = sum of 18:2n-6.
[7] n-6 : n-3 FAR = (C18:2n-6) ÷ (C18:3n-3).

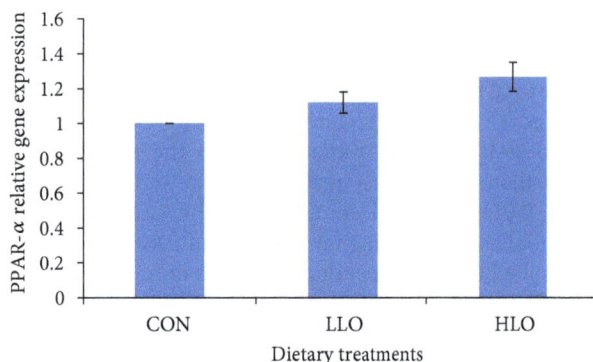

FIGURE 1: Comparison of PPAR-α relative gene expression in the subcutaneous adipose tissue of Boer goats fed diets with different levels of linseed oil. Values were normalized with a housekeeping gene, β-actin. Then, treated samples were expressed relative to gene expression of CON group. Values are means ± 1 standard error bar. HLO: high LO; LLO: low LO; CON: without LO.

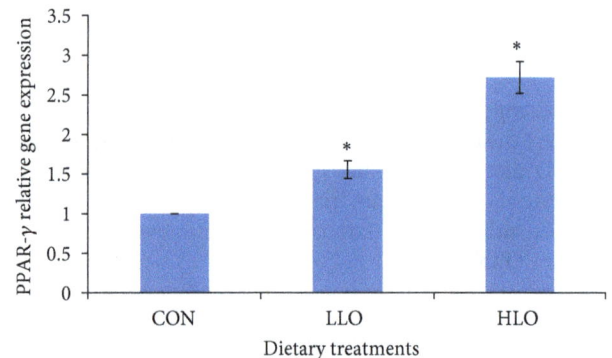

FIGURE 2: Comparison of PPAR-γ relative gene expression in the subcutaneous adipose tissue of Boer goats fed diets with different levels of linseed oil. Values were normalized with a housekeeping gene, β-actin. Then, treated samples were expressed relative to gene expression of CON group. Values are means ± 1 standard error bar. Values indicated by the ∗ show significant difference compared with the CON group ($P < 0.05$). HLO: high LO; LLO: low LO; CON: without LO.

all treatment groups ($P > 0.05$, Figure 1) indicating that different LO levels in the diet had no effect on the upregulation of the PPAR-α gene expression. The current study showed that the LO altered the PPAR-γ expression in the goat tissues where increasing the LO increased the PPAR-γ expression in the HLO and LLO treatment groups compared to the CON group.

The SCD gene expression showed a significant ($P < 0.05$, Figure 3) reduction in the HLO group compared to the CON group suggesting that the SCD gene was downregulated by the HLO and LLO treatment. This effect was more pronounced in the LO group.

Effect of Linseed Oil Dietary Supplementation on Fatty Acid Composition and Gene Expression in Adipose Tissue of Growing Goats

185

TABLE 4: Fatty acid composition (percentage of total identified fatty acids) of the subcutaneous adipose tissue in Boer goats fed diets with different levels of linseed oil[1].

Fatty acids	Treatment diets			SEM	P value	
	HLO	LLO	CON		Linear	Quadratic
C10:0, capric	0.13	0.13	0.12	0.01	0.744	0.890
C12:0, lauric	2.96	3.52	3.66	0.26	0.661	0.051
C14:0, myristic	5.80	4.59	5.15	0.19	0.506	0.793
C14:1, myristoleic	0.73	0.57	0.57	0.02	0.066	0.086
C15:0, pentadecanoic	0.73	0.81	0.90	0.07	0.368	0.738
C15:1, pentadecenoic	0.17	0.27	0.29	0.03	0.710	0.110
C16:0, palmitic	23.39	22.55	20.54	0.32	0.493	0.008
C16:1n-7 palmitoleic	3.43	3.79	4.40	0.20	0.078	0.418
C17:0, margaric	1.11	1.24	1.19	0.09	0.810	0.609
C17:1, margaroleic	0.95	0.89	0.83	0.06	0.200	0.213
C18:0, stearic	7.97	7.48	7.28	0.31	0.349	0.646
C18:1n-9, oleic	39.76	40.17	39.23	0.38	0.274	0.145
C18:1trans-11, vaccenic	2.67	2.79	3.25	0.10	0.014	0.171
C18:2n-6, linoleic	4.06	5.72	6.76	0.27	0.004	0.001
CLA cis-9, trans-11	0.62	0.77	1.05	0.06	0.007	0.002
CLA cis-12, trans-10	0.10	0.13	0.17	0.01	0.031	0.013
C18:3n-3, α-linolenic	1.59	0.86	0.48	0.10	0.001	0.005
C20:4n-6, arachidonic	2.34	2.88	3.55	0.12	0.002	0.001
C20:5n-3, eicosapentaenoic	0.27	0.11	0.08	0.03	0.523	0.882
C22:5n-3, docosapentaenoic	0.55	0.37	0.26	0.03	0.132	0.002
C22:6n-3, docosahexaenoic	0.68	0.37	0.20	0.05	0.004	0.001
SFA[2]	42.09	40.32	38.85	0.48	0.983	0.114
UFA[3]	57.91	59.68	61.15	0.48	0.913	0.214
MUFA[4]	47.69	48.47	48.58	0.48	0.234	0.249
n-3 PUFA[5]	3.10	1.72	1.03	0.19	0.004	0.001
n-6 PUFA[6]	6.40	8.60	10.31	0.32	0.002	0.001
Total CLA[7]	0.72	0.89	1.23	0.07	0.008	0.002
n-6 : n-3 FAR[8]	2.07	5.02	10.03	0.79	0.001	0.001
UFA : SFA	1.38	1.48	1.57	0.03	0.981	0.252
PUFA : SFA ratio	0.23	0.26	0.29	0.01	0.062	0.494

HLO: high LO; LLO: low LO; CON: without LO.
[1] The data are expressed as the percentage of total identified fatty acids.
[2] SFA = sum of C10:0 + C12:0 + C14:0 + C15:0 + C16:0 + C17:0 + C18:0.
[3] UFA = sum of C14:1 + C16:1 + C17:1 + C18:1n-9 + C18:2 + C18:3 + C20:4, C22:6, C20:5n-3 + C22:5-3 + C22:6n-3.
[4] MUFA = sum of C14:1 + C16:1 + C17:1 + C18:1n-9.
[5] n-3 PUFA = sum of C18:3n-3 + C20:5n-3 + C22:5n-3 + C22:6n-3.
[6] n-6 PUFA = sum of 18:2n-6 + 20:4n-6.
[7] Total CLA = sum of CLA cis-9 trans-11 + CLA cis-12 trans-10.
[8] n-6 : n-3 FAR = (C18:2n-6 + C20:4n-6) ÷ (C18:3n-3 + C20:5n-3 + C22:5n-3 + C22:6n-3).

4. Discussion

4.1. Adipose Tissue FA Composition. The FA composition of adipose tissue is determined by *de novo* lipogenesis, desaturation, dietary lipids composition and the difference in the utilization of various FA by the animal body. The proportions of the predominant FA in the goat subcutaneous adipose tissue in this study were similar to the FA proportions of the subcutaneous adipose tissue of beef heifers fed different plant oil sources as reported by Noci et al. [16].

The linear increase in subcutaneous adipose tissue concentrations of C18:1 trans-11 by decreasing dietary LO may be explained by the ability of ruminal bacteria to synthesis this isomer from C18:2n-6 [12]. Harfoot et al. [31] reported that C18:1 trans-11 was the primary end product in the rumen rather than C18:0 when C18:2n-6 was supplied in higher concentrations. The concentrations of C18:2n-6 that ranged from 4.06 to 6.76% were somewhat similar to the 4.78% reported for the subcutaneous adipose tissue of Hanwoo steers fed linseed [32] but higher than the 1.43%

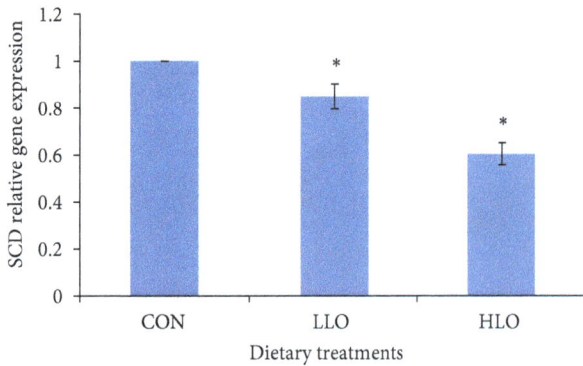

FIGURE 3: Comparison of SCD relative gene expression in the subcutaneous adipose tissue of Boer goats fed diets with different levels of linseed oil. Values were normalized with a housekeeping gene, β-actin. Then, treated samples were expressed relative to gene expression of CON group. Values are means $\pm$ 1 standard error bar. Values indicated by $*$ show significant difference compared with the CON group ($P < 0.05$). HLO: high LO; LLO: low LO; CON: without LO.

reported in heifers fed different vegetable oils [16]. The concentration of C18:2n-6 tended to decrease ($P < 0.01$) with increasing dietary LO, possibly because of its partial conversion to C18:0 in the rumen. The decrease in C20:4n-6 when the dietary LO was increased to 1.30% which coincided with the major decrease in dietary concentrations of C18:2n-6 was also reported by Igarashi et al. [33] for rat adipose tissue. The quadratic increases in the concentration of n-3 PUFA (C18:3n-3, C22:5n-3, and C22:6n-3) coincided with the major increase in dietary concentration of C18:3n-3 of the HLO diet. These increases are in agreement with the results of Kim et al. [34] who fed cattle with linseed. Jerónimo et al. [35] also reported that the concentration of C22:5n-3 and C22:6n-3 increased in ruminant intramuscular fat when they were fed high levels of linseed oil rather than sunflower oil. The capacity of conversion of C18:3n-3 to health promoting n-3 LC-PUFA is limited in humans [36] stressing the importance for its dietary supply. Increasing the LO in the diet resulted in a partial substitution of n-6 PUFA by n-3 PUFA in membranes. A possible reason can be the competition between C18:2n-6 and C18:3n-3 for the same desaturation and elongation enzymes which affect the conversion to LC-PUFA derivatives. Generally, a positive relationship has been reported between the concentrations of dietary C18:2n-6 and adipose tissue CLA cis-9 trans-11 in grazing heifers fed diets supplemented with plant oil-enriched concentrates [16]. The C18:1 trans-11 isomer is an intermediate product in the microbial biohydrogenation of dietary C18:1n-9, C18:2n-6, and C18:3n-3 [12]. The C18:1 trans-11 concentration in the subcutaneous adipose tissue, which could be converted to CLA cis-9 trans-11, tended to increase linearly with decreasing dietary LO. Likewise, the concentrations of CLA cis-9 trans-11 increased linearly ($P < 0.01$) as the goats consumed diets of decreasing LO, with the concentration of CLA cis-9 trans-11 increasing dramatically in goats fed the CON diet. This result may be explained by the

increase in C18:2n-6 in goats fed the diet without the addition of LO, indicating that more C18:2n-6 was isomerized to CLA cis-9 trans-11 and hydrogenated to C18:1 trans-11 in the rumen of goats fed the higher C18:2n-6 leading to a higher deposition in the adipose tissue. Jerónimo et al. [35] reported that when linseed oil fed to lambs was replaced with soybean oil which contained high amounts of C18:2n-6 as the fat supplement, the concentrations of C18:1 trans-11 as well as CLA cis-9 trans-11 in intramuscular fat were increased. Therefore, the synthesize of these isomers from C18:2n-6 may have been more efficient than that from C18:3n-3. At least for forage-based diets, the well-established mainstream pathway for the ruminal biohydrogenation of C18:2n-6 is straightforward with an initial isomerization with the formation of CLA cis-9 trans-11 and its reduction to C18:1 trans-11. The C18:2n-6 supplementation would therefore result in more rumen-derived CLA cis-9 trans-11 and less diverse biohydrogenation-derived FA compared with LO. The content of CLA cis-9 trans-11 in meat decreased with increasing the dietary LO, confirming the previous results of Bessa et al. [9]. In addition Noci et al. [16] observed that the CLA cis-9 trans-11 content in intramuscular fat was higher in heifers supplemented with sunflower oil as a source of C18:2n-6 than with LO.

The n-6 : n-3 FAR is highly influenced by the FA composition of the diet fed to the animals [37]. Lowering the ratio of n-6 to n-3 FA in food products have been recommended to prevent or modulate certain diseases in humans [38]. The n-6 : n-3 FAR in food should range between 1 and 4 [39]. The n-6 : n-3 FAR found in the HLO treatment (2.07) was within this range. In the HLO treatment, only 48.51% of n-3 PUFA were n-3 LC-PUFA. This can be important considering that the health benefits of n-3 PUFA are mostly associated with the n-3 LC-PUFA as the metabolism of C18:3n-3 in humans is limited [36]. The PUFA : SFA and n-6 : n-3 FAR are indices used to evaluate the nutritional value of fat for human consumption. Increasing the PUFA content of the diet, by including sources rich in either n-6 or n-3 PUFA, generally improves the PUFA : SFA ratio [40]. This was also observed in the present trial, and in all diets where the PUFA : SFA ratio was always lower than 0.29, which is the minimum value recommended for the human diet.

4.2. Adipose Tissue Gene Expression. Previous studies have also revealed that the PUFA activated the PPAR efficiently, although the very LC-PUFA such as erucic acid and nervonic acid and short-chain FA (<Cl0) cannot activate the PPAR-α and PPAR-γ because these FA are exclusively metabolized in the peroxisomes [41]. The PPAR-α responds to changes in dietary fat by activating the expression of various enzymes involved in fatty acyl CoA formation and hydrolysis, FA elongation and desaturation, and FA oxidation [42].

The significant differences in the mRNA expression of PPAR-γ in the subcutaneous adipose tissue between the different dietary treatments suggest that the principal pathway through which FA act to modulate the expression of lipogenic genes is through the altered expression of PPAR-γ. Al-Hasani and Joost [43] also showed that increasing the dietary LO in the rodent diet can increase PPAR-γ activity in target tissues which is associated with increased insulin sensitivity. An alternative explanation is that the effects of FA on PPAR-γ signaling are mediated through changes in PPAR-γ activity rather than changes in gene expression, which was not measured in this study, since FA have been shown to act as natural ligands of the PPAR-γ gene in other studies [44]. This study also showed that the PPAR-γ contributes to the regulation of subcutaneous adipose tissue gene expression. As PPAR-γ is related to the expression regulation of several gene-encoding proteins involved in adipocyte metabolism, it could also be a candidate gene affecting the fat deposition, including intramuscular and subcutaneous adipose tissue deposition [45]. The results of the current study clearly support that there is a relationship between gene expression of PPAR-γ and the intake of the n-3 PUFA. The increase in n-3 PUFA of the adipose tissues in the present study increased the PPAR-γ gene expression. Increasing the dietary LO might stimulate PPAR-γ target gene expression such as lipoprotein lipase (LPL), fatty acid transport protein (FTTP), and acyl-CoA synthase (ACS) [46, 47].

To the best of our knowledge, this is the first study that examined the effects of different dietary LO levels on the PPAR-α, PPAR-γ, and SCD mRNA expression in goat tissues. This study demonstrated for the first time that the dietary LO inhibits the expression of the gene that codes for the critical enzyme required to desaturate vaccenic acid to CLA in the goat adipose tissue. Furthermore, there is evidence from the present study that the degree of inhibition of transcription for this gene was related to the dietary LO level.

The expression of SCD is known to be strongly modulated by several nutrients such as FA, carbohydrates and hormones [18, 22, 48], and cholesterol [49]. Similarly, Daniel et al. [50] also showed a reduction in SCD gene expression in the adipose and liver tissues of lambs fed forage compared with a concentrate-based diet. The downregulation in mRNA levels was probably due to the high concentration of C18:3n-3 in the forage compared with the concentrate diet. Alpha-linolenic acid (C18:3n-3) also inhibited SCD gene expression in the mouse adipocytes [18, 51]. Given that C18:3n-3 is the predominant essential FA in grass [52], these results have implications for strategies to further augment concentrations of CLA in the tissue of goats reared on pasture. Waters et al. [22] and Igarashi et al. [18] found a negative relationship between SCD gene expression and n-3 PUFA in beef cattle and the rat, respectively, which is supported by the present finding where downregulation of SCD occurred for the HLO dietary treatment group with the highest concentration of α-linolenic acid. Bellinger et al. [53] showed that feeding a mixture of n-3 PUFA, α-linolenic acid (eicosapentaenoic acid), EPA, and docosahexaenoic acid (DHA) resulted in a 50% suppression of SCD mRNA in the rat liver. Nutrients, especially FA, have been shown to regulate SCD at both the enzyme activity [51] and transcriptional level [54]. In human cell lines, the transcription of SCD is under the control of two transcription factors, namely, PPAR-α and PPAR-γ [54]. The present study also examined the gene expression of these two transcription factors as affected by the different dietary LO, and while there was downregulation of the SCD gene expression by increasing the dietary LO, the PPAR-γ gene

expression was significantly increased in the adipose tissue. Actually, the PPAR is a key regulator of SCD which mediates its transcriptional activation [55].

Nutritionists strongly recommend an increased human consumption of CLA and n-3 PUFA [56–58]. The CLA in human tissues may be synthesized through the tissue desaturation of vaccenic acid by SCD [59] and thus may be increased by vaccenic acid in the human diet. Results of the present study have important implications with regards to ingesting n-3 PUFA which may have negative effects on the de novo synthesis of CLA in the human muscle through potential reductions in the SCD gene expression as shown in this and many other studies. However, because a positive relationship of n-6 PUFA and particularly the n-6 : n-3 FAR with SCD gene expression exists, a correct balance of dietary n-3 PUFA appears to be of critical importance to achieve optimal SCD gene expression levels and, in turn, CLA production in the adipose tissue. A ruminant product that naturally contains CLA and n-3 PUFA could be a good regular source of these important FA and a good alternative to more expensive nutritional supplements. The present study has important implications for the establishment of dietary strategies to augment the concentration of both CLA and n-3 PUFA in ruminant tissues. However, further work is required to determine the biochemical and molecular mechanisms controlling the synthesis and deposition of n-3 PUFA and CLA in the goat tissues to optimize the effects of n-3 PUFA on the PPAR-α, PPAR-γ, and SCD gene expression. This will provide better strategies to consistently produce nutritionally enhanced chevon.

5. Conclusion

Increasing the linseed oil in the goat diet had increased the total n-3 PUFA due to the high levels of C18:3n-3 in linseed oil, resulting in a reduced n-6 : n-3 FAR of the subcutaneous adipose tissue. However, the synthesis of EPA, docosapentaenoic acid (DPA), and DHA from dietary C18:3n-3 seems to be limited, and thus the EPA, DPA, and DHA enriched goat meat would contribute only a small amount compared to the recommended daily intake for human diet. The results indicate that maximum of tissue CLA cis-9 trans-11 concentration was observed with the low C18:3n-3 in the diet which contained low amounts of linseed oil and it decreased linearly by increasing the C18:3n-3.

The present study investigated how changes in the dietary FA affect the mRNA level expression of genes related to fat metabolism in the subcutaneous adipose tissue in Boer goats. The results showed that different dietary fat led to different FA profiles in the adipose tissues and levels of PPAR-α, PPAR-γ, and SCD gene expression. Goats fed treated OPF-based diets with high n-3 PUFA showed an upregulation of the PPAR-γ and downregulation of the SCD gene expression compared to the goats fed with low n-3 PUFA.

Thus, the data indicate that utilization of diets with the highest α-linolenic acid is a valid approach to obtain goat meat enriched with n-3 PUFA. Increasing the n-3 PUFA in the diet using linseed oil decreased the n-6 : n-3 FAR in the subcutaneous adipose tissue of growing Boer goats. These changes would likely improve the health status of the chevon produced.

Acknowledgments

The authors are very grateful to the Faculty of Veterinary Medicine, Universiti Putra Malaysia. This research was supported by the Malaysian Government E-Science Grant no. 05-01-04-SF0200.

References

[1] D. I. Givens, "The role of animal nutrition in improving the nutritive value of animal-derived foods in relation to chronic disease," *Proceedings of the Nutrition Society*, vol. 64, no. 3, pp. 395–402, 2005.

[2] K. W. J. Wahle, S. D. Heys, and D. Rotondo, "Conjugated linoleic acids: are they beneficial or detrimental to health?" *Progress in Lipid Research*, vol. 43, no. 6, pp. 553–587, 2004.

[3] R. J. B. Bessa, P. V. Portugal, I. A. Mendes, and J. Santos-Silva, "Effect of lipid supplementation on growth performance, carcass and meat quality and fatty acid composition of intramuscular lipids of lambs fed dehydrated lucerne or concentrate," *Livestock Production Science*, vol. 96, no. 2-3, pp. 185–194, 2005.

[4] Y. M. Goh, *Dietary manipulations using oil palm (Elaeis guineensis) fronds to increase the unsaturated fatty acid content of mutton under tropical conditions [Ph.D. thesis]*, Universiti Putra Malaysia, Selangor, Malaysia, 2002.

[5] M. Ebrahim, *Production of omega-3 enriched chevon through diets supplemented with oil palm (Elaeis guineensis) fronds [M.S. thesis]*, Universiti Putra Malaysia, Selangor, Malaysia, 2009.

[6] J. Santos-Silva, I. A. Mendes, P. V. Portugal, and R. J. B. Bessa, "Effect of particle size and soybean oil supplementation on growth performance, carcass and meat quality and fatty acid composition of intramuscular lipids of lambs," *Livestock Production Science*, vol. 90, no. 2-3, pp. 79–88, 2004.

[7] G. Demirel, A. M. Wachira, L. A. Sinclair, R. G. Wilkinson, J. D. Wood, and M. Enser, "Effects of dietary n-3 polyunsaturated fatty acids, breed and dietary vitamin E on the fatty acids of lamb muscle, liver and adipose tissue," *The British Journal of Nutrition*, vol. 91, no. 4, pp. 551–565, 2004.

[8] A. M. Wachira, L. A. Sinclair, R. G. Wilkinson, M. Enser, J. D. Wood, and A. V. Fisher, "Effects of dietary fat source and breed on the carcass composition, n-3 polyunsaturated fatty acid and conjugated linoleic acid content of sheep meat and adipose tissue," *The British Journal of Nutrition*, vol. 88, no. 6, pp. 697–709, 2002.

[9] R. J. B. Bessa, S. P. Alves, E. Jerónimo, C. M. Alfaia, J. A. M. Prates, and J. Santos-Silva, "Effect of lipid supplements on ruminal biohydrogenation intermediates and muscle fatty acids in lambs," *European Journal of Lipid Science and Technology*, vol. 109, no. 8, pp. 868–878, 2007.

[10] S. L. Cooper, L. A. Sinclair, R. G. Wilkinson, K. G. Hallett, M. Enser, and J. D. Wood, "Manipulation of the n-3 polyunsaturated fatty acid content of muscle and adipose tissue in lambs," *Journal of Animal Science*, vol. 82, no. 5, pp. 1461–1470, 2004.

[11] N. S. Kelley, N. E. Hubbard, and K. L. Erickson, "Alteration of human body composition and tumorigenesis by isomers of conjugated linoleic acid," *Modern Dietary Fat Intakes in Disease Promotion*, pp. 121–131, 2010.

[12] C. G. Harfoot and G. P. Hazelwood, "Lipid metabolism in the rumen," in *The Rumen Microbial Ecosystem*, pp. 382–426, Elsevier Science Publishing, London, UK, 1997.

[13] J. M. Griinari, B. A. Corl, S. H. Lacy, P. Y. Chouinard, K. V. V. Nurmela, and D. E. Bauman, "Conjugated linoleic acid is synthesized endogenously in lactating dairy cows by $\delta9$-desaturase," *Journal of Nutrition*, vol. 130, no. 9, pp. 2285–2291, 2000.

[14] A. de la Torre, D. Gruffat, D. Durand et al., "Factors influencing proportion and composition of CLA in beef," *Meat Science*, vol. 73, no. 2, pp. 258–268, 2006.

[15] A. P. Moloney, C. Kennedy, F. Noci, F. J. Monahan, and J. P. Kerry, "Lipid and colour stability of m. longissimus muscle from lambs fed camelina or linseed as oil or seeds," *Meat Science*, vol. 92, no. 1, pp. 1–7, 2012.

[16] F. Noci, P. French, F. J. Monahan, and A. P. Moloney, "The fatty acid composition of muscle fat and subcutaneous adipose tissue of grazing heifers supplemented with plant oil-enriched concentrates," *Journal of Animal Science*, vol. 85, no. 4, pp. 1062–1073, 2007.

[17] A. P. Simopoulos, "Evolutionary aspects of diet, the ω-6/ω-3 ratio and genetic variation: nutritional implications for chronic diseases," *Biomedicine and Pharmacotherapy*, vol. 60, no. 9, pp. 502–507, 2006.

[18] M. Igarashi, K. Ma, L. Chang, J. M. Bell, and S. I. Rapoport, "Dietary n-3 PUFA deprivation for 15 weeks upregulates elongase and desaturase expression in rat liver but not brain," *Journal of Lipid Research*, vol. 48, no. 11, pp. 2463–2470, 2007.

[19] H. Sampath and J. M. Ntambi, "Polyunsaturated fatty acid regulation of genes of lipid metabolism," *Annual Review of Nutrition*, vol. 25, pp. 317–340, 2005.

[20] T. Hajar, Y. M. Goh, M. A. Rajion et al., "Omega 3 polyunsaturated fatty acid improves spatial learning and hippocampal peroxisome proliferator activated receptors (PPARα and PPARγ) gene expression in rats," *BMC Neuroscience*, vol. 13, no. 1, article 109, 2012.

[21] D. E. Graugnard, P. Piantoni, M. Bionaz, L. L. Berger, D. B. Faulkner, and J. J. Loor, "Adipogenic and energy metabolism gene networks in Longissimus lumborum during rapid post-weaning growth in Angus and Angus × Simmental cattle fed high-starch or low-starch diets," *BMC Genomics*, vol. 10, article 142, 2009.

[22] S. M. Waters, J. P. Kelly, P. O'Boyle, A. P. Moloney, and D. A. Kenny, "Effect of level and duration of dietary n-3 polyunsaturated fatty acid supplementation on the transcriptional regulation of $\Delta9$-desaturase in muscle of beef cattle," *Journal of Animal Science*, vol. 87, no. 1, pp. 244–252, 2009.

[23] J. M. Ntambi, "Regulation of stearoyl-CoA desaturase by polyunsaturated fatty acids and cholesterol," *Journal of Lipid Research*, vol. 40, no. 9, pp. 1549–1558, 1999.

[24] NRC, *Nutrient Requirements of Small Ruminant*, National Academy Press, Washington, DC, USA, 6th edition, 2007.

[25] AOAC, *Official Methods of Analysis*, edited by K. Herlick, Association of Official Analytical Chemists, Arlington, Va, USA, 15th edition, 1990.

[26] P. J. van Soest, J. B. Robertson, and B. A. Lewis, "Methods for dietary fiber, neutral detergent fiber, and nonstarch polysaccharides in relation to animal nutrition," *Journal of Dairy Science*, vol. 74, no. 10, pp. 3583–3597, 1991.

Effect of Linseed Oil Dietary Supplementation on Fatty Acid Composition and Gene Expression in Adipose
Tissue of Growing Goats

189

[27] J. Folch, M. Lees, and G. H. Sloane Stanely, "A simple method for the isolation and purification of total lipides from animal tissues," *The Journal of Biological Chemistry*, vol. 226, no. 1, pp. 497–509, 1957.

[28] M. A. Rajion, J. G. McLean, and R. N. Cahill, "Essential fatty acids in the fetal and newborn lamb," *Australian Journal of Biological Sciences*, vol. 38, no. 1, pp. 33–40, 1985.

[29] M. Ebrahimi, M. A. Rajion, Y. M. Goh, and A. Q. Sazili, "Impact of different inclusion levels of oil palm (*Elaeis guineensis* Jacq.) fronds on fatty acid profiles of goat muscles," *Journal of Animal Physiology and Animal Nutrition*, vol. 96, no. 6, pp. 962–969, 2012.

[30] E. Dervishi, C. Serrano, M. Joy, M. Serrano, C. Rodellar, and J. H. Calvo, "The effect of feeding system in the expression of genes related with fat metabolism in semitendinous muscle in sheep," *Meat Science*, vol. 89, no. 1, pp. 91–97, 2011.

[31] C. G. Harfoot, R. C. Noble, and J. H. Moore, "Factors influencing the extent of biohydrogenation of linoleic acid by rumen micro-organisms in vitro," *Journal of the Science of Food and Agriculture*, vol. 24, no. 8, pp. 961–970, 1973.

[32] C. M. Kim, J. H. Kim, Y. K. Oh et al., "Effects of flaxseed diets on performance, carcass characteristics and fatty acid composition of Hanwoo steers," *Asian-Australasian Journal of Animal Sciences*, vol. 22, no. 8, pp. 1151–1159, 2009.

[33] M. Igarashi, F. Gao, H. W. Kim, K. Ma, J. M. Bell, and S. I. Rapoport, "Dietary n-6 PUFA deprivation for 15 weeks reduces arachidonic acid concentrations while increasing n-3 PUFA concentrations in organs of post-weaning male rats," *Biochimica et Biophysica Acta*, vol. 1791, no. 2, pp. 132–139, 2009.

[34] C. M. Kim, J. H. Kim, T. Y. Chung, and K. K. Park, "Effects of flaxseed diets on fattening response of Hanwoo cattle: 2. Fatty acid composition of serum and adipose tissues," *Asian-Australasian Journal of Animal Sciences*, vol. 17, no. 9, pp. 1246–1254, 2004.

[35] E. Jerónimo, S. P. Alves, J. A. M. Prates, J. Santos-Silva, and R. J. B. Bessa, "Effect of dietary replacement of sunflower oil with linseed oil on intramuscular fatty acids of lamb meat," *Meat Science*, vol. 83, no. 3, pp. 499–505, 2009.

[36] G. C. Burdge and P. C. Calder, "α-linolenic acid metabolism in adult humans: the effects of gender and age on conversion to longer-chain polyunsaturated fatty acids," *European Journal of Lipid Science and Technology*, vol. 107, no. 6, pp. 426–439, 2005.

[37] K. Raes, S. De Smet, and D. Demeyer, "Effect of dietary fatty acids on incorporation of long chain polyunsaturated fatty acids and conjugated linoleic acid in lamb, beef and pork meat: a review," *Animal Feed Science and Technology*, vol. 113, no. 1–4, pp. 199–221, 2004.

[38] A. P. Simopoulos, "Importance of the omega-6/omega-3 balance in health and disease: evolutionary aspects of diet," *Healthy Agriculture, Healthy Nutrition, Healthy People*, vol. 102, pp. 10–21, 2011.

[39] A. P. Simopoulos, "Omega-6/omega-3 essential fatty acid ratio and chronic diseases," *Food Reviews International*, vol. 20, no. 1, pp. 77–90, 2004.

[40] L. A. Sinclair, "Nutritional manipulation of the fatty acid composition of sheep meat: a review," *Journal of Agricultural Science-Cambridge*, vol. 145, no. 5, pp. 419–434, 2007.

[41] H. Osmundsen, J. Bremer, and J. I. Pedersen, "Metabolic aspects of peroxisomal β-oxidation," *Biochimica et Biophysica Acta*, vol. 1085, no. 2, pp. 141–158, 1991.

[42] A. Pawar and B. J. Donald, "Unsaturated fatty acid regulation of peroxisome proliferator-activated receptor alpha activity in rat primary hepatocytes," *The Journal of Biological Chemistry*, vol. 278, no. 38, pp. 35931–35939, 2003.

[43] H. Al-Hasani and H. G. Joost, "Nutrition-/diet-induced changes in gene expression in white adipose tissue," *Best Practice and Research: Clinical Endocrinology and Metabolism*, vol. 19, no. 4, pp. 589–603, 2005.

[44] F. Chiarelli and D. Di Marzio, "Peroxisome proliferator-activated receptor-γ agonists and diabetes: current evidence and future perspectives," *Vascular Health and Risk Management*, vol. 4, no. 2, pp. 297–304, 2008.

[45] P. Tontonoz and M. S. Bruce, "Fat and beyond: the diverse biology of PPARγ," *Annual Review of Biochemistry*, vol. 77, no. 1, pp. 289–312, 2008.

[46] S. P. Kaplins'kyĭ, A. M. Shysh, V. S. Nahibin, V. I. Dosenko, V. M. Klimashevs'kyĭ, and O. O. Moĭbenko, "Omega-3 polyunsaturated fatty acids stimulate the expression of PPAR target genes," *Fiziolohichnyĭ Zhurnal*, vol. 55, no. 2, pp. 37–43, 2009.

[47] S. É. Michaud and G. Renier, "Direct regulatory effect of fatty acids on macrophage lipoprotein lipase: potential role of PPARs," *Diabetes*, vol. 50, no. 3, pp. 660–666, 2001.

[48] J. M. Ntambi and M. Miyazaki, "Regulation of stearoyl-CoA desaturases and role in metabolism," *Progress in Lipid Research*, vol. 43, no. 2, pp. 91–104, 2004.

[49] H. J. Kim, M. Miyazaki, and J. M. Ntambi, "Dietary cholesterol opposes PUFA-mediated repression of the stearoyl-CoA desaturase-1 gene by SREBP-1 independent mechanism," *Journal of Lipid Research*, vol. 43, no. 10, pp. 1750–1757, 2002.

[50] Z. C. T. R. Daniel, R. J. Wynn, A. M. Salter, and P. J. Buttery, "Differing effects of forage and concentrate diets on the oleic acid and conjugated linoleic acid content of sheep tissues: the role of stearoyl-CoA desaturase," *Journal of Animal Science*, vol. 82, no. 3, pp. 747–758, 2004.

[51] A. M. Sessler, N. Kaur, J. P. Palta, and J. M. Ntamb, "Regulation of stearoyl-CoA desaturase 1 mRNA stability by polyunsaturated fatty acids in 3T3-L1 adipocytes," *Journal of Biological Chemistry*, vol. 271, no. 47, pp. 29854–29858, 1996.

[52] R. J. Dewhurst, N. D. Scollan, M. R. F. Lee, H. J. Ougham, and M. O. Humphreys, "Forage breeding and management to increase the beneficial fatty acid content of ruminant products," *Proceedings of the Nutrition Society*, vol. 62, no. 2, pp. 329–336, 2003.

[53] L. Bellinger, C. Lilley, and S. C. Langley-Evans, "Prenatal exposure to a maternal low-protein diet programmes a preference for high-fat foods in the young adult rat," *The British Journal of Nutrition*, vol. 92, no. 3, pp. 513–520, 2004.

[54] J. Eeckhoute, O. Frédérik, S. Bart, and L. Philippe, "Coordinated regulation of PPARγ expression and activity through control of chromatin structure in adipogenesis and obesity," *PPAR Research*, vol. 2012, Article ID 164140, 9 pages, 2012.

[55] C. W. Miller and J. M. Ntambi, "Peroxisome proliferators induce mouse liver stearoyl-CoA desaturase 1 gene expression," *Proceedings of the National Academy of Sciences of the United States of America*, vol. 93, no. 18, pp. 9443–9448, 1996.

[56] R. J. Deckelbaum and C. Torrejon, "The omega-3 fatty acid nutritional landscape: health benefits and sources," *The Journal of Nutrition*, vol. 142, no. 3, pp. 587S–591S, 2012.

[57] J. MacRae, L. O'Reilly, and P. Morgan, "Desirable characteristics of animal products from a human health perspective," *Livestock Production Science*, vol. 94, no. 1-2, pp. 95–103, 2005.

[58] A. Dilzer and Y. Park, "Implication of conjugated linoleic acid (CLA) in human health," *Critical Reviews in Food Science and Nutrition*, vol. 52, no. 6, pp. 488–513, 2012.

[59] A. M. Turpeinen, M. Mutanen, A. Aro et al., "Bioconversion of vaccenic acid to conjugated linoleic acid in humans," *American Journal of Clinical Nutrition*, vol. 76, no. 3, pp. 504–510, 2002.

19

Lovastatin Production by *Aspergillus terreus* Using Agro-Biomass as Substrate in Solid State Fermentation

Mohammad Faseleh Jahromi,[1] **Juan Boo Liang,**[2] **Yin Wan Ho,**[1] **Rosfarizan Mohamad,**[3] **Yong Meng Goh,**[4] **and Parisa Shokryazdan**[1]

[1] *Laboratory of Industrial Biotechnology, Institute of Bioscience, Universiti Putra Malaysia, 43400 Serdang, Malaysia*
[2] *Laboratory of Animal Production, Institute of Tropical Agriculture, Universiti Putra Malaysia, 43400 Serdang, Malaysia*
[3] *Faculty of Biotechnology and Biomolecular Sciences, Universiti Putra Malaysia, 43400 Serdang, Malaysia*
[4] *Faculty of Veterinary Medicine, Universiti Putra Malaysia, 43400 Serdang, Malaysia*

Correspondence should be addressed to Juan Boo Liang, jbliang@putra.upm.edu.my

Academic Editor: Anuj K. Chandel

Ability of two strains of *Aspergillus terreus* (ATCC 74135 and ATCC 20542) for production of lovastatin in solid state fermentation (SSF) using rice straw (RS) and oil palm frond (OPF) was investigated. Results showed that RS is a better substrate for production of lovastatin in SSF. Maximum production of lovastatin has been obtained using *A. terreus* ATCC 74135 and RS as substrate without additional nitrogen source (157.07 mg/kg dry matter (DM)). Although additional nitrogen source has no benefit effect on enhancing the lovastatin production using RS substrate, it improved the lovastatin production using OPF with maximum production of 70.17 and 63.76 mg/kg DM for *A. terreus* ATCC 20542 and *A. terreus* ATCC 74135, respectively (soybean meal as nitrogen source). Incubation temperature, moisture content, and particle size had shown significant effect on lovastatin production ($P < 0.01$) and inoculums size and pH had no significant effect on lovastatin production ($P > 0.05$). Results also have shown that pH 6, 25°C incubation temperature, 1.4 to 2 mm particle size, 50% initial moisture content, and 8 days fermentation time are the best conditions for lovastatin production in SSF. Maximum production of lovastatin using optimized condition was 175.85 and 260.85 mg/kg DM for *A. terreus* ATCC 20542 and ATCC 74135, respectively, using RS as substrate.

1. Introduction

Lovastatin is a potent drug for lowering the blood cholesterol and it was the first statin accepted by United States Food and Drug Administration (USFDA) in 1987 as a hypercholesterolemic drug [1]. It is a competitive inhibitor of HMG-CoA reductase, which is a key enzyme in the cholesterol production pathway [2]. Lovastatin is a secondary metabolite during the secondary phase (idiophase) of fungi growth [3]. This product can be produced by cultures of *Penicillium* species [4], *A. terreus* [5–7], *Monascus* species [8, 9], *Hypomyces*, *Doratomyces*, *Phoma*, *Eupenicillium*, *Gymnoascus*, and *Trichoderma* [10]. Although the ability of different groups of fungi for production of lovastatin was reported in many studies, only production of this compound by *A. terreus* was

commercialized (for manufacture of high quantity of lovastatin for used as anticholesterol drag) [11]. Microorganisms are able to produce lovastatin in SSF or submerged culture [5, 7, 12–15]. Experiment showed that quantity of lovastatin production in SSF is significantly higher than submerged culture [5]. Different substrates were used for lovastatin production in SSF, including sorghum grain, wheat bran, rice, and corn [5, 7]. These substrate materials are normally expensive and are competing with food or feed ingredients for human and livestock. On the other hand, large quantity of agro-industrial biomass such as RS and OPF are produced globally particularly in the tropical countries. These agro-biomass are often burned away for disposal, causing huge environmental concerns, with only some remaining being used as roughage feed for ruminant livestock. These biomasses are, however, potential substrates for growth of microorganisms and production of biomaterials.

Over the last 250 years, the concentration of atmospheric methane (CH_4) increased by approximately 150% [16], with agricultural activities contributing 40% of the total anthropogenic source, of which 15 to 20% is from enteric fermentation in ruminants [17]. On the other hand, ruminal CH_4 production accounts for between 2 to 15% of dietary energy loss for the host animals [18]. Because of the negative effects on environment and the host animal nutrition, mitigation of enteric CH_4 emission in ruminant livestock had been extensively researched, including the use of various mitigating agents such as ionophores [19], organic acids [20], fatty acids [21], methyl coenzyme M reductase inhibitors [22], vaccine [23], and oil [24]. However, these technologies have limited application primarily because they, besides suppressing CH_4 also, decrease nutrients digestibility (such as oil and fatty acids), have negative effect on human and animal health (antibiotics), or are not economically acceptable (methyl coenzyme M reductase inhibitors and vaccine).

Wolin and Miller [25] showed significantly reduction in growth and activity of methanogenic Archaea using lovastatin without any negative effect on cellulolytic bacteria that was due to the effect of this drug on inhibition the activity of HMG-CoA reductases in the archaeal microorganisms. It is uneconomical to use pure lovastatin as a feed additive for the mitigation of CH_4 production in ruminants. Production of this component using low-cost substrate and process for being used as animal feed additive were the main objective of present study.

Thus, the primary objective of this study was to investigate the efficacy of two strains of A. terreus (ATCC 20542 and ATCC 74135) for production of lovastatin using RS and OPF as substrates. In addition, the effects on nitrogen source, mineral solution, moisture, incubation time, pH, inoculum size, particle size, and incubation time on lovastatin production were investigated.

2. Materials and Methods

2.1. Substrate. RS and OPF were collected from the local fields in the state of Selangor, Malaysia. The materials were ground and sieved through mesh size 6 to obtain particles size of about 3.4 mm and dried in oven at 60°C for 48 h and used in SSF studies.

2.2. Microorganism and Preparation of Spore Suspension. A. terreus, ATCC 20542 and ATCC 74135, used in this study was obtained from the American Type Culture Collection (ATCC). They were maintained on potato dextrose agar (PDA) slants at 32°C for 7 days, stored at 4°C, and subcultured every two weeks. For the preparation of spore suspension, 10 mL of sterilized 0.1% Tween-80 solution was added to the 7-day old culture slants of the fungi at the end of incubation, surface of the culture was scratched with sterilized loop, and the Tween-80 solution containing spores was transferred into 100 mL Schott bottle containing the same solution and agitated thoroughly using a shaker to suspend the spores. The number of spores was measured

using a hemocytometer and adjusted to approximately 10^7 spores/mL for use as inoculum throughout the study.

2.3. Solid State Fermentation. This study consisted of two subexperiments. In the first, the efficacy of lovastatin production by two strains of A. terreus (ATCC 20542 and ATCC 74135) using two types of agro-biomass (RS and OPF) as substrate was examined. In addition, soybean meal, urea and ammonium sulphate were used as nitrogen sources and the need to supplement mineral to enhance the fermentation process was studied. In the next subexperiment, fermentation conditions were optimized for maximum lovastatin production in SSF. The procedure of SSF for each subexperiments was described below. Both subexperiments were conducted in triplicate. Presented data are Mean ± Standard Deviation.

Subexperiment 1: Effect of Substrate, Nitrogen, and Mineral Solution. Solid state fermentation was carried out in 500 mL Erlenmeyer flasks containing 20 g of the respective substrate (RS or OPF). The moisture content of the substrate were adjusted with mineral solution (KH_2PO_4: 2.1 g/L, $MgSO_4$: 0.3 g/L, $CaCl_2$: 0.3 g/L, $FeSO_4$: 0.11 g/L, $ZnSO_4$: 0.3 g/L) or distilled water to produce a moisture content of approximately 75%. pH of all the solutions were adjusted to 6 before adding into the solid substrate. For the study on effect of nitrogen source on lovastatin production, 1% urea, 1% ammonium sulphate, or 10% of soybean meal was added in to the solid culture. The contents in the flasks were autoclaved for 15 min at 121°C and after cooling the flasks to room temperature, 10% inoculum were added and the contents of the flasks were thoroughly mixed. The flasks were incubated at 32°C for ten days.

Subexperiment 2: Optimization the Fermentation Condition. Since the production of lovastatin using RS as substrate was significantly higher than that for OPF, a follow-up experiment was conducted to optimize several factors (pH, temperature, particle size, inoculums size, and initial moisture content) known to affect SSF process for production of lovastatin by both strains of A. terreus using only RS as substrate. The second experiment consisted of five subexperiments, each evaluating the effect of one of the above five factors on the SSF process with the remaining factors being constant. For study on the effect of pH, because of the difficulty of adjusting the pH of solid sample in SSF, pH of solution was adjusted using 1 M Sodium hydroxide and 1 M hydrogen chloride to pH 5, 6, 7, and 8 (before adding in the substrate). Incubation temperatures between 25 to 42°C, five different particle sizes: <425 μm (mesh no. 40), 425–600 μm (mesh no. 30), 600–1400 μm (mesh no. 14), 1.4–2 mm (mesh no. 10), and 2–3.35 mm (mesh no. 6), inoculums size of 5, 10, and 15% and initial moisture contents at 50, 66, and 75% were investigated.

2.4. Extraction and Determination of Lovastatin. At the end of fermentation, the solid culture was dried at 60°C for 48 h and 0.5 g of the dry culture was extracted with 15 mL methanol and shaking in a shaker for 60 min at 220 rpm [5].

TABLE 1: Effect of nitrogen source and mineral solution on lovastatin production in solid state fermentation by *A. terreus* using rice straw (RS) and oil palm frond (OPF) as substrates.

Treatments	Lovastatin production (mg/kg DM)	
	ATCC 20542	ATCC 74135
RS	154.48 ± 22.88^a	157.07 ± 1.92
RS plus mineral	119.35 ± 16.59^{ab}	146.09 ± 7.49
RS plus mineral and urea	96.3 ± 3.02^{bc}	118.26 ± 6.38
RS plus mineral and soybean meal	126.36 ± 22.84^{ab}	139.63 ± 25.45
RS plus mineral and ammonium sulphate	66.19 ± 3.39^c	129.39 ± 21.90
Significant	**	NS
OPF	7.84 ± 0.14^b	13.09 ± 2.22^b
OPF plus mineral	9.66 ± 0.03^b	7.2 ± 1.55^b
OPF plus mineral and urea	21.25 ± 4.66^b	17.58 ± 3.03^b
OPF plus mineral and soybean meal	70.17 ± 4.84^a	63.76 ± 14.00^a
OPF plus mineral and ammonium sulphate	10.06 ± 0.31^b	11.16 ± 3.41^b
Significant	**	**

NS: not significantly different.
**Significantly different at 1% level.
a,b,c indicating that means for each substrate within column are significantly different.

After filtration with membrane filter (0.2 μm), the concentration of lovastatin in the filtrate was assayed using HPLC (Waters, USA, 2690) attached with an ODS column (Agilent, 250 × 4.6 mm i.d., 5 μm). The mobile phase consisted of acetonitrile and water (70:30 by volume) contained 0.5% acetic acid. The flow rate was 1 mL/min. The photo diode array (PDA) detection range was set from 210 to 400 nm and lovastatin was detected at 237 nm. The sample injection volume was 20 μL, and the run time was 12 min. Since two forms of lovastatin (lactone and β-hydroxyl) are normally present in the fermented culture, they were separately determined in the HPLC. Commercial lovastatin (mevinolin K, 98%, HPLC grade, M2147, Sigma, USA) used as standard is in the lactone form. And β-hydroxyl lovastatin was produced from the lactone form using the method of Friedrich et al. [26]. Briefly, to prepare β-hydroxyl acid, lactone lovastatin was suspended in 0.1 M NaOH and heated at 50°C for 1 h in a shaking incubator. Subsequently, the mixture was adjusted to pH 7.7 with 1 M HCl, filtered through 0.2 μm filters and used as standard for HPLC. The retention times of the hydroxyl and lactone forms of lovastatin were 6.668 and 10.898 min, respectively. Different concentrations ranged from 0.5 to 500 ppm of lovastatin were used as standard and standard curve of lovastatin.

2.5. Scanning Electron Microscope. Microscopy analysis was conducted using a Scanning Electron Microscope (SEM) to determine the morphological growth of *A. terreus* on the surface of RS. The samples were dried (60°C for 48 h), cut into 1 mm size, and affixed to a metal SEM stub and sputter coated in gold using SEM coating unit (BAL-TEC SCD 005 Spotter coater). The coated specimens were viewed using Environmental Scanning Electron Microscope (Phillips XL30, Germany) at accelerating voltage of 15–25 KV.

2.6. Chemical Analysis. Neutral detergent fibre (NDF) and acid detergent fibre (ADF) were determined by the detergent system [27]. Acid detergent lignin (ADL), crude protein (CP), and ash were determined using the method described by AOAC [28]. Hemicelluloses content was estimated as the difference between NDF and ADF, while cellulose content was the difference between ADF and ADL.

2.7. Statistical Analysis. All of the experiments were done with 3 replicate. Individual culture flasks were considered as experimental units. Data were analyzed as a completely randomized design (CRD) using the general linear model (GLM) procedure of SAS 9.2 [29] with a model that included treatment effects and experimental error. All multiple comparisons among means were performed using Duncan's new multiple range test ($\alpha = 0.05$).

3. Results and Discussion

3.1. Lovastatin Determination. Lovastatin was quantified as its β-hydroxyl and lactone forms (Figure 1). The β-hydroxyl of lovastatin is the more active form of this drug but is unstable [5], thus preparation for its standard solution was prepared freshly from the lactone form according to Friedrich et al. [26]. Because the hydroxyl form of lovastatin is not stable, lactone form is normally the primary lovastatin detected in the fermented products. The quantities of lovastatin reported in Table 1 were the combination of the two forms, but results of this study show that β-hydroxyl form is the dominant lovastatin in the solid cultures.

It was reported that the conditions needed for the conversion of lactone into the β-hydroxyl form are high pH (e.g., by addition of NaOH), heating to 50°C, and naturalization by acid. Since none of the above conditions was applied in this study, the β-hydroxyl form of lovastatin present in the extract in this study is believed to be a direct product of SSF and not due to conversion from the lactone form. Hydroxyl lovastatin has been reported to be the more active form of this drug [26], and its efficacy for inhibition of HMG-CoA reductase will be validated in subsequent experiment.

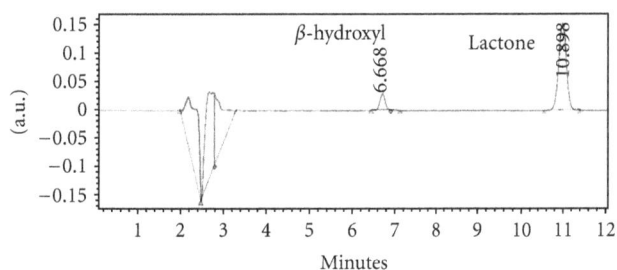

FIGURE 1: HPLC chromatogram of lovastatin in β-hydroxyl and lactone forms.

3.2. Effect of Substrate. In the first experiment, the efficacy of two strains of *A. terreus* on the production of lovastatin in SSF using RS and OPF as substrate was investigated. Lovastatin productions by both strains using RS and OPF as substrates are shown in Table 1. Quantity of lovastatin production by both strains of *A. terreus* in the RS samples was higher than OPF. The highest lovastatin production was 154.48 and 157.07 mg/kg DM, respectively, for *A. Terreus* ATCC 20542 and ATCC 74135 using RS as substrate without additional nitrogen and mineral supplementation.

Although the two strains of *A. terreus* tested were capable to produce lovastatin from RS and OPF, results of this study showed that RS is a more suitable substrate, producing significantly higher lovastatin than OPF.

Lovastatin can be produced from 9 molecules of acetyl-CoA during the fermentation (Figure 2) [30]. Since acetate production by microbial activity is correlated with carbohydrate fermentation, source of carbohydrate is important for production of this product. Lignocelluloses including cellulose, hemicelluloses, and lignin are the main components of agricultural biomass, therefore, for production of acetate, these macromolecules must first be hydrolyzed into their subunits such as glucose, xylose, in the presence of the appropriate enzymes, such as cellulase, hemicellulase, pectinase, and cellulobiase. The resulted monomers can then be used during the fermentation process of fungi and production of acetyl-CoA which can be used as substrate for lovastatin production (Figure 2). Therefore, the microorganisms selected for SSF must be able to degrade the lignocelluloses by producing sufficient amount of the appropriate enzymes to hydrolyze the respective lignocelluloses fractions. Production of cellulase (such as betaglucosidase, endoglucanase, and cellobiohydrolase) and hemicellulases (mainly xylanase) enzymes by *A. terreus* and their effects on lignocelluloses degradation have been well documented [31–33]. However, there is no known data on production of lignin degradation enzyme by *A. terreus*.

RS has two main advantages over OPF as substrate for *A. terreus* in SSF. Lignin content of RS is half that of OPF and its hemicelluloses content is higher than OPF (Figure 3). The lower production of lovastatin recorded for OPF using the two *A. terreus* could be partly due to the absence of enzyme to degrade the lignin component of the biomass. Reports also showed that *A. terreus* has high ability to produce hemicellulase enzymes such as xylanase [34], making RS

which contains high hemicelluloses more appropriate as substrate for lovastatin production.

3.3. Effect of Mineral Solution and Nitrogen Source. Effects of mineral solution and nitrogen on lovastatin production by *A. terreus* are shown in Table 1. Addition of mineral solution has no significant effect on quantity of lovastatin production by both strains of *A. terreus*. Although addition of nitrogen source has no significant effect on lovastatin production by *A. terreus* ATCC 74135 ($P > 0.05$), it negatively affected lovastatin production by *A. terreus* ATCC 20542 ($P < 0.01$) mainly in the ammonium sulphate treatment. Supplementation of minerals and nitrogen sources (especially soybean meal) increased the of total lovastatin to about 64 mg/kg DM in *A. terreus* ATCC 74135 using OPF, but it is still half of that compared to RS (Table 1).

Minerals and nitrogen are essential growth nutrients for microorganisms [15]. In submerged fermentation, these nutrients must be added into the medium to sustain the growth of the microorganisms [34]. Agricultural biomass such as RS contain mineral components [35, 36]. Ash content of RS and OPF which is indicator of mineral content of these by products was shown in Figure 3. This quantity can supply the request of *A. terreus* for growth and lovastatin production and additional mineral solution has no effect on enhancement of the lovastatin production by this fungus.

Although nitrogen source is important for fermentation, but carbon : nitrogen (C : N) ratio is more important for lovastatin production. Negative effect of additional nitrogen source on lovastatin production was reported in previous study [37]. In the study of growth of *A. terreus*, and lovastatin production, the above authors reported that type of carbon and nitrogen sources and C : N ratio in the medium has important effect on lovastatin production and by increased the C : N ratio (from 14.4 to 23.4 and 41.3) the ability of *A. Terreus* for lovastatin production was increased. Use of a slowly metabolized carbon source (lactose) in combination with either soybean meal or yeast extract under nitrogen-limited conditions gave the highest lovastatin production. Effect of C : N ratio on lovastatin production was also studied by Bizukojc and Ledakowicz [6] who reported that higher C : N ratio can provide better fermentation condition for lovastatin (mevinolinic acid) production (Figure 4). Results of the chemical analysis showed that the availability of N in RS (CP = 4.2%) is higher than OPF (CP = 3.2%) (Figure 3). On the other hand, stranger bands in the cell wall polymers of OPF in comparison to the RS due to the higher lignin (Figure 3) can suppress the availability of nitrogen for fungal activity in OPF. It can be concluded that RS without additional nitrogen has better C : N ratio in comparison to the samples that contain additional nitrogen sources. In contrast, to achieve the suitable C : N ratio in the OPF culture, additional nitrogen source is requested.

3.4. Effect on Lignocellulose Reduction. Although the main objective of this study was to investigate the ability of *A. terreus* to produce lovastatin using biomass as substrate, ability of this fungus to reduce the lignocelluloses content

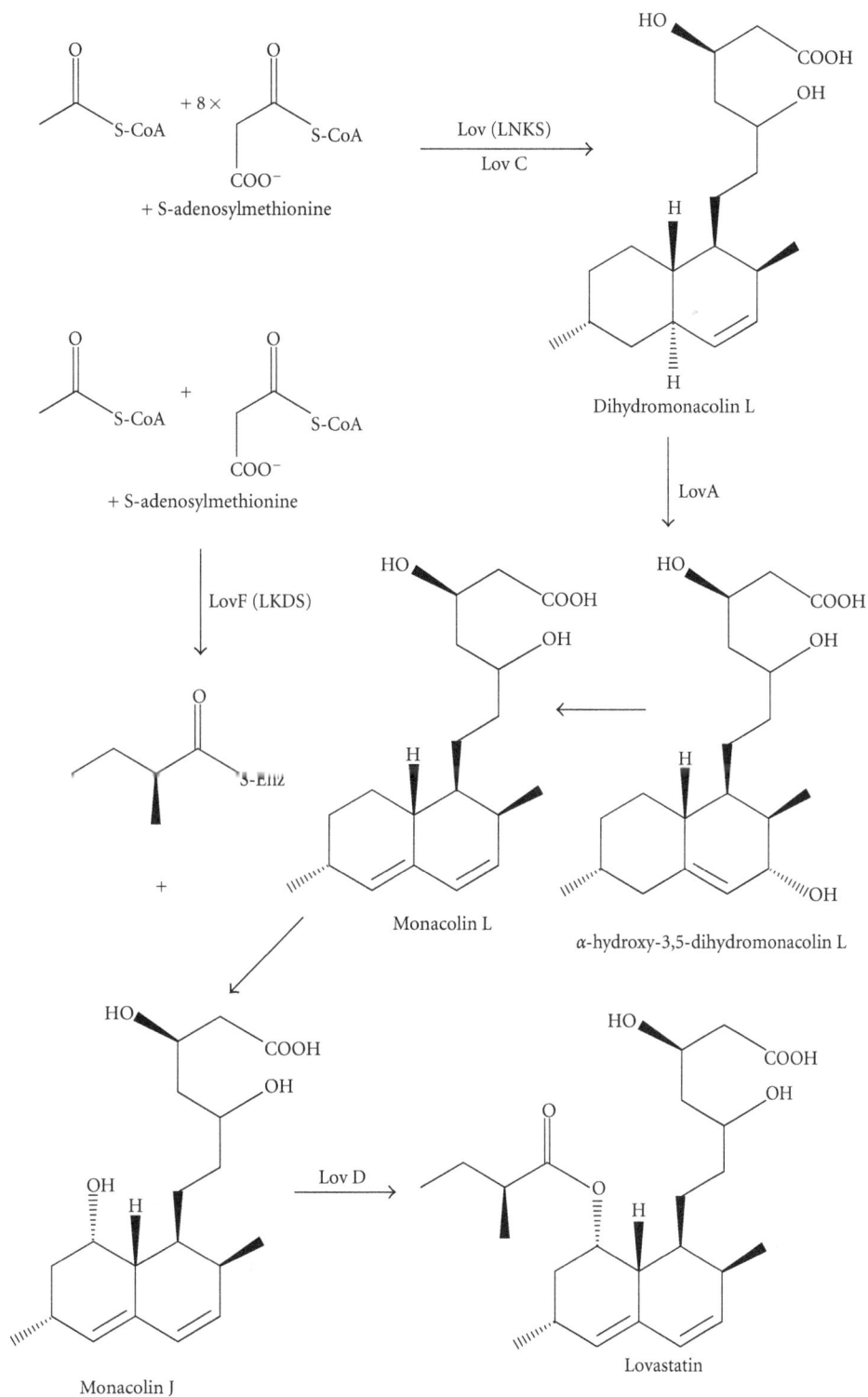

FIGURE 2: Lovastatin production pathway [30].

of RS and OPF was also important, particularly when these materials would be considered as feed for ruminants. Effect of the two strains of *A. terreus* on reduction of lignocelluloses content of RS and OPF is shown in Table 2. Results show that (i) additional mineral solution has no effect ($P > 0.05$) to enhancing the ability of *A. terreus*

to reduce lignocelluloses content of RS and OPF but (ii) nitrogen supplement increases the ability of the fungi to degrade cellulosic materials. Both nitrogen solution, urea and ammonium sulphate significantly increased lignocellulosic components reduction. *A. terreus* has higher ability for reduction of lignocelluloses content of RS compared to OPF.

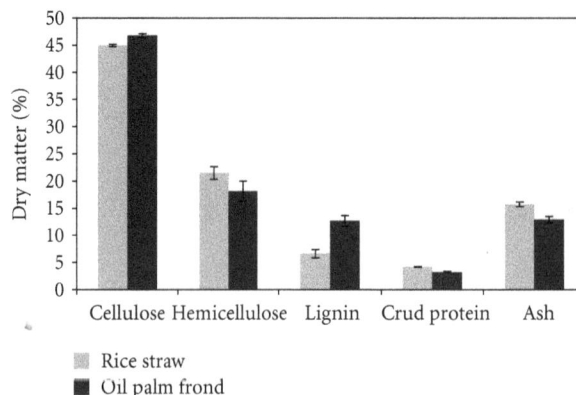

FIGURE 3: Lignocellulose, crude protein, and ash content of RS and OPF.

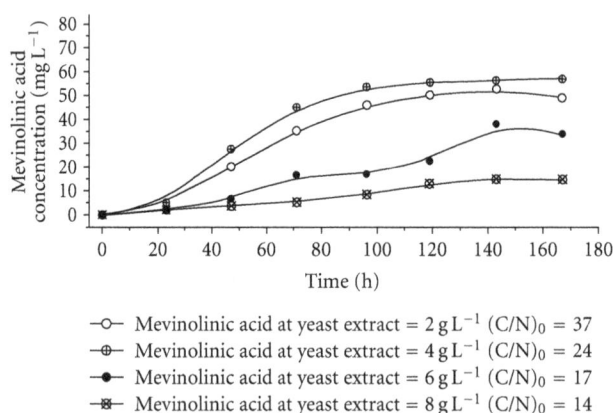

FIGURE 4: The evolution of mevinolinic acid (lovastatin) production in the batch culture at different initial yeast extract concentrations and C:N ratio content [6].

Although the effect of ammonium sulphate on lignocelluloses content reduction was higher than urea, this difference was not significant (Table 2). Because of its affordable price, urea is the most widely used nonprotein nitrogen in ruminant feed [38]. Thus, it is used as the nitrogen source in the following experiment for *A. terreus*.

Based on the higher yield of lovastatin using RS as substrate, the follow-up study for optimization of influencing factors (pH, temperature, particular size, inoculum size and initial moisture content and incubation time) was examined using the two strains of fungi on RS alone with addition of urea.

3.5. Effect of Initial pH. Study on initial pH (Figure 5) indicated that this factor has no significant effect on lovastatin production by both strains of *A. Terreus* ($P > 0.05$); however, pH 6 produced the highest lovastatin; 67.88 and 85.49 mg/kg DM by *A. terreus* ATCC 20542 and *A. terreus* ATCC 74135, respectively. There was no significant difference in lovastatin production for pH of 5 to 8; however, pH higher than 7 or lower than 6 had negative effect on lovastatin production. Kumar et al. [39] showed that pH 5.8–6.3 provided optimum conditions for lovastatin production by *A. terreus* in the batch process. Other researchers [40–42] also reported

optimum pH of within the range of 5–7 using various fungi for statin production. The above information thus suggested that optimum initial pH for lovastatin production in SSF is near neutral pH with small variations which could be due to the types of substrate and microorganism used in the fermentation process. On the other hand, [43] reported that by controlling pH and slowly adding the carbon source lovastatin yield could increase five folds.

3.6. Effect of Temperature. One of the important factors effecting microbial activity and thus biomaterial production is incubation temperature. Results of this study suggest that 25°C is optimum temperature for lovastatin production in SSF by the two strains of *A. terreus* with maximum production of 171.61 and 202.93 mg/kg DM for *A. terreus* ATCC 20542 and *A. terreus* ATCC 74135, respectively (Figure 6). Increasing the incubation temperature to higher than 25°C has negative effect on lovastatin production and *A. terreus* ATCC 74135 being more sensitive to the temperature change. Similar incubation temperature (25°C) was also reported previously [43, 44] for lovastatin production. Optimum temperature of 29.5°C for lovastatin production had been reported by Panda et al. [40] using *Monascus purpureus* and *Monascus Ruber* while Kumar et al. [39] showed that 28°C is

TABLE 2: Effect of SSF on cellulose and hemicellulose contents of RS and OPF (% of dry mater).

Treatments	A. terreus ATCC 20542		A. terreus ATCC 74135	
	Cellulose	H-cellulose	Cellulose	H-cellulose
Nonfermented RS	44.97 ± 0.22^{ab}	21.50 ± 1.16^{a}	44.97 ± 0.22^{a}	21.50 ± 1.16^{a}
Fermented RS (FRS)	46.83 ± 2.12^{a}	18.06 ± 1.20^{bc}	45.12 ± 0.23^{a}	13.13 ± 0.79^{b}
FRS plus mineral	45.80 ± 0.67^{ab}	19.63 ± 0.80^{ab}	44.91 ± 0.72^{a}	12.38 ± 0.76^{b}
FRS plus mineral and urea	40.70 ± 2.52^{c}	16.96 ± 1.78^{dc}	41.97 ± 0.61^{b}	9.21 ± 1.05^{c}
FRS plus mineral and ammonium sulphate	42.72 ± 3.69^{bc}	15.11 ± 1.65^{d}	40.97 ± 0.51^{c}	7.79 ± 2.50^{c}
Significant	*	**	**	**
Nonfermented OPF	46.76 ± 0.34^{a}	18.13 ± 1.83	46.76 ± 0.34^{a}	18.13 ± 1.83
Fermented OPF (FOPF)	43.22 ± 2.59^{b}	19.36 ± 0.31	46.23 ± 1.17^{a}	15.66 ± 1.05
FOPF plus mineral	43.04 ± 1.28^{b}	15.82 ± 2.92	45.34 ± 0.95^{ab}	15.01 ± 0.82
FOPF plus mineral and urea	42.37 ± 1.92^{b}	16.90 ± 0.15	43.40 ± 0.89^{b}	15.40 ± 0.62
FOPF plus mineral and ammonium sulphate	44.66 ± 0.54^{ab}	18.30 ± 1.80	43.59 ± 1.69^{b}	15.00 ± 1.51
Significant	*	NS	*	NS

NS: not significantly different.
*Significantly different at 5% level.
**Significantly different at 1% level.
a,b,c indicating that means within column are significantly different.

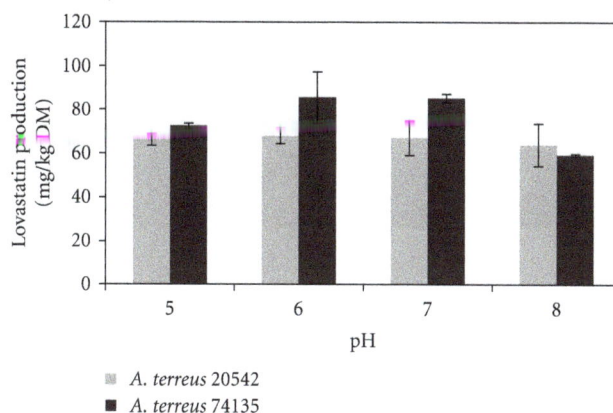

FIGURE 5: Effect of pH on lovastatin production by two strains of A. terreus ($P > 0.05$).

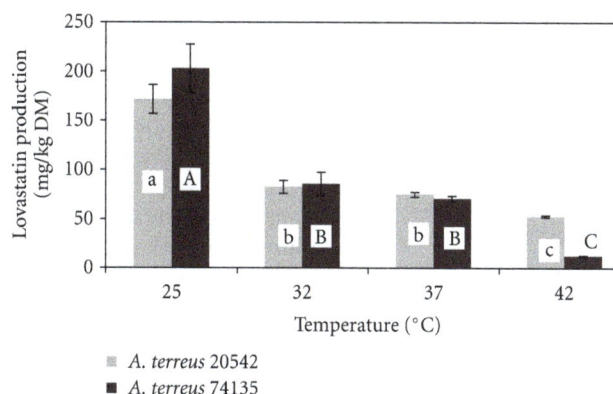

FIGURE 6: Effect of incubation temperature on lovastatin production by A. terreus ($P < 0.01$). (a, b, and c) indicate differences among means between samples for A. terreus ATCC 20542. (A, B, and C) indicate differences among means between samples for A. terreus ATCC 20542.

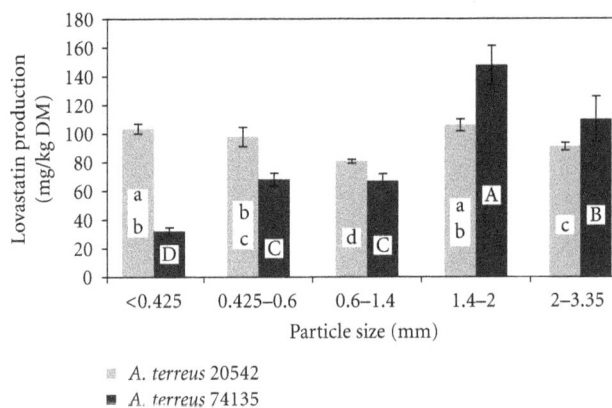

FIGURE 7: Effect of particle size on lovastatin production by two strains of A. terreus ($P < 0.01$). (a, b, c, and d) indicate differences among means between samples for A. terreus ATCC 20542. (A, B, C, and D) indicate differences among means between samples for A. terreus ATCC 20542.

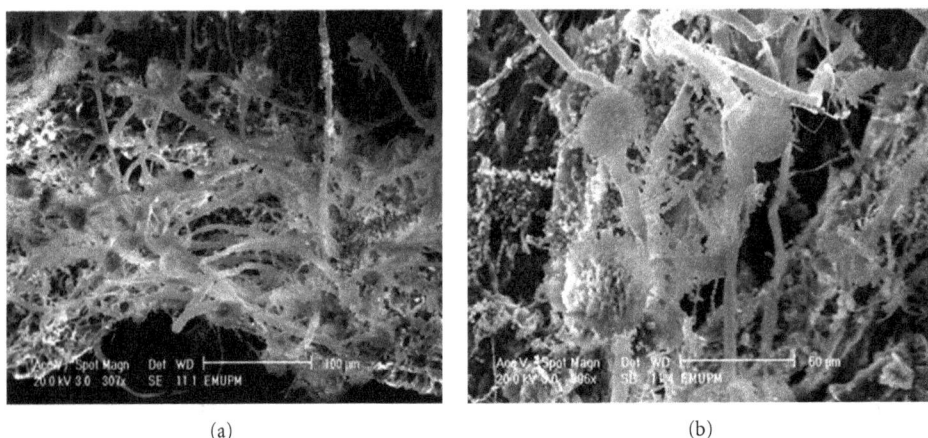

FIGURE 8: Scanning electron micrographs of A. Terreus ATCC 20542 (a) and A. Terreus ATCC 75135 (b) on the surface of RS. High concentration of mycelium and spore could have negative effects on air flow in the solid culture.

optimum incubation temperature for lovastatin production by A. terreus.

3.7. Effect of Particle Size. Five different particle sizes of RS were used to study the effect of this factor on lovastatin production. The results suggest optimum particle size of RS for lovastatin production was between 1.4 to 2 mm and A. terreus ATCC 74135 is more sensitive to changes in particle size (Figure 7). The above results are differed with that of Valera et al. [41], who found increasing particle size of wheat bran as substrate (from 0.4 to 1.1 mm) in SSF resulted in reduction of lovastatin production. They further reported an interaction effect between particle size and moisture content of solid material on lovastatin production.

There are two opposing effects of particle size on SSF process at any given moisture content. The first is small particle size that increases surface area of solid materials for the attachment and growth of the fungi. The second is smaller particle size that reduces interspace between particles and gas phase oxygen transfer and thus reduces the growth potential of the aerobic microorganisms. In addition, growth

and multiplication of microorganisms on the surface of solid materials further reduce the gas phase space and make it even more difficult for air transfer between solid particles. Effect of high micellium content on reduction of the gas phase space of RS culture are shown in Figure 8. High concentration of mycelium and spores present between the solid particles reduce the flow of air in the culture and could reduce the available oxygen for growth of A. terreus.

On the other hand, some reports that indicate reduction of particle size enhanced production of lovastatin in SSF [15]. Wei et al. [7] reported that grounding rice through 20 mesh size (840 μm) has positive effect but further reduction using 40 mesh size (420 μm) reduced lovastatin production compared with the natural size of rice grain. The formation of lovastatin is strictly dependent on the oxygen supplied; however, Bizukojc and Ledakowicz [6] reported that aeration rate up to 0.308 vvm is preferred for lovastatin biosynthesis as higher aeration has negative effect on lovastatin production.

3.8. Effect of Moisture. Moisture content had significant effect on lovastatin production ($P < 0.01$) (Figure 9). Results

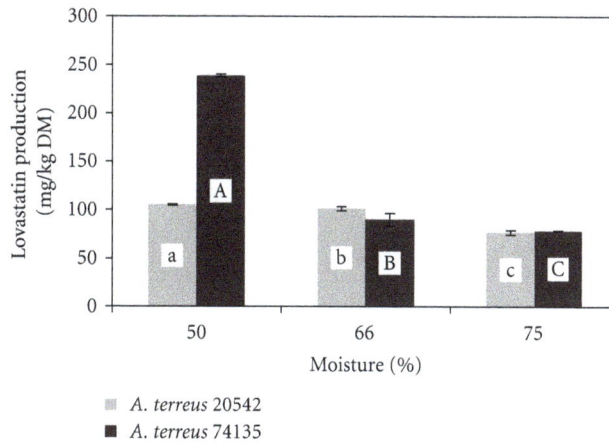

FIGURE 9: effect of moisture content on lovastatin production by two strains of *A. terreus* ($P < 0.01$). (a, b, and c) indicate differences among means between samples for *A. terreus* ATCC 20542. (A, B, and C) indicate differences among means between samples for *A. terreus* ATCC 20542.

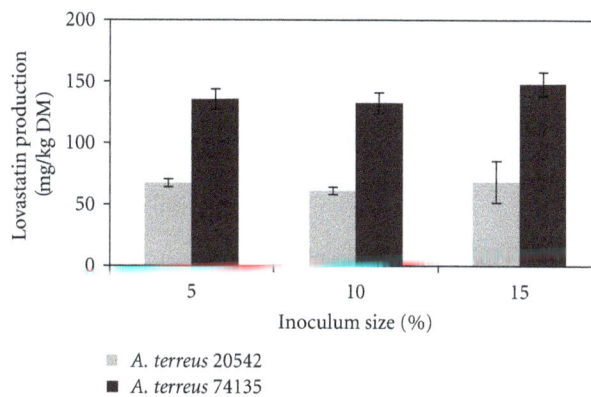

FIGURE 10: Effect of inoculums size on lovastatin production by *A terreus* ($P > 0.05$).

of this study suggest that 50% moisture content is optimum for lovastatin production by two strains of *A. Terreus*. Results also showed that *A. Terreus* ATCC 74135 is more sensitive to the moisture and at 50% moisture production of lovastatin by this strain can be up to 238.74 mg/kg DM. Other studies [41, 42] reported that slightly higher (58 to 60%) initial moisture produced maximum yield of statin using other fungi. High moisture content resulted in aggregation of substrate particles, reduction of aeration and leading to anaerobic conditions [45].

3.9. Effect of Inoculum Size. Figure 10 shows that different inoculum sizes did not affect ($P > 0.05$) lovastatin production by the two strains of *A. Terreus*. Previous study showed no significant different in lovastatin production using 5×10^7 to 10×10^7 spores/mL (wihin the ranged used in this study), but use of lower than 5×10^7 spores/mL can depress the production of lovastatin [15].

3.10. Effect of Incubation Time. To study the effect of incubation time on lovastatin production, optimal conditions for all the other factors (pH = 6, temperature = 25°C, inoculums size = 10% and moisture = 50%) obtained

earlier were applied. Results of the study suggested that maximum production of lovastatin was achieved on day 12 with lovastatin production of 175.85 mg/kg DM for *A. terreus* ATCC 20542 but day 8 with lovastatin production of 260.85 mg/kg DM for *A. terreus* ATCC 74135 (Figure 11). In both fungi cultures, the concentration of lovastatin increased until day 8 with no significant effect on lovastatin production thereafter. Pansuriya and Singhal [15] reported that lovastatin production by *A. terreus* in SSF using wheat bran increased until day 3 of fermentation with no further enhancement thereafter. The shorter duration of the above study compared to the present study could be due to higher quality of substrate (rice bran versus RS). The yields of lovastatin obtained in this study were about 20% of those reported [5, 42] using wheat bran and groundnuts oil cake as substrates. The lower lovastatin yield recorded in this study, using agro-biomass as compared to those using high-energy grains and oil seeds, was acceptable.

4. Conclusion

Results of this study suggest that RS is the better substrate than OPF for lovastatin production in SSF. Although

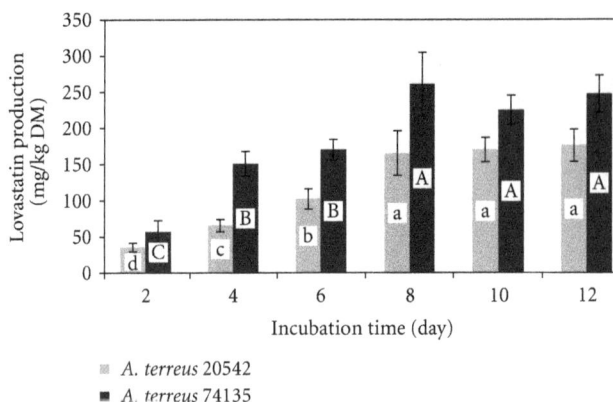

FIGURE 11: Lovastatin production by *A. terreus* in optimum condition at different time of incubation ($P < 0.01$). (a, b, c, and d) indicate differences among means between samples for *A. terreus* ATCC 20542. (A, B, and C) indicate differences among means between samples for *A. terreus* ATCC 20542.

additional nitrogen source has no benefit for improvement the lovastatin production, it enhances lignocelluloses degradation by the fungi. Since one of the main objectives of this study was to evaluate the use of the fermented RS as ruminant feed, urea was supplemented as nitrogen source in the fermentation process. Results of optimization experiment indicate that pH 6, 25°C incubation temperature, 10% inoculums size, 50% moisture content, and 8 days fermentation are the best conditions for maximum lovastatin production in SSF using RS as substrate with *A. terreus* ATCC 74135 recorded higher lovastatin production of 260.85 mg/kg DM after 8 days fermentation.

The present study provides a new insight for production of lovastatin and/or other similar biomaterials of high value from agro-biomass, which otherwise may be sources of pollutant to the environment. Furthermore, the lovastatin enriched fermented RS has the potential to be used as antimethanogenesis feed supplement for reduction of enteric methane production in the ruminant animals. The above suggestion needs further investigations.

Acknowledgment

This study was supported by the Fundamental Research Grant Scheme (FRGS 1/2010 UPM) of the Department of Higher Education Malaysia.

References

[1] J. A. Tobert, "Lovastatin and beyond: the history of the HMG-CoA reductase inhibitors," *Nature Reviews Drug Discovery*, vol. 2, no. 7, pp. 517–526, 2003.

[2] A. W. Alberts, J. Chen, and G. Kuron, "Mevinolin: a highly potent competitive inhibitor of hydroxymethylglutaryl-coenzyme A reductase and a cholesterol-lowering agent," *Proceedings of the National Academy of Sciences of the United States of America*, vol. 77, no. 7, pp. 3957–3961, 1980.

[3] K. Gupta, P. K. Mishra, and P. Srivastava, "A correlative evaluation of morphology and rheology of *Aspergillus terreus* during lovastatin fermentation," *Biotechnology and Bioprocess Engineering*, vol. 12, no. 2, pp. 140–146, 2007.

[4] A. Endo, M. Kuroda, and Y. Tsujita, "ML 236A, ML 236B, and ML 236C, new inhibitors of cholesterogenesis produced by *Penicillium citrinum*," *Journal of Antibiotics*, vol. 29, no. 12, pp. 1346–1348, 1976.

[5] N. Jaivel and P. Marimuthu, "Optimization of lovastatin production in solid state fermentation by *Aspergillus terreus*," *Optimization*, vol. 2, no. 7, pp. 2730–2733, 2010.

[6] M. Bizukojc and S. Ledakowicz, "Biosynthesis of lovastatin and (+)-geodin by *Aspergillus terreus* in batch and fed-batch culture in the stirred tank bioreactor," *Biochemical Engineering Journal*, vol. 42, no. 3, pp. 198–207, 2008.

[7] P. L. Wei, Z. N. Xu, and P. L. Cen, "Lovastatin production by *Aspergillus terreus* in solid-state fermentation," *Journal of Zhejiang University*, vol. 8, no. 9, pp. 1521–1526, 2007.

[8] S. A. Sayyad, B. P. Panda, S. Javed, and M. Ali, "Optimization of nutrient parameters for lovastatin production by *Monascus purpureus* MTCC 369 under submerged fermentation using response surface methodology," *Applied Microbiology and Biotechnology*, vol. 73, no. 5, pp. 1054–1058, 2007.

[9] T. Miyake, K. Uchitomi, M. Y. Zhang et al., "Effects of the principal nutrients on lovastatin production by *Monascus pilosus*," *Bioscience, Biotechnology and Biochemistry*, vol. 70, no. 5, pp. 1154–1159, 2006.

[10] A. Endo, K. Hasumi, and A. Yamada, "The synthesis of compactin (ML-236B) and monacolin K in fungi," *Journal of Antibiotics*, vol. 39, no. 11, pp. 1609–1610, 1986.

[11] A. L. Demain, *Novel Microbial Products For Medicine and Agriculture, Vol. 1*, Elsevier Science, 1989.

[12] L. S. T. Lai, C. C. Pan, and B. K. Tzeng, "The influence of medium design on lovastatin production and pellet formation with a high-producing mutant of *Aspergillus terreus* in submerged cultures," *Process Biochemistry*, vol. 38, no. 9, pp. 1317–1326, 2003.

[13] W. F. Ruddiman and J. S. Thomson, "The case for human causes of increased atmospheric CH_4 over the last 5000 years," *Quaternary Science Reviews*, vol. 20, no. 18, pp. 1769–1777, 2001.

[14] S. Suryanarayan, "Current industrial practice in solid state fermentations for secondary metabolite production: the Biocon India experience," *Biochemical Engineering Journal*, vol. 13, no. 2-3, pp. 189–195, 2003.

[15] R. C. Pansuriya and R. S. Singhal, "Response surface methodology for optimization of production of lovastatin by solid

state fermentation," *Brazilian Journal of Microbiology*, vol. 41, no. 1, pp. 164–172, 2010.

[16] R. K. Pachauri and A. Reisinger, Eds., *Intergovernmental Panel on Climate Change, in Summary For Policymaker of Synthesis Report*, IPCC, Cambridge, UK, 2007.

[17] P. J. Crutzen, I. Aselmann, and W. Seiler, "Methane production by domestic animals, wild ruminants, other herbivorous fauna, and humans," *Tellus B*, vol. 38, no. 3-4, pp. 271–284, 1986.

[18] A. R. Moss, "Measuring Methane Production from Ruminants," in *Methane: Global Warming and Production by Animals*, H. P. S. Makkar and P. E. Vercoe, Eds., p. 105, Chalcombe Publications, Canterbury, UK, 1993.

[19] F. X. Wildenauer, K. H. Blotevogel, and J. Winter, "Effect of monensin and 2-bromoethanesulfonic acid on fatty acid metabolism and methane production from cattle manure," *Applied Microbiology and Biotechnology*, vol. 19, no. 2, pp. 125–130, 1984.

[20] S. A. Martin, "Manipulation of ruminal fermentation with organic acids: a review," *Journal of Animal Science*, vol. 76, no. 12, pp. 3123–3132, 1998.

[21] F. Dohme, A. Machmüller, A. Wasserfallen, and M. Kreuzer, "Ruminal methanogenesis as influenced by individual fatty acids supplemented to complete ruminant diets," *Letters in Applied Microbiology*, vol. 32, no. 1, pp. 47–51, 2001.

[22] S. Y. Lee, S. H. Yang, W. S. Lee, H. S. Kim, D. E. Shin, and J. K. Ha, "Effect of 2-bromoethanesulfonic acid on in vitro fermentation characteristics and methanogen population," *Asian-Australasian Journal of Animal Sciences*, vol. 22, no. 1, pp. 42–48, 2009.

[23] Y. J. Williams, S. Popovski, S. M. Rea et al., "A vaccine against rumen methanogens can alter the composition of archaeal populations," *Applied and Environmental Microbiology*, vol. 75, no. 7, pp. 1860–1866, 2009.

[24] N. Mohammed, N. Ajisaka, Z. A. Lila et al., "Effect of Japanese horseradish oil on methane production and ruminal fermentation in vitro and in steers," *Journal of Animal Science*, vol. 82, no. 6, pp. 1839–1846, 2004.

[25] M. J. Wolin and T. L. Miller, "Control of rumen methanogenesis by inhibiting the growth and activity of methanogens with hydroxymethylglutaryl-SCoA inhibitors," *International Congress Series*, vol. 1293, pp. 131–137, 2006.

[26] J. Friedrich, M. Zuzek, M. Bencina, A. Cimerman, A. Strancar, and I. Radez, "High-performance liquid chromatographic analysis of mevinolin as mevinolinic acid in fermentation broths," *Journal of Chromatography A*, vol. 704, no. 2, pp. 363–367, 1995.

[27] P. J. Van Soest, J. B. Robertson, and B. A. Lewis, "Methods for dietary fiber, neutral detergent fiber, and nonstarch polysaccharides in relation to animal nutrition," *Journal of Dairy Science*, vol. 74, no. 10, pp. 3583–3597, 1991.

[28] AOAC, *Official Methods of Analysis*, AOAC: Association of Official Analytical Chemists, Washington, DC, USA, 15th edition, 1990.

[29] SAS, "SAS Institute Inc., 2008. SAS Online Doc 9. 2," SAS Institute Inc., Cary, NC, USA, 2008.

[30] Z. Jia, X. Zhang, Y. Zhao, and X. Cao, "Enhancement of lovastatin production by supplementing polyketide antibiotics to the submerged culture of *Aspergillus terreus*," *Applied Biochemistry and Biotechnology*, vol. 160, no. 7, pp. 2014–2025, 2010.

[31] J. Gao, H. Weng, D. Zhu, M. Yuan, F. Guan, and Y. Xi, "Production and characterization of cellulolytic enzymes from the thermoacidophilic fungal *Aspergillus terreus* M11 under solid-state cultivation of corn stover," *Bioresource Technology*, vol. 99, no. 16, pp. 7623–7629, 2008.

[32] J. Gao, H. Weng, Y. Xi, D. Zhu, and S. Han, "Purification and characterization of a novel endo-β-1,4-glucanase from the thermoacidophilic *Aspergillus terreus*," *Biotechnology Letters*, vol. 30, no. 2, pp. 323–327, 2008.

[33] G. Emtiazi, N. Naghavi, and A. Bordbar, "Biodegradation of lignocellulosic waste by *Aspergillus terreus*," *Biodegradation*, vol. 12, no. 4, pp. 259–263, 2001.

[34] N. B. Ghanem, H. H. Yusef, and H. K. Mahrouse, "Production of *Aspergillus terreus* xylanase in solid-state cultures: application of the Plackett-Burman experimental design to evaluate nutritional requirements," *Bioresource Technology*, vol. 73, no. 2, pp. 113–121, 2000.

[35] S. Ikeda, Y. Yamashita, and I. Kreft, "Mineral composition of buckwheat by-products and its processing characteristics to konjak preparation," *Fagopyrum*, vol. 16, pp. 89–94, 1999.

[36] M. Antongiovanni and C. Sargentini, "Variability in chemical composition of straws," *Options Mediterraneennes Serie Seminaires*, vol. 16, pp. 49–53, 1991.

[37] J. L. Casas López, J. A. Sánchez Pérez, J. M. Fernández Sevilla, F. G. Acién Fernández, E. Molina Grima, and Y. Chisti, "Production of lovastatin by *Aspergillus terreus*: effects of the C : N ratio and the principal nutrients on growth and metabolite production," *Enzyme and Microbial Technology*, vol. 33, no. 2-3, pp. 270–277, 2003.

[38] J. X. Liu, Y. M. Wu, X. M. Dai, J. Yao, Y. Y. Zhou, and Y. J. Chen, "The effects of urea-mineral lick blocks on the liveweight gain of local Yellow cattle and goats in grazing conditions," *Livestock Research For Rural Development*, vol. 7, no. 2, 1995.

[39] M. S. Kumar, S. K. Jana, V. Senthil, V. Shashanka, S. V. Kumar, and A. K. Sadhukhan, "Repeated fed-batch process for improving lovastatin production," *Process Biochemistry*, vol. 36, no. 4, pp. 363–368, 2000.

[40] B. P. Panda, S. Javed, and M. Ali, "Optimization of fermentation parameters for higher lovastatin production in red mold rice through co-culture of *Monascus purpureus* and *Monascus ruber*," *Food and Bioprocess Technology*, vol. 3, no. 3, pp. 373–378, 2010.

[41] H. R. Valera, J. Gomes, S. Lakshmi, R. Gururaja, S. Suryanarayan, and D. Kumar, "Lovastatin production by solid state fermentation using *Aspergillus flavipes*," *Enzyme and Microbial Technology*, vol. 37, no. 5, pp. 521–526, 2005.

[42] N. S. Shaligram, S. K. Singh, R. S. Singhal, G. Szakacs, and A. Pandey, "Compactin production in solid-state fermentation using orthogonal array method by *P. brevicompactum*," *Biochemical Engineering Journal*, vol. 41, no. 3, pp. 295–300, 2008.

[43] M. Manzoni and M. Rollini, "Biosynthesis and biotechnological production of statins by filamentous fungi and application of these cholesterol-lowering drugs," *Applied Microbiology and Biotechnology*, vol. 58, no. 5, pp. 555–564, 2002.

[44] A. Endo, K. Hasumi, and S. Negishi, "Monacolins J and L, new inhibitors of cholesterol biosynthesis produced by *Monascus ruber*," *Journal of Antibiotics*, vol. 38, no. 3, pp. 420–422, 1985.

[45] R. P. Tengerdy, "Solid substrate fermentation," *Trends in Biotechnology*, vol. 3, no. 4, pp. 96–99, 1985.

Permissions

The contributors of this book come from diverse backgrounds, making this book a truly international effort. This book will bring forth new frontiers with its revolutionizing research information and detailed analysis of the nascent developments around the world.

We would like to thank all the contributing authors for lending their expertise to make the book truly unique. They have played a crucial role in the development of this book. Without their invaluable contributions this book wouldn't have been possible. They have made vital efforts to compile up to date information on the varied aspects of this subject to make this book a valuable addition to the collection of many professionals and students.

This book was conceptualized with the vision of imparting up-to-date information and advanced data in this field. To ensure the same, a matchless editorial board was set up. Every individual on the board went through rigorous rounds of assessment to prove their worth. After which they invested a large part of their time researching and compiling the most relevant data for our readers. Conferences and sessions were held from time to time between the editorial board and the contributing authors to present the data in the most comprehensible form. The editorial team has worked tirelessly to provide valuable and valid information to help people across the globe.

Every chapter published in this book has been scrutinized by our experts. Their significance has been extensively debated. The topics covered herein carry significant findings which will fuel the growth of the discipline. They may even be implemented as practical applications or may be referred to as a beginning point for another development. Chapters in this book were first published by Hindawi Publishing Corporation; hereby published with permission under the Creative Commons Attribution License or equivalent.

The editorial board has been involved in producing this book since its inception. They have spent rigorous hours researching and exploring the diverse topics which have resulted in the successful publishing of this book. They have passed on their knowledge of decades through this book. To expedite this challenging task, the publisher supported the team at every step. A small team of assistant editors was also appointed to further simplify the editing procedure and attain best results for the readers.

Our editorial team has been hand-picked from every corner of the world. Their multi-ethnicity adds dynamic inputs to the discussions which result in innovative outcomes. These outcomes are then further discussed with the researchers and contributors who give their valuable feedback and opinion regarding the same. The feedback is then collaborated with the researches and they are edited in a comprehensive manner to aid the understanding of the subject.

Apart from the editorial board, the designing team has also invested a significant amount of their time in understanding the subject and creating the most relevant covers. They scrutinized every image to scout for the most suitable representation of the subject and create an appropriate cover for the book.

The publishing team has been involved in this book since its early stages. They were actively engaged in every process, be it collecting the data, connecting with the contributors or procuring relevant information. The team has been an ardent support to the editorial, designing and production team. Their endless efforts to recruit the best for this project, has resulted in the accomplishment of this book. They are a veteran in the field of academics and their pool of knowledge is as vast as their experience in printing. Their expertise and guidance has proved useful at every step. Their uncompromising quality standards have made this book an exceptional effort. Their encouragement from time to time has been an inspiration for everyone.

The publisher and the editorial board hope that this book will prove to be a valuable piece of knowledge for researchers, students, practitioners and scholars across the globe.

List of Contributors

Carlos Barreiro, Juan F. Martın, and Carlos Garcıa-Estrada
Proteomics Service of INBIOTEC, Instituto de Biotecnologıa de Leon (INBIOTEC), Parque Cientıfico de Leon, Avenida. Real, no. 1, 24006 Leon, Spain

Yuichi Endo, Daisuke Iwaki, Yumi Ishida, Minoru Takahashi and Teizo Fujita
Department of Immunology, Fukushima Medical University School of Medicine, 1-Hikarigaoka, Fukushima 960-1295, Japan

Misao Matsushita
Department of Applied Biochemistry, Tokai University, Hiratsuka, Kanagawa 259-1292, Japan

Jennifer L. Behan and Yvonne E. Cruickshank
School of Life, Sport and Social Sciences, Edinburgh Napier University, Sight hill Campus, Edinburgh EH11 4BN, UK

Gerri Matthews-Smith and Malcolm Bruce
School of Nursing, Midwifery and Social Care, Edinburgh Napier University, Edinburgh EH11 4BN, UK

Kevin D. Smith
School of Life, Sport and Social Sciences, Edinburgh Napier University, Sight hill Campus, Edinburgh EH11 4BN, UK
Community Drug Problem Service, Spittal Street Centre, Edinburgh EH3 9DU, UK

Mior Ahmad Khushairi Mohd Zahari
Department of Process and Food Engineering, Faculty of Engineering, Universiti Putra Malaysia, Serdang, 43400 Selangor, Malaysia
Faculty of Chemical and Natural Resources Engineering, Universiti Malaysia Pahang, Lebuhraya Tun Razak, Kuantan, 26300 Pahang, Malaysia

Mohd Noriznan Mokhtar
Department of Process and Food Engineering, Faculty of Engineering, Universiti Putra Malaysia, Serdang, 43400 Selangor, Malaysia

Jailani Salihon
Faculty of Chemical and Natural Resources Engineering, Universiti Malaysia Pahang, Lebuhraya Tun Razak, Kuantan, 26300 Pahang, Malaysia

Mohd Ali Hassan
Department of Process and Food Engineering, Faculty of Engineering, Universiti Putra Malaysia, Serdang, 43400 Selangor, Malaysia
Department of Bioprocess Technology, Faculty of Biotechnology and Bio molecular Sciences, Universiti Putra Malaysia, Serdang, 43400 Selangor, Malaysia

Hidayah Ariffin
Department of Bioprocess Technology, Faculty of Biotechnology and Bio molecular Sciences, Universiti Putra Malaysia, Serdang, 43400 Selangor, Malaysia

Yoshihito Shirai
Department of Biological Functions and Engineering, Graduate School of Life Science and Systems Engineering, Kyushu Institute of Technology, 2-4 Hibikino, Wakamatsu-ku, Kitakyushu, Fukuoka 808-0196, Japan

Yoshikazu Takada
Department of Dermatology and Biochemistry and Molecular Medicine, University of California Davis School of Medicine, Research III Suite 3300, 4645 Second Avenue, Sacramento, CA 95817, USA

David C. Kilpatrick
Scottish National Blood Transfusion Service, National Science Laboratory, Ellen's Glen Road, Edinburgh EH17 7QT, UK

James D. Chalmers
MRC Centre for Inflammation Research, University of Edinburgh, 47 Little France Crescent, Edinburgh EH16 4TJ, UK

Edward F. Plow, Loic Doeuvre and Riku Das
Department of Molecular Cardiology, Joseph J. Jacobs Center for Thrombosis and Vascular Biology, Cleveland Clinic, 9500 Euclid Avenue, NB50, Cleveland, OH 44195, USA

Angels Diaz-Ramos, Anna Roig-Borrellas, Ana Garcia-Melero and Roser Lopez-Alemany
Biological Clues of the Invasive and Metastatic Phenotype Research Group, (IDIBELL) Institut d'Investigacions Biomediques de Bellvitge, L'Hospitalet de Llobregat, 08908 Barcelona, Spain

Jia Li and Hongfeng Yuan
Department of Ophthalmology, Research Institute of Surgery and Daping Hospital, Third Military Medical University, Chongqing, 400042, China

Changqing Li
Department of Pharmacy, Chongqing Red Cross Hospital, Chongqing 400020, China

Fang Gong
Department of Neonatology, Yong Chuan Affiliated Hospital, Chongqing Medical University, Chongqing 402160, China

Meili Gao
Department of Biological Science and Engineering, Key Laboratory of Biomedical Information Engineering of the Ministry of Education, School of Life Science and Technology, Xian Jiaotong University, Xi'an, Shaanxi 710049, China

Yongfei Li
School of Materials and Chemical Engineering, Xian Technological University, Xian, Shaanxi 710032, China

Xiaochang Xue
State Key Laboratory of Cancer Biology, Department of Bio pharmaceutics School of Pharmacy, Fourth Military Medical University, Xian, Shaanxi 710032, China

Xianfeng Wang
Department of Anesthesiology, Wake Forest School of Medicine, Winston-Salem, NC 27157, USA

Jiangang Long
Institute of Mitochondrial Biology and Medicine, Key Laboratory of Biomedical Information Engineering of the Ministry of Education, School of Life Science and Technology, Xian Jiaotong University, Xian, Shaanxi 710049, China

Ibrahim Guillermo Castro-Torres and Margarita Virginia Saavedra-Velez
Facultad de Quimica Farmaceutica Biologica, Universidad Veracruzana, Xalapa, Veracruz, Mexico

Elia Brosla Naranjo-Rodriguez
Laboratorio de Neurofarmacologia, Departamento de Farmacia, Facultad de Quimica, Universidad Nacional Autonoma de Mexico, Mexico, DF, Mexico

Miguel Angel Dominguez-Ortiz and Janeth Gallegos-Estudillo
Laboratorio de Productos Naturales, Instituto de Ciencias Basicas, Universidad Veracruzana, Xalapa, Veracruz, Mexico

Riccardo Montioli, Barbara Cellini, Mirco Dindo, Elisa Oppici and Carla Borri Voltattorni
Section of Biological Chemistry, Department of Life Sciences and Reproduction, University of Verona, Strada Le Grazie 8, 37134 Verona, Italy

Jacek Rutkowski and Bogumil Brzezinski
Department of Biochemistry, Faculty of Chemistry, Adam Mickiewicz University, Grunwaldzka 6, 60-780 Poznan, Poland

Motoe Taniguchi
Department of Functional Biology, Kanagawa Dental College, 82 Inaokacho, Yokosuka 238-8580, Japan
Department of Maxillofacial Diagnostic Science, Kanagawa Dental College, 82 Inaokacho, Yokosuka 238-8580, Japan

Yuko Mikuni-Takagaki
Department of Functional Biology, Kanagawa Dental College, 82 Inaokacho, Yokosuka 238-8580, Japan

Hiroshi Mori
Department of Maxillofacial Diagnostic Science, Kanagawa Dental College, 82 Inaokacho, Yokosuka 238-8580, Japan

Junko Iizuka and Yoshiharu Mukai
Department of Oral Medicine, Kanagawa Dental College, 82 Inaokacho, Yokosuka 238-8580, Japan

Yukio Hirata and Yukari Murata
Department of Dental Sociology, Kanagawa Dental College, 82 Inaokacho, Yokosuka 238-8580, Japan

Yumi Ito
Yokohama Training Center, Kanagawa Dental College, 3-31-6 Turuyacho, Kanagawa-ku, Yokohama 221-0835, Japan
Department of Pathology, Tsurumi University School of Dental Medicine, Yokohama 230-8501, Japan

Mariko Iwamiya
Clinical Laboratory, Kanagawa Dental College, 82 Inaokacho, Yokosuka 238-8580, Japan

Xuehua Li, Tatiana Syrovets and Thomas Simmet
Institute of Pharmacology of Natural Products and Clinical Pharmacology, Universitat Ulm, Helmholtzstraße 20, 89081 Ulm, Germany

Jan Eric Jessen and Johann Orlygsson
Faculty of Natural Resource Sciences, University of Akureyri, Borgir, Nordurslod 2, 600 Akureyri, Iceland

Yang Shi, Shuqun Hu, Qingwei Song and Shengcai Yu
Department of General Surgery, The Affiliated Hospital of Xuzhou Medical College, Xuzhou, Jiangsu 221002, China

Xiaojun Zhou, Jun Yin, Lei Qin and Haixin Qian
Department of General Surgery,The First Affiliated Hospital of Soochow University, Suzhou, Jiangsu 215006, China

Y. M. Goh
Department of Veterinary Preclinical Sciences, Faculty of Veterinary Medicine, Universiti Putra Malaysia, 43400 Serdang, Selangor, Malaysia
Institute of Tropical Agriculture, Universiti Putra Malaysia, 43400 Serdang, Selangor, Malaysia

M. Ebrahimi and M. A. Rajion
Department of Veterinary Preclinical Sciences, Faculty of Veterinary Medicine, Universiti Putra Malaysia, 43400 Serdang, Selangor, Malaysia

A. Q. Sazili
Department of Animal Science, Faculty of Agriculture, Universiti Putra Malaysia, 43400 Serdang, Selangor, Malaysia

J. T. Schonewille
Division of Nutrition, Department of Farm Animal Health, Faculty of Veterinary Medicine, Utrecht University, P.O. Box 80151, 3508 TD Utrecht, The Netherlands

Mohammad Faseleh Jahromi, Yin Wan Ho and Parisa Shokryazdan
Laboratory of Industrial Biotechnology, Institute of Bioscience, Universiti Putra Malaysia, 43400 Serdang, Malaysia

Juan Boo Liang
Laboratory of Animal Production, Institute of Tropical Agriculture, Universiti Putra Malaysia, 43400 Serdang, Malaysia

Rosfarizan Mohamad
Faculty of Biotechnology and Bio molecular Sciences, Universiti Putra Malaysia, 43400 Serdang, Malaysia

Yong Meng Goh
Faculty of Veterinary Medicine, Universiti Putra Malaysia, 43400 Serdang, Malaysia